HANDBOOK OF ADHESIVES AND SEALANTS IN CONSTRUCTION

HANDBOOK OF ADHESIVES AND SEALANTS IN CONSTRUCTION

Joseph S. Amstock
President, Professional Adhesive and Sealant Systems
Huntingdon Valley, Pennsylvania

McGRAW-HILL
New York San Francisco Washington D.C. Auckland Bogotá
Caracas Lisbon London Madrid Mexico City Milan
Montreal New Delhi San Juan Singapore
Sydney Tokyo Toronto

Library of Congress Cataloging-in-Publication Data

Amstock, Joseph S.
    Handbook of adhesives and sealants in construction / Joseph Amstock.
       p.   cm.
    Includes bibliographical references.
    ISBN 0-07-001616-X
    1. Sealing compounds.   2. Adhesives.   I. Title.

TP988.A47   2000
691'.99—dc21   00-058670

McGraw-Hill
A Division of The **McGraw·Hill** *Companies*

Copyright © 2001 by The McGraw-Hill Companies, Inc. Printed in the United States of America. Except as permitted under the United States Copyright Act of 1976, no part of this publication may be reproduced or distributed in any form or by any means, or stored in a data base or retrieval system, without the prior written permission of the publisher.

1 2 3 4 5 6 7 8 9 0   DOC/DOC   9 0 2 1 0 9 8 7 6

ISBN 0-07-001616-X

The sponsoring editor for this book was Larry Hager and the production supervisor was Sherri Souffrance. It was set in Times Roman by Lone Wolf Enterprises, Ltd.

Printed and bound by Maple-Vail Book Mfg Group.

This book is printed on recycled, acid-free paper containing a minimum of 50% recycled, de-inked fiber.

McGraw-Hill books are available at special quantity discounts to use as premiums and sales promotions, or for use in corporate training programs. For more information, please write to the Director of Special Sales, Professional Publishing, McGraw-Hill, Two Penn Plaza, New York, NY 10121-2298. Or contact your local bookstore.

Information contained in this work has been obtained by The McGraw-Hill Companies, Inc. ("McGraw-Hill") from sources believed to be reliable. However, neither McGraw-Hill nor its authors guarantee the accuracy or completeness of any information published herein and neither McGraw-Hill nor its authors shall be responsible for any errors, omissions, or damages arising out of use of this information. This work is published with the understanding that McGraw-Hill and its authors are supplying information but are not attempting to render engineering or other professional services. If such services are required, the assistance of an appropriate professional should be sought.

---

*To Vicki, John, Nigel,
Graham and Ayrton*

CONTENTS

Preface   xvii
About the Author   xix

Chapter 1 Introduction to Sealants and Adhesives   1.1

    Historical Background of Adhesives / 1.2
    Required Properties of Joint Sealants / 1.3
    Available Materials / 1.9
    Sealants / 1.12
    Mastics / 1.12
    Thermoplastics Applied / 1.12
    Thermoplastics / 1.13
    Thermosetting, Chemically Curing / 1.13
    Accessory Materials / 1.14
    Primers / 1.14
    Bond Breakers / 1.14
    Backup Materials / 1.14
    Preformed Seals / 1.14
    Rigid Waterstops and Miscellaneous Seals / 1.15
    Flexible Waterstops / 1.15
    Gaskets and Miscellaneous Seals / 1.15
    Strip (Gland) Seals / 1.16
    Compression Seals / 1.16
    Flexible Foam (Impregnated) / 1.16
    Flexible Foam (Non-Impregnated) / 1.17
    Tension/Compression Seal Systems / 1.17
    Introduction to Adhesives / 1.17
    Anaerobic / 1.18
    Conclusion / 1.18
    References / 1.20

Chapter 2 Epoxy Adhesives and Sealants   2.1

    Definition of Epoxy Resin / 2.4
    History / 2.4
    Chronology / 2.5
    Characteristics that Justify the Use of Epoxies / 2.6
    Epoxy—The Intermediate Resin / 2.7
    Resins Derived from Epichlorohydrin (EPI) / 2.8
    Applications of Epoxy Resins / 2.9
    Typical End-Use Applications / 2.10
    Summary / 2.11
    References / 2.12

Chapter 3 Acrylic Adhesives and Sealants   3.1

  Introduction / 3.1
  Development / 3.3
  Polymer Selection / 3.3
  Mixing Equipment / 3.4
  Raw Materials / 3.4
    Binder / 3.4
    Pigments, Fillers and Extenders / 3.3
    Dispersant / 3.4
    Plasticizer / 3.4
  Wetting Agent / 3.4
  Freeze-Thaw Stabilizer / 3.5
  Adhesion Promoter / 3.5
  Surface Drying Regulator / 3.5
  Defoamer / 3.5
  Acrylic Latex-Based Caulk / 3.5
  Solvent-Release Acrylic Caulks / 3.6
  Formulation and Physical Data / 3.6
  Clear Waterborne Caulks and Sealants / 3.7
  Anaerobic and Modified Acrylic Adhesives / 3.14
  Solvent Release Sealant Systems / 3.18
  Formulations / 3.19
  Other Formulas / 3.20
  Typical Applications / 3.20
    Vertical Mastics / 3.20
    Aqueous Acrylic Caulk / 3.21
    Acrylic Latex Caulk / 3.22
  Advantages Over Oil-and/or Solvent-Containing Caulks / 3.23
  Advantages Over Polyvinyl Acetate Latex Caulks / 3.24
  Characteristics of Acrylic Latex Polymers / 3.24
  Mixing Instructions / 3.24
  Areas of Application and Use / 3.25
  References / 3.26

Chapter 4 Anaerobics and Cyanoacrylates   4.1

  Introduction / 4.1
  Anaerobics / 4.1
  Basic Chemistry / 4.2
    Monomers / 4.3
    Initiators / 4.3
    Accelerators / 4.4
    Stabilizers / 4.4
    Form Modifers / 4.4
  Types and Forms / 4.4
  Processing / 4.6
  Cyanocrylates / 4.6
  Chemical Composition / 4.7
  Manufacture of 2-Cyanoacrylate Esters / 4.8
  Bonding Action / 4.10
  Storage Stability / 4.11
  Summary / 4.11
  References / 4.11

Chapter 5 Fluorcarbons and Hot Melt Adhesives and Sealants   5.1

Introduction / 5.1
Hot Melt Composition / 5.2
Chemical Structure / 5.4
   Ethylene Vinyl Acetate / 5.4
   Vinyl Acetate Ethylene / 5.4
   Low-Density Polyethylene / 5.6
   C30 Alkenyl Succinic Anhydride (C30ASA) / 5.7
Critical Properties / 5.7
   Viscosity / 5.7
   Melt Index (MI) / 5.9
   Density / 5.10
   Elongation / 5.10
   Tensile Modulus / 5.10
   Hardness / 5.10
   Softening Point / 5.11
Typical End-Use Formulation / 5.11
Chemical Suppliers / 5.12
Equipment Suppliers / 5.13
Summary / 5.13
References / 5.14

Chapter 6 Asphaltic, Oleoresinous, and Oil-Based Putty   6.1

Introduction / 6.1
   Oleoresinous Caulks and Sealants / 6.1
   Water Emulsion Sealants / 6.2
   Plastic Sealants / 6.2
   Hot-Poured Bituminous Sealant / 6.2
Typical Formulations / 6.3
Applications / 6.4
Summary / 6.6

Chapter 7 Concrete, Cements, Grouts, and Mortars   7.1

Introduction / 7.1
Architectural Concrete—History / 7.1
Understanding Concrete as a Building Component / 7.5
   Concrete / 7.5
   Masonry / 7.5
   Mortar / 7.5
   Portland Cement / 7.5
Composition and Hydration of Cement / 7.6
Mortars / 7.7
Portland Cement Plaster / 7.8
Brass / 7.9
Cracking in Floors / 7.10
Causes of Floor Cracking / 7.10
How Various Joints Work / 7.10
   Isolation Joints / 7.11
   Control Joints / 7.11
Construction Joints / 7.11
How Joints are Formed / 7.11
   Isolation Joints / 7.11
   Control Joints / 7.14
   Construction Joints / 7.15
Spacing and Layout of Joints / 7.15

Cracking in Concrete Repairs—A Complicated Process / 7.15
Understanding Cracking / 7.20
Visualizing Strength / 7.23
Shrinkage / 7.25
Keeping Cracks to a Minimum / 7.25
Having the Right Concrete / 7.25
   Expansion Tests / 7.27
   Expansion Rate / 7.27
   Interground Versus Expansive Components / 7.27
   Experience / 7.27
Joint Solutions / 7.28
Basic Joint Types / 7.28
   Construction Joints / 7.28
   Isolation Joints / 7.29
   Shrinkage-Contraction Joints / 7.29
Typical Problems / 7.29
Filling the Joint / 7.29
   Rigid Epoxy Filler / 7.30
   Semi-rigid Epoxy Filler / 7.31
   Elastomeric (Soft) Fillers / 7.32
Leave In-Place Forms / 7.33
Strengthening the Joint Edge / 7.33
   Rapid Epoxy Filler / 7.33
Hard Aggregate / 7.33
Conventional Steel-Protected Joints / 7.33
Factory-Produced Steel Angle Joints / 7.34
Gypsum Concrete / 7.35
Plaster and Water / 7.35
Aggregate and Portland Cement / 7.36
Quality Control / 7.37
In-Service Aspects / 7.37
Construction and Installation / 7.37
Forensic Study of Failures / 7.38
Grout Specifications / 7.38
Height Change / 7.39
Short-Form Specifications / 7.40
Summary / 7.40
Reference / 7.40
Further Reading / 7.41

Chapter 8 Butyl Sealants   8.1

Introduction / 8.1
Basic Chemistry / 8.2
Summary / 8.11
References / 8.11
Further Reading / 8.11

Chapter 9 Permapol ® Polymers   9.1

Introduction / 9.1
   Permapol P2 / 9.2
   Permapol P3 / 9.2
Chemistry / 9.2
Typical Physical Properties / 9.3
Summary / 9.4
References / 9.4

Chapter 10 Neoprene, Hypalon ®, and Nitrile   10.1

Introduction / 10.1
Neoprene / 10.2
Types of Neoprene Sealants and Adhesives / 10.3
   Neoprene Type KNR / 10.3
   Neoprene AC / 10.3
   Neoprene AD / 10.3
   Neoprene AD-G / 10.4
   Neoprene AF / 10.4
   Neoprene AH / 10.4
   Neoprene FB / 10.5
   Neoprene Type GN and GNA / 10.5
   Neoprene Type GRT / 10.5
   Neoprene Type WRT / 10.5
   Neoprene WHV / 10.5
   Neoprene Type CG / 10.5
Types of Nitrile Sealants and Adhesives / 10.6
Hypalon ® / 10.7
Types of Hypalon ® Sealants and Adhesives / 10.7
Summary / 10.16
References / 10.16

Chapter 11 Polysulfide and LP ™ Polymers   11.1

Introduction / 11.1
Mercaptan-Terminated Liquid Polymers / 11.1
LP ™ Polysulfide Basic Characteristics / 11.3
Curing Mechanisms / 11.3
Types and Uses of Polysulfide Sealants / 11.5
Design Characteristics for LP ™ Polysulfide Sealants / 11.8
Typical Formulations / 11.12
Summary / 11.15
References / 11.16

Chapter 12 Polyurethane (Urethane)   12.1

Introduction / 12.1
Simple Urethane Reaction—Polyols / 12.2
Polyurethane Chemistry / 12.3
   One-Component Sealants / 12.3
The Basics of Urethane and Polyurethane Reactions / 12.4
   Urethane Reaction / 12.4
   Polyurethane Reaction / 12.4
   Isocyanates / 12.5
   Prepolymers / 12.7
   Polyester Polyols 12.7
   Auxiliary Components / 12.9
Advantages and Limitations of Urethane Sealants / 12.9
Suggested Starting Formulations for Construction Sealants / 12.9
   Polyurethane-Based Insulating Glass Sealant / 12.10
   One-Part Pourable Joint Sealer / 12.13
Other Formulations / 12.13
   P-Toluene Sulfonyl Isocyanate (PTSI) / 12.13
   Poly bd-Based Resin Sealants / 12.17
Summary / 12.21
References / 12.21

Chapter 13 Silicones   13.1

Introduction / 13.1
History and Chemistry / 13.2
Types of Silicone Sealants and Applications / 13.4
The Age of Chemical Fasteners / 13.6
   Structural Glazing Applications / 13.7
   Highway, Bridge, and Airfield Sealants [4] / 13.8
   Preformed Silicone Profiles / 13.14
   Remedial Caulking with Silicone Sealants / 13.17
Suggested Starting Formulations of Silicone / 13.23
Summary / 13.24
References / 13.24

Chapter 14 New Polymeric Systems   14.1

Introduction / 14.1
Polymeric Systems / 14.1
Summary / 14.4
References / 14.4

Chapter 15 Firestops and Flameproofing   15.1

Introduction / 15.1
Current Model Building Codes / 15.2
Elements of Firestop Technology / 15.2
Opening Size / 15.3
Penetrant Combustibility / 15.6
A Firestop—Basics to a Better Understanding / 15.7
Material Performance / 15.8
Ease of Application / 15.8
Manufacturer's Commitment / 15.9
Firestop Systems—An Introduction / 15.9
Firestopping Sealants / 15.10
   Mortars, Caulks, Intumescents / 15.10
Firestops—Life Safety Protection / 15.11
Firestop Materials and Compounds / 15.11
FlameSafe ® Firestop Sealants / 15.12
Nelson CLK ™ Firestop Sealants / 15.12
SpecSeal ® Series ES Elastomeric Sealant / 15.12
Nelson FSC ™ Fire Protective Coatings / 15.12
FlameSafe ® Firestop Coatings / 15.12
Pensil ® 300 Sealant / 15.13
Specifications / 15.27
Typical Application / 15.32
Summary / 15.32
References / 15.33

Chapter 16 Gaskets, Foamed and Solid Tapes   16.1

Introduction / 16.1
Types of Tapes, Gaskets, Etc. / 16.1
Application Techniques / 16.6
Gaskets / 16.8
Properties and Evaluation / 16.9
Structural Gaskets / 16.11

Summary / 16.16
References / 16.16

Chapter 17 Formulary for Sealants and Adhesives   17.1

Introduction / 17.1
The Markets / 17.1
The Formulary / 17.3
Summary / 17.14
References / 17.14

Chapter 18 Specifications, Testing, and Quality Assurance   18.1

Introduction / 18.1
Selecting Sealants, Manufacturers and Applicators / 18.2
Description of Team Members / 18.3
   Specification Writer / 18.3
   Training Architects as Specifiers / 18.4
   Owner / 18.5
   Architect-Engineer [A-E] / 18.5
   Contractor / 18.6
   Specialty Contractor / 18.6
   Estimators / 18.7
Manufacturers Require Specifications / 18.7
Attorneys Need Specifications / 18.7
Identifying Problems / 18.8
Making Recommendations for Repairs / 18.8
Preparation of Quality Specifications / 18.8
Competitive Bidding / 18.9
Negotiated Method / 18.10
Sealant Validation / 18.10
Construction Documents Technology Manual / 18.11
Types of Standards / 18.12
   United States Federal Specifications / 18.12
ASTM Standards Important to High Performance Sealants / 18.14
Other Pertinent ASTM Standards (United States) / 18.20
Canadian Standards for High Performance Sealants / 18.25
Other Canadian Standards: Caulking and Sealing Compounds / 18.27
American Concrete Institute Standards / 18.28
Preformed Sealants and Tapes / 18.29
American Association of State Highway and Transportation Officials
   Specification / 18.30
U.S. Army Corps of Engineers / 18.31
American Architectural Manufacturers Association / 18.31
State of California / 18.33
   City of Los Angeles / 18.33
Construction Specifications Institute / 18.33
Quality Assurance / 18.33
   Scope and Purpose / 18.33
Summary / 18.34
References / 18.34

Chapter 19 Joint Design, Joint Details, and Installation   19.1

Why Joints are Required / 19.2
Why Sealing is Needed / 19.3

Joint Design as Part of Overall Structural Design / 19.4
Types of Joints and their Function / 19.5
  Contraction (Control) Joints / 19.5
  Expansion (Isolation) Joints / 19.5
  Construction Joints / 19.6
  Combined and Special-Purpose Joints / 19.6
  Hinge Joints / 19.6
  Sliding Joints / 19.6
  Cracks / 19.7
Joint Configurations / 19.7
Joint Details / 19.7
Structures / 19.17
Slabs on Grade, Highway and Airports / 19.17
Construction and Installation Considerations / 19.18
Installation of Sealants / 19.18
Joint Configuration with Sealing in Mind / 19.19
Preparation of Joint Surfaces / 19.19
Inspection of Readiness to Seal / 19.21
Priming, Installation of Backup Materials and Bond-Breakers / 19.21
Installation of Field-Molded Sealants, Hot Applied / 19.22
Installation of Field-Molded Sealants, Cold-Applied / 19.22
Installation of Compression Seals / 19.23
Installation of Preassembled Devices / 19.25
Installation of Waterstops / 19.26
Installation of Gaskets / 19.26
Installation of Fillers / 19.27
Neatness and Cleanup / 19.27
Safety Precautions / 19.27
Summary / 19.28
References / 19.28

Chapter 20 How Sealants Function — 20.1

Introduction / 20.1
Classification of Sealants / 20.1
Behavior of Sealants in Butt Joints / 20.2
Malfunction of Sealants / 20.2
Behavior of Sealants in Lap Joints / 20.7
Effect of Temperature / 20.7
Shape Factor in Field-Applied Sealants / 20.9
Function of Bond-Breakers and Backup Materials / 20.9
Function of Fillers in Expansion Joints / 20.10
Functions of Primer / 20.12
Summary / 20.12
References / 20.12

Chapter 21 Joint Movement and Design Schematics — 21.1

Introduction / 21.1
Determination of Joint Movements and Locations / 21.1
Selection of Butt-Joint Widths for Field-Applied / 21.2
Selection of Butt-Joint Shape for Field-Applied Sealants / 21.8
Selection of Size of Compression Seals for Butt Joints / 21.10
Limitations on the butt Joint Widths and Movements for the Various
    Types of Sealants / 21.10
Lap Joint Sealant Thickness / 21.11
Shape and Size of Rigid Waterstops / 21.11

Shape and Size of Gaskets and Miscellaneous Seals / 21.16
Measurement of Joint Movements / 21.16
Means of Measuring Joint Movements / 21.16
Corresponding Measurement of Temperature and Moisture Control / 21.17
Survey of Joint Sealant Performance / 21.17
References / 21.18

Chapter 22 Performance, Defects, Repair and Maintenance of Sealants   22.1

Questionable Performance / 22.1
Repair of Concrete Defects and Replacement of Sealants / 22.2
Saw-Cutting Joints in Concrete / 22.2
Why Saw-Cut Joints / 22.2
Where to Saw-Cut / 22.2
When to Saw-Cut / 22.3
Unacceptable Raveling / 22.4
  Uncontrolled Cracking / 22.5
Equipment Affects Timing / 22.6
How to Saw-Cut / 22.7
Depth of Cut / 22.7
Saw-Cut Joint Sequence / 22.7
Joint Curing / 22.8
Preparing Cracks for Sealing / 22.9
Reservoir or Dam / 22.9
Creating a Dam / 22.10
Cleaning the Dam / 22.11
Normal Maintenance / 22.12
Sealing in the Future—Concluding Remarks / 22.12
  What is Possible Now / 22.12
  Advancements Still are Required / 22.13
  Educating the Public / 22.13
  Future Codes, Standards, Recommended Practices and Specifications / 22.13
New Developments / 22.14
  Field-Applied Sealants / 22.14
  Polyvinyl Chloride Coal Tar / 22.14
  Specifications and Warranties / 22.14
  Installation of Hot-Applied Sealants / 22.18
  Preformed Sealants / 22.18
  Compression Seals / 22.18
Joint Face Armoring and Mechanical Locking of Seals / 22.18
Modular Compression Seal Systems / 22.21
Adhesion Lubricants / 22.21
Tension-Compression Seal Improvements / 22.24
Strip (or Gland) Seals / 22.24
Trends in Contraction Joint Practice for Slabs on Grade, Highway, and Airports / 22.28
Two-Stage Joints for Buildings / 22.28
Application and Performance / 22.32
Summary / 22.34
References / 22.34

Chapter 23 Trade Organizations, Publications, Government and Suppliers   23.1

Notes / 23.22

Chapter 24 Glossary of Terms   24.1

    References / 24.76

Appendix A Temperature Conversion Chart   A.1

Appendix B Greek Alphabet   B.1

Appendix C Periodic Table of Elements   C.1

Appendix D Conversion Chart   D.1

Appendix E Equivalents   E.1

Appendix F Conversions (Metric and Color), Properties and Comparisons of Building Materials   F.1

Appendix G Key to Symbols   G.1

Appendix H Estimates and Requirements   H.1

Appendix I Concrete Composition   I.1

Index   IN.1

PREFACE

This book deals with sealant and adhesive materials used in the construction industry.

It is intended as an educational tool for designers, architects, engineers, contractors, and manufacturers of construction adhesives and sealants, as well as the many students of architecture studying waterproofing techniques and proper installation.

It is the author's intention to treat the complicated chemistry, manufacturing process, surface preparation, and proper installation of joint sealants under a variety of field conditions, so the uninitiated will have an improved understanding on this subject.

In most cases, architects, builders, and building owners are interested in the performance aspects of adhesives and sealants, not necessarily the chemistry or the physical properties. This book contains a wealth of information about the performance of a variety of generic materials. The author has organized the information so it can be used systematically, leading the reader from the source, to need, design, specific formulations, and, finally, installation.

It would be impossible to thank everyone who has assisted in the production of this book. Virtually every manufacturer of adhesives and sealants, and accessories and equipment, has contributed through specific technical data, published papers, and other relevant materials.

A special thanks to Elf Atochem of North America, Rohm and Haas Co., Shell Oil Products Co., Sealant, Waterproofing and Restoration Institute, Bayer Corp., and the Construction Specifiers Institute. I am especially grateful to my friends throughout the industry who discussed many of the problems via the telephone.

I would like to express my sincerest thanks to Gary J. Amstock, my son, for his unselfish contributions of hundreds of hours spent preparing the charts, graphs, figures, and illustrations to make this book a useful tool.

Joseph S. Amstock

ABOUT THE AUTHOR

Joseph S. Amstock is a consulting chemical engineer with more than 45 years experience in the caulk, sealants, glazing, and waterproofing industries designed for construction. He was associated with Products Research and Chemical Corp. and Bostik, Inc.

He has served on various committees in the American Society for Testing and Materials, Building Research Institute, Sealed Insulating Glass Manufacturers Association, and Sealant Waterproof and Restoration Institute, and he was on the board of directors of the National Wood Window and Door Association. During his career, he was assigned to Mexico City, Mexico; Brussels, Belgium; and Galway, Ireland.

He now operates his own consulting firm, Professional Adhesive and Sealant Systems.

Professional organizations and trade journals in several countries have published more than 21 technical papers credited to him. In addition, he has written chapters about sealants for John Wiley and Sons, as well as McGraw-Hill. In 1997, he was Editor-in-Chief for the *Handbook of Glass in Construction*, published by McGraw-Hill, a book that since has been translated into Spanish, *Manual del Vidrio en la Construcion*.

CHAPTER 1
INTRODUCTION TO SEALANTS AND ADHESIVES

This chapter is an introduction to the functional properties of sealants, adhesives, and other accessory materials used in construction. Readers of this book will primarily be architects, design engineers, contractors, and specification writers who are interested in the performance properties of sealants and adhesives, their applications and uses. The book will help the professional choose the right material for the job.

An adhesive is a substance capable of holding two or more surfaces together in a strong, often permanent bond [1]. Natural adhesives have been known for centuries. Before the widespread use of synthetic resins, the terms *adhesive* and *glue* were considered synonymous because the earliest commercial adhesives were glues and were generally of animal or marine origin. The term glue is still used widely, and such expressions as *glue line*, *glue pot* and *glue spreader* occupy prominent places in our vocabulary, even though they do not necessarily refer to animal glues. The term glue generally implies a substance of a sticky character, although many of the modern adhesives are not sticky or tacky. Adhesive, however, is a more inclusive term, encompassing many types of synthetic resins, cellulosic resins, natural resins, and glues of animal origin. Adhesives can also be classified by their reaction to heat (thermoplastic and thermosetting) and by their ability to remain rigid or to stretch (elastomer adhesives).

HISTORICAL BACKGROUND OF ADHESIVES

The history of adhesives to the present is largely the history of animal glues and their applications. The romance of the art of gluing has been vividly portrayed by Darrow [2], who reviews Egyptian, Roman, Renaissance, and modern references to the application of glues, including the development of vegetable adhesives from cassava flour. Egyptian murals and veneered caskets in our museums indicate the early use of the art of gluing. We can assume that the Egyptians, who acquired many of their arts from earlier cultures, merely applied these techniques whose origins are lost in the shadows of history.

The first commercial glue plant was founded in Holland in 1690, and the first glue factory in the United States in 1808 by Elijah Upton, founder of the American Glue Company of Boston. It is noteworthy that up to the present day, the majority of glue applications were made in the manufacturer of furniture. In reviewing books about the art of furniture manufacture, the use of glues is mentioned often. Except for the introduction of rubber and pyroxlin cements a hundred years ago, adhesives technology advanced little until the twentieth century. In the last few decades, natural adhesives have been improved, and a few synthetics have been developed in the laboratories. Large-scale modern application of adhesives is now essential to the construction industry in such materials as plywood, and laminated beams (wall partitions in floors and ceilings). The greatest use of adhesives is in the manufacture of wood products such as plywood.

A sealant is a substance, generally of polysulfide, polyurethane, polyvinyl chloride, acrylic, or silicone used to weatherseal or caulk an opening or for sealing expansion/contraction joints in building structures.

This book is designed to aid our understanding of the nature of sealants and adhesives and specifically their performance in practice and their evolution over the past 50 years. In later chapters, the chemical composition and specific physical properties of generic sealants and adhesives will be discussed along with specific physical properties. In the twentieth century, many natural adhesives have been modified or replaced by synthetics—most of these are polymers (polymerization) that provide great strength and flexibility. The so-called superglues, described in Chapter 4 are polymers called cyanoacrylates.

Because of their physical limitations, many adhesive and sealant materials perform well only in joints of small initial width and subsequent movement. The configuration of the joint, the process by which it is constructed (formed) and access for installation of the adhesive and sealant also impose restrictions on the type of material that might be suitable for a particular application. In actual practice, environmental conditions often dictate additional performance requirements beyond those needed to accommodate movements alone. Selection of the most appropriate material for a particular application is not a simple matter in view of all the variables involved.

Once an understanding is gained of the basic properties of the materials required, then available materials can be classified and related as to their suitability in various types of joints. This information is conveniently displayed in a series of tables and is cross-referenced in later figures that illustrate the details of various joint applications in concrete structures.

This early portion of the chapter discusses field-molded sealing materials used where one surface of the finished joint is open to permit the sealing operation [3]. Sealants used for these applications are listed in Table 1.1. The joint design for an expansion (isolation) joint may consist of a filler strip below the area where the sealant will be placed, bond breaker material to separate the caulking from adhering substrate, and backup materials to support the material from sagging. The appurtenant materials are listed in Table 1.2. Preformed materials used in joints open on at least one surface and materials used as waterstops and gaskets are listed in Table 1.3.

Table 1.4 shows some of the current uses to which the various sealants and adhesives are put and consideration of storage and handling for installation. In cross-referencing types of material, the Roman numeral system is used as in Tables 1.1 and 1.4. Individual field-molded sealant materials are identified by letters A, B, C, and so on as in Table 1.1.

Individual preformed sealant compounds are identified by numbers as given in Table 1.3.

REQUIRED PROPERTIES OF JOINT SEALANTS

For satisfactory performance a sealant and adhesive must:

1. be an impermeable material;
2. deform to accommodate the movement and rate of extension occurring at the joint;
3. sufficiently retain its original properties and shape if subjected to cyclical deformations;
4. adhere to various substrates. This means for all sealants and adhesives, except those preformed sealants that exert a force against the concrete surfaces or are mechanically interlocked within anchorage, the sealant must bond to the substrates and not fail in adhesion (lose its bond to the substrate specified) nor peel at corners or other local areas of high stress.
5. In addition, sealants and adhesives must not internally rupture (that is, fail in cohesion);
6. resist flow due to gravity (or fluid pressure) or acceptable softening at higher service temperatures;
7. not harden or become unacceptably brittle at lower temperatures;
8. not be adversely affected by aging, weathering or other factors for a reasonable service life under the range of temperatures and other conditions that occur.

TABLE 1.1 Material Used for Sealants in Joints Open on at Least One Surface

Group		Field-molded				Preformed
		Thermoplastics		Thermosetting		Compression
Type	I. Mastic	II. Hot-applied	III. Cold-applied	IV. Chemically curing	V. Solvent release	VI. Seal
Composition	(A) Drying oils (B) Non-drying oils (C) Low-melt-point asphalt (D) Polybutenes (E) Polyisobutylenes or combination of D&E All used with fillers such as asbestos fiber or siliceous materials. All contain 100% solids, except D & E, which may contain solvent.	(F) Asphalts (G) Rubber asphalts (H) Pitches (I) Coal tars (J) Rubber coal tars All contain 100% solids. (W) Hot-applied PVC coal tar	(K) Rubber asphalts (L) Vinyls (M) Acrylics (K) Contains 70-80% solids (L) (M) Contain 75-90% solids. All contain solvent. (K) May be an emulsion (60-70% solids).	(N) Polysulfide (O) Polysulfide coal tar (P) Polyurethane (Q) Polyurethane coal tar (R) Silicones (S) Epoxy (N) (R) contain 95-100% solids. (O) (Q) (S) contain 90-100%solids. (P) Contains 75-100% solids. (N) (P) (R) May be either 1- or 2-component system. (O) (Q) (S) 2-component system	(T) Neoprene (U) Butadiene styrene (V) Chlorosulfonated polyethylene (T) (V) Contain 80-90% solids. (U) contains 85-90% solids. (R) Silicones	(3) Neoprene rubber
Colors	(A) (B) Varied (C) Black only (D) (E) Limited	Black only	(K) Black only (L) (M) Varied	(N) (R) (S) Varied (O) (P) Limited (Q) Black only	(T) Limited (V) Varied	Black exposed surfaces may be treated to give varied colors
Setting or curing	Noncuring, remains viscous, A & B form skin on exposed surface.	Noncuring, sets upon cooling. Softens on warming. Hardens on cooling. (W) Resistant	Noncuring, sets on release of solvent or evaporation of water. (M) Remains soft except for surface skin.	2-component system catalyst 1-component moisture pickup from the air	Release of solvent	
Aging and weathering resistance	Low	Moderate (W) High resistance to weather	Moderate	High	High	High
Increase in hardness in relation to (1) age	High	High to moderate (W) No hardness	High	(S) High (N) (O) (P) (Q) (R) Moderate	High	Low
or (2) Low temp.	High	High to moderate (W) No hardness	High	(S) (N) (O) (P) (Q) (R) Low	High	Low

(Continued)

TABLE 1.1 Material Used for Sealants in Joints Open on at Least One Surface *(Continued)*

Group		Fieldmolded					Preformed
		Thermoplastics		Thermosetting			Compression
Type	I. Mastic	II. Hot-applied	III. Cold-applied	IV. Chemically curing	V. Solvent release		VI. Seal
Recovery	Low	Moderate (W) High	Low	(N) (O) Moderate (P) (Q) (R) High (S) LowResistance	Low Low		High Moderate
Moderate to water	(P) (Q) (R) (S) High	Moderate	High (N) (O) Moderate				
High resistance to indentation and intrusion of solids	Low	Low at high temperatures (W) High	Low at high temperatures	High	Low		High
Shrinkage after installation	High	Varies (W) None	High	Low	High		None
Resistance to chemicals	High except to solvents and fuels	(F) (G) High except to solvents and fuels (H) (I) (J) High and fuel resistant (W) High	(K) High except to solvents and fuels (L) (M) High except to alkalis and oxidizing acids	(N) (P) Low to solvents, fuels, oxidizing acids (O) (Q) Low to solvents, but moderate fuel resistance (R) Low to alkalis (S)High	Low to solvents, fuels and oxidizing acids		High
Modulus at 100% elongation	Not applicable	Low	Low	(R) (O) (P) (Q) Low (R) High and low (S) Not applicable	Moderate		
Allowable extension and compression	± 3%	± 5% (W) ± 25% extension	± 7%	± 25% except (S) less > ± 50% (R)	± 7%		Must be compressed at all times to 45–85% of its original width
Other properties	(A) (B) D) (E) non-staining (D) (E) pick up dirt; use in concealed location only	Due to softening in hot weather, usable only in horizontal joints (W) No flow at elevated temperatures	(K) Usable in inclined joints	(N) (P) (R) (S) Non-staining	(U) (V) Non-staining (V) Good vapor and dust sealer		
Unit first cost	(A) (B) (C) Very low (D) (E) Low	(F) (G) (H) (I) (J) Very low (W) Medium	(K) Very low (L) Low (M) High	(O) (Q) (R) High (N) (P) (S) Very high	(T) (U) (V) Low		(3) High

1.5

TABLE 1.2 Preformed Materials Used for Fillers and Backup

Composition type	Uses and governing properties	Installation
(11) Natural rubber (a) Sponge (b) Solid	Expansion joint filler. Readily compressible and good recovery. Solid rubber may function as filler but primarily intended as gasket.	High pliability may cause installation problems. Weight of plastic concrete may precompress it. In constructive joints attach to first placement with adhesive.
(12) Neoprene or butyl sponge tubes	Backup. Where resilience is required in large joints. Check for compatibility with sealant as to staining.	Compressed into joints with hand tools.
(13) Neoprene or butyl sponge rods	Backup. Used in narrower joints, e.g. contraction joints in canal linings and cover-slabs and pavements. Check for compatibility with sealant as to staining.	Compressed into joints with hand tools or roller.
(14) Expanded foam, ethylene vinyl acetate	(a) Expansion joint fillers. Readily compressible, good recovery, nonabsorptive.	Must be rigidly supported for full length during concreting.
	(b) Backup. Compatible with most sealants	Compressed into joint with hand tools.
(15) Expanded polyethylene, polyurethane and polystyrene rigid foams	Expansion joint filler. Useful to form a gap but after significant compression will not recover.	Support in place during concreting. In construction joints attach to first placement. Sometimes removed after concreting where no longer needed.
(16) Bituminous or resin-impregnated corkboard	Expansion joint filler. Readily compressible and resilient. Not compatible and must be isolated from most nonasphaltic sealants.	Support in place during concreting, or attach to preceding placement. Boards easily damaged by careless handling.
(17) Bentonite or dehydrated cork	Filler with self-sealing properties. Absorption of water after installation causes material to swell. Cork can be compressed.	Cork available in moisture-proof liners that require removal before installation. Bentonite in powder form, loose or within cardboard liners.
(18) Wood—Cedar, redwood, pine, chipboard, untreated fiberboard	Expansion joint filler, has been widely used in the past. Swells when water is absorbed. Not as compressible as other fillers and less recovery. Natural woods should be knot-free.	Rigid and easily held alignment during concreting.
(19) Bituminous-impregnated fiberboard	Expansion joint filler. Widely used. Resilient cane fiber used. Has moderate recovery after compression. Should not be compressed more than 50% or bitumen-extruded, which may damage sealant.	Reasonably rigid to hold alignment during concreting or placed against preceding placement.

(Continued)

TABLE 1.2 Preformed Materials Used for Fillers and Backup (*Continued*)

Composition type	Uses and governing properties	Installation
(20) Metal or plastic	(a) Expansion joint filler. Hollow compressible thin-gauge box. Used only in special applications.	Installed as for wood fiberboard materials.
	(b) Backup. Foil, inert to sealants, but shape irregular.	Crumbled and placed in joint.
(21) Glass fiber, mineral wool	(a) Expansion joint filler. Made in board form by impregnating with bitumen or resins. Easily compressed.	Installed as for wood or fiberboard materials.
	(b) Backup. Inert without impregnation so as not to damage sealant.	In mat form or packed loose material or yarn.
(22) Oakum, jute, manila yarn and rope, and piping upholstery cord	The traditional materials for packing joints before installation sealants. Where used as backup should be treated with oils.	Packed in joint to required depth.
(23) Portland cement, grout or mortar	Used in joints in precast units and pipes, and sometimes behind waterstops, to fill the remaining gap when no movement is expected.	Bed (mortar) Inject (grout)

In addition, depending on the specific service conditions, the sealant may be required to resist one or more of the following: intrusion of foreign material, wear, indentation, pickup or attack by chemicals present. Further requirements might be that the sealant has a specific color, resists change of color or is nonstaining to the substrate.

Finally, the sealant must not deteriorate when stored for a reasonable time prior to use and it must be relatively easy to handle and install and be free of substances harmful to the user and substrate or other materials that abut (refer to Chapter 19). In certain locations, regulations may restrict the use of sealants that contain solvents deemed to be polluting.

Chapter 18 lists current specifications that include methods of testing and use for sealants and adhesives used in construction. Those listed are sponsored by the United States Army, Navy, Air Force and General Services Administration (federal); American Society of Testing and Materials (ASTM); American Association of State Highway Transportation Officials (AASHTO); and the Hydro-Electric Power Commission of Ontario (Canada). These specifications are in wide general use and copies can be obtained from the addresses listed in Chapter 23. Local specifications, too numerous to mention, also exist, and can be determined through individual architectural or engineering firms.

TABLE 1.3 Preformed Materials Used for Waterstops, Gaskets, and Miscellaneous Sealing Purposes

Composition and type	Properties significant to application	Available in	Uses
(1) Butyl–conventional rubber-cured	High resistance to water, vapor, and weathering. Low permanent set and modulus of elasticity formulations possible, giving high cohesion and recovery. Tough. Color–black, can be painted.	Beads, rods, tubes, flat sheets, tapes, and purpose-made shapes.	Waterstops, combined crack inducer and seal, pressure-sensitive dust and water-sealing tapes for glazing and curtain walls.
(2) Butyl–raw, polymer-modified with resins and plasticizers	High resistance to water vapor and weathering. Good adhesion to metals, glass, plastics. Moldable into place but resists displacement, tough and cohesive. Black, can be painted.	Beads, tapes, gaskets, grommets.	Glazing seals, lap seams in metal cladding, curtain-wall panels.
(3) Neoprene–conventional rubber-cured	High resistance to oil, water, vapor, and weathering. Low permanent set. Color–basically black, but other surface colors can be incorporated.	Beads, rods, tubes, flat sheets, tapes, purpose-made shapes. Either solid or open- or closed-cell sponges.	Waterstops, glazing seals, insulation, and isolation of service lines. Tension-compression seals, compression seals, gaskets.
(4) PVC (polyvinyl-chloride): Thermoplastic extrusions or moldings	High water and vapor, but only moderate chemical resistance. Low permanent set end modulus of elasticity formulations possible, giving high cohesion and recovery. Tough. Can be softened by heating for splicing. Color–pigmented black, brown, green, etc.	Beads, rods, tubes, flat sheets, tapes, purpose-made shapes.	Waterstops, gaskets, combined crack inducer and seal.
(5) Poly—isobutylene, noncuring	High water, vapor resistance. High flexibility at low temperatures. Flows under pressure, surface pressure sensitive, high adhesion. Sometimes used with butyl compounds to control degree of cure. Color–black, gray, white.	Beads, tapes, grommets, gaskets.	Gaskets, glazing seals, curtain-wall panels, acoustical partitions.
(6a) SBR (Styrene butadiene rubber) (6b) NBR (Nitrite butadiene rubber) Polyisoprene, polydiene–conventional rubber cure	High water resistance. NBR has high oil resistance.	Beads, rods, flat sheets, tapes, gaskets, grommets, purpose-made shapes. Either solid or cellular sponges.	Waterstops, gaskets for pipes. Insulation and isolation of service lines.
(7) Polyurethane foam impregnated with polybutylene	Low recovery at low temperature, can be installed in damp joints. Color–black, gray.	Rods, flat sheets (strips), open-cell sponges.	Gaskets, compression seals.

(Continued)

TABLE 1.3 Preformed Materials Used for Waterstops, Gaskets, and Miscellaneous Sealing Purposes (*Continued*)

Composition and type	Properties significant to application	Available in	Uses
(8) Natural rubber–cured (vulcanized)	High water resistance but deteriorates when exposed to air and sun. Low resistance to oils and solvents. Now largely superseded by synthetic materials. Color– Black.	Purpose-made shapes.	Waterstops, gasket for pipes.
(9) Metals (a) Copper (b) Steel (stainless) (c) Lead (d) Bronze	For waterstops (a) Ductile and flexible, but work hardens under flexing and fractures. (b) Rigid, must be V- or U-corrugated to accommodate any movement and should be anchored. (c) Deforms readily but inelastic to deformation under movement.	Flat and preshaped strips, lead molten or yarn.	(a) (b) Waterstops (c) Protection for joint edges in floors (d) Panel dividers in floor toppings
(10) Rubber asphalts	Natural rubber 8, butyl 1, or neoprene 3 digested in asphalt. High viscosity, some elasticity moldable into place.	Beads, rods, flat sheets (strip), (IIG IIIK), gasket form.	As alternatives to hot- or cold-applied rubber asphalts (IIG IIIK). Gasket for pipes.

AVAILABLE MATERIALS

No single material has the properties necessary to fully meet every application. It is, therefore, a matter of selecting material that is economically and physically acceptable.

For many years, oil-based mastics, bituminous compounds, and metallic materials were the only sealants available. For many applications these traditional materials did not perform well, and in recent years there has been the development of many types of elastomeric sealants whose behavior is largely elastic rather than plastic and which are flexible rather than rigid at normal service temperatures. Elastomeric materials are available as field-molded and preformed sealants. Though initially more expensive, preformed sealants might be cheaper in the long run because they usually have a longer service life. Furthermore, they can seal joints where considerable movements occur that otherwise could not possibly be sealed by traditional field-molded materials. This has opened new engineering and architectural possibilities to the design engineer of various structures.

I have made an attempt in this book to list or discuss every attribute of every sealant. My discussion is limited to those features I consider important so that a suitable choice can be made by the designer, specifier, and end user.

TABLE 1.4 Uses for Field-Molded and Preformed Sealants[1,4,5,6]

Type of application		I. Mastics	II. Hot-applied	III. Cold-applied	IV. Chemical cure	V. Solvent release	VI. Seal	VII. Waterstops	VII. Gaskets	IX. Misc.
			Thermoplastics		*Thermosetting*		*Compression*			
			Field-molded					*Preformed*		
Structures not under fluid pressure; e.g., buildings, bridges, storage bins, retaining walls Note 3	Caulking & glazing	A B D E		L M	N P R S	T U V	3		1 3 4 5	1 2
	Precast panels	A B D E		M	N P R	T V	3		1 3 4 5 7	1 2 10
	Walls (vertical joints)	A B D E			N P R	T V	3 7		7	10
	Roof deck (horiz. joints)	A B D E	F G W		N O P Q	T V	3 7		7	10
	General floors				N O P Q		3			1 4 9d
	Industrial floors		G H W	K	N P Q		3			1 3 9c
	Floors w/ oil&solvents Services		H I J W		N O Q S		3			1
									3 8	11
	Bridges		G W		N O Q		3 7			10 3 Note 2
Containers subject to fluid pressure; e.g., water containing or excluding structures Note 3	Canal linings	C	G W	K			3	1 4		
	Precast pipes	C	G W				3	1 3 4 6 8 9a 9b	1 3 4 5 6 8	10
	Tanks & monolithic pipe	C	G W	K			3	1 3 4 6		10
	Swimming pools		G W				3	4 9a 9b		
	Dams									
	Walls & floors w/ water outside		G W	K			3	1 3 4 6 9b		3 Note 2
Pavements	Walkways	F G W		K	N O P Q		3			
	Highways	G W			N O Q		3			
	Airports	G W			O Q		3			
	Areas w/ fuel spillage	H J W			O Q		3			
Grouting nonworking cracks				K	S					2 3
Suitable in above applications where joint movement is:	None or very small*	A B C D E	F G H I J	K L M	N O P Q R	T U V	3	1 3 4 6 8 9a 9b	1 3 4 5 6 7 8	1 2 9c 9d
	Small**		F G H I J	K L M	N O P Q R	T U V	3	1 3 4 6 8 9a 9b	7	
	Large**				N O P Q R		3	3 4		3 Note 2
	Very large**									

(Continued)

* Contraction joints ** Expansion joints

TABLE 1.4 Uses for Field-Molded and Preformed Sealants *(Continued)*

Type of application	Field-molded						Preformed			IX. Misc.
	Thermoplastics		Thermosetting			Compression				
	I. Mastics	II. Hot-applied	III. Cold-applied	IV. Chemical cure	V. Solvent release	VI. Seal	VII. Waterstops	VII. Gaskets		
Storage life: Limited (l) Over 1 year (o) Emulsions are damaged by freezing	A B C D E(o)	F G H I I(o)	K L M(o)	N O P Q R S(l)	T(o) U V(l)	3 7(o)	1–9(o)	1–8(o)	1–11(o)	
Installation: Knife or trowel (k) Insert (i), heat & pour (h) Mix if 2-component (m) Note 5 Hand gun (g), pressure gun (p) Preposition (pp)	A B(k)(g)(p) C(k)(g) D E (g)(p)	F G H I (h) (w) (h)	K L (g)(k)(p) M(g) preheat to 100°F (40°C)	N O P Q R S (m)(k)(g)(p)	T U V (g)(k)(p)	3(i)	1–9(pp)	1–8(pp)	1234 9d 10(pp) 9c(i)(h) 11(pp)	

Notes to table 1.4

Note 1: Table 1.4 is only a general guide. Before deciding on a material-specific application, all circumstances, particularly the joint movement expected, a suitable joint design and joint detail must be considered.

Note 2: 3 Refers to tension-compression seals described in Section 3.6.

Note 3: Certain sealants may contain substances toxic to potable water or foodstuffs. Check local or national restrictions that may govern use in areas exposed to these.

Note 4: Certain materials are equally suitable for both vertical and horizontal joints. Others are not and while they might stay in place in horizontal joints, they would sag or flow out of vertical joints in hot weather. Asphalt and rubber-asphalt materials are examples of these. Some materials are available in two grades. One known as nonsag or gun-grade athermotropic is suitable for vertical joints. The other known as self-leveling or pour grade is intended for use in horizontal joints.

Note 5: Pot life (time material is still usable after mixing) is limited and correct proportioning and mixing is critical with 2-component materials.

Note 6: Field-molded sealants furnished as follows:

Liquid in drums, cans or cartridges A B D E
Liquid in drums C K W
Liquid in drums or cans O Q
Liquid in cans P S
Liquid in cans or cartridges L N R T U V
Liquid in cartridges M
Solid in cakes for melting F G H I J

SEALANTS

Mastics

These compounds are composed of a viscous liquid rendered immobile by the addition of fibers and fillers. They do not usually harden, set, or cure after application but, instead, form a skin on the surface exposed to the atmosphere. Mastics are Type I and listed in Table 1.1 (A) or (B) drying or nondrying oils (including oleoresinus compounds), (C) low melting point asphalts (D) polybutenes, (E) polyisobutylene or a combination of these materials. With any of these, a variety of fillers can be used, including fibrous talc or finely divided calcareous or siliceous materials. The functional extension-compression range for these materials is approximately +3 percent. They can be used where only very small joint movements are anticipated, and economy of first cost outweighs that of maintenance or replacement. With aging, most mastics tend to harden with the increasing depth of oxidation and loss of volatiles, thus reducing their serviceability. Polybutene and polyisobutylene mastics have a somewhat longer service life than do the older type mastics. The main use of mastics is in caulking and glazing.

Thermoplastics Applied

These are materials that become soft on heating and set to a given hardness on cooling, usually without chemical change. They are generally black and are listed under Type II in Table 1.1. They include (F) asphalts, (G) rubber asphalts, (H) pitches, (I) coal tars, and (J) rubber coal tars. Thermoplastics are useable over an extension-compression range of +5 percent. This limit is directly influenced by service temperatures and aging characteristics of specific materials. Though initially cheaper than some other sealants, their effective life is shorter in practice. They tend to lose elasticity and plasticity with age, to accept rather than reject foreign materials, and extrude from the joints that close tightly or that have been overfilled. Physical properties may be affected adversely by overheating during installation.

Those with an asphaltic base are softened by hydrocarbons, such as oils, gasoline, or jet fuel spillage. Tar-based materials are fuel- and oil-resistant and are preferred for service stations, refueling and vehicle parking areas, airfield aprons and holding pads.

The use of sealant types F, G, H, I, and J are restricted to horizontal joints because they would run out of vertical joints during installation or subsequently evaporate in warm weather. They are used in building roof decks and containers. They have been widely used in pavement joints in the past, but chemically cured, thermoset, field-molded sealants or compression seals are now preferred.

The Type II (W) polyvinylchloride coal tars listed in Table 1.1 have the following enhanced characteristics and properties:

1. They do not flow at elevated service temperature.
2. They are resilient.

3. They have good resistance to weathering and aging.
4. They are resistant to jet fuels or other similarly aggressive chemicals.
5. The allowable extension and compression is up to ± 25 percent.
6. The unit cost is medium.

Polyvinylchloride coal tar sealants are commonly used in pavement and canal liner joints.

Thermoplastics

As noted previously, thermoplastics are long-chain high polymers, whose chains become entangled. They become more flexible with heat, allowing deformation of the mass. They become less flexible as they cool, leading to increased rigidity and hardness, ultimately to brittleness. These materials can be softened and hardened repeatedly by heating and cooling, hence the term *thermoplastic*. Cold-applied solvent or emulsion type thermoplastics are set either by the release of solvents or the breaking of emulsions on exposure to air. They can be heated to a temperature not exceeding 120° F (49° C) to facilitate application, but usually they are handled at ambient temperatures. Release of solvent or water as a consequence of heating can cause shrinkage and increased hardness, resulting in the reduction of permissible joint movement and serviceability. Type III products listed in Table 1.1 include (K) rubber asphalts, (L) vinyl, (M) acrylics, and (X) modified butyl rubbers are available in a variety of colors. Their maximum extension/compression ranges ± 7 percent. Heat softening and cold hardening, however, can reduce this figure.

These materials are restricted in use to joints with small movements. Rubber asphalts listed in Table 1.1 are Type III (K) and are used in canal linings, tanks and fillers for cracks. Type III (L) vinyl, (M) acrylics, and (X) modified butyl rubbers are mainly used in buildings for caulking and glazing.

Thermosetting, Chemically Curing

Sealants in this class are either one- or two-component systems that cure by chemical reaction from the liquid form in which they are applied to a solid state. Type IV sealants listed in Table 1.1 are (N) polysulfide, (O) polysulfide coal tar, (P) polyurethane (urethane), (Q) polyurethane coal tar, (R) silicone, and (S) epoxy-based materials. The properties that make them suitable as sealants for a wide range of uses are their resistance to weathering and ozone, flexibility and resilience at both high and low temperatures, and inertness to a wide range of chemicals, including, for some, solvents and fuels. In addition, the abrasion and indentation resistance of urethane sealants is above average. Thermosetting, chemically curing sealants have expansion/compression ranges of: silicone +100/-50 percent, polyurethane ± 25 percent, polysulfide ± 25 percent, and epoxy-based materials less than 25 percent. Silicone sealants remain more flexible over a wider temperature range than other field-molded liquid sealants. If substrate conditions are clean and otherwise suitable, then thermosetting, chemically curing sealants can stand greater movements than other field-molded sealants and they generally have a much greater service life.

ACCESSORY MATERIALS

Primers

Where primers are required, a suitable priming material compatible with the sealant is usually supplied. In the case of hot-poured, field-molded sealants, the primers are usually high viscosity bituminous or tar cut back with solvent. To overcome damp surfaces, wetting agents may be included in primer formulations, or materials may be used that preferentially wet such surfaces, such as polyamide-cured coal-tar epoxies.

Bond Breakers

Many backup materials do not adhere to sealants and thus, separate bond breakers are not needed where these are used. Polyethylene tape, coated papers, and metal foils often are used where a separate bond breaker is needed.

Backup Materials

These materials tend to limit the depth of the sealant; limit displacement and traffic and fluid pressure; facilitate tooling and shaping and may serve as a bond breaker to prevent the sealant from bonding to the back of the joint. Suitable preformed materials are listed in Table 1.2. In selecting backup material, the recommendations of the sealant manufacturer should be followed to ensure compatibility.

The backup material preferably should be compressible within itself so that the sealant is not forced out as the joint closes, and it should recover as the joint opens. Care must be taken to select the correct width and shape of the material so that it is compressed about 50 percent after installation. Stretching, twisting, or braiding of tube or rod stock should be avoided. Backup material and fillers containing bitumen or volatile materials should not be used with thermosetting, chemical curing field-molded sealants, since they can migrate to, and/or be absorbed at joint interfaces, impairing adhesion.

PREFORMED SEALS

Tables 1.3 and 1.4 cover preformed sealants for two applications, distinguished by how they are installed in the work and their subsequent accessibility. Traditionally, preformed sealants have been subdivided into two classes, flexible and rigid. Most rigid performed sealants are metallic; examples are metal waterstops and flashing. Flexible preformed sealants usually are made from natural or synthetic rubbers, polyvinyl chloride and similar materials, and are used for waterstops, gaskets, and other sealing purposes. Preformed equivalents of certain materials, i.e., rubber asphalts, usually categorized as field-molded, are available as a convenience to handling and installation. In recent times, however, a new and very important use of preformed sealants has been made in the form of strip (gland) seals. (See strip gland seals later in this chapter.) Flexible seals can be installed where the joint is open on at least one surface after the remaining concreting operations are complete and can be replaced in service if necessary.

Rigid Waterstops and Miscellaneous Seals

Rigid waterstops are made of steel, copper, and, occasionally, lead. The stiffness of steel waterstops can lead to cracking in adjacent concrete. Steel waterstops are primarily used in dams and other heavy construction projects. Stainless steel may be desirable in particularly corrosive environments. After welding, annealing of steel sometimes is required to improve flexibility at the weldment. Copper waterstops are used in dams and general construction. They are highly resistant to corrosion but must be handled with care to avoid damage. For this reason and cost, flexible waterstops are often used instead.

Flexible Waterstops

The types of materials suitable and in use as flexible waterstops are shown in Table 1.3. Butyl, neoprene, and natural rubbers have good extensibility and resistance to water and chemicals and they can be formulated to give good recovery and fatigue resistance.

Polyvinyl chloride (PVC) compounds, however, are probably the most widely used now. PVC is not quite as elastic as the rubbers, recovers more slowly from deformation, and is susceptible to oils, but grades with sufficient flexibility (especially important at low temperatures) can be formulated. PVC has the advantage of being thermoplastic and, hence, can be spliced easily on the job or custom-configured for joint intersections. Flexible waterstops, as shown in Table 1.4, are used widely as the primary sealing system in dams, tanks, monolithic pipe lines, flood walls, and swimming pools, to keep water in, and, in buildings below grade or in earth-retaining walls, to keep the water out.

Gaskets and Miscellaneous Seals

Gaskets and materials in the form of a thick ribbon (tape) are sealants widely used with glazing and for precast concrete panels in certain walls. Gaskets also are used extensively at joints between precast pipes and where mechanical joints are needed in service lines. Suitable materials are listed in Table 1.3 and uses in Table 1.4. The sealing action is obtained either because the sealant is compressed between the joint faces (gaskets) or because the surface of the sealant, as in the case of polyisobutylene, is pressure-sensitive and thus adheres. Chapter 17 will explore these types of materials in depth.

Strip (Gland) Seals

These sealing systems are essentially exposed flexible waterstops and are used frequently in bridge expansion joints, either in single units, or serially in modular systems. Neoprene, natural rubber and EDPM (ethylene propylene diene monomer) natural rubber are the main materials in these applications where the seals are anchored at the ends and configured so that they can fold or flex as the joint opens and closes [3].

Compression Seals

Preformed compression seals are compartmentalized and extruded to the required configuration from elastomeric compounds, most commonly neoprene and EDPM. For effective sealing, sufficient contact pressure must be maintained at the joint face, which means the seal must always be in some degree of compression. Compression is accomplished by internal webs that fold and flex to accommodate movement, yet keep the side faces of the seal in contact with the joint faces. Good resistance to compression set (that is, the material must recover to its original shape and size sufficiently when released) is required to obtain these characteristics.

Lubricants are used to facilitate installation of compression seals. For machine installation, additives are needed to make the lubricant thixotropic (giving it increased fluidity during agitation). Special lubricant adhesives that both prime and bond have been formulated for use where improved seal-to-joint face contact is required.

Compression seals are effective joint sealants over a wide range of temperatures in almost all applications. Seals can be used individually or as components for modular systems.

Flexible Foam (Impregnated)

Another type of preformed compression seal is polybutylene-impregnated foam (usually a flexible, open-cell polyurethane compounded with coal tar). This material has found limited application in structures such as buildings and bridges, but its recovery at low temperatures is too slow to follow joint movements and when highly compressed, the impregnated flexible foam exudes and stains the pavement. This generally limits the application to joints where less than ± 10 percent extension/compression occurs at low temperatures, or +20 percent where the temperature is above 50° F (10° C). The material often must be bonded to the joint faces. (See Chapter 16 for more information about impregnated flexible foam.)

Flexible Foam (Non-Impregnated)

One type of sealant in this category is a cross-linked, closed-cell ethylene vinyl acetate expanded-foam material that exhibits good chemical resistance properties to most mild, non-oxidizing acids and alkalines. The material, which can fit any shape or size required, is heat-welded into sheets and cut to length. Heat welding can be done on the job site, to either fabricate lengths or make alterations, with a polytetrafluroethylene-coated (PTFE) iron. An adhesive compatible with ethylene vinyl acetate is used to bond the sealant to the joint face. Based on manufacturer's literature, the allowable movement should be less than 50 percent of the nominal width. Although it has some tension capability, it is preferable that it not be used in tension. (See Chapter 16 for more information about this type of sealant.)

Tension/Compression Seal Systems

Tension/compression seal systems are composed of relatively massive, molded-block-style elastomeric material, commonly neoprene or EDPM, in which a metal bridging plate can be incorporated either at its surface or embedded within. The elastomeric element is anchored to both joint faces, and movement is accommodated by a combination of grooves and shear deformation of the elastomeric component [4]. When this system is used in bridge deck expansion joints, the elastomeric element must be tough and abrasion-resistant against traffic loads of wear.

INTRODUCTION TO ADHESIVES

This second reaction describes the various classes of adhesives used in construction. These classes represent an agreement among adhesive experts on the basic categories of adhesives available today. In order to define these classes, some examples of system formulation are given in the description that will follow. In Chapters 2, 3, and 4, we offer some formulations used in the construction industry and will attempt to show specific areas of application.

From a practical standpoint, it sometimes is difficult to differentiate an adhesive from a sealant. Sealants are adhesives that usually have a considerably lower strength than most structural adhesives. Sealants usually have a greater elongation and are applied in thick glue lines. Table 1.5 is an aid to architects, design engineers, and contractors in the selection of adhesives for use in assembly applications. Because of the wide variety of potential adhesive applications and properties, this data must necessarily be supplemented with knowledge gained by the user through a more intimate association with the particular problem. The intention of this guide is to be useful to the designer of bonded assemblies involving a broad spectrum of substrates rather than zeroing in on specifics. Many formulations are available, particularly in the categories of epoxies and solvent or bodied-solvent cements for thermoplastics. The success of any application requires quality adhesives that must be employed according to the manufacturer's instructions.

Anaerobic

Anaerobic adhesives are generally esters of alkylene glycols and either acrylic or methacrylic acid. The formulations contain hydroperoxides that under ordinary circumstances would polymerize acrylates quite rapidly. Oxygen, however, inhibits the initiation of the polymerization by scavenging free radicals and thus preventing reaction. When the adhesive is placed in a glue line and the accessibility of oxygen is restricted, polymerization proceeds quite quickly. This type of adhesive is appropriate for non-permeable substrates and has been used extensively in machine fasteners.

The typical system might be formulated from polyethylene glycol dimethyl acrylate and cumene hydroperoxide. This system would be stored in a permeable container (usually polyethylene) that allows oxygen to pass through the container wall and inhibit the polymerization. This adhesive system might be modified further by the addition of thickening agents or plasticizers that would allow a gradation into an anaerobic sealant. Such formulations give the methacrylates a wide range of strength and characteristics. A number of other structural modifications are also possible to alter the flexibility and elongation of dimethacrylates. These acrylate-based adhesives are limited by the environmental and chemical resistance of conventional acrylate plastics, although the difunctional monomers might exhibit some improvement in solvent resistance over linear acrylate polymers.

This adhesive type along with cyanoacrylates is the fastest growing class of adhesives. Convenience in use and quick cure has caused the displacement of many of other types of adhesives.

CONCLUSION

The architect or design engineer should be aware of the movements in the structures being designed. All types of joints will be discussed in detail later in this book.

TABLE 1.5 Typical Properties of the 5 Most Widely used Chemically Reactive Structural Adhesives

Property	Anaerobic	Modified Acrylic	Cyanoacrylate	Epoxy	Polyurethane
Substrates bonded	Metals, glass thermosets	Most smooth non-porous	Most non-porous metals and plastics	Most	Most smooth porous
Service temperature range °F (°C)	−67 to 300 (−55 to 149)	−100 to 250 (−73 to 250)	−67 to 175 (−55 to 79)	−67 to 250 (−55 to 121)	−25 to 175 (−157 to 79)
Impact resistance	Fair	Good	Poor	Poor	Excellent
Tensile shear Ksi (MPa)	2.50 17.50	3.70 25.90	2.70 18.90	2.20 15.40	2.20 15.40
Peel strength piw (N/m)	10 (1750)	30 (5250)	3 (<525)	3 (<525)	80 (14,000)
Heat cure or mixing required	No	No	No	Yes	Yes
Solvent resistance	Excellent	Good	Good	Excellent	Good
Moisture resistance	Good	Good	Poor	Excellent	Fair
Odor	Mild	Strong	Moderate	Mild	Mild
Toxicity	Low	Moderate	Low	Moderate	Moderate
Flammability	Low	High	Low	Low	Low
Gap, limitation inch (mm)	0.025 (0.635)	0.030 (0.762)	0.010 (0.254)	None	None

REFERENCES:

1. Kinloch, A.J. 1987. *Adhesion and Adhesives*. Danbury, Connecticut: Grolier Electronic Publishing, Inc.
2. Darrow, F.L. 1930. *Story of an Ancient Art*. (n.p.) Perkins Glue Company.
3. ACI 504R-90, *Guide for Sealing Joints in Concrete Structures*, American Concrete Institute, 1990.
4. Skiest, Irving. 1962. *Handbook of Adhesives*. New York: Reinhold Publishing Co.

CHAPTER 2
EPOXY ADHESIVES AND SEALANTS

Like unsaturated polyesters, epoxies (EP) are thermosetting and can be defined as any molecule containing more than one a-epoxy group (whether situated internally, terminally, or on cyclic structures) capable of being converted to a useful thermoset form. The term is used to indicate the resins in both thermoplastic (uncured) and thermoset (cured) states [1]. Because of similarities, epoxies and unsaturated polyesters often are used for the same purposes, but the somewhat more complex curing and fabricating process and higher cost of epoxies mean they are generally employed in building when a polyester will not do. Epoxy resins, also known as epoxides, are monomers (low modulus) and prepolymers (high modulus) that further react with curing agents to yield the desirable flexible, semi-rigid, or rigid thermosetting plastics. Epoxy resin selection is usually based on performance properties, reactivity, handling characteristics, availability, and cost. The curing agent, also known as the hardener, chemically brings about the change from liquid, paste, or mortar consistency to a solid plastic. It is in this state that the system is usually used—there is limited usage in the uncured, non-cross-linked state.

Few industries exist in which epoxy formulations are not found. Epoxy adhesives and sealants can be formulated to meet the many physical property requirements. Table 2.1 lists typical data for three two-component sealants—a general-purpose sealant, a fast-curing sealant, and a high-performance material. They have found particularly wide use in the construction industry associated with concrete. They form the basis of the concrete industry's rehabilitation and preservation product lines. Different formulations are employed for weatherability, heat, flame, chemical, and electrical resistance and are widely used as structural adhesives. Although epoxies are typically brittle, they can be modified, as shown by elongation of the high-performance sealant. The table also shows that their gel time and cure schedule can be manipulated for fast or slow cure.

TABLE 2.1 Physical and Mechanical Properties of Two-Component Pastes

Property	General purpose	Fast setting	High performance
Color when mixed	Cream	Gray	Gray
Viscosity, Pa's			
Resin	50	260	1
Hardener	35	160	100R/71H
Mixed	45	250	100R/100H
Mix ratio			
Weight/weight	100R/80H*	100R/100H	1.36
Volume/volume	100R/100H	100R/100H	0.97
Specific gravity			
Resin	1.17	1.48	1 hr
Hardener	0.92	1.44	—
Gel time	2 hrs	4 min	5 days/77°F (25°C) or 2 hrs/190°F (88°C)
Cure schedule	24 hrs/77°F (25°C) or 30 min/212°F (100°C)	4 hrs/77°F (25°C)	
Aluminum lap shear strength, Mpa (ksi) (a)			
At -75° F (-60° C)	20 (2.9)	10 (1.5)	29 (4.2)
At 77° F (25° C)	18 (2.6)	20 (2.9)	31 (4.5)
At 180° F (82° C)	<2 (0..3)	<8 (1.2)	18 (2.6)
At 250° F (121° C)	—	—	6.9 (1.0)
T-peel strength, N/mm, lbf/in.)	—	—	2 (11)
Chemical resistance MPA (ksi) gasoline (90 days) JP-4 jet fuel, 7 days	17 (2.4)	—	34 (4.9)
Other substrate, lap-shear, Mpa (ksi) At 77°F (25°C)	23 (3.3) (copper)	—	10 (1.5) (polyetherimide)

* R = resin; H = hardener

Because of their brittleness, unreinforced epoxies cannot be used structurally, but when reinforced with glass fibers and other fibrous, laminar, or particulate reinforcements, they provide excellent structural materials. Adhesion to many substrates is excellent. Consequently, the most common building applications for reinforced epoxies are reinforced plastics, laminates, crack fillers, industrial and decorative (i.e., terrazzo) flooring, and adhesives for metals, masonry, and concretes. A major use is heavy-duty protective coatings. Epoxy formulations are widely used as flooring systems, thin-film and build-up coatings, penetration sealers (with or without decorative surface films), epoxy sand-filled grouts, patching compounds, and mortar overlays. They provide excellent anti-skid surfaces, chemical resistance, and weatherability.

TABLE 2.2 Physical and Mechanical Properties of One-Component Epoxy Sealants

Property	Epoxy-film tape	Thermosetting hot-melt	Epoxy paste
Appearance	Black film on carrier	Semi-green solid	Red-brown paste
Viscosity	—	Extrude at 140-175° F (60-80° C)	300 Pa's
Specific gravity	1.4	1.35	1.44
Elongation at break,%	—	1-2	7
Sag, mm (in.)	0, 0.08 in (2 mm), layer/390° F (200° C)	0, 0.2 in 0(6 mm), bead/340° F (171°C)	< 0.12 in (6 mm) or 0.12 in, 340° F (171° C)
Cure schedule	24 hrs/77° F (25° C) or 30 min/212° F (100° C)	4 hrs/77° F (25° C)	5 days 77° F (25° C) or 2 hrs/190° F (88° C)
Shelf life	3 mos/77° F (25° C)	3 mos/77° F (25° C)	3 mos/77° F (25° C)
Cold rolled steel, lap-shear strength, Mpa (ksi)(a)			
At −22° F (−60° C)	21 (3.0)	23 (3.3)	23 (3.3)
At 77° F (25° C)	16 (2.3)	20 (2.9)	22 (3.2)
At 180° F (82° C)	15 (2.2)	17 (2.5)	19 (2.8)
Salt spray endurance 500 hrs Mpa (ksi)	15 (2.2)	17 (2.5)	18 (2.6)

Table 2.2 shows the typical physical properties for one-component epoxies. This table compares epoxy film tape, thermosetting hot-melt, and epoxy paste. Because the hardener is already incorporated in the main resin in one-component sealants, their shelf lives are considerably lower than two-component sealants or adhesives.

Epoxy adhesives are perhaps the most versatile of the structural adhesives. Although generally characterized as being strong but brittle, they can be formulated to be more flexible without loss of tensile strength. They are able to bond a variety of substrates efficiently and can be formulated to cure either at room temperature or elevated temperatures, under dry or wet conditions. Epoxies can be more expensive than other adhesives but they are quite competitive on a cost-per-formance basis.

As seen in Table 2.3, the shelf life at room temperature for one-component epoxy adhesives and caulks is limited. The stability for each component of two-component materials is often one or more years. This table shows typical properties for epoxy-syntactic sealants. Partly because of the high levels of glass micro-balloons present as filler in the syntactic material, they can be expected to possess good sag resistance.

TABLE 2.3 Physical Properties of Epoxy Syntactic Sealants

Property	One-component	Two-component
Appearance	White paste	Off-white paste

Viscosity
  Resin
  Hardener, Pa's
  Mixed
	—
	—
	—
		Paste
		0.400
		Paste

Extrudability, 0.125 in (3.2 mm) nozzle @ 80 psi (550 kPa), in³/mi, (cm³/min)	150 (25)	350 (21)
Specific gravity	1.0	0.7
Work life	8 hrs 77° F (25° C)	30 min 77° F (25° C)
Cure schedule	1 hr/340° F (171° C)	24 hrs 77° F (25° C)
Shelf life	3–6 mos −40° F (-40° C)	6 mos 77° F (25° C)
Aluminum lap-shear strength, Mpa (ksi)(a) At 77° F (25° C)	6.8 (0.99)	8.6 (2.6)

Compressive strength, Mpa (ksi)
  At 77° F (25° C)
  At 340° F (171° C)
	138 (20)
	69 (10)
		38 (5.5)

DEFINITION OF EPOXY RESIN

For the purpose of this book, epoxy resin is defined as any molecule containing more than one a-epoxy group (whether situated internally, terminally, or on cyclic structures) capable of being converted to a useful thermoset form. The term indicates the resins in both the thermoplastic (uncured) and thermoset (cured) state [2].

HISTORY

Epoxy resins are prepared commercially by:

1. The dehydrohalogenation of the chlorohydrin prepared by the reaction of epichlorohydrin with a suitable di- or polyhydroxyl material or other active-hydrogen-containing molecule.
2. The reaction of olefins with oxygen-containing compounds such as peroxides or per-acids.
3. The dehydrohalogenation of chlorohydrins prepared by routes other than route one.

CHRONOLOGY

- Schrade cites the first commercial attempt to prepare resins from epichlorohydrin in 1927.
- Dr. Pierre Castan, Switzerland, and Dr. S.O. Greenlee, United States, shared credit for the synthesis of the first materials designated as epoxy resins.
- In 1936, Dr. Castan produced a low-melting, amber-colored resin, which then was reacted with phthalic anhydride to produce a thermoset compound. Dr. Trey Freres, Switzerland, shared in the developments.
- In 1939, Devoe-Reynolds explored the epichlorohydrin-bisphenol-A synthesis for the production of new resins.
- By the end of the 1930s, there was considerable work in several countries pointing toward the synthesis of epoxy resins and exploring chemical reactions that would later assume importance.
- In 1948, from a base of almost zero, production of the diglycidyl ether of bisphenol-A resins increased considerably: 20 million pounds in 1954, 30 million pounds in 1957, and 110 million pounds in 1965. After 1965, the production increase began to stabilize at 10 percent per year.
- Because of their high adhesive strength, excellent film-forming abilities, and easy application, they currently dominate the industry in the areas of rehabilitation and preservation, in applications such as crack repair, control joint sealing, patching, bonding, and surface sealing.

Epoxy systems that are properly formulated to be compatible with concrete, wood, steel, and other structural building materials have earned acceptance in a complex building material market. Epoxy formulations typically are considered to be trade secrets. Since the 1970s, some companies, however, have received chemical and/or applications patents in the United States, Canada, and other countries. Even though the percentage of epoxy resins consumed each year in construction is smaller than in other industries, the demands on formulation, application, and cure are probably more strenuous. In most industries, the epoxy formulation can be helped during the curing process by application of heat. This is not so in construction and industrial applications. Each epoxy formulation must be capable of curing on its own under warm or cold conditions of application. The curing agents are formulated to provide long or short pot lives under predetermined hot, warm, or cold temperature curing ranges. These formulations use the following terminology:

- Slow cure—80° to 140° F (27° to 60° C)
- Normal cure—40° to 140° F (5° to 60° C)
- Rapid cure—0° to 140° F (-18° to 60° C)

Super rapid cure formulations have a pot life of only 30 seconds. The curing reaction is exothermic. Exotherm is defined as the increase of temperature of the compound above the cure temperature due to energies released during the polymerization process. Most epoxy formulations require this exotherm to generate the reactive process of the cure. The exotherm can reach temperatures higher than 392° F (200° C). Cold temperature formulations do not rely on the exotherm process to cure; they depend on a combination of chemical and exotherm reactions under isothermal conditions. The curing process is irreversible. Once the resin has been used, it cannot be converted into its original form.

Epoxy formulations during the design stage are categorized by lowest-temperature cure, moisture content of substrate (dry, moist, or underwater abilities of application and adherence), solid content, reactivity, and consistency or viscosity (liquid, paste, or mortar) for a specific use. Conditions of application, the desired form of protection, and the physical properties (such as compressive, tensile and flexural strengths, the modulus of elasticity, and the coefficient of thermal linear expansion, etc.) determine the compatibility of a substrate. After cure and use are determined, the epoxy resin and curing agent types are selected. Then it is a relatively routine task to develop the formulation to meet the desired goal.

Ambient-cured epoxy systems are a mixture of two components: the epoxy resin (component A), and the curing agent (component B). Neither component is stable or commercially useful separately. The components are manufactured separately and not combined until ready for use by the person applying the material. Epoxy resins, also known as epoxides are monomers (low modulus) or prepolymers (high modulus) epoxies that further react with curing agents to yield the desired flexible, semi-rigid, or rigid thermosetting plastics. Epoxy resin selection usually is based on performance properties, reactivity, handling characteristics, availability and cost. The term *epoxy* is used to indicate the resins in both the thermoplastic (uncured) and the thermoset (cured) states. The curing agent, also known as the *hardener*, chemically brings about the change from the liquid, paste, or mortar consistency to a solid plastic. It is in this semi-cured state that the system usually is used; in the uncured, non-cross-linked state there is limited usage.

CHARACTERISTICS THAT JUSTIFY THE USE OF EPOXIES

The characteristics that justify use of epoxies demonstrate their usefulness as structural adhesives in building. They are as follows:

- Solventless liquids at room temperatures
- Almost no shrinkage during curing
- No volatile products evolved during hardening
- Thermosetting

- Gap-filling qualities
- Water, including boiling water, resistance
- Alkali, acid, and solvent resistance
- Good mechanical properties
- Electrical insulation
- Thermal shock resistance (freeze/thaw cycle)
- Fatigue resistance
- Adhesion to most substrates

EPOXY—THE INTERMEDIATE RESIN [3]

The epoxide or ethoxylene group:

$$\underset{H_2C - CH ---}{\overset{O}{\diagup \diagdown}}$$

is a highly reactive three-member ring. Both basic and acidic materials cause it to open. With primary and secondary amines, a gylcidal ether reacts to give an amine alcohol.

$$RNH_2 + H_2C \underset{\diagdown \ \diagup}{-} CH \cdot CH_2 \, O - \rightarrow$$
$$O$$
$$RNH \cdot CH_2 \cdot CHOH \cdot CH_2 -$$

Under the influence of other bases, anhydrides, and Lewis acids, an external or internal epoxide may polymerize to a short-chain polyether.

$$n \, (R^1 \, CH \underset{\diagdown \diagup}{-} CHR^2) \xrightarrow{cat}$$
$$O$$
$$- \, (-CHR^1 - CHR^2 - O-) - n$$

In addition to catalyzing the polymerization of epoxides, anhydrides react with epoxides to form ether alcohols. A small amount of water or alcohol is required to initiate the reaction.

RESINS DERIVED FROM EPICHLOROHYDRIN (EPI)

The most widely used epoxy intermediates are those made by the alkaline condensation of epichlorohydrin (epi) and bisphenol-A (bis) to give the resins of the structural formula, as shown in Figure 2.1. The commercial bis-epi resins are, essentially, mixtures of polymers in which the average n varies from zero to approximately 20. Resins in which the average n is 2 or more are solid at room temperature. It is the liquid bis-epi resins, however, that are of the greatest interest to adhesive formulators. In the synthesis of epoxies, liquid resins are derived by using an excess of epichlorohydrin with respect to bisphenol-A, in the presence of sodium hydroxide as diglycidyl ether of bisphenol-A (DGEBA), possessing a viscosity of 11 to 15 Pa•s at 77° F (25° C).

Side reactions in the preparation form species other than epoxide, such as hydrolyzable chlorine, bound chlorine, and α-glycol, each of which is illustrated in Figure 2.2. Moreover, the monomer content (DGEBA) is usually about 70 percent to 80 percent of the composition, with oligomeric species and side products (such as those shown above) representing the difference.

Solid epoxies can be prepared by chain extension of the liquid resin bisphenol-A in catalytic polymerization, using an excess of the former to provide epoxide terminal groups, as shown in Figure 2.3.

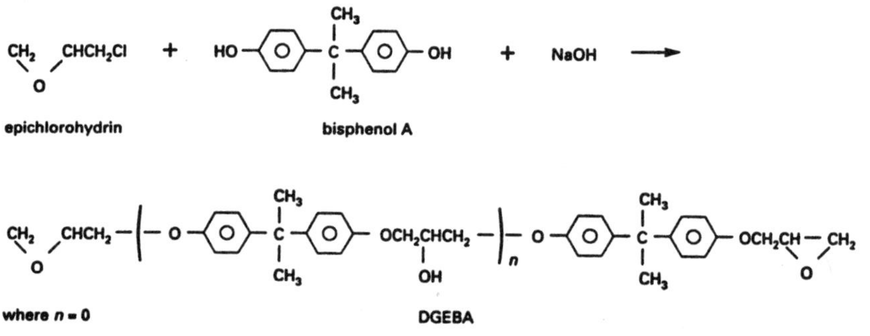

FIGURE 2.1 Diglycidyl ether of bisphenol-A (epoxy) synthesis.

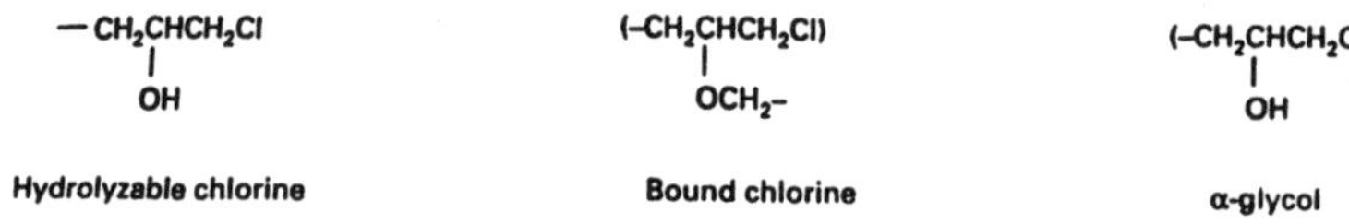

FIGURE 2.2 Side reactions.

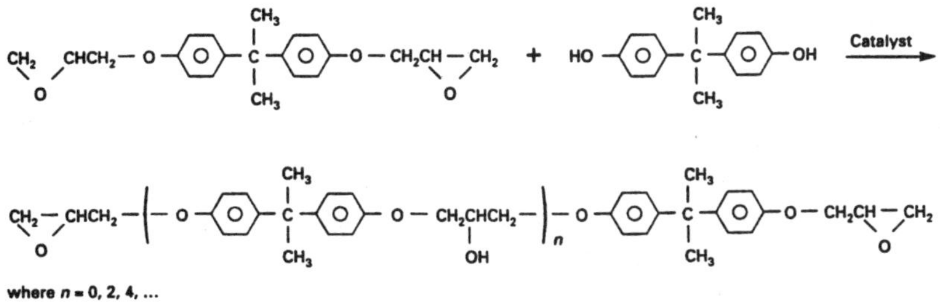

FIGURE 2.3 Solid epoxy preparation.

APPLICATIONS OF EPOXY RESINS

Because of their versatility, epoxy resins are used in thousands of construction and industrial applications. The range of applications includes:

- Adhesives for honeycomb structures and for concrete topping compounds
- Caulking compounds to repair plastic and metals used in buildings
- Caulking and sealant compounds in buildings and highway construction applications, and areas where high orders of chemical resistance are required
- Laminating resins
- Epoxy-based solution coatings, used as maintenance and product finishes, masonry finishes, structural steel coatings, tank linings and concrete-floor paints, gym and floor varnishes
- Decorative floor applications; chemically resistant mortars and floor topping compounds and lightweight, chemically resistant foams

Primarily used as industrial and construction materials, epoxy resins were introduced into the consumer and building trades in the early 1960s as specialty-packaged adhesives and some epoxy-based paints, coatings, and adhesives have since gained wide acceptance on those levels. It therefore should be acknowledged that the basic formulator no longer can be considered the supreme authority on all details of end-use. Specialized uses have led to the evolution of end-use specialists in formulating companies. Structural adhesives and sealants demand use that requires specialized products and expertise.

EPOXY ON-SITE APPLICATIONS

The structural and chemical technology utilized in epoxy formulations is considered an art as well as a science. Performance characteristics and years of successful applications have created an appeal for epoxies from both the performance and economic points of view.

TYPICAL END-USE EPOXY FORMULATIONS

The data presented are starting formulations and are intended to give the reader the basic background about what can be accomplished with a little research. For many general-purpose applications (such as metal to metal, wood to glass, concrete to concrete), where service environments are not unusual and temperatures do not vary markedly from normal ambient, the selection of the adhesive or sealant formulas can be based on cost and convenience. Specialized uses, however, have led to the evolution of specialized formulating companies. These companies compound the diluents, flexibilizers and fillers and develop hardeners to suit the end-use application. The following formulas suggest some of these specialties (Figures 2.4-2.9).

Every formulator has recipes for epoxy patches and sealing compounds; these are only a few of the favorites.

Part A		Part B	
Liquid epoxy resin, EEW-188	100	Jeffamine D-230	25
		N-aminoethylpiperazine (AEP)	4
		Nonyphenol	12
Typical physical properties			
Resin/curing blend volume		2:1	
Physical and exothermic data			
Brookfield viscosity, cps @ 77°F (25°C)		1600	
Gel time, minutes (200 g mass)		28.0	
Peak exotherm temperature, F & C		366.8°F (186°C)	
Time to peak temperature, minutes		36.0	

FIGURE 2.4 Decorative aggregate. (*Courtesy of Texaco Chemical Co., Austin, Texas 78761.*)

Part A		Part B	
Epoxy resin, EEW-188	100	Jeffamine D-230	25
		Nonyphenol	16
		N-aminoethylpiperazine (AEP)	4
		#3 sand/silica flour, 3/1 pbw	1600
Typical physical properties			
Sand/binder volume		8:1	
Matrix exotherm data			
Brookfield viscosity, cps @ 77°F (25°C)		1250	
Gel time, minutes (200 g mass)		42.6	
Peak exotherm temperature, F & C		384.8°F (196°C)	
Time to peak temperature, minutes		50.4	
Compressive strength at failure, psi		> 12,000	

FIGURE 2.5 Sand-filled flooring. (*Courtesy of Texas Chemical Co., Austin, Texas 78761.*)

Part A	
ERL 3794	100
Aluminum powder	60
Santocel 54	5–10
Part B	
MDA (methylenedianiline)	28.5

The above hot-melt mix is cast in ½-inch diameter tubes and allowed to harden at room temperatures. The sticks will remain usable for more than five weeks at room temperature, or much longer when refrigerated.

The part to be soldered must first be heated sufficiently to melt the end of the stick under slight pressure, then additional heat is applied cautiously to effect cure.

FIGURE 2.6 Stick solder. (*Courtesy of Union Carbide Plastics Co.*)

Part A		No. 1	No. 2
Araldite 6020		100	100
Part B			
Thiokol LP-3		50	50
DMAPA (dimethyl amonopropylamine)		10	—
DMP 30		—	6

(for resilient bonds and good low-temperature characteristics)

FIGURE 2.7 Tough general purpose adhesive.

Part A	Pbw
Thiokol LP 3	100
Hydrite clay 121	140
Trimethylaminomethyl phenol	20
Toluene	65
Part B	
DER 331	200
Hydrite clay 121	105
Toluene	5
Basic physical test data	
Tensile strength, psi	450
Flexural strength, psi	345
Compressive strength, psi	4350

This formulation was used to bond old concrete to new concrete after seven days at 80°F (26.67°C)

FIGURE 2.8 Epoxy concrete adhesive.

Part A	
Epotuf 6140	100
Part B	
Thiokol LP 3	50
DMP 30 and DMP 10[a]	7.5
Calcium carbonate[b]	120

DMP 10 is slower than DMP 30, hence more DMP 10 should be used in the summer and more DMP 30 in the winter.

[a] Rohn & Haas Co.
[b] Diamond Alkali Co.

FIGURE 2.9 Concrete bonding adhesive.

SUMMARY

Epoxy resin technology had its genesis in research conducted in the United States and Europe just before World War II. The first resins—the reaction products of epichlorohydrin and bisphenol-A—were produced commercially in 1947. In a little more than 10 years, they achieved a production volume of 30 million pounds, and then 60 million pounds six years later.

In the late 1950s, a number of new epoxy resins, different from the earlier diglycidyl ethers, were introduced and by the end of 1960 at least 25 distinct types of resins were available commercially. At this point, the term epoxy resin became generic. It is now applied to a wide array of materials.

Epoxy resins belong to the thermosetting class of plastics and bear similarities to materials such as phenolics and polyesters. Epoxy resins offer great versatility, low shrinkage, good chemical resistance, and outstanding adhesion. They can bond nearly any substrate, and therefore have a large share of the construction adhesive market.

REFERENCES:

1. Wilkes, Joseph, A. 1988. *Encyclopedia of Architecture, Design, Engineering and Construction.* New York: John Wiley & Sons.
2. Lee, Henry and Kris Neville. 1967. *Handbook of Epoxy Resins.* New York: McGraw-Hill Book Company.
3. Skiest, Irving. 1962. *Handbook of Adhesives.* New York: Reinhold Publishing Company.

CHAPTER 3
ACRYLIC ADHESIVES AND SEALANTS

INTRODUCTION

Acrylics comprise a family, but most building applications are based on *polymethyl methacrylate* (PMMA). It has excellent optical clarity, color stability, and exceptional weatherability and it is lightweight. Today, manufacturers widely use acrylic sealants and adhesives in applications including automotive, electronics, medical and, most important, construction. Acrylics are used in building primarily for glazing, lighting, and decorative features where weather resistance is necessary. Another use is as a sealant for curtain-wall panels. Many companies, however, restrict or ban the use of (meth)acrylic acids, which are the main adhesion-promoting constituents of acrylic adhesives and sealants.

Water-based acrylic sealants and adhesives have become well established as alternatives to solvent-based systems. A few raw material suppliers developed programs demonstrating the feasibility of these sealants and adhesives, created modifying resins and elastomers with the desired performance properties, and developed formulations for specific end-use applications in the construction industry.

Acrylate and methyl methacrylate esters are unsymmetrically substituted ethylene and can be represented by the generic formula:

$$\begin{array}{c} H \quad R \\ \diagdown \quad \diagup \\ C = C \\ \diagup \quad \diagdown \\ H \quad\quad COOR \end{array}$$

in which $R = H$ for acrylates and $R = CH_3$ for methacrylate. The identity and nature of the R and R-' groups determine the properties of both the monomers and the polymers [1]. Polymers of this class are noted for their clear color and the stability of their properties upon aging under severe service conditions. The properties of the polymers can be made to vary from those of extremely tacky sealants and adhesives, to rubbers, tough plastics, and even hard powders. The acrylics used for sealants are tailor-made copolymers of acrylic and/or methacrylic esters and other monomers. Acrylic and methacrylate monomers are extremely versatile building blocks.

Specifically, acrylic sealants are primarily based on ethyl, butyl, and 2-ethyl hexyl acrylate monomers, as well as small quantities of methyl methacrylate, acrylic, and/or methacrylic acids, and other specialty monomers.

Historically, the development of the acrylates and methacrylates proceeded slowly. Acrylic acid, (propenoic acid) was first prepared by the air oxidation of acrolein in 1843 [2], and methacrylic acid was first obtained in 1865 by dehydration and hydrolysis of ethyl α-hydroxyisobutyrate. Methyl and ethyl acrylates were prepared in 1873, but were not observed to polymerize at that time [3]. In 1880, poly (methyl) acrylate was reported by G.W.A. Kahlbaum, who noted that on dry distillation up to 320° F (169° C), the polymer did not depolymerize. Acrylates and methacrylates first received serious attention from Otto Rohm, who observed the remarkable properties of acrylic polymers while preparing for his doctoral dissertation in 1901. A quarter of a century elapsed, however, before he was able to translate his observations into commercial reality. He obtained a U.S. patent on the sulfur vulcanization of acrylates in 1912 [4]. Based on the continuing work in Rohm's laboratory, the first limited production of acrylates began in 1927 by Rohm and Haas Company, in Darmstadt, Germany. Use of this class of compound has grown in the United States to a total of more than 800,000 tons in 1997.

In 1933, a significant breakthrough occurred with the development of the use of acrylics for safety glass. This work was carried out by Rohm and Haas Co. and I.E. DuPont de Nemours and Co. in the United States, by Rohm and Haas Co. in Germany, and in England by Imperial Chemicals Industry Ltd.

The uniqueness of methyl methacrylate as a plastic component accounts for its industrial and construction use and, in this capacity, it far exceeds the combined volume of all the other methacrylates. Today, manufacturers widely use acrylic adhesives in applications including automotive, construction, and medical. Many companies, however, restrict or ban the use of (meth)acrylic acids, which are the main adhesion-promoting constituent in acrylic adhesives. The major automobile manufacturers in the United States mandated that no acrylic acid would be used in the manufacturing of their vehicles by the end of 1997.

New low-odor, noncorrosive adhesives manufactured in Torrington, Conn., are free of (meth)acrylic acid. They maintain their structural integrity while meeting increasingly stringent hazardous materials requirements and eliminating corrosion in applications, including specific sealing applications.

In the 1960s, Lord Corp. was the first manufacturer to develop an acrylic adhesive based on methyl methacrylate [5]. Today, we call this a *first-generation adhesive* (FGA). In 1968, Loctite Corp. received a patent for an anaerobic structural adhesive that was nonflammable—an advance over previous formulations. Manufacturers required scrupulously clean surfaces for bonding on metal and glass, using almost no plastics.

In 1972, DuPont Co. patented and immediately licensed a second-generation acrylic (SGA). Its curing chemistry was similar to that in two-part anaerobic adhesives, but it was based on methacrylate, as were first-generation acrylics. The creation of SGAs (also called toughened or modified acrylics) offered enormous promise. The SGAs appeared to be poised to take over the entire multi-million dollar, structural-bonding market. Then, a megatrend occurred in society—increased consciousness of worker safety issues and environmental pollution issues.

DEVELOPMENT

The formulations listed throughout this chapter are offered as suggested starting points for the manufacturer to use in creating a satisfactory commercial product. Although the suggested formulations exhibit good performance properties and stability, it is recognized that they cannot always be followed exactly, because of raw material availability or for economic reasons. Therefore, one of the purposes of this chapter is to aid the sealant manufacturer in developing an acceptable formulation based on the latest acrylic polymers.

POLYMER SELECTION

The selection of a binder is the most important decision to be made before manufacture of a sealant, caulk, or adhesive. Many major companies such as Rohm and Haas and duPont produce a variety of acrylic emulsions specifically designed for manufacturing latex caulk and adhesives that also exhibit significantly better performance properties than butyl, PVA, or oil-based systems, and at a moderate price.

MIXING EQUIPMENT

To attain maximum ingredient dispersion in the sealant formulations and to fully use the suggested formulations in this chapter, the use of mixers that are capable of high shear at relatively low mixing speed is suggested. This type of mixer will yield a creamy, smooth (not grainy) caulk after approximately 1½ hours of mixing time. Many mixers are equipped with a vacuum to give an air- and bubble-free mix, a desirable feature. Mixers that do not generate adequate shear can produce caulks with poor extender dispersion. This can result in the formation of aggregates, the separation and exudation of ingredients, and the loss of stability. Care should be taken to keep the mixer covered and sealed during operation to prevent the loss of liquid, and consequently, caulk surface skinning, along with the formulation of agglomerates in the finished product.

RAW MATERIALS

Binders

Binders are acrylic emulsion polymers and serve as the fundamental ingredients that are responsible for the caulk's particular physical and chemical properties.

Pigments, Fillers, and Extenders

These solid sealant ingredients contribute to the desired color, rheological properties, performance properties, and bulk. Examples are calcium carbonates, titanium dioxide, and clay.

Dispersants

These additives help to disperse the pigment particles to obtain a stable, homogeneous product.

Plasticizers

The placticizer permits the formulation of soft, flexible caulks with good elongation and recovery.

Wetting Agents

The wetting agent improves mechanical stability, lowers sealant consistency, and enhances package storage stability.

Freeze-Thaw Stabilizers

Ethylene glycol is added to latex caulk formulations to improve sealant freeze-thaw stability.

Adhesion Promoters

Glycidoxypropyl trimethoxy silane is an example of a material that enhances the wet adhesion properties to glass and glazed ceramic tile substrates.

Surface Drying Regulators

The addition of mineral thinner to latex caulk formulations retards skin formation to allow adequate wet-edge and tooling properties.

Defoamers

Defoamers prevent excessive air-entrapment during mixing.

ACRYLIC LATEX-BASED CAULK

Latex sealants and caulks are formulated from acrylic polymers, fillers, and surfactants dispersed in water. Additives include silanes to promote adhesion, ethylene glycol for freeze-thaw stability, and other agents to control the rheological properties.

Acrylic latex caulks should be limited almost exclusively to indoor applications. They are thermoplastic, and in fact are so stiff in the cold that they cannot be applied at temperatures below 40° F (4.4° C). Maximum joint movement is ± 7.5 percent to ± 12.5 percent of the joint width, although we will present formulas later in this chapter that are capable of meeting Class A requirements. Shrinkage is generally high (20 percent to 30 percent), tear resistance is poor, water resistance is low, and there is little recovery from extension. Latex caulks also freeze at temperatures below zero, and are susceptible to rain washout in the first 12 to 18 hours after application [6].

For indoor use, these sealants are fast-skinning and can be painted in 30 minutes to an hour. They are nonstaining, do not harden with age, and will adhere to glass, ceramics, drywall, gypsum plaster, and most metals without priming. Concrete, stucco, masonry, plastics, and most woods, however, must be primed before caulking. Latex sealants also can be applied to damp surfaces or substrates. They are inexpensive and have, for all practical purposes, replaced oil-based caulks for interior residential and light construction applications [7].

SOLVENT-RELEASE ACRYLIC CAULKS

Solvent-release acrylics are based on acrylic acid combined with fillers, catalysts, plasticizers, and solvents. Solvent acrylics are better suited than latex caulks for exterior use. Their odor during cure is so noxious that they cannot be used in closed areas, occupied buildings, or near foodstuffs, which will absorb the odor. Movement potential is $\pm$ 10 percent to $\pm$ 12.5 percent, although recent formulations are available to meet the $\pm$ 25 Class A requirements. Adhesion to most surfaces is excellent with nominal cleaning and no priming required.

Solvent acrylics have good weathering characteristics, resist ultraviolet and ozone deterioration, are nonstaining and color stable. Water resistance, however, is poor and shrinkage is relatively high. They also have low recovery and are thermoplastic. Some sealants must be heated at the job site for application in cold temperatures. Those that require heating are products with lower range solvent content, which means less shrinkage and hardening. Surface skinning will take up to 36 hours or more, during which time the caulking picks up dirt, noticeably discoloring the lighter-colored compounds.

Acrylics are used as binders for both aqueous and solvent-based caulks and sealants.

FORMULATION AND PHYSICAL DATA

The use of waterborne, clear (translucent and transparent) caulks and sealants continues to grow in popularity in the sealant industry. These products were introduced little more than a decade ago, yet today they account for about one-fifth of the total sales in the sealant and caulk market [8]. A new product promises to help clear-latex caulks and sealants expand the overall market. Rhoplex® 2620 emulsion is an innovative acrylic polymer designed as a binder in clear sealant formulations. It provides all the features characteristic of acrylic binders typically employed in clear latex caulks: comparatively low cost, easy handling and cleanup, and the ability to be painted. It also offers a major advantage: the capacity to pass the performance standards for Class A architectural sealants under Federal Specification TT-S-00230C.

This capability means that sealants based on this emulsion have a balance of flexibility and elasticity necessary to tolerate, without failing, a $\pm$ 25 percent change in the size of the joint. Consequently, they can be employed successfully in many applications demanding interior and exterior applications where no conventional latex caulks and sealants—clear or pigmented—previously have been used.

Other latex emulsions will be included that allow sealant formulators to produce clear sealants of exceptional clarity with good performance. They also allow the compounder to produce a flexible sealant without the addition of an external plasticizer. Clear sealants are growing in popularity among formulators and users alike. For formulators, clear sealants eliminate some problems associated with color matching, reproducing the exact color in every batch, and carrying slow-moving inventory. For consumers, clear sealants allow the natural beauty of complex-colored substrates such as wood or glazed tile to show through.

All water-based clear sealants go on white and ultimately cure clear. Most clear sealants based on Rhoplex® reach their clarity faster than clear sealants based on other types of latex emulsions. Based on sealants applied to ½ inch × ¾ inch pine channels, these clear sealants usually dry in about 10 days, whereas clear sealants based on other emulsion polymers can take more than 30 days to develop their ultimate clarity.

CLEAR WATERBORNE CAULKS AND SEALANTS

In the past, caulks were pigmented white almost universally to match the white finishes prevalent on architectural substrates. Today, although white is still the most common color for building exteriors, it no longer dominates. Current tastes increasingly run to the use of earth tones to match multiple finishes on outdoor surfaces. Caulks can be pigmented to match these colored substrates, but this approach presents difficulties. Color-blending caulks is a complicated, labor-intensive process that requires considerable formulating expertise. Matching colors is also a very subjective matter: a formulator can achieve what he thinks is a perfect match only to have the architect or client reject the color.

In many cases, the construction industry has found that an unpigmented caulk constitutes a better alternative on any substrate that is not white. Because these formulations contain nothing to make them opaque and give them color, they are translucent to transparent. Joint sealant materials actually are visible through these clear products. Consequently, the caulking material does not clash with a substrate; instead, it takes on the appearance of that material.

Unpigmented sealants are not a new phenomenon; non-aqueous products of this type have been available for years. Waterborne formulations based on acrylic emulsions, however, offer significant advantages over silicone and other non-aqueous sealants. The aqueous acrylic sealants are less expensive than translucent silicone sealants, and, as latex systems, they are easy to work with, dry rapidly, and can be cleaned up with soap and water. They also can be painted and generally are clearer than any of the silicone sealants on the market.

For more rigorous interior and exterior applications, the construction industry generally prefers products that can meet all the performance requirements for Class A architectural sealants under Federal Specification TT-S 00230C. Traditional latex sealants did not satisfy these criteria, so they usually were not specified for the more difficult applications. Innovations in polymer design and a modified approach to polymer synthesis made it possible for Rohm and Haas chemists to create a binder that provides the balance of cohesive and adhesive strength necessary to tolerate a high degree of joint movement.

To achieve this optimal balance of cohesive and adhesive strength, the developers of this emulsion devised a product with superior mechanical properties. Table 3.1 outlines a typical formulation of this binder that displays an excellent blend of tensile strength and elongation at each of three different test temperatures. Table 3.2 outlines these typical mechanical values. Many latex caulks have a marginal extensibility under these conditions. For the formulation shown in Table 3.1, however, a sample will stretch to almost seven times its original length before breaking. Table 3.3 shows typical values that meet Federal Specification TT-S-00230C, Class A requirements.

These special considerations were uppermost in the minds of the chemists who developed the starting formulation TC-20-3, in Table 3.1. Each ingredient in this formulation was selected specifically for its ability to help produce a high-performance clear sealant with Class A Aymar durability.

Again, ingredients should be added to the formulation in the order listed to avoid stability problems and to facilitate their assimilation. For the same reason,

TABLE 3.1 Clear Sealant Formula TC-20-3

Materials		Pounds	Gallons
Rhoplex® 2620[a]		726.42	85.52
Sodium lauryl sulfate (Stepan WA-100)[b]	premix	1.12	0.13
Water		10.00	1.20
Kathon® LX 1.5%[a]		0.60	0.07
Propylene glycol	premix	7.48	0.86
Ethylene glycol		7.48	0.80
White mineral oil[c]		49.88	6.87
Adhesion promoter (Silane Z-6040)[d]		4.04	0.45
Ammonium hydroxide (28% NH_3)	premix	7.50	0.97
Water		42.64	5.12
Skane® M-8[a]		0.50	0.06
Fumed silica (Cab-O-Sil M-5)[e]		24.94	1.37
Ultraviolet absorber (Tinuven 1130)[f]		0.50	0.05
Hindered amine light stabilizer (Tinuven 765)[f]		0.50	0.05
Add materials to a Sigma blade mixer in the order listed and mix for 30 minutes.			
Total volume, gal	100.54		
Percent solids by weight (theory)	60.16		
Percent solids by volume (theory)	57.46		
Density @ 25° C, lb/gal	8.79		

[a] Rohn and Haas Co., Philadelphia, Pa.
[b] Stepan Chemical Co., Northfield, Ill.
[c] Amoco 35 USP, Amoco Oil Co., Ellicott City, Md.; Drakkeol 35, Penreco, Butler, Pa. (or equivalent)
[d] Dow Corning Corp., Midland, Mich.
[e] Cabot Corp., Tuscola, Ill.
[f] Ciba Geigy Corp., Hawthorne, N.Y.

TABLE 3.2 Typical Mechanical Properties* of TC-20-3 Sealant†

Tensile strength		psi
25° C	max.	40
	@ break	39
0° C	max.	> 118‡
	@ break	> 118
− 18° C	max.	670
	@ break	670
Elongation		%
25° C	max. tensile	1145
	@ break	1145
0° C	max. tensile	>1700
	@ break	>1700
− 18° C	max. tensile	695
	@ break	695
Recovery after 25% extension		%
25° C		92

* ½" gauge length, 0.2 in./min crosshead speed
† ⅛" slabs cured 1 week at 25°C + 2 weeks at 50°C
‡ Test was terminated at 1700% elongation (118 psi tensile)

it is advisable to premix some of the components before they are charged to the mix. One premix should consist of sodium lauryl sulfate, water, and Kathon LX preservative. The second should be composed of mineral oil and adhesion promoter. Ammonium hydroxide and water comprise the third premixture.

Formulators should be aware that the presence of mineral oil in the mix will only inhibit the generation of foam, not completely suppress it. The agitation of the mixer inevitably will introduce some air into the formulation. Consequently, after all the ingredients have been blended thoroughly enough to produce a homogeneous mixture, the formulation should be de-aerated by operating the mixer under vacuum during the last 5 to 10 minutes of the formulating process. The vacuum necessary to draw the entrained air from the sealant will vary from batch to batch, depending on the size of the batch and the type of equipment used.

Table 3.4 is a formula based on another type of acrylic for knife-grade and cartridge (AS-50-1) applications. It is composed of relatively high solids, for a solvent-based system.

Tables 3.5a, 3.5b, and 3.5c present typical physical and mechanical properties based on the above-referenced formula.

The formula in Table 3.6 is the one acrylic copolymer emulsion that allows sealant formulators to produce both clear sealants of exceptional clarity *and* clear and pigmented sealants for good performance.

TABLE 3.3 Typical Performance of TC-20-3 in Federal Specification TT-S-00230C, Class A

Standard	Requirements	TC-20-3
Slump		
Vertical at 50°C	$3/16"$ max.	None
Vertical at 25°C	$3/16"$ max.	None
Horizontal at 50°C	No deformation	None
Horizontal at 25°C	No deformation	None
Extrusion rate	$<= 45$ sec. ($1/2"$ orifice)	< 5
Shore A hardness	15 min., 50 max.	10
Weight loss	10% max.	40%
Cracking	None allowed	None
Chalking	None allowed	None
Tack-free time, hrs	72 max.	$<= 7$
Stain	None allowed	None
Aymar durability, Class A		
Glass	$<= 1\frac{1}{2}$ sq in. total loss in	Pass
Aluminum	bond and cohesive areas	Pass
Concrete		Pass
Wet adhesion, pli*		
Glass	$=> 5$	> 13 FD
Aluminum	$=> 5$	> 13 FD
Concrete	$=> 5$	13 C
Adhesion after UV exposure, pli*		
Glass	$=> 5$	15 LC/C

Note: Shore A hardness, Aymar durability, and peel adhesion were determined after curing one week at 25° C plus two weeks at 50° C.

* Adhesion failure modes:
    FD= Fabric delamination   C = Cohesive   LC - Light cohesive failure

Some of the important benefits are:

- Best ultimate clarity available
- Shortest time to reach ultimate clarity
- Outstanding blush resistance
- Excellent resistance to dirt pick-up
- One binder for both clear and pigmented formulations

Tables 3.7 and 3.8 provide chemical and physical properties for the formulation given in Table 3.6. Clear sealants temporarily can turn white or blush when they are exposed to moisture. Clear sealants based on this polymer have excellent resistance to blushing and can be immersed in water for 20 hours without doing so.

TABLE 3.4 Cartridge and Knife-Grade Formulation (AS-50-1)

Material		Pounds	Gallons
Rhoplex 1950 (63%)	} premix	387.4	44.55
Kathon® LX (1.5%)		1.3	0.15
Octylphenoxy polyethoxy ethanol, 70% active (T-DET 407)		10.0	1.09
Ethylene glycol		7.5	0.81
Thickener (Nastrosol 250 MXR)		3.8	0.34
Dispersant (KTPP)		1.3	0.06
Tamol® 850		1.5	0.15
Calcium carbonate (Camel Tex)		726.5	32.18
Titanium dioide (Ti-Pure R-901)		16.1	0.48
Mix the above for 1¼ hours, then incorporate:			
Rhoplex 1950 (63%)		129.1	14.85
Mineral spirits (Varsol #1)		29.3	4.47
Mix for 10 minutes, then add:			
Defoamer (Nopco NXZ)		1.1	0.15
Mix for for an additional 5 minutes.			
Total volume gallons		99.30	
% solids by weight, (theory)		82.26	
% plasticizer on binder		0.0	
Pigment binder ratio		2.28/1	
Pigment/(binder + plasticizer) ratio		2.28/1	
PVC		47.05	
Volume solids (theory)		69.93	
Density, pounds/gallon @ 25°C		13.24	

TABLE 3.5a Application Properties and Stability Caulk Formulation AS-50-1

pH	
Initial	7.9
Aged*	7.4
Consistency† (sec)	
Initial	6.1
Aged	6.1
Extrusion rate† (g/sec)	
Initial	45.1
Aged*	45.2
Channel cracking‡	
Cure = 1 wk @ 50° C	No cracking
Vertical channel slump§ (mm)	
Room temp.	1.0

* Caulk heat aged 30 days @ 50° C
† Consistency and extrusion rate testing temperature = 25°C
‡ White pine, 6" × ½" × ⅜" channel
§ Aluminum, 6" × ½" × ⅜" channel

TABLE 3.5b Mechanical Properties* Caulk Formulation AS-50-1

Tensile strength† (psi)	
Maximum	21.3
@ break	11.6
% elongation†	
@ maximum psi	335
@ break	680
Flexibility‡	
−15°F, ½" dia. mandrel 180° bend	Pass
Hardness, Shore A	
30 days @ 25° C/50% RH	42

* 25°C/50% RH
† Cure time = 2 weeks @ 25°C/50% RH
‡ Cure time = 2 weeks @ 25°C/50% RH, then 1 week @ 50°C

TABLE 3.5c Rhoplex® 1950—Adhesion of Formulation AS–50–1*

Substrate	Adhesion† (pli)
Glass‡	3A
Aluminum	15 CP
Birch plywood	7A

* Cure = 1 week @ 25°C/50% RH +2 weeks @ 50°C
† Failure modes: A = Adhesive CP = Cohesive peak; pli = pounds per linear inch.
‡ Silane adhesion promoters (such as Silane Z-6040, Dow Chemical Co.) can be added to the formulation to improve adhesion to glass. However, the long-term shelf stability of silanes in the AS–1950–1 formulation is limited.

TABLE 3.6 Formulation TC-28-1*

Material		Pounds	Gallons
Rhoplex 928		844.2	95.39
Water		8.0	0.98
Ammonium hydroxide (28% NH$_3$)	} premix	6.0	0.78
Mix for 5 minutes			
Sodium laurel sulfate (Stepanol, WA-100)		1.2	0.14
Mineral spirits (Texanol)		12.9	1.63
Ethylene glycol		2.5	0.27
Defoamer (Nopcon NXZ)		3.2	0.43
Mix for 5 minutes			
Acrysol® TT-615 (add slowly)		11.0	1.26
Mix for 5 minutes			
Adjust the pH to 7.5-7.8 with 28% NH$_4$OH (add slowly), mix an additional 10 minutes, and then de-air.			
Total		889.0	100.86
Typical values			
% total solids (weight)		59.4	
% total solids (volume)		56.9	
Density (lbs/gal)		8.8	

* For sealants with enhanced freeze/thaw stability, increase the ammonium hydroxide level to 6.5 pounds, reduce the Texanol level to 6.5 pounds, and increase the glycol level (propyl glycol preferred) to 8.9 pounds.

Because this polymer can produce both clear and pigmented sealants, it offers formulators cost savings by:

- Reducing the number of raw materials
- Simplifying inventory management and reordering
- Eliminating the need for an extra bulk storage tank

Some minor changes in ingredients are required to have a pigmented system as shown in Table 3.9, formulation AS-28-1.

ACRYLIC ADHESIVES AND SEALANTS

TABLE 3.7 Clear Sealant Performance Formulation TC-28-1*

Property	TC-28-1
Adhesion to glass, pli	
18 days air dry	1.8A
+1 day at 120° F	11A
+2 days at 120° F	16A
*Flexibility, 1" mandrel**	
Passes, ° F	30
Time to clear, days†	10
Ultimate clarity†	Clear
Tack (cured 1 mo. @ 77° F†)	Low
Water blushing†	
1 hour soak	None

* 1/4" thick wet slab, cured two weeks @ 50°C
† 3/4 × 1/2" unpainted pine channel, cured at 25°C/50% RH

TABLE 3.8 Clear Sealant Dry Adhesion Formulation TC-28-1*

Substrate	TC-28-1 pli
Anodized aluminum	25 FT
Formica® laminate	10–15 A
Polycarbonate	34 FT
PMMA (Plexiglass® sheet)	16A
Vinyl siding	24 FT
Ceramic tile	10 A
Glass	13–17 A
Birch plywood	32 FT
Construction plywood	8–24 SF
Pine clapboard	18–31 SF
Cedar clapboard	7–10 SF
Concrete	4–5 SF

Key:
A = Adhesive   FT = Fabric tear
C = Cohesive   SF = Substrate
FD = Fabric delamination failure

* Cured one week at 50°C.

TABLE 3.9 Formulation AS-28-1*

Materials	Pounds	Gallons
Rhoplex 928	390.0	44.1
Water	20.0	2.4
Octyphenoxy polyethoxy ethanol, OPE-40, 70% active	7.0	0.76
Kathon® LX, 1.5% } premix	1.3	0.15
Water	1.3	0.15
Ethylene glycol	12.5	1.34
Tamol® 850	1.5	0.15
HEC thickener (Natrosol 250 MXR)	1.75	0.15
Potassium tripoly phosphate (KTPP)	1.3	0.06
Plasticizer (Santicizer 160, Benzoflex 9-88, or equivalent)	121.0	14.6
Mineral spirits (Varsol #1)	25.0	3.8
Adhesion promoter (Silane Z-6040)	0.6	0.07
Calcium carbonate (Drikalite)	300.0	13.3
Calcium carbonate (Atomite)	300.0	13.3
Pigment (Ti-Pure R-901)	10.0	0.32
Ammonium hydroxide (28% NH_3)	1.5	0.19
Total	1196.0	95.02
Typical values		
Pigment/binder ratio	2.5	
Pigment/binder and plasticizer solids	1.67	
% plasticizer on polymer solids	49.6	
% total solids (weight)	81.5	
% total solids (volume)	71.6	
Density (lbs/gal)	12.6	

* Source: Rohn and Haas

Tables 3.10, 3.11, and 3.12 are typical formulas based on A-920 emulsion that will provide properties that meet performance requirements of TT-S-00230, Class A, formulas AS-920-A1, AS-920-B1, and AS-920-C1.

ANAEROBIC AND MODIFIED ACRYLIC ADHESIVES

Soaring energy and labor costs and the need to improve productivity have led product managers to review the advantages and disadvantages of using high-performance, convenience-engineering adhesives as substitutes for mechanical fastening methods. Two relatively new classes of these structural anaerobics and modified acrylics have begun to achieve acceptance as assembly tools because of

TABLE 3.10 Formulation AS-920-A1

Materials	*Pounds per 100 gallons*
Rhoplex A-920 (62%)	468.5
Premix	
Water	1.5
Octylphenoxy polyethoxy ethanol, 70% active (T-DET-0407)	11.4
Kathon® LX, 1.5%	1.5
Ethylene glycol	14.3
Tamol® 850	1.4
Skane® M-8	1.1
Defoamer (Nopco NXZ)	2.0
Premix	
Plasticizer (Santicizer S-160 or equivalent)	144.9
Thickener (Natrosol 250HR)	3.4
Premix	
Mineral spirits (Varsol #1)	11.4
Adhesion promoter (Silane Z-6040)	0.68
Dispersant (KTPP)	1.4
Calcium carbonate (Drikalite)	512.2
Pigment (TiPure R-901)	11.4
Ammonium hydroxide (28% NH_3)	2.3
Total weight	1189.4

Add the ingredients in the order listed and mix in a Sigma-type mill for about 1.5 hours. Final pH should be adjusted to 7.8/8.0 using 28% ammonium hydroxide. Vacuum and de-air for about 5 minutes at about 25 in Hg.

Typical values	
% solids wt	82.0
Pigment binder ratio	1.8
% plasticizer on polymer solids	50.0
Density (lbs/gal)	11.9
pH	7.8-8.0

TABLE 3.11 Formulation AS-920-B1

Materials	Pounds per 100 gallons
Rhoplex A-920 (62%)	396.5
Premix	
Water	47.6
Octylphenoxy polyethoxy ethanol, 70% active (T-DET-0407)	9.7
Kathon® LX, 1.5%	1.3
Ethylene glycol	12.1
Tamol® 850	1.2
Skane® M-8	1.0
Defoamer (Nopco NXZ)	1.9
Premix	
Plasticizer (Santicizer S-160 or equivalent)	122.6
Thickener (Natrosol 250HR)	2.4
Premix	
Mineral spirits (Varsol #1)	9.7
Adhesion promoter (Silane Z-6040)	0.58
Dispersant (KTPP)	1.2
Calcium carbonate (Drikalite)	654.5
Pigment (TiPure R-901)	9.7
Ammonium hydroxide (28% NH_3)	2.1
Total weight	1274.1

Add the ingredients in the order listed and mix in a Sigma-type mill for about 1.5 hours. Final pH should be adjusted to 7.8-8.0 using 28% ammonium hydroxide. Vacuum de-air for about 5 minutes at about 25 in Hg.

Typical values

% solids wt	82.0
Pigment binder ratio	2.7
% plasticizer on polymer solids	50.0
Density (lbs/gal)	12.7
pH	7.8-8.0

their abilities to cure rapidly at room temperature without mixing or metering and to adhere to a variety of substrates. Although they are not found extensively in construction applications, they should be mentioned with a brief list of some of their properties [9].

Basic to the high performance of modern engineering adhesive systems is the ability to provide both the rigidity required for high tensile strength and the toughness and flexibility required for high peel and impact resistance. Although approaching these requisites via different chemical routes, the aerobic adhesives and modified acrylic adhesives are each capable of producing this somewhat paradoxical combination.

Anaerobic structural adhesives represent the culmination of an evolutionary process that began a number of years ago with the familiar anaerobic locking and

TABLE 3.12 Formulation AS-920-C1

Materials	Pounds per 100 gallons
Rhoplex A-920 (62%)	346.6
Premix	
Water	80.2
Octylphenoxy polyethoxy ethanol, 70% active (T-DET-0407)	8.4
Kathon® LX, 1.5%	1.1
Ethylene glycol	10.6
Tamol® 850	1.0
Skane® M-8	0.8
Defoamer (Nopco NXZ)	1.7
Premix	
Plasticizer (Santicizer S-160 or equivalent)	86.0
Thickener (Natrosol 250HR)	2.5
Premix	
Mineral spirits (Varsol #1)	8.4
Adhesion promoter (Silane Z-6040)	0.51
Dispersant (KTPP)	1.0
Calcium carbonate (Drikalite)	808.4
Pigment (TiPure R-901)	8.4
Ammonium hydroxide (28% NH_3)	1.7
Total weight	1367.3

Add the ingredients in the order listed and mix in a Sigma-type mill for about 1.5 hours. Final pH should be adjusted to 7.8-8.0 using 28% ammonium hydroxide. Vacuum de-air for about 5 minutes at about 25 in Hg.

Typical values

% solids wt	82.5
Pigment binder ratio	3.8
% plasticizer on polymer solids	4.0
Density (lbs/gal)	13.7
pH	7.8-8.0

sealing adhesives. The initial products, although ideally suited to improving the strength and reliability of threaded fasteners and sealing leakage, did not possess the flexibility and toughness required of true structural adhesives. Today, sophisticated combinations of methacrylate and urethane chemistry provide the requisite properties while maintaining the anaerobic characteristics of stability by air, and cure in the absence of air.

Modified acrylic adhesives are also the result of lengthy developmental efforts. Desired properties in these products are achieved by using solutions of polymeric rubbers in low molecular weight mono-methacrylate monomers, most often methyl methacrylate. Recent concerns about plant and factory problems arising from the use of moderately toxic, flammable methyl methacrylate have led to the development of a modified acrylic engineering adhesive that minimizes toxicity and flammability through the use of an higher molecular weight monomethacrylate. To differentiate this latest modified acrylic from more conventional modern acrylics, the two systems will be referred to respectively as non-volatile and volatile modified acrylics. As indicated in Table 3.13, both structural anaerobics and modified acrylics are 100 percent solids, nonsolvent systems. The methyl methacrylate base of most volatile modified acrylics, however, does introduce significant plant handling problems. The structural anaerobics and nonvolatile modified acrylics, based on methacrylate monomers of lower vapor pressure, are considerably less toxic (particularly through inhalation) and less flammable. Because of these factors, they are generally less difficult to handle in a plant environment.

Key factors in the curing process, during which the uncured liquid is transformed into a solid polymeric adhesive, include the speed of cure, the gap or joint through which the cure will process, and the substrates capable of being bonded. Table 3.14 presents a summary of curing data for anaerobic and modified acrylic engineering adhesives. Although a number of variables influence cure speed (substrates, joint size, choice of activator), the time to reach fixture or handling strength is generally in the following order: nonvolatile modified acrylics, anaerobic, and volatile modified acrylics. Once fixture has occurred, there is little difference in the time required to reach 50 percent and 100 percent of ultimate strength.

Table 3.15 provides a summary of cured performance properties of the engineering systems on prepared substrates. The data clearly highlight the superior

TABLE 3.13 Properties of Uncured Anaerobics and Modified Acrylics

	Anaerobic	Volatile acrylic	Nonvolatile acrylic
Toxicity	Very low	Moderate	Low
Flashpoint, °F	> 200	40-60	> 200
Solvents	None	None	None

TABLE 3.14 Curing Anaerobics and Modified Acrylics

	Anaerobic	Volatile acrylic	Nonvolatile acrylic
Fixture time, min*			
2 mils	1-5	5-12	1-2
20 mils	5-10	15-20	4-5
50% strength, hr	4-8	4-8	4-8
100% strength, hr	24	24	24
Maximum gap tolerance, mils	30-40	30-40	30-40
Recommended surfaces	Clean/prepared metals, glass ceramics	Oily and *as-received* metals, most plastics, concrete	

* Range for a variety of prepared and unprepared metal and plastic substrates

TABLE 3.15 Cured Anaerobics and Modified Acrylics on Prepared Metal Substrates

	Anaerobic	Volatile acrylic	Nonvolatile acrylic
Tensile shear–steel, psi* (ASTM D 1002)			
2 mils	4500	5000	2700
20 mils	3000	3200	1000
Tensile shear–aluminum, psi† (ASTM D 1002)			
2 mils	3300	3900	2600
20 mils	2200	3000	1000
Impact–steel, ft lb/in. (ASTM D 950)	25	12	11
T-Peel–steel, lb/in. (ASTM D 903)	60	38	28
Thermal operating range, °F	−60 to 400	−40 to 300	−40 to 250
Hot strength, % RT (ASTN D 1151)			
200° F	56	38	30
300° F	49	12	11
350° F	18	0	0
Solvent resistance (ASTM D 896)			
% retention (188° F/14 days)			
Motor oil	100+	79	66
Gasoline	55	51	22
Water	55	46	61
Phosphate esters	98	93	75

* Degreased, sand-blasted steel
† Degreased aluminum

performance characteristics of structural anaerobics when the metal substrate has been prepared specifically for bonding operations. Table 3.15 also illustrates the decrease in tensile shear on going from volatile to nonvolatile modified acrylics.

Tensile shear values for modified acrylics on a variety of substrates, for which structure the anaerobics generally are not recommended, are given in Table 3.16. Comparative strengths of the two systems on a received, unprepared test specimens mirror differences tabulated for prepared metal substrates. With the exception of epoxy-glass and polycarbonate, where the volatile modified acrylics show clear advantage, there is little differential for a variety of plastics.

SOLVENT RELEASE SEALANT SYSTEMS

Man has used solvent-based adhesives and sealants since antiquity. Prominent examples include the pine oleoresin to assemble writing brushes and other simple articles in China. From these beginnings, the chemical industry developed products that are more specialized. A solution of polybeta pinene tackifier and natural rubber coated on proper substrate gives a high-performance, pressure-sensitive tape, even by today's standards. Solutions of rosin, polymerized rosin and rosin ester tackifiers with natural rubber (later through SBR and other synthetic elastomers) produced adhesives for myriad uses, including packaging, laminating,

TABLE 3.16 Tensile Shear Strengths of Volatile and Nonvolatile Modified Acrylics

Substrate	Volatile acrylic	Nonvolatile acrylic
As-received steel–2 mils	4500	2000
Epoxyglass		
2 mils	2900	1800
20 mils	2000	1000
ABS		
2 mils	700*	600*
20 mils	500*	500*
Acrylic–2 mils	600*	900*
Nylon–2 mils	300	300
Phenolic–2 mils	1200*	1100*
Polyester–2 mils	1200	1100
Polycarbonate–2 mils	1000	300*
Polystyrene–2 mils	300*	300*

* Substrate failure

construction, product assembly, and tapes and labels. Using similar formulations, the development of EVA polymers, butyl rubber, and block copolymers made hot melt adhesives and sealants possible. Large-scale cracking of petroleum fractions made available not only byproduct unsaturates such as isoprene for synthetic rubber and block copolymers, but piperylene and other fractions for synthetic hydrocarbon tackifiers. This, in turn, made synthetic adhesives and sealants possible without using natural products or derivatives.

As the 1990s ended perhaps the most prominent trend in the adhesive and sealant industry was a decline in use of solvent-borne materials, principally for environmental reasons. Because of economics and performance, however, some solvent-borne systems, including solvent-based acrylic sealants, survive.

FORMULATIONS

Acrylic sealant and adhesive formulations are usually based on polymers of acrylate ester mixtures, most commonly the methyl, ethyl, butyl, and 2-ethylhexyl derivatives. These are low Tg material, flexible, tacky, and soft at normal temperatures. Adjusting the ratios of the esters and other monomers allows tailoring the polymer for a specific end use. For example, addition of methyl methacrylate or other methacrylates increases the tensile strength, at the same time raising the Tg. Inclusion of small amounts of functional monomers, such as methacrylatic acid or hydroxyethyl methacrylate, can have a profound effect on rheology and adhesion to certain surfaces.

Polymerization of the chosen monomer can follow many different schemes. One procedure starts out by charging a certain amount of monomer mixture, solvent, and catalyst to an appropriate reactor.

OTHER FORMULAS

This section will cover a few formulations based on solution acrylic polymers suggested by Schnee-Morehead, Inc. [10]. These are acrylic polymer solutions of varying solids content in toluene. Tables 3.17 and 3.17a are based on a 68 percent solid polymer in both clear and pigmented formulae, whereas Table 3.18 is a higher-solids pigmented formula.

Tables 3.19, 3.20, and 3.21 list the typical performance properties for these formulations.

TYPICAL APPLICATIONS

Vertical Mastics

Exterior elastomeric masonry coatings serve both in new construction and to bridge and cover flaws and cracks in older structures. The vertical mastic is applied in thick films (usually from a water-based system), which form an elastic membrane when the water evaporates. The membrane remains elastic even at low temperatures, is nontacky, and resists dirt pickup at high temperatures. Con-

TABLE 3.17 Solvent Release Sealant Formulations Based on Solution Acrylic Polymers, Clear

Materials	% by weight
Clear	
SM ARO* acrylic polymer (68% NVM in toluene)	80.00
SM 7603 acrylic polymer (83% NVM in xylene)	—
Drilalite† calcium carbonate	—
Duramite† calcium carbonate	—
Aerosil 200‡	6.30
Thixatrol ST§	—
Toluene	13.60
Xylene	—
MEMO¶ Silane	—
Glymo¶ Silane	0.10

* Schnee-Morehead, Inc., Irving, Texas
† ECC International, Sylacauga, Ala.
‡ Degussa Corp., Ridgefield Park, N.J.
§ Rheox, Inc. Hightstown, N.J.
¶ Hüls America, Inc., Piscataway, N.J.

TABLE 3.17a Solvent Release Sealant Formulations Based on Solution Acrylic Polymers, Pigmented

Materials	% by weight
White #1	
SM ARO* acrylic polymer (68% NVM in toluene)	48.70
SM 7603 acrylic polymer (83% NVM in xylene)	—
Drilalite† calcium carbonate	47.00
Duramite† calcium carbonate	—
Aerosil 200‡	—
Thixatrol ST§	1.90
Toluene	2.30
Xylene	—
MEMO¶ Silane	—
Glymo¶ Silane	0.10

* Schnee-Morehead, Inc., Irving, Texas
† ECC International, Sylacauga, Ala.
‡ Degussa Corp., Ridgefield Park, N.J.
§ Rheox, Inc. Hightstown, N.J.
¶ Hüls America, Inc., Piscataway, N.J.

TABLE 3.18 Solvent Release Sealant Formulations Based on Solution Acrylic Polymers

Materials	% by weight
White #2	
SM ARO* acrylic polymer (68% NVM in toluene)—	
SM 7603 acrylic polymer (83% NVM in xylene)	42.78
Drilalite† calcium carbonate	—
Duramite† calcium carbonate	46.40
Aerosil 200‡	—
Thixatrol ST§	4.26
Toluene	—
Xylene	6.46
MEMO¶ Silane	0.10
Glymo¶ Silane	—

* Schnee-Morehead, Inc., Irving, Texas
† ECC International, Sylacauga, Ala.
‡ Degussa Corp., Ridgefield Park, N.J.
§ Rheox, Inc. Hightstown, N.J.
¶ Hüls America, Inc., Piscataway, N.J.

TABLE 3.19 Typical Performance Properties
(Cured 28 days @ 77°F—7 days @ 120°F)

Property		Clear	
% NVM by weight		60.70	
Density, lbs/gal ASTM D 1475		8.64	
Pigment/polymer		0.11/1	
Slump (inch)		> 0.1	
Extrusion rate (sec/g) 60 psi, 0.104" orifice		< 1	
Tack-free time (minutes) ASTM D 2377		30	
Mechanical/properties ASTM D 412			
Maximum tensile strength (psi)		20	
% elongation @ break		> 1000	
% recovery		80	
Shore A hardness ASTM C 661		20	
180° F peel adhesion (pli/% COH) ASTM C 794			
		Dry/Wet	
Glass		9/100	0/AF
Anodized aluminum		9/100	8/100
Wood		9/100	4/AF

ventional thin paint films must be relatively hard at normal-use temperature in order to achieve the toughness, abrasion resistance, and low degree of permeability that is required. Consequently, conventional paint films tend to be brittle at low temperatures, but they are adequate for the task for which they were designed. Vertical caulks and mastics provide low-cost, lightweight, and easy construction or repair of building walls with high insulation value. Vertical caulk development had its origin in Europe after World War II, when the need for new, rapid, and inexpensive methods to facilitate the rebuilding of Europe advanced the technology.

Aqueous Acrylic Caulk

Applying this type of caulk to the exterior joints around windows and doors in both residential and commercial structures can be a messy operation when ordinary butyl rubber or linseed oil-based compounds are used. Moreover, it is likely

TABLE 3.20 Typical Performance Properties

Property (Cured 28 days @ 77°F—7 days @ 120°F)	White #1	
% NVM by weight	82.10	
Density, lbs/gal ASTM D 1475	12.05	
Pigment/polymer	1.42/1	
Slump (inch)	> 0.1	
Extrusion rate (sec/g) 60 ps, 0.104" orifice	1	
Tack-free time (minutes) ASTM D 2377	30	
Mechanical/properties ASTM D 412		
Maximum tensile strength (psi)	30	
% elongation @ break	> 1000	
% recovery	80	
Shore A hardness ASTM C 661	35	
180° F peel adhesion (pli/% COH) ASTM C 794	Dry	Wet
Glass	10/100	0/AF
Anodized aluminum	9/100	7/100
Wood	9/100	4/100

TABLE 3.21 Typical Performance Properties

Property (Cured 28 days @ 77°F—7 days @ 120°F)	White #2	
% NVM by weight	86.30	
Density, lbs/gal ASTM D 1475	11.85	
Pigment/polymer	1.43/1	
Slump (inch)	> 0.1	
Extrusion rate (sec/g) 60 ps, 0.104" orifice	5	
Tack-free time (minutes) ASTM D 2377	180	
Mechanical/properties ASTM D 412		
Maximum tensile strength (psi)	80	
% elongation @ break	> 500	
% recovery	50	
Shore A hardness ASTM C 661	50	
180° F peel adhesion (pli/% COH) ASTM C 794	Dry	Wet
Glass	6/100	6/100
Anodized aluminum	6/100	6/100
Wood	6/100	4/100

that the result of this labor will be disappointing. In several years, after the new caulk has weathered, discolored, and lost its flexibility, the task must be performed again. A new type of sealant, however, promises to solve these problems.

Acrylic Latex Caulk

This new class of compounds is known as acrylic latex caulk. Now available commercially, it is being stocked by building product supply houses and hardware stores. The brand names are endless. The caulks also are suitable for semirigid building construction where only slight to moderate joint movement is anticipated. These systems are based on high solids, thermoplastic polymers that exhibit long-lasting flexibility, outstanding resistance to ultraviolet light, rapid skin-over and cure-through time, plus the convenience and ease of application typical of an aqueous product.

ADVANTAGES OF ACRYLIC LATEX CAULKS OVER OIL-CONTAINING AND/OR SOLVENT-CONTAINING CAULKS

- Easy application and convenient cleanup make acrylic latex caulks attractive both esthetically and economically. These caulks are similar to acrylic latex in that they: are not sticky, messy, or difficult to apply by gun, work, or tool. Water used for tooling also can be used as a solvent to wipe the uncured caulk from the sash, glass, hands, and clothing. Less time and effort are required to make a neat job and to clean up after the job is through.
- Acrylic latex caulks can be applied over damp surfaces. Oil- and solvent-based caulks and sealants tend to curl away from wet substrates. When applied to damp surfaces, acrylic sealants do not suffer any loss in adhesion, which permits exterior caulking shortly after rain.
- Latex paints, with their desirable combination of easy application, fast drying, and convenient cleanup, can be applied with excellent results over acrylic sealants within 30 minutes after the caulk is gunned into the joint. The ability to accept quickly and hold exterior latex paint is a characteristic not shared by competitive solvent-containing sealants. Latex paint applied over oil, butyl, polysulfide, and polyurethane materials can fail by cracking almost immediately. Silicone sealants do not accept latex paints at all. Short- or long-term adhesion of latex paints to acrylic caulk is excellent.
- Acrylic latex caulks have short tack-free time, under normal conditions less than 20 minutes, a valuable asset in preventing early dirt pickup. Acrylic caulks also cure rapidly, usually within 48 hours after application. Cure-through time in an unusually deep joint is about 72 hours.
- Acrylic latex-based caulks exhibit exceptional retention of their initial flexibility, elongation, and recovery properties. These systems show outstanding resistance to hardening, embrittling, and cracking with age and exposure. Generally, acrylic caulks exhibit improved properties compared to butyl and oil-based linseed caulks.
- Acrylic caulks can be formulated in a variety of colors, including bright whites and pastels. Because of the clarity of the acrylic polymer, minimum amounts of prime pigments are necessary to produce true colors. Because these polymers permit high pigment loading, it is possible to formulate them in a broad cost-performance range.
- Color stability of acrylic latex caulks is excellent. They display the high resistance to degradation from ultraviolet light that is typical of acrylic polymers. Unlike oil, butyl, and polysulfide sealants that turn yellow on exposure, acrylic systems retain their original shade even after extended Fade-O-Meter and Weather-O-Meter exposure. Because of the exceptional color retention of the acrylic caulk, painting can be eliminated when the color of the caulk that is selected matches adjacent trim.

- The acrylic-based caulks are one-part systems that have excellent package or shelf stability. If formulated and handled as recommended, these compounds will not separate on standing even when stored for long periods. In addition, these materials are non-staining. They can be used on wood, brick, concrete, marble, or any other commonly used building materials without danger of discoloring the substrate.

ADVANTAGES OVER POLYVINYL ACETATE LATEX CAULKS

- Compared to earlier latex caulk systems based on polyvinyl acetate emulsion polymers, acrylic latex sealants show superior flexibility, elongation, and recovery properties, and better retention of flexibility on exterior exposure.
- The superior pigment binding of acrylic latex polymers permits the formulation of sealants with higher solids. Due to high solids in these formulations, there is less caulk shrinkage after application.
- Completely nonclumping sealant formulations can be prepared in both gun-grade and squeeze-tube grade consistencies. Consequently, caulk application is easier and caulked joints have a smoother, more even appearance.
- Acrylic latex caulks possess outstanding freeze-thaw stability and excellent package or shelf stability. Recommended formulations successfully pass at least five freeze-thaw cycles without noticeable change in appearance and performance. Packaged caulks have been stored for one year without changing.

CHARACTERISTICS OF ACRYLIC LATEX POLYMERS

As supplied, acrylic polymers generally are opaque, white aqueous emulsions containing 100 percent acrylic polymer at 55 percent solids. Films of these polymers are clear, flexible, and resistant to damage from ultraviolet light. The sealants or caulks adhere well to a variety of porous and nonporous substrates. The pigment binding ability of the polymer is excellent and permits very high pigment loading in compounds.

Table 3.22 summarizes some of the important properties of an acrylic emulsion. These properties are typical for all the formulations found in the text.

MIXING INSTRUCTIONS

Sealants and caulking compounds based on acrylic latex polymers are easy to prepare. No premixing, grind, or letdown is necessary. Formulations can be made by consecutive, single-step addition of the required ingredients to a single con-

TABLE 3.22 Properties of an Acrylic Emulsion

Properties	Values
Appearance	Milky white liquid
Solids content	55 ±0.5%
pH, as packed	4.5 ±0.25%
Weight per gallon	8.9 lbs
Emulsifying system	Anionic
Specific gravity at 77° F (25° C)	1.07
Brookfield viscosity (LVT viscometer #2 spindle 60 rpm)	< 300 cps
Minimum film-forming temperature	< 36° F (< 2° C)
Tukon hardness of polymer film (KHN)	< 1

tainer, with appropriate mixing. It is difficult to produce off-grade batches, provided ingredients are added in the order listed and in the amounts recommended in the formulation. The use of closed mixing vessels is suggested so those solids do not climb through the water evaporation phase during mixing. Closed sigma-blade mixers, Ross mixers, Pony mixers, Ribbon blenders, and similar types of high-shear, low-speed agitation equipment are suitable for preparing acrylic latex-based caulks.

It is important to allow mixing to continue for at least 1½ hours. Although it is possible to achieve an apparently homogeneous batch of sealant in the mixer in ½ hour, it is not sufficient time for complete dispersion of all the ingredients in a manner that will prevent separation during prolonged storage.

AREAS OF APPLICATION AND USE

Surveys indicate that acrylic latex and some solvent acrylic systems are used in the home and more than 50 percent are used in outdoor applications. Typical areas of usage are:

1. Bathrooms and kitchens
   a. Bath and shower seals
   b. Secure and seal basins
   c. Finish tiles and tile grout
   d. Finish around bathroom fixtures, furniture, etc.
2. Interior rooms
   a. Door jambs
   b. Skirting boards
   c. Fill gaps between walls and ceilings
   d. Seal in window frames
   e. Fill cracks in plaster

3. Exterior
   a. Seal in window frames
   b. Seal around doors
   c. Seal around extractor fans, chimneys, and air conditioners
   d. Seal around columns, timber posts, and brickwork
   e. Seal, ship, lap, and repair flashing and general roof leaks
   f. Replace faulty pointing
   g. Repair gutters and drain pipes
   h. Repair asbestos roofs, etc.

REFERENCES

1. Kine, B.B. and R.W. Novak. 1984. "Rohm and Haas Company Acrylic and Methacrylic Ester Polymers" in *Encyclopedia of Polymer Science and Engineering*. (1). 2nd ed. New York: John Wiley and Sons, Inc.
2. Redtenbacker, J. 1843. Ann. 47. (n.p.)125.
3. Caspary, W. and B. Tollens. (n.d.) Ann 167. (n.p.) 241.
4. US. Patent 1,121,134, 12/15/1914, O. Rohm.
5. Bachmann, Andrew G. (n.d.) *Advances in Acrylic-Adhesive Technology*. Torrington Connecticut: Dymax Corporation Adhesives & Sealants Industry.
6. Knolopka, K.M., Lomax, J. and D. K. Speck. (n.d.) Resin Review, (XLI) No.3.
7. Beal, C. 1990. "Caulking and Sealants, Selecting and Specifying." The Construction Specifier. (March).
8. Hauser, Dr. M. and Dr. J. T. Loft, J.T. 1980. "Anaerobic and Modified Acrylic Adhesives." (Loctite Corp.) Adhesives Age. (12).
9. Hauser, Dr. M. and Dr. J. T. Loft, J.T. 1980. "Anaerobic and Modified Acrylic Adhesives." (Loctite Corp.) Adhesives Age. (12).
10. Courtesy of Schnee-Morehead, Inc., Irving, Texas.
11. Bachmann, Andrew G., (n.d.) Dymax Corporation Adhesives and Sealants Industry.

CHAPTER 4
ANAEROBICS AND CYANOACRYLATES

INTRODUCTION

This chapter presents a brief description of the classes of sealants and adhesives used in this book. Anaerobics and cyanoacrylates sometimes are referred to as superglues and represent an agreement of several technologists and the author on the basic categories of sealants and adhesives used in the construction industry today. From a practical viewpoint, it is difficult to differentiate between an adhesive and a sealant. Sealants are adhesives that usually have considerable lower strength than most other adhesives. In general, they have higher elongation and are applied in thick glue lines [1]. Like their sister systems, anaerobics, cyanoacrylate sealants, and adhesives cure in the absence of oxygen. They are best suited for bonding nonporous materials such as plastics, metal, and glass [2].

ANAEROBICS

Anaerobic adhesives are usually esters of alkylene glycols and either acrylic or methacrylic acid. The formulations contain hydroperoxides that under ordinary circumstances would polymerize acrylates quite readily. Oxygen, however, inhibits the initiation of the polymerization by scavenging free radicals and thus preventing reaction. When the adhesive is placed in a glue line and accessibility of oxygen is restricted, polymerization proceeds quite quickly. This class of adhesive is quite appropriate for a wide variety of nonpermeable substrates and has been used extensively in machine fasteners [3].

Anaerobics are single component monomeric liquids that cure by free-radical polymerization; the typical system might be formulated from polyethylene glycol dimethacrylate and cumene hydroperoxide. This system would be stored in a permeable container (usually polyethylene), which allows oxygen to pass through the container wall and inhibit the polymerization. This adhesive system might be modified further by the addition of thickening agents or plasticizers, which would allow a gradation into an anaerobic sealant. Such formulations give the methacrylates a wide range of strength and characteristics. A number of other structural modifications can alter the flexibility and elongation of dimethacrylates. These sealants and adhesives are available in machinery and structural grades; the adhesives provide high tensile and shear strengths in flat assemblies, such as curtainwall laminates.

The environmental and chemical resistance of conventional acrylate plastics limits these acrylate-based adhesives and sealants, although the difunctional monomers can exhibit some improvement over linear acrylate polymers in their resistance to solvents. These types, along with cyanoacrylates, are the fastest-growing class of sealants and adhesives. Their convenience in use and quick cure results in a displacement of a number of other types of adhesives.

Anaerobic adhesive and sealants are applied to three types of surfaces: active, inactive, and inhibiting. Active surfaces include clean metals and thermoset plastics and result in the fastest cure. Inactive surfaces include some metals that are slow to cure. Inhibiting surfaces include bright plating, chromates, oxides and certain anodizes. Primer and heat must be used with inhibiting surfaces. Anaerobics attack some plastics and rubbers. Specific plastics such as, Teflon, polyolefins, and nylon can be bonded with anaerobics [4].

Average cure speeds can range from fast (five minutes to two hours) to moderate (two to six hours) to slow (six to 24 hours) at room temperature in the absence of primers. Heat accelerates the cure. In general, anaerobic sealants and adhesives have a service temperature limit of 300° F (149° C) and are moderately priced. Structural anaerobics have high structural strength and good moisture-, and solvent- and temperature-resistance. Structural sealants and adhesives are based on anaerobic chemistry and are best known for bonding small, rigid components because the cured products tend to be relatively hard and inflexible.

BASIC CHEMISTRY

Anaerobic formulations still contain the methacrylate functional monomers and/or resins and free-radical polymerization initiators found in the early formulations. Since 1990, a typical formula has comprised some or all of these components: monomers (and/or resins), initiators of free radical polymerization, accelerators for the initial process, stabilizers, thickeners, and other form and performance modifiers, and adhesion promoters. Each component in the formulation is listed below.

Monomers

The selection in anaerobic formulations is a major determinant of cured adhesive properties. Most of the monomers used are methacrylate. Dimethacrylates of poly-alkoxy glycols are useful bases of anaerobic formulations. Their formulas are shown in Figure 4.1.

Initiators

Hydroperoxides are used in most anaerobic formulations. Cumene hydroperoxide is a frequently cited initiator. The hydroperoxide provides stability in the formula, yet it is suitably reactive along with accelerators when in contact with substrates and in the absence of oxygen. The distinctive feature of an anaerobic adhesive, in contrast to the larger set of acrylic adhesives, is that they exist in one part, with a

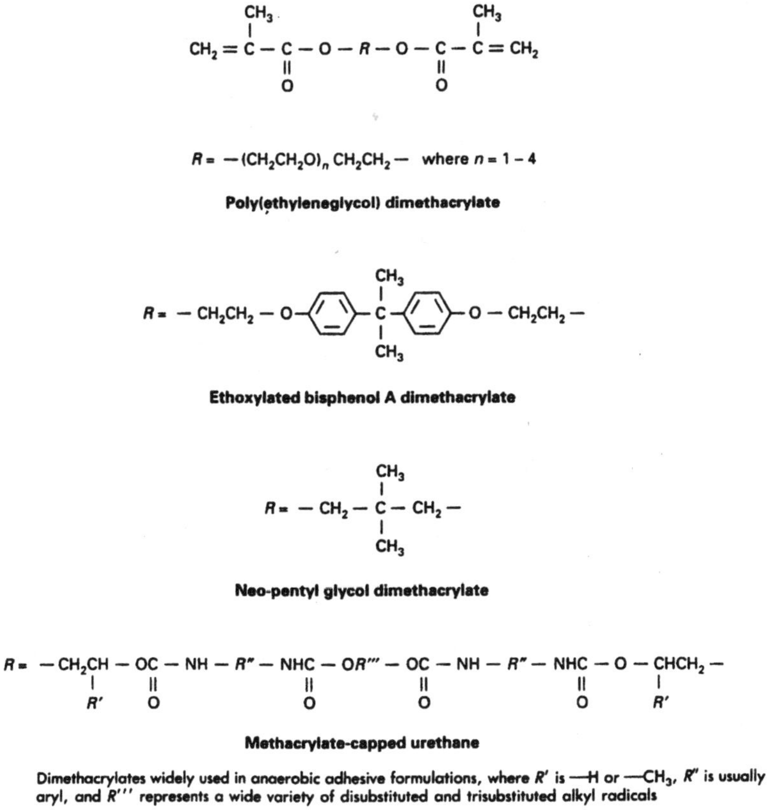

$$CH_2 = \overset{CH_3}{\underset{}{C}} - \underset{\overset{\|}{O}}{C} - O - R - O - \underset{\overset{\|}{O}}{C} - \overset{CH_3}{\underset{}{C}} = CH_2$$

$R = -(CH_2CH_2O)_n CH_2CH_2-$ where $n = 1-4$

Poly(ethyleneglycol) dimethacrylate

$$R = -CH_2CH_2 - O - \langle \text{Ar} \rangle - \overset{CH_3}{\underset{CH_3}{C}} - \langle \text{Ar} \rangle - O - CH_2CH_2 -$$

Ethoxylated bisphenol A dimethacrylate

$$R = -CH_2 - \overset{CH_3}{\underset{CH_3}{C}} - CH_2 -$$

Neo-pentyl glycol dimethacrylate

$$R = -\underset{R'}{\overset{}{C}}H_2CH - \underset{\overset{\|}{O}}{OC} - NH - R'' - NHC - \underset{\overset{\|}{O}}{O}R''' - \underset{\overset{\|}{O}}{OC} - NH - R'' - NHC - \underset{\overset{\|}{O}}{O} - \underset{R'}{\overset{}{C}}HCH_2 -$$

Methacrylate-capped urethane

Dimethacrylates widely used in anaerobic adhesive formulations, where R' is —H or —CH_3, R'' is usually aryl, and R''' represents a wide variety of disubstituted and trisubstituted alkyl radicals

FIGURE 4.1 Chemistry of anaerobic adhesive and sealants.

fine balance of stability and reactivity. The careful choice of inhibitor is just one element used to create that fine balance.

Accelerators

Anaerobically curing adhesives and sealants require accelerators for curing that provides positive fixturing. Saccharin (benzoic sulfimide) and many compounds containing acidic NH groupings have proved to be accelerators and/or initiators, especially when used in combination with aromatic amines. Combinations such as saccharin N, N-dimethyl-*p*-toluidine and saccharin with tetrahydroquinoline are widely used as accelerators.

Stabilizers

Stabilizers are critical to the formulation of useful anaerobic adhesive products. Phenolic and quinoidal stabilizers often are used. Chelators are used to trap traces of transition metals that could cause initiator decomposition with consequent preemptive curing.

Form Modifiers

Form modifiers include dyes, fluorescing agents, thickeners, and pigments.

Performance Modifiers

Plasticizers and adhesion promoters are included in this group. Plasticizers generate lower compressive strength and adhesion promoters enhance adhesion. Other performance modifiers include lubricants, such as polytetrafluoroethylene (PTFE) powder, and fillers such as mica.

TYPES AND FORMS

The chemistry of anaerobic sealants and adhesives allows for the formulation of many functional types for specific applications. The construction market limits their use; however, they can be used as:

- Thread-locking adhesives
- Thread-sealing adhesives
- Porosity sealants
- Structural adhesives

Table 4.1 shows a variety of adhesives, some geared to construction, that compete with anaerobics in specific applications. Cost factors of anaerobic products frequently are formulated using expensive raw materials. Extensive and specific quality-control systems are necessary to guarantee performance. Prices, as late as 1999, are typically $26 per pound ($52.63 per 250 kg.), but when anaerobics replace mechanical alternatives, savings are 2 to 10 times the cost of the adhesive. Table 4.2 outlines the uncured properties of anaerobic adhesives and sealants and Table 4-3 lists the cured properties.

TABLE 4.1 Adhesives that Compete with Anaerobics in Given Applications

Application area	Competing adhesive	Advantages	Limitations
Porosity sealing	Sodium silicate solutions	Very low cost	Slow cure of water-soluble sealants
	Styrene-thinned polyesters	Low cost	Odor and toxicity
	Heat-accelerated methacrylates	Simple	Bleed out, caused by heating
Flange sealing	Room temperature curing	Good temperature resistance and gap filling	Tendency to shim
	Solvent base resins	Low cost	Initially present solvent results in poor water-resistance
Bonding	Epoxy adhesives and sealants	Lower cost, can gap-fill	One-part or two-part, requiring heat cure
	Acrylic adhesives and sealants	Wider range of substrates	Odor and flammability

TABLE 4.2 Uncured Properties of Anaerobic Adhesives and Sealants

Properties	Typical value range	Comments
Visual features	Colorless to highly pigmented	A fluorescing agent may be added to aid on-part inspection.
Viscosity	Water-thin liquid to viscous pastes	The rheology is chosen for the application involved.
Vapor pressure, psi (Pa)	0.0015–0.0045 (10–30)	The major contribution is usually minor solvent residue in the raw material.
Density, g/cm^2	1–1.3	The concentration of fillers determines the level of density.
Flash point, °F (°C)	175–300°F (80–150°C)	
Curing characteristics, response to radox/ absence of air Fixturing on steel, min° Full cure on steel, hr	5–300 6–100	Heat, light, and activators can be used.

TABLE 4.3 Properties of Cured Anaerobic Sealants and Adhesives

Properties	Typical values	Comments
Coefficient of thermal expansion, 10^{-4}/K	1	
Thermal conductivity, Btu in/hr ft^2 °F (W/m.K)	0.7 (0.1)	
Specific heat, Btu/lb °F (kj/kgK)	0.7 (0.3)	
Elongation @ break, %	0.10–10	Most have low (1%) elongation @ break.
Adhesive strength Shear on steel, ksi (Mpa)	0.15–5.8 (1–40) 0.15–4.3 (1–30)	Products are developed to have Tensile on steel, ksi (Mpa) values appropriate to applications.
Hot strength @ RT, % @ 212° F (100° C) @ 300° F (150° C)	30–100 15–100	Some products contain B-staging additives, which increase in strength upon heating.
Strength retained after aging for 1000 hrs, % @ 212° F (100° C) @ 300° F (150° C)	80–100 20–100	Some products contain B-staging additives, which increase in strength upon heating.

PROCESSING

Considerations about shelf life, pot life storage requirements, substrates, geometry and cure requirements, the use of primers and activators, and application equipment are discussed below.

Shelf life and pot life depend not only on the formulation, but also on the package used. As the term anaerobic implies, there is a strong dependence on access to oxygen in the air as a stabilizing factor. A package stability of months, afforded by large air-impermeable containers become *years* in small, thin-walled, low-density polyethylene bottles and tubes.

A unique advantage of anaerobic adhesives and sealants is the stability of the one-component formulation in air. Stability in dispensing equipment also is good if contact with redox-active metals is avoided.

CYANOACRYLATES

Cyanoacrylates derive their unusual behavior from a high polarization of the unsaturated bond of the acrylate. They polymerize quickly at room temperature without a catalyst and are suitable for a number of adherends because of the presence of a number of basic materials on most substrates that may serve as activators for the ionic polymerization. Cyanoacrylates can be modified by adding thickeners and other polymers, thereby creating cyanoacrylate pastes.

Development of these adhesives began in the early 1950s by the Tennessee Eastman Company and became commercially available in 1958. Since then, many other companies worldwide, both formulators and distributors, have entered the field of cyanoacrylates.

CHEMICAL COMPOSITION

Cyanoacrylate esters are 1,1-disubstituted ethylene with the following basic structure:

$$CH_2 = C \begin{cases} CN \\ COOR \end{cases}$$

Their reaction is due to the ability of the electron withdrawing ester and nitrile groups to share negative charges. Three resonance structures are shown in Figure 4.2. Ethyl and methyl groups are the most commonly used because of their low price, high bond-strengths, and availability.

The adhesives have been used widely for thread-locking applications. Most manufacturers warn that caution should be used in handling because of the danger of bonding one's skin to the adhesive. Some users have found that the cyanoacrylates are subject to hydrolytic degradation when exposed to moisture. They are thermoplastic when cured, and, consequently, are limited in temperature capability and chemical resistance. Table 4.4 outlines the commercially available 2-cyanoacrylate esters, their uses and properties. Of these, ethyl and methyl are the most commonly used because of their price, bond strength, and availability.

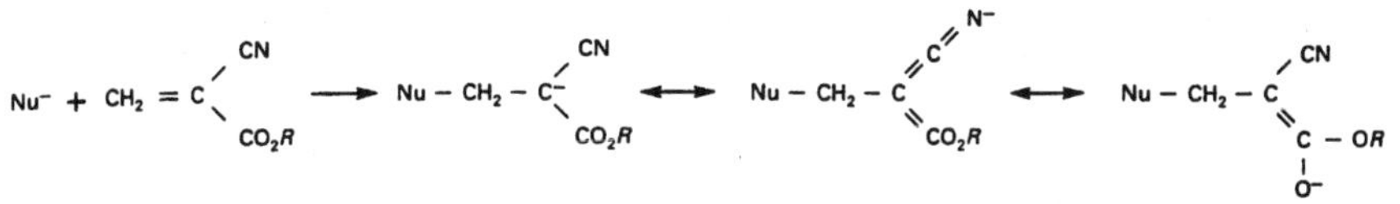

FIGURE 4.2 Resonance of the anion of cyanoacrylate esters, where Nu^- refers to a negatively charged nucleophile.

TABLE 4.4 Commercially Available Two-Cyanoacrylate Esters

Ester	Radical group, R	Properties and uses
Methyl	$-CH_2$	Strongest bond on metals; good solvent and temperature resistance.
Ethyl	$-CH_2CH_3$	General purpose; forms strong bonds on metals, plastics and rubbers.
Allyl	$-CH_2CH=CH_2$	Use temperature above 212° F (100° C).
Butyl	$-CH_2CH_2CH_2CH_3$	Plastics and rubbers, improved flexibility; less irritating vapor than that of lower cyanoacrylates.
2-methoxyethyl	$-CH_2CH_2OCH_3$	Low odor; improved flexibility; nonfogging.
2-ethoxyethyl[3]	$-CH_2CH_2OCH_2CH_3$	Low odor; improved flexibility; nonfogging.
2-methoxy-1-methylethyl	$-CHCH_2OCH_2CH_3$	Low odor; improved flexibility; nonfogging.

MANUFACTURE OF 2-CYANOACRYLATE ESTERS

Several methods are used to synthesize cyanoacrylate monomers. The most important industrial methods use the sequences:

- Base-catalyzed Knoevenagel reaction of formaldehyde with a cyanoacrylate ester to produce the 2-cyanoacrylate ester, which spontaneously polymerizes
- Thermal depolymerization under acidic conditions to regenerate the monomer
- Addition of stabilizer, thickeners, and other ingredients to produce a commercial adhesive or sealant
- Purification of the monomer by one or more distillations

The first step, the reaction of formaldehyde with the cyanoacrylate ester, preferably is carried out using an excess of the ester. The catalyst is usually a secondary amine such as piperdine. Two likely pathways for this reaction are via the hydroxymethyl derivative and the Mannich base, as shown in Figure 4.3.

The 2-cyanoacrylate esters are unstable in the presence of amines and undergo anionic polymerization. The excess cyanoacetate ester used in the Knoevenagel reaction acts as a chain transfer agent, reducing the molecular weight of the polymer and lowering the reaction viscosity. The two end-groups are derived from the cyanoacetate ester, as shown in Figure 4.4.

The second step is depolymerization of the polymer by an unzipping mechanism. A cyanoacrylate polymer readily depolymerizes because the ceiling temperature is low enough that other thermal reactions do not readily occur. The ceiling temperature is determined by the relative stabilities of the monomer and polymer. Structural features that either stabilize the monomer or lower the stability of the polymer, reduce the ceiling temperature. In the case of cyanoacrylates,

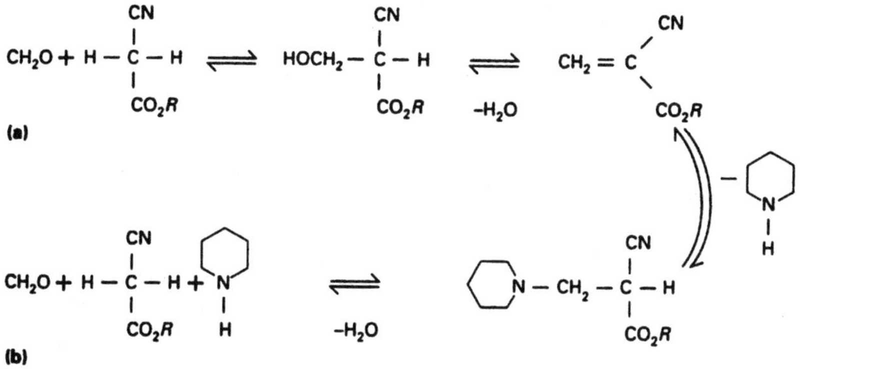

FIGURE 4.3 Knoevenagel reaction in cyanoacrylate esters.

Pathways for the Knoevenagel reaction in cyanoacrylate esters. (a) Via the hydroxymethyl derivative. (b) Via the Mannich base

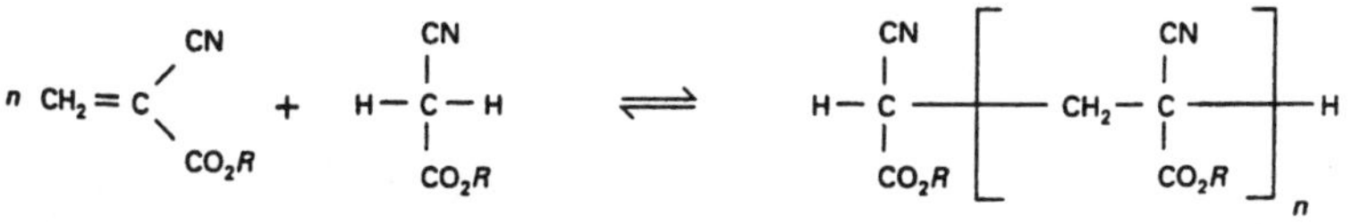

FIGURE 4.4 Polymer based on cyanoacrylate esters.

Structure of the polymer based on cyanoacrylate esters

a feature that stabilizes the monomer is conjunction of the double bond. Stearic acid strain from the interactions of the neighboring side groups lowers the stability of the polymer.

The depolymerization step is carried out at high temperatures—300° to 375° F (150° to 190° C)—and high vacuum—1 to 5 torr (135 to 665 Pa)—in the presence of inhibitors that prevent repolymerization. The 2-cynaoacrylate ester distills over and is collected onto a nonvolatile acid anhydride or strong acid, such as hydroquinone, in the receiver and free radical trap. A byproduct of the depolymerization is a 2,4 dicyanoglutarate, as shown in Figure 4.5. In the third step, the monomer is separated from this material, as well as from excess polymerization inhibitors, by one or more distillations. The fourth step is covered by the addition of additives, stabilizers, thickeners, and other commercial products used to produce adhesives.

Figure 4.6 illustrates the electrical charge on the propagating polymer depending on which inhibitor is used.

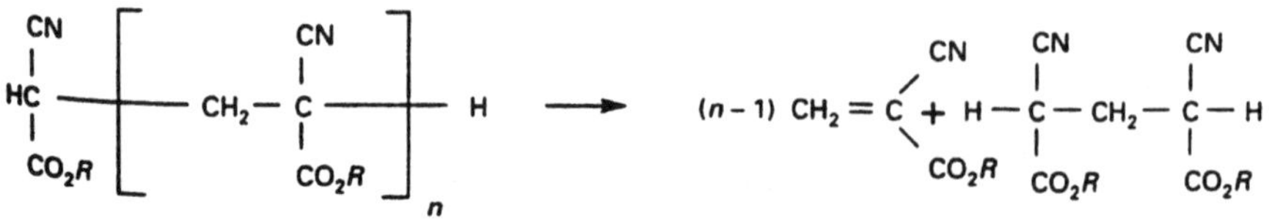

FIGURE 4.5 Depolymerization of the polymer based cyanoacrylate esters.

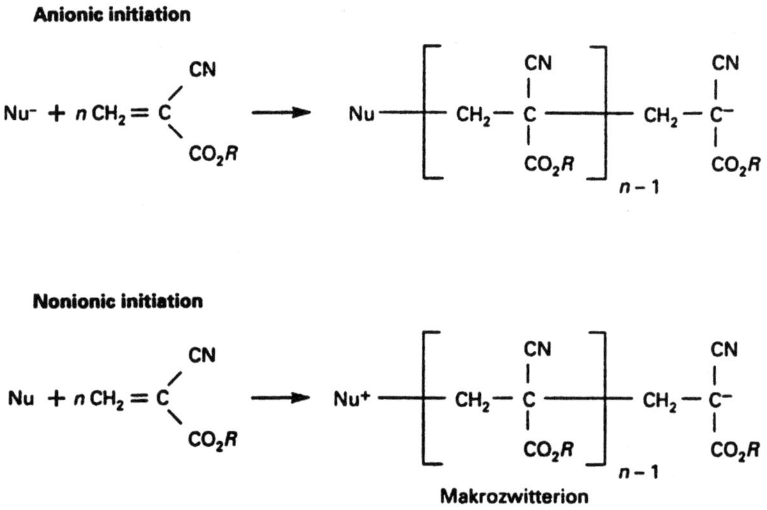

FIGURE 4.6 Mechanisms of initiation in the polymerization of cyanoacrylate esters.

BONDING ACTION

The bonding action of cyanoacrylates depends on the presence of either alkaline sites or adsorbed water on the surface of the substrate. Glass has an alkaline surface and bonds rapidly. Rubber and plastics contain additives, such as antioxidants and vulcanization agents, which can migrate to the surface and initiate polymerization. Metal has a bonding surface that is an oxide or hydroxide. As stated in Table 4.1, cyanocrylates can be used in construction applications where bonding to metals, in particular, is involved.

SURFACE PREPARATION

Cyanoacrylates provide the best bonds on clean, dry, grease-free surfaces. In many cases, a simple solvent wipe is sufficient. To obtain the highest bond strengths, which in many cases exceed the internal strength of the substrate, a more elaborate treatment is necessary.

STORAGE STABILITY

Cyanoacrylates are prone to both polymerization and hydrolysis. They should be stored in inert containers that have low permeability to moisture. Packages include polyethylene bottles and tubes made of aluminum, tin, or polyethylene. The stability of an adhesive or sealant is improved greatly by storing it at reduced temperatures, preferably in a freezer. The container should be allowed to warm to room temperature before it is opened to prevent moisture from condensing and penetrating the adhesive. Most manufacturers offer a six-month to one-year shelf life when stored at room temperature.

SUMMARY

The chemistry of anaerobic formulations differs in composition form and cured properties, depending on the particular application. All, however, are one-part adhesives that maintain a fine balance between package stability and reactivity in contact with metal surfaces. This reactivity causes a redox-triggered hardening process in the absence of oxygen. A heat and surface activator can be used to augment this curing mechanism. Anaerobic adhesives and sealants can be formulated in many physical fluid forms, from water-thin liquids to thixotropic pastes. The end-use largely determines the choice.

The U.S. Food and Drug Administration has not approved the use of cyanoacrylates for medical applications because the material contains an excess of toxic ingredients, such as formaldehyde. Cyanoacrylates, however, have been used for medical applications in other countries. They have limited application in the construction market because of their cost and the difficulty in handling.

REFERENCES

1. The International Plastics Selector, Inc. 1979-80. Adhesives (desktop data bank). San Diego, Calif.
2. Catena, W. 1993. "Top Anaerobic Function Depends on Adhesive, Air, and Metal Balance," National Styrach and Chemical Co. Adhesives Age. June.
3. Catena, W. 1993. "Top Anaerobic Function Depends on Adhesive, Air, and Metal Balance," National Styrach and Chemical Co. Adhesives Age. June.
4. Melody, D.P. 1990. Anaerobics, Loctite International. Connecticut.

CHAPTER 5
FLUOROCARBONS AND HOT MELT ADHESIVES AND SEALANTS

INTRODUCTION

The adhesive and sealant industries have had a remarkable growth over the past 25 years. Hot melts, in particular, grew from 100 million pounds in 1970 to more than 500 million pounds in 1995. Their U.S. dollar value was just short of $2 billion or 7 percent of the adhesive and sealant markets. The rapid growth of hot-melt adhesives and sealants is the result of several factors and has produced many advantages.

A hot-melt sealant or adhesive often denotes a lower molecular weight, less viscous material than thermoplastic. They are 100 percent nonvolatile and are solid at room temperature [1]. Then they are melted, heated usually to 300° to 400° F (148.89° to 204.44° C), and applied in a molten state. On cooling, they solidify. The thermoplastic nature (melting when heated and solidifying when cooled) is inherent in hot-melt adhesives and sealants. Many thermoplastic materials are used in the molten state, including wax, polyethylene, asphalt, rubber-wax blends, and ethylene-vinyl acetate-reinforced wax materials. These products are rather simple, with limited requirements in adhesion and end-use performance. Hot-melt adhesives and sealants differ from other liquids and hot-applied systems in that they set by cooling rather than by chemical curing or by evaporation of a liquid carrier.

Hot melts must be melted. When molten, they must be fluid to apply. The required fluidity or viscosity, while very broad, is limited by the application equipment. The viscosity limits imposed by equipment guide the composition of hot-melt systems, which consist of the essential ingredients, the polymer and a diluent [2]. The cooling produces rapid bonding. In addition, hot-melt adhesives and sealants can be reused simply by reheating; they can be formulated to bond with a wide range of substrates and perform a gap-filling function. Hot melts have excellent moisture resistance and because they are 100-percent solid, transportation and storage problems are minimized, and energy costs are low.

In spite of their many advantages, hot-melt adhesives and sealants do have some limitations. Because they are thermoplastic materials, heat resistance can be poor, particularly with ethyl vinyl acetate (EVA) formulations. Additionally, the strength of hot melts does not match that of thermosetting materials. Hot-melt materials also tend to produce surface bonds with less substrate penetration than solvent-borne adhesives and sealants. Table 5.1 lists several advantages and limitations for hot melts.

Since hot melts bond almost immediately, they are useful for high-speed production applications. One of the major markets for this type of adhesive is packaging, including case sealing and composite cans. Hot-melt sealants are also designed as insulating glass sealants with regular or partially cured butyl as the base. One of the newer approaches is the use of 100 percent solid, thermoplastic compounds that are heated to their lower viscosity and then applied as hot melts. In the insulating glass industry, these products are known as hot-flow sealants.

HOT MELT COMPOSITION

A typical hot-melt material is composed of three components:

- Polymer (30 percent to 40 percent)
- Tackifying resin (30 percent to 40 percent)
- Petroleum wax (20 percent to 30 percent)

TABLE 5.1 Advantages and Limitations of Hot Melts

Advantages	Limitations
1. Speed of bond formation	1. Limited toughness at useable viscosities
2. Fast set time	2. Limited adhesion
3. Increase of production	3. Lack of molten properties, such as tack, range, and wetting ability
4. Easy automation of hot-melt dispensing equipment	4. Produces low viscosity
5. Gap filling	5. Poor mechanical properties
6. Water insensitivity	
7. No disposal costs and elimination of recovery costs	
8. Barrier properties	
9. Wide formulating latitude to meet cost and performance criteria	
10. Reduced maintenance and cleanup costs	

Polymer forms the backbone of the sealant or adhesive, controlling its strength and toughness. Tackifying resin contributes wetting and tack and wax lowers the melt viscosity and controls the setting speed. Antioxidants, fillers, plasticizers, and blowing agents can be used to enhance other properties.

Backbone polymers used in hot-melt sealants and adhesives include EVA copolymers, low-density polyethylene (LDPE), atactic polypropylene (APP), ethylene ethyl acetate (EEA), polyamides, polyesters, and thermoplastic elastomers (TPE). EVAs are the most widely used polymers for hot melts because they are compatible with both amorphous tackifying resins and crystalline waxes. A blend of butyl rubber serves as the elastomeric binder with amorphous polypropylene and various resins to give thermoplastic properties. Blends of butyl, ethylene propylene rubber, and precured butyl also can be used; but in all cases, formulations must be tailored carefully to avoid excessive melt viscosity. Polymers, such as LDPE and EVA, and fillers and plasticizers, such as polybutene, are used to modify compound properties. These compounds are applied by thermal pumping and dispensing equipment designed for bulk hot-melt materials, or by a semiportable heat extruder gun. Applications involve mainly plant assembly of windows and various curtain-wall panels, line sealing of prefabricated buildings, and adhesive and sealant end-uses where the compound can flow into place and both adhere to the substrates and seal out the elements.

The backbone polymer usually is modified with a low molecular weight noncrystalline, polymeric material known as a *tackifying resin*. Tackifying resins are derived from rosin or natural terpenes, or are synthesized from hydrocarbons, and range from hard, brittle materials to semiliquids. They are selected based on their chemical polarity, softening point, heat stability, and color. Tackifying resins can range from nonpolar aliphatic hydrocarbons to highly polarized rosin esters. Polarity in tackifying resins affects substrate adhesion, polymer compatibility, and, to a lesser extent, flexibility and adhesive viscosity.

The primary function of the wax component of the hot-melt materials is to control the open time of the total system by reducing its viscosity so it can be applied easily to a surface. Paraffin waxes, derived from petroleum residue, are straight-chain, saturated hydrocarbons with large crystalline structures. They are characteristically hard, brittle materials with a sharp melting point and excellent resistance to gas and moisture permeation. Microcrystalline waxes used for adhesives and sealants are relatively soft, flexible materials with a broader melting range. They have better adhesive properties, but poorer barrier properties, than paraffin. Higher melting-point coating grades of microcrystalline waxes are harder and less flexible than the laminating grades. Blends of paraffin and microwaxes are often used to improve adhesive properties. In place of the petroleum waxes, high-melting-point synthetic waxes are used to frequently provide high-temperature properties and greater cohesive strength.

Microcrystalline waxes and synthetic waxes are selected on the basis of melting point, 190° to 195° F (87.78° to 90.56° C), respectively, oil content, hardness (flexibility), and color. Melting point affects the heat resistance and open time of the adhesive and also can affect adhesion. High-melting-point waxes increase the adhesive strength of hot-melt sealants and adhesives.

An antioxidant often is added to hot melts to stabilize viscosity and prevent formation of a surface skin or *char* in the mixing and application equipment. Most polymers and tackifying resins, however, already contain an antioxidant. If additional stabilization is needed, a hindered phenolic antioxidant is the preferred choice for adhesives made with ethylene vinyl acetate and low-density polyethylene.

CHEMICAL STRUCTURE

Ethylene Vinyl Acetate

EVA copolymers are characterized, like all EVA resins, by melt index and the vinyl acetate content-incorporated (VA-incorporated) in the polymer. EVA resins range in VA-incorporated content from 6 percent to 33 percent by weight. The midrange EVAs, with a VA-incorporated content of 19 percent to 28 percent, are used most widely for adhesives and coatings. The 28-percent VA-incorporated copolymers are used widely in wax-based, hot-melt adhesives for product-assembly applications with excellent moisture- and gas-barrier properties. Higher VA-incorporated copolymers are preferred if improved adhesion to nonporous substrates is required. Lower VA-incorporated materials offer greater hardness and blocking resistance at lower cost.

Vinyl Acetate Ethylene

VAE copolymers have high VA-incorporated levels of 40 percent to 51 percent. They are polar and, for the most part, noncrystalline thermoplastics that are useful in waxless, hot-melt adhesives and sealants, solvent-based pressure-sensitive adhesives, and clear coatings. The high vinyl acetate content of VAE copolymers improves their solubility in organic solvents and broadens their compatibility with polar modifying resins. The range of EVA and VAE adhesives and sealants available from Quantum [3] is shown in Figure 5.1.

Both EVA and VAE copolymers are produced from the same technology used by Quantum to manufacture LDPE. Vinyl acetate monomer is copolymerized with ethylene to produce a chain of carbon atoms with acetoxy groups distributed randomly along the chain, as shown in Figure 5.2.

The acetoxy groups reduce crystallinity and increase flexibility in the resulting polymer. The acetoxy groups increase the polymer density, however, because the molecular weight of vinyl acetoxy is higher than that of ethylene. The concentration of the ethylene might be higher in one portion of the polymer than in others because acetoxy groups are scattered randomly along the ethylene chain. Where

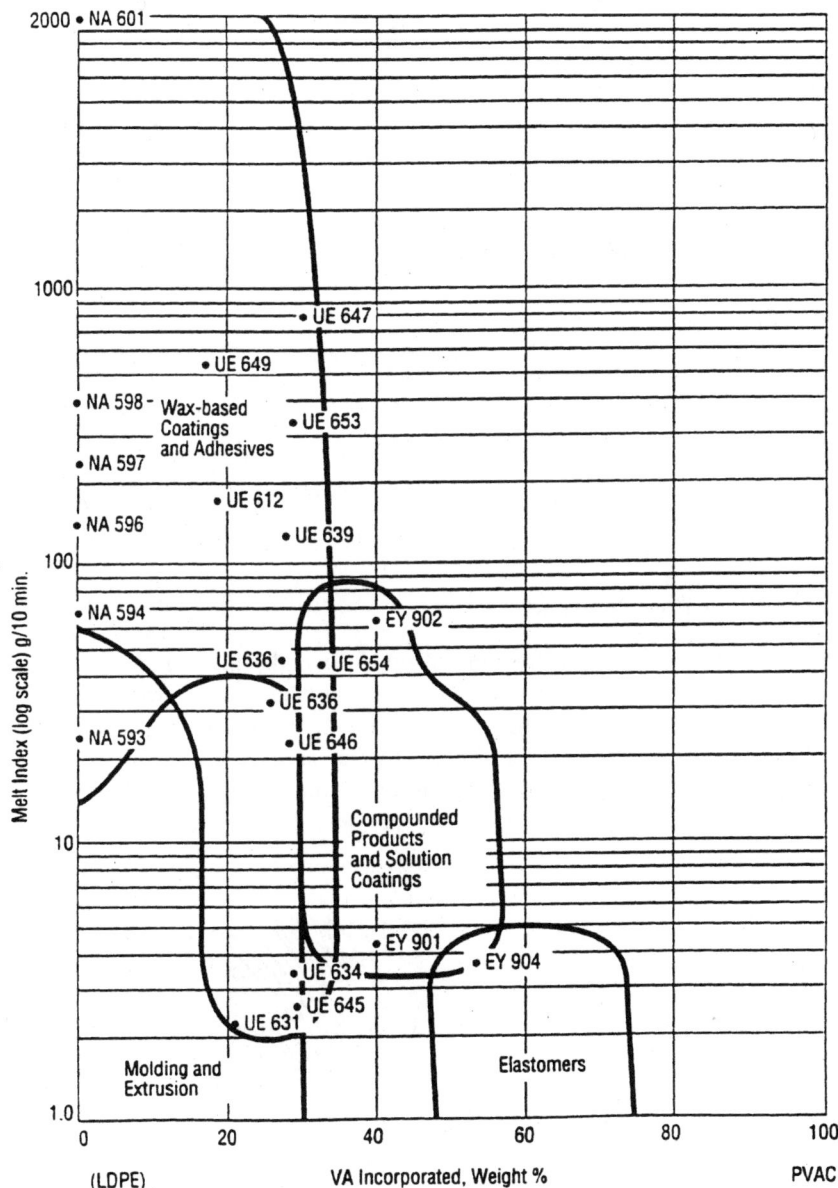

FIGURE 5.1 Application Areas for Quantum Polymers.

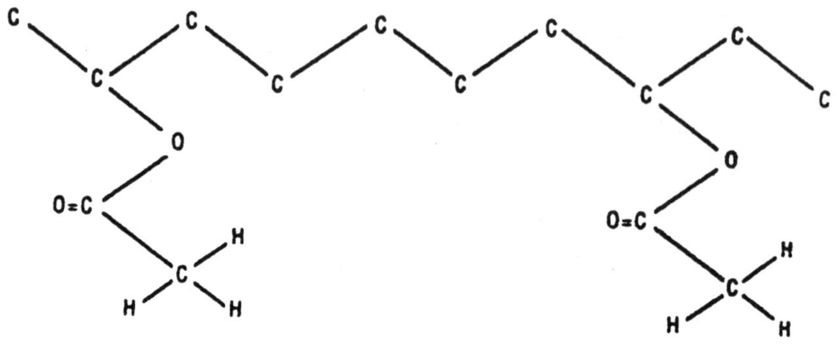

FIGURE 5.2 Part of an ethylene vinyl acetate copolymer chain.

the concentration of acetoxy groups is high, a polar, noncrystalline, or amorphous region occurs. In Figure 5.3, the crystalline portion of the EVA polymer is represented by the letter A; the amorphous region is represented by the letter B.

Low-Density Polyethylene

The free radical, bulk polymerization of ethylene under high pressure produces LDPE. This process produces a polymer consisting of chains of carbon atoms with attached hydrogen atoms. The chains can contain many long or short side branches, which account for the term branched polyethylene, as shown in Figure 5.4.

The overall size of each molecule determines its molecular weight. The average weight of all the molecules in the polyethylene chain is indicated by its melt

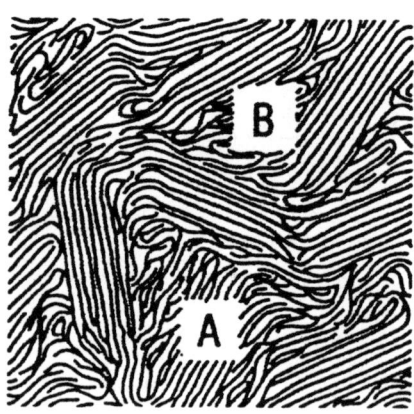

FIGURE 5.3 Part of a branched polyethylene chain.

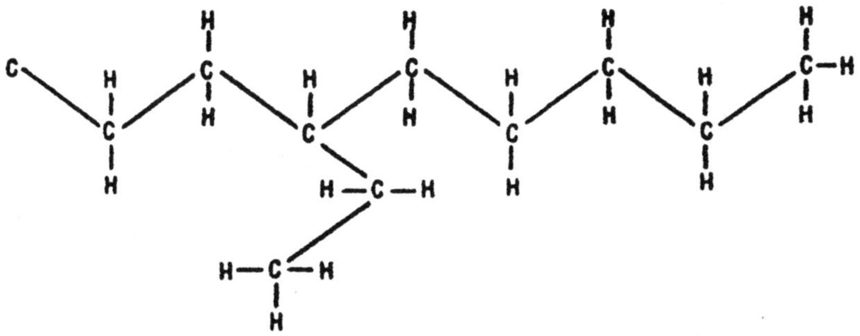

[Chemical structure diagram]

FIGURE 5.4 Crystalline portion (A) and amorphous portion (B) of ethylene polymer.

index: the higher the average molecular weight, the lower the melt index and vice versa. Since melt index is a measurement of the flow behavior of a polymer under standard conditions of temperature, pressure, and orifice size, average molecular weight is an important polymer characteristic.

The amount of branching off the main polymer chain affects the density of the polymer. Branching inhibits the formation of ordered or crystalline regions in the polymer as it solidifies from a melt. Thus, less-crystalline polymers have lower densities than polymers with fewer chain branches and greater crystallinity. Crystallinity contributes strength and heat resistance to the polymer. Branched or amorphous regions give the polymer greater flexibility and adhesion.

C30 Alkenyl Succinic Anhydride (C30ASA)

C30ASA is a new hot-melt-adhesive raw material that functions as both a viscosity modifier and a compatibilizing agent. It also promotes greater interaction among the adhesive components and builds better new adhesive properties with improved performance. This unique product is compatible with EVAs, LDPEs and other more common hot melts. As with most raw materials, the formulas have dozens of variations. For our purposes, it is not practical to publish all the starting material, but Figure 5.5 lists some hot melts containing C30ASA with microwax and a couple with ethylene copolymer blends.

CRITICAL PROPERTIES

Viscosity

The concentration of molecular weight of the polymer is the strongest factor in determining the hot-melt viscosity. The molten viscosity, considered to be the

Raw materials	Parts by weight			
	1	2	3	4
Elvax 460	150	150	150	150
C_{30}ASA	100	—	100	—
Microwax	—	100	—	100
Foral 85	150	—	—	—
Statac	—	—	—	—
Piccolyte S-100	—	150	—	—
Piccolyte A-100	—	—	—	—
CKM 2432	—	—	—	—
Isoterp 95	—	—	150	—
Piccotex 75	—	—	—	150
Piccovar L-60	—	—	—	—
Antioxidant	2	2	2	2

Physical properties	Values			
Ring and ball softening point, °C ASTME E 28-67	91	92	92	92
Melt viscosity, °F, Brookfield HAF Thermosel				
325	39.0	41.5	44.0	27.2
350	26.6	27.0	37.5	18.6
375	16.7	17.0	20.0	13.0
400	12.2	12.0	18.0	9.2
Tensile strength, psi ASTM D 1708				
Yield	500	939	1176	798
Ultimate	923	1048	683	704
Elongation, %	429	336	81	168

FIGURE 5.5 Hot melt adhesive with C_{30}ASA.

most distinctive property of a given product, is the most common and most reliable measure of identification and is an indicator of its uniformity. To help visualize the magnitude of viscosities for hot melts, several common liquids are listed below with their approximate viscosities at room temperature.

Liquid	Viscosity, cP
Water	1
Kerosene	10
Light weight motor oil	100
Glycerin	1,000
Corn Syrup	10,000
Molasses	100,000

The standard method of measuring the viscosity uses the Brookfield Viscometer with a Thermosel Chamber (ASTM D 3236). This meter is a low-shear, rotational viscometer with a controllable temperature source, capable of viscosity measurements from room temperature up to 500° F (260° C). Apparent viscosity, expressed in centipoises (cP) or milli-Pascal-sec (mPa-s), is meaningless without an accompanying test temperature.

Most copolymers are too high in viscosity to be measured with a Brookfield viscometer. Brookfield viscosity of a low-melt index polymer can be determined by measuring the viscosity of blends of polymer diluted with wax at several concentrations. The viscosity of a polymer then can be approximated by extrapolating the wax concentration to zero. Brookfield viscosity for LDPE resins and EVA copolymers also can be calculated from the extrusion conditions used for determining melt index, according to the following formula:

$$\text{Viscosity, cP @ 374° F (190° C)} = \frac{\sim 9,500,000}{\text{Melt index}}$$

From Brookfield viscosity measurements of blends of EVA copolymers and waxes at various temperatures, the following relationship between temperature and viscosity has been determined:

$$\ln \eta = \frac{A}{T} + B$$

where (η) = Brookfield viscosity, cP
T = absolute temperature, K
A, B = constants

In practical terms, this equation means the melt viscosity of the polymer approximately doubles for every 50° F (10° C) decrease in temperature. In addition to flow behavior, melt index is an indicator of a number of other important properties. As the melt index decreases or the molecular weight increases, the softening point, tensile strength, elongation, stiffness, and hardness of the polymer all increase.

Melt Index (MI)

Melt index is inversely proportional to average molecular weight and melt viscosity. The higher the MI, the lower the molecular weight and viscosity of the polymer. Melt index is measured with an extrusion plastometer or melt indexer at standard extrusion conditions of 374° F (190° C), with an applied weight of 2.16 kg (ASTM D 1238 Condition 190/2.16). The melt-index value is the weight of polymer that is extruded through a standard orifice as dg/min or g/10 min.

Density

Density in polyethylene momopolymers is a measure of crystallinity that controls the stiffness, strength, and barrier properties of the plastic. The higher the density of polyethylene hot melts, the greater the stiffness, and the higher the yield strength, and the better the resistance to permeation by gases and water vapor.

Density is measured by two common methods, either in a density-gradient column or by liquid displacement. In the gradient-column method (ASTM D 1505), used primarily with polyethylene, a glass column is filled with a liquid that gradually increases in density from the top to the bottom. A small sample placed in the column sinks to the point where the density is the same as that of the liquid surrounding it. The density, or level of the sample, is read from a calibrated scale on the column. In the displacement method (ASTM D 792), a compression-molded sample of polymer first is weighed in air and then in an inert liquid of known density. Density is calculated from the sample weights and liquid density.

Tensile Strength

Tensile strength is the maximum stress a specimen will sustain in a tensile test to failure, expressed as pounds per square inch.

Elongation

Elongation is the increase in length of a specimen stretched to failure, expressed as a percent of the original gauge length. Both test methods use ASTM D 638, using a 0.075 in. thick dumbbell-shaped specimen, cut from a compression molded plaque. Samples are pulled in a constant cross-head rate testing machine, using a jaw separation rate of 20 in/min.

Tensile Modulus

Polymer stiffness can be determined in a number of ways, but tensile modulus is the easiest and most common measure. Modulus is the ratio of the applied stress to the resulting deformation or strain at any point on the stress/strain curve. The test method is ASTM D 638 and is expressed as pounds per square inch.

Hardness

Hardness with plastics, sealants, and adhesives often is used as a simple measurement of stiffness and durometer. The lower the hardness, the more flexible the polymer. Hardness is the resistance of the polymer to indentation by a blunted needle. *Shore A or D Durometer* (ASTM D 2240) is used in this type of determination.

Softening Point

Softening point is the temperature at which a molded polymer disk becomes soft enough that the weight of a small steel ball causes it to sag one inch. ASTM E 28, also known as the ring-and-ball softening test, is the test method followed. The softening point of a polymer is important in hot-melt adhesives and sealants because this property influences heat resistance, blocking, application temperature, and open time or solidification point.

TYPICAL END USE FORMULATIONS

The following formulations are starting positions that chemists, engineers, or architects might use to begin to determine what is best for them or the construction project. Figure 5.6 shows an insulating glass, hot-melt sealant that can be used for the primary seal of a dual-sealed system. A more simplified formulation that employs a butyl mastic as the barrier sealant between the spacer and glass is shown in Figure 5.6a. Figure 5.7 is a typical EVA-based, hot-melt general-purpose adhesive. Another general-purpose material, described in Figure 5.8, is based on a Kraton SIS block copolymer. In a flooring application, one might use

Raw materials	Parts by weight
Exxon butyl 065	184
Afax 500 HLO/Vestoplast x3632	72
Black pigment, PZ 700 black 35	2.5
Calofort S	240
Elvax 210/EVA 28-400	80
Stabelite ester 10	216
Hyvis 2000	72
Silane A1120	12

Physical properties	Values
Viscosity, cp	300,000 to 500,000
Ball and ring softening point	311° to 338°F (155° to 170°C)
Appearance	Smooth, lump-free
Gassing, 356°F (180°C)	Material should show no signs of gassing
Shear adhesion, psi A1 to A2	32 (2.5kg/cm²)

FIGURE 5.6 Hot melt butyl insulating glass sealant.

Raw materials	Parts by weight
Vistanex LM-MH	100
Mistron vapor talc	48
N 990 (MT) carbon black	2

FIGURE 5.6a Insulating glass extruded mastic.

Raw materials	Parts by weight
duPont Elvax 220	300
Polyterpene resin	500
Microcrystalline wax, 185°	40
Antioxidant	1.0–2.0

FIGURE 5.7 EVA-based hot melt adhesive.

Raw materials	Parts by weight
Krato SIS block polymer	60
Polyterpene tackifier	120
Shell process oil	20–40
Antioxidant	1.0–2.0

FIGURE 5.8 Elastomeric sealant.

Raw materials	Parts by weight
SBR 1018	18.6
SBR 1009	81.4
Petroleum hydrocarbon (bp 116°–136°C)	407
Calcium carbonate	151
Hard clay	232
Antioxidant	2.3
Polymerized resin	244

FIGURE 5.9 Ceramic tile cement.

Raw materials	Parts by weight
Partially crosslinked butyl rubber	100
Silene 732 D silica	50
N326 (HAF-LS) carbon black	140
Parapol 950 polybutene	140
Hecolyn resin	30

This product is manufactured in a single-step process using a kneader mixer.

FIGURE 5.10 Resilient sealing tape.

Raw materials	Parts by weight
Exxon butyl 065	120
Chlorobutyl 1066	80
Calcium carbonate, coated	100
N326 (HAF-LS) carbon black	200
Parapol 950 polybutene	230
Zinc oxide	4
Magnesium oxide	0.8
Stearic acid	0.8

This is a two-stage process:
1. Banbury masterbatch: butyl, chlorobutyl, CaCO, ZnO, MgO, stearic acid. Mix 5 minutes at 324°F (164.22°C).
2. Kneader mix: Masterbatch plus other ingredients.

FIGURE 5.10a Resilient sealing tape.

a quick-setting vinyl floor tile PSA, as illustrated in Figure 5.9. Two resilient tape formulations in Figures 5.10 and 5.10a offers the end-user tapes that can be used for sealing glass and insulating glass units to both inside and outside of the window, as illustrated in Figure 5.11.

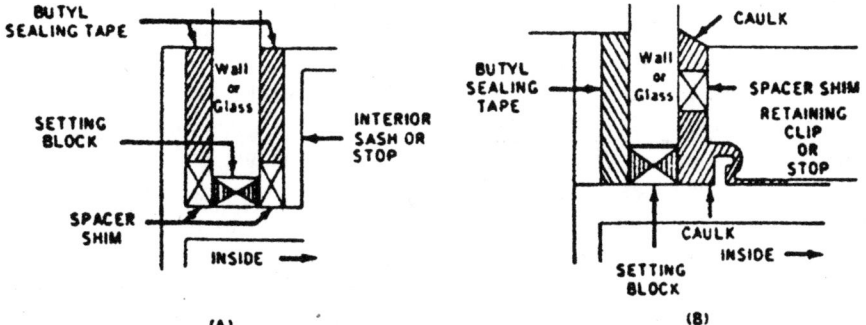

Typical architectural tape installations.

FIGURE 5.11 Architectural tape Installation.

CHEMICAL SUPPLIERS

Shell Chemicals Co.
E.I. duPont de Nemours & Co., Inc.
USI Chemicals Co.
Union Carbide Co.
Exxon Chemicals
Dow Chemical
Eastman Chemicals Products

EQUIPMENT SUPPLIERS

Nordson Corporation, Amherst, Ohio 44001
Meltex Corporation, Peachtree City, Georgia 30269
Accumeter Laboratories, Marlborough, Massachusetts 01752
Boulton-Emerson, Inc., Lawrence, Massachusetts 01842
Grayco/LTI, Monterey, California 93940

SUMMARY

The adhesion and sealant industries in the United States will continue to grow and remain attractive markets, despite their complex niche orientation and the continued evolution of both technology and market demands. The principal use for hot melts in sealants is in the manufacture of insulating glass units, with an estimated annual consumption of 30 million pounds. Although large volumes of sealants and adhesives have penetrated the construction market, hot melts have not gained much in this conservative industry. The application techniques of hot melts are not amenable to on-site application. Therefore, most of the hot melts used in construction are for prefabricated panels and other materials, specifically self-adhesive flooring, shutters, and the assembly of window casings.

Companies that maintain a long-term commitment to marketing and technology development and focus on specific customer needs will be able to capitalize on many opportunities and ultimately will emerge as winners in the marketplace.

REFERENCES

1. Cagle, C.V. 1973. *Handbook of Adhesive Bonding*. New York: McGraw-Hill, Inc.
2. Exxon Chemical Co., Polymers Group, Houston, TX.
3. Quantum Chemical Corp. 1992. *Adhesives, Sealants and Coatings Manual*.

CHAPTER 6
ASPHALTIC, OLEORESINOUS, AND OIL-BASED PUTTY

INTRODUCTION

Oil-and resin-based sealants, caulks, and putties were the primary adhesive materials before the development of the curtain wall in the late 1940s, and before the advent of polysulfide sealants. At that time, there was a long history of using unmodified bituminous materials as joint and crack sealers. Beginning with oil-based caulks, some of the most frequently used sealants of this type are analyzed in this chapter.

Oleoresinous Caulks and Sealants

These generally cure by surface oxidation and have bases of linseed oil, fish oil, soybean oil, tung oil, castor oil, or marine oil. Some of the oils can be drying oils, which quickly oxidize and harden. Additives of metallic naphthenates are included to obtain various degrees of cure-through. Modifiers commonly are used to permanently plasticize; these softer caulks are not recommended in relatively nonmoving joints with up to ± 5 percent joint movement, but hardening oil-based caulks have a maximum movement capability of ± 2 percent. The formulation of good oil- and resin-based caulks parallels the formulation of other caulks and sealants in that there is a proper ratio of filler to binder and that the sealants or caulks must set or cure within a desirable period of time to produce a product that will meet existing standards. Table 6.1 offers the advantages and limitations of oil-based caulks.

TABLE 6.1 Advantages and Limitations of Oil-Based Caulks

Advantages	Limitations
1. Can remain plastic	1. Little or no recovery
2. Tools easily	2. Very little flexibility
3. Applies easily	3. Can crack and harden
4. Single component	4. Little movement, ± 2% to ± 5%
5. No primer required	5. Not speced for moving joints
6. Lowest cost of all caulks	6. Slow cure rate
7. 10-year longevity	7. Can stain substrates
8. Low shrinkage	8. Few good caulks available
9. Fast skinning	9. Up to 20% shrinkage

Requirements for movement are much lower for oil-based caulks, and the raw material is cheaper, which is why these materials find a use in their segment of the market. Much of this type of material is used in glazing factory casement-type windows and meets ASTM C 669. Other sealants, such as silicone or solvent-based acrylics, are much preferred. A good reason for the oil-based variety of this type of caulk is the much lower price.

Other desirable features are the fast skinning and reasonable flexibility that good caulks give. Flexibility comes with the use of polybutene. Caulks and sealants for use in building joints also exhibit fast skinning, but then remain flexible within the mass of caulk for long periods of time. None of the referenced standards gives a hardness requirement.

Water Emulsion Sealants

Emulsified in water, they cure to rubber-like substances when the water is driven off at room or elevated temperatures.

Plastic Sealants

These are based on a combination of polybutenes and polyisobutylenes compounded with nondrying vegetable oils that remain as a permanent plastic compound.

Hot-Poured Bituminous Sealant

Straight asphalt, or asphalt modified caulks with reclaimed rubber and/or mineral fillers, is included in this group. The material is supplied in a semisolid form and must be melted in order to pour it into the crack or joint. In the case of straight or mineral-filled asphalt, the melting can be done in small direct-fired carts. This type of sealant usually is employed for small repairs and maintenance work, particularly by municipalities. Rubber-bitumen compounds, however, must be heated indirectly in an oil bath, with the temperature controlled at 400° to 410° F (205° to 210° C) to avoid scorching the rubber. This sealant is the choice for large construction and maintenance jobs.

Also available are asphaltic cement, cutback asphalts, filled asphalts, coal tar, and emulsions. The principal advantages to these unmodified bitumens are that they are readily available and cost relatively little. Using any of these caulks is better than nothing, but they rarely can be considered effective sealants, because their life expectancy runs from a few months in severe climates to only two years in especially mild environments. It might be best to regard these materials as joint fillers, since they have many shortcomings as sealants. These materials become brittle at low temperatures. Some become too soft and are too easily penetrated at high temperatures. The coal tar will flow. The greatest shortcoming might be that asphalts simply do not have a significant potential to bond with portland cement concrete. In fact, asphalts often are considered to be incompatible with concrete.

Sealants based on oleoresinous materials, asphalt, and various types of oil are generally nonhardening. Table 6.2 offers a few materials, their coverage costs, and some basic uses. These soft-setting sealants stay wet after application (plasticizers continuously come to the surface) and never truly dry. They generally cannot be depended upon to perform joining functions, although some formulations are used as adhesives in very low-stress joints. These sealants are characterized by the mastic type of paste usually applied to seams with a trowel or brush. Many consistencies and formulations are available, so it is difficult to generalize about them. For example, soft-setting natural resin and oil-based sealers are available in drying and nondrying types. The drying types stay soft inside but develop a protective skin; then, on drying, remain permanently tacky.

Some nondrying sealants are formulated in such a consistency that they can be packaged in tape form. These are not adhesive-backed and can be easily thumbed into place. For purposes of this chapter, they are not considered to be sealing tapes, but simply another example of a heavy-consistency, nonhardening sealant.

TYPICAL FORMULATIONS

A good quality gun-grade, oil-based caulk might have the general formula listed in Figure 6.1. A knife-grade compound would contain a less-viscous oil and have 20 to 25 percent of the selected oil. Less-fibrous fillers would be used with slightly more ground calcium carbonate filler and very little solvent. More polybutene can be added for improved knifing ability. A second typical formula listed in Figure 6.2 uses a different oil base. The oils dry and oxidize because of the cross-linking of the conjugated double bonds in the presence of a catalyst. These types are applied by caulking gun and usually can be painted over in 24 hours. They typically are used in consumer applications.

TABLE 6.2 Typical Sealant Applications and Representative Costs

Sealant base	Cost per gallon	Uses
Nonhardening		
Oleoresinous	2–7	Sealing of concrete joints, masonry copings, and tile. Glazing single lights of glass.
Asphalt and bituminous	1–5	Sealing fraying surfaces, metal joints, silos, air conditioners. Caulking of expansion and con traction joints.
Polybutene	3–8	General construction-type caulking and glazing. Seal between dissimilar metals.
Hardening		
Oleoresin, asphalt, and bituminous	1–7	Same general uses as nonhardening. Hardening-type materials should be used where pressure limits exceed the limitations of the nonhardening formulations.

Raw materials	Parts by weight
Boiled linseed oil	25–30
Calcium carbonate	45–50
Polybutene	5–10
Colorants	3–5
Thixotropic agents	2–4
Dryers	0.1–0.3
Mineral spirits	10–12

Physical properties	Values
Consistency	Heavy mastic
Movement capability, %	± 5
Hardness, Shore A	30 after 1 year
Cracking	None after 300 hours in weatherometer
Weight loss, %	< 20

FIGURE 6.1 Starting formulation for an oil-based caulk

APPLICATIONS

Oil- and resin-based caulks are furnished in both gun and knife grade and may be delivered to the job site in tenth of a gallon cartridges or one- to five-gallon pails. Building supply or hardware stores offer this system in smaller sizes. The materials are easy to apply and require no priming except for quick dusting. Priming may be desirable with wood substrates.

Raw materials	Parts by weight	
	1	2
Blown soybean oil	30.0	17.1
Polybutene	15.0	7.7
Soya fatty acid	2.0	—
Tall oil fatty acid	—	1.1
Calcium carbonate	35.6	56.0
Fiberous talc	8.4	14.0
Cobalt naphthenate, 6% Co	0.5	0.3
Mineral spirits	8.5	3.8
45–50		
3–5		
2–4		
0.1–0.3		
10–12		

Physical properties	Values
Consistency	Heavy mastic
Movement capability, %	± 5
Hardness, Shore A	30 after 1 year
Cracking	None after 300 hours in weatherometer
Weight loss, %	< 20

FIGURE 6.2 Typical oleoresinous (oil-based) sealant, gun grade.

Applicators and contractors use the caulks to glaze wood and metal sash, door and window frames, interior cracks sealing, lap joint sealing of interior ductwork, copings, and other noncritical moving areas. The homeowner will use the caulk for the same applications, and for the exterior would use a solvent acrylic-based sealant for better movement flexibility. The best available compounds for a 10-year product life are ASTM specifications C-570 for oil- and resin-based caulking for building construction, and specification C-669 for glazing back bedding and face glazing of metal sash cover good quality caulk.

Some outmoded specifications are still on the books, but in many cases have been replaced by the following:

- TT-C-00598c for oil- and resin-based types
- TT-G-410E for back bedding and face glazing
- TT-P-0079b for linseed oil type
- TT-P-781a for putty and elastic compounds for glazing metal sash

SUMMARY

Oil- and resin-based sealants and caulks are nonelastic systems designed for joints with little or no movement. They have some shrinkage and hardness will gradually increase, because they are based on drying oils. These are some of the lowest-priced caulks on the market and are available in limited colors and quality grades. Easy to apply, they have no storage or handling problems; in addition, they do not stain or require any joint cleaning or priming. Joint movement capability is maximum ± 5 percent. During the past 25 years or more, this material has lost a considerable market share to acrylics, polysulfides, polyurethanes, and silicones. There is no potential growth for oil-based caulks; the market has bottomed out.

CHAPTER 7
CONCRETE, CEMENTS, GROUTS, AND MORTARS

INTRODUCTION

Cement is a natural material with adhesive and cohesive properties that make it capable of bonding mineral fragments into a compact whole. This is achieved by bonding fine and coarse aggregate particles with cement paste, which is a mixture of cement and water [1]. Concrete is one of the four most important materials used in civil engineering and building. No type of steel, aluminum, or plastic panels applied to the outside of buildings can match the value of large exterior wall sections of precast concrete in speed of erection, permanence, or versatility of form and function. In many respects, concrete is superior to steel. It can be cast in any desired shape and water will not rust it, so it requires no painting. It is much more fireproof than steel and is used in many buildings to make them fire-resistant; and it can speed up construction when precast units, such as beams, columns, and wall sections are used.

The use of cementing materials goes back to ancient Egypt and Rome, but the invention of modern portland cement usually is attributed to Joseph Aspdin, a builder in Leeds, England, who obtained a patent for it in 1824. Today, the annual world production of portland cement is around 700 million metric tons.

ARCHITECTURAL CONCRETE—HISTORY [2]

Roman engineers used concrete in 200 B.C.; however, it was largely ignored during the following centuries and did not re-emerge as a viable building material until the Industrial Revolution. Table 7.1, is a chronological record of the advancement of architectural concrete [3].

TABLE 7.1 Architectural Concrete Time Line

55 B.C. Roman Amphitheater, Pompei, Italy	1894 Church of St. Jean de Mont Martre (exposed arches and vaults), A. de Baudot, France	1916 Concrete-frame house, R. Van't Hoff, Utrecht, The Netherlands
27 B.C. Cement cures in water, Rome, Italy	1895 Factory at Tourcoing (concrete structural frame), F. Hennebique, Tourcoing, France	1917 Orly Airport hangars (parabolic vaults), E. Fressinet, Orly, France
A.D. 1-199 Roman Colosseum, Basilica of Constantine, Pantheon Dome, Rome, Italy	1901-1908 Arch bridges (continuity of two-way slab and "mushroom" column), R. Maillart, Switzerland	1918 Selma cargo ship (lightweight concrete boat), R. J. Wig, United States
1121 Reading Abbey, Reading, UK	1902 Rue Franklin Apartments (concrete frame), A. Perret, Paris, France	1919 Futurism, Sant'Elia, Italy
1756 Eddystone Lighthouse, J. Smeaton, Cornwall, UK		1920 Five Points of Architecture, Le Corbusier, Second International Exposition
1796 Roman cement patent, J. Parker, London, UK	1902 Ingalls Building (reinforced-concrete skyscraper), Ferro-Concrete Co., Cincinnati, Ohio	1922 Notre Dame du Raincy (vaults and columns exposed), A. Perret, Raincy, France
1824 Portland cement patent, J. Aspdin, London, UK	1904 Cité Industriale (reinforced concrete, planned for use in all building types), T. Gamier, Paris, France	1928 Horticultural Hall (elliptical arches exposed), Easton/Robertson, Westminster, UK
1825–1843 Thames Tunnel, Rotherlide, UK	1905 Rue de Ponthieu garage (exposed concrete frame), A. Perret, France	
1854 Reinforced concrete patent, W. Wilkinson, London, UK		1931 Villa Savoye (concrete columns, slabs, spiral stairs), Le Corbusier, Poissy, France
1856 Reinforced concrete patent, F. Coignet, Paris, France	1907 Tilt-up wall construction, R. Aiken, Zion City, Illinois	1932 Concrete pumped by compressed air
1867 Reinforced concrete patent for wire mesh in plate slabs and arches, J. Monier, Paris, France	1908 Ingersol Terrace (tract houses cast-in-place with modular forms), T. Edison, Union, New Jersey	1934 Empire swimming pool (hinged arches with stiffening ribs), O. Williams, London, UK
1871 Portland cement, D. Saylor, United States	1909 La Mouche Slaughterhouse, T. Garnier, France	1935 Hippodrome Raceway (cantilevered shell vaults), Torroja, Madrid, Spain
1875 Ward's Castle, R. Mook, Port Chester, New York	1909 Transit-mixed concrete	
1889 Bridge (reinforced concrete arch), E. Ransome, San Francisco, California	1912 Centenary Hall (ribbed dome), A. Perret, Breslau, Germany	1938 Swiss Pavilion, Le Corbusier, Paris, France

TABLE 7.1 Architectural Concrete Time Line (*Continued*)

1938 Air-entrained concrete sidewalks, N.Y. Department of Transportation, New York, New York	**1954** Church of the Miraculous Virgin (double-curved vaults), F. Candela, Mexico City, Mexico	**1963** Yale Art and Architectural Building (bush-hammered and reeded texture), P. Rudolph, New Haven, Connecticut
1939-1943 Ferrocement boats, P.L. Nervi	**1957** Cabero warehouse (folded-plate roof), F. Candela, Vallejo, Mexico	**1963** New Haven parking garage (continous-board form), P. Rudolph, New Haven, Connecticut
1939 Cement Hall (barrel vault), R. Maillart, Zurich, Switzerland	**1957** Kips Bay Apartments (concrete warehouse), I.M. Pei, New York	
1950 Johnson's Wax Laboratory (cantilevered floors from hollow reinforced core), F. L. Wright, Racine, Wisconsin	**1958** UNESCO headquarters, Breuer/Nervi, Paris, France	**1965** Salk Institute (smooth-form finish), L. Kahn, La Jolla, California
1950 Exhibition Hall (ribbed dome, ferrocement), P. Nervi, Turin, Italy	**1959** Monastere Sainte-Marie de la Tourette (board form and exposed aggregate), Le Corbusier, Lyons, France	**1967** Marina City (concrete high rise), B. Goldberg, Chicago, Illinois
1951 Cosmic Ray Pavilion (thin shell parabolas), F. Candela, University City, Mexico	**1960** Kurashiki City Hall (New Brutalism), K Tange, Kurashiki, Japan	**1967** Dallas City Hall, New York University Towers (board and fin-formed textures), I.M. Pei
1951-1960 Secretariat, Courts of Justice, Parliament (rough-cast concrete, wood and steel forms), Le Corbusier, Chandigarh, India	**1960** Hydraulic-pumped concrete in mobile trucks	**1978** National Gallery of Art, I.M. Pei, Washington, D.C.
	1960 TWA Terminal (shell roof), E. Saarinen, New York, New York	**1985** River City, B. Goldberg, Chicago, Illinois
1952 Unite d'Habitation (board-formed columns, spandrels, soffits), Le Corbusier, Marseilles, France	**1962** Air-traffic control towers, I. M. Pei/Severud, Chicago, Illinois	**1986** Javits Convention Center, I. M. Pei, New York
	1962-1983 National Assembly Building, L. Kahn, Dacca, Bangladesh	

Advancements in reinforced concrete design made by A. deBaudot, F. Hennebique, R. Maillart, and A. Perret showed that the material was capable of supporting structural loads and was an alternative to newly developed structural iron products. The standardization of concrete soon enabled engineers to design with confidence. Dramatic examples of structural concrete technology are the interior vaults of the church of St. Jean de Montmartre, designed by A. de Baudot and built in 1894, and the long-span bridges of Switzerland, designed by R. Maillart and built in the early 1900s.

Portland cement first was manufactured in the United States in 1871. Since then, the United States has led the way in the engineering, manufacturing, and use of concrete and masonry in the building and construction fields. Ernest Ransome and several structural engineers in the United States developed the technology of reinforced concrete for grain silos and low-rise factories.

At the turn of the century, a concern for comprehensive planning caused the architects of the early modern movement to break from the traditional emphasis on applied ornamentation and present straightforward, unadorned designs that reflected the machine-age aesthetic (modular and factory mass-produced), with structural frames exposed in the final finish. This is best illustrated in the Cité Industriale presentation by T. Granier in 1904. In this project, factories, housing, and public buildings were to be made of exposed reinforced concrete. This material was pioneered for industrial construction because the flat slab allowed more space from floor to ceiling (with flat plate design, dropped beams were not required); allowed greater daylight into the building (punched openings in bearing masonry were eliminated); did not require additional treatment for fireproofing (as did steel structures); and with some additional attention, a reasonable finish could be attained without adding another layer of construction (i.e., terra cotta cladding or brick). (See Chapter 15 for more information about fireproofing.)

In 1920, Le Corbusier's Five Points of Architecture exposition advanced Garnier's artistic direction and Ransome's engineering advancements, and established reinforced concrete slabs and columns as a basic tenet of modern architecture.

Unfortunately, during World War II, the demand for steel (armaments) surpassed the demand for concrete, with the result that steel technology rapidly upgraded and was better prepared for the building boom after the war. Theoretical work, however, continued, re-emphasizing the principles first professed in the early 1900s by modern architects. R. Maillart designed the Cement Hall, a thin-shell arch that dramatically presented another possibility of concrete, in 1939. This cast-in-place apartment building with rough, board-formed concrete surfaces became the most influential concrete building in its time. It soon was followed by Le Corbusier, whose work in Chandigrah, introduced this low-energy technology to developing nations with spectacular results. In the United States, I.M. Pei took the building type from its association with utilitarian factories and brought it to a higher sophistication. The Earth Sciences building at the Massachusetts Institute of Technology (MIT) exemplifies architectural concrete design and construction as follows:

1. Design with construction orientation
2. Architectural input into form-work design
3. Concern for modular form work
4. Specification of cement color
5. Construction of a mock-up

UNDERSTANDING CONCRETE AS A BUILDING COMPONENT

Portland cement is universally considered the most important masonry material used in modern construction. Its numerous advantages make it one of the most economical, versatile, and universally used construction materials available. It is used for buildings, bridges, sewers, culverts, foundations, footings, piers, retaining walls, and pavements. A concrete structure, either plain or reinforced, is singular among the many systems of modern construction. In its plastic state, concrete can be handled readily and placed in forms and cast into any desired shape. Quality concrete work produces structures that are lasting in appearance and require comparatively little maintenance. Before we proceed, some basic terms need definition.

Concrete

Concrete can be defined simply as a mixture of aggregates (sand, gravel, or crushed stone) held together by cement.

Masonry

Masonry refers to any type of construction involving the laying of substantial units—blocks, stones, concrete blocks, or tile—with or without a cementing agent to hold them together.

Mortar

Mortar is the cementing agent used when placing units such as blocks and bricks together. Technically speaking, mortar might be described as a type of concrete.

Portland Cement

Portland cement usually is made from calcareous material, such as limestone or chalk, and from alumina- and silica-bearing material such as clay or shale. The manufacturing process essentially consists of grinding the raw materials, mixing them intimately in specified proportions, and burning them in a large rotary kiln at a temperature of approximately 2500° F (1350° C). The material sinters and partially fuses into a ball known as a clinker. The clinker is cooled and ground to a fine powder. Gypsum is added to control the speed of setting when the cement is mixed with water. There is no typical portland cement manufacturing plant, but all cement manufacturing processes are basically the same as outlined in Figure 7.1.

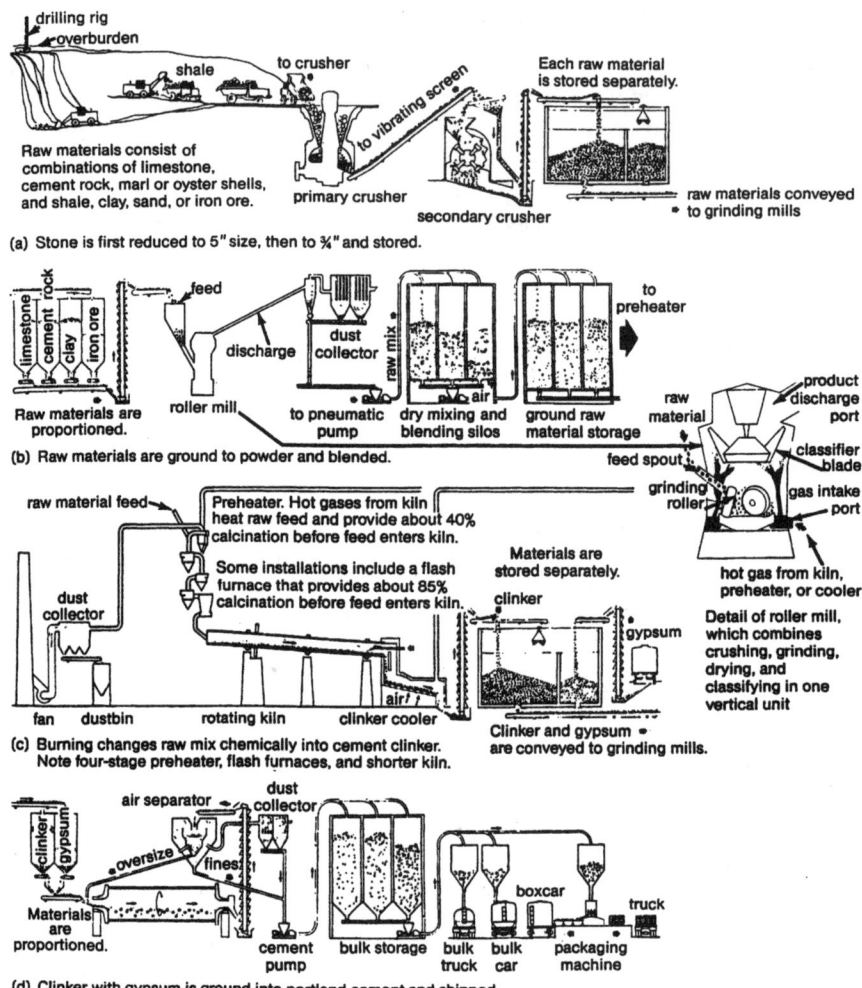

FIGURE 7.1 Portland cement manufacturing process.

COMPOSITION AND HYDRATION OF CEMENT

The main compounds in portland cement and their typical percentage content are as follows:

Tricalcium silicate	$3CaO \cdot Si_2O_3$	55%
Dicalcium silicate	$2CaO \cdot Si_2O_3$	25%
Tricalcium aluminate	$3CaO \cdot Al_2O_3$	9%
Tetracalcium aluminoferrite	$4CaO \cdot Al_2O_3 \cdot Fe_2O_3$	8%

Although all cement is basically the same, different cements are manufactured to meet different physical and chemical requirements for specific applications. Differences between cements are the result of variations in the type and quantity of raw materials used in the manufacturing process. The American Society of Testing and Materials (ASTM) identifies eight types of portland cement, as shown in Table 7.2

Concrete mixtures can also be divided into different classes. One important factor that distinguishes a particular mixture is the type of portland cement used. Concrete mixtures also differ with respect to the proportions of cement, water, and aggregates used to create them. Further differences can be created by the inclusion of various admixtures, such as air-entraining agents or chemicals, which help concrete to set up properly in extremely cold weather. Proportions of materials used in concrete are illustrated in Figure 7.2.

MORTARS [4]

In the late nineteenth century, stronger mortars were created by sweetening the lime with small amounts of portland cement. The addition of cement caused the mortars to harden more quickly, which allowed thicker joints and more rapid placement of masonry units. This was a great step forward for masonry construction. The process of adding cement to mortar mixtures eventually led to the development of masonry cement—a factory-prepared combination of materials that produces the properties desired in a mortar.

TABLE 7.2 Types of Portland Cement

Type I	A general-purpose portland cement suitable for all uses where the special properties of other types are not required. Typical uses include pavements, sidewalks, buildings, bridges, and concrete blocks.
Type II	A specific type of cement used for structures in water or soil containing moderate amounts of sulfate. Also used to moderately control heat buildup in large piers, heavy abutments, and heavy retaining walls.
Type III	A high-early-strength cement with a short curing time. Used when forms need to be removed as soon as possible or when a structure must be put into service quickly. Can be used in cold weather because it permits a reduction in the controlled curing time.
Type IV	A special cement used for constructing dams and other massive concrete structures. Generally not available.
Type V	A cement designed to resist chemical attack by soil and water high in sulfates.
Types IA, IIA	These cements are used to make air-entrained concrete. They have the same and IIIA properties as Types I, II, and III, except that they have small quaniti-ties of air-entraining materials combined with them.
White portland	A portland cement made from raw materials containing little or no iron or cementmanganese (the substances that give cement its gray color). Used primarily in stucco, terrazzo, cement paint, finish-coat plaster, tile grout, and decorative concrete.

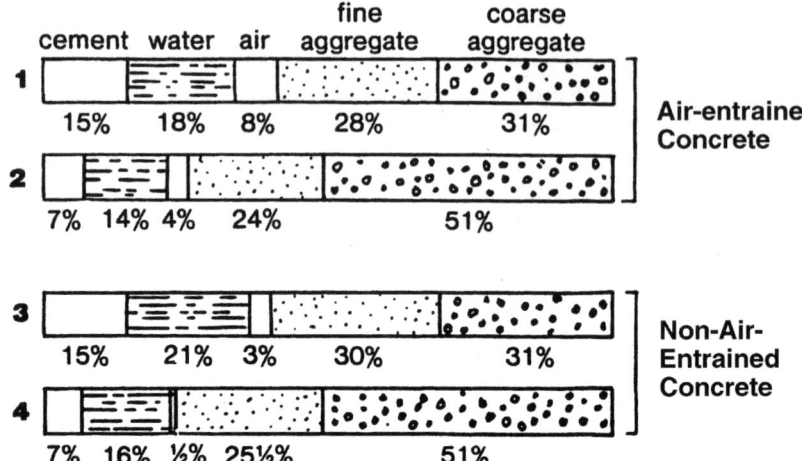

Range in proportions of materials used in concrete, by absolute volume. Bars one and three represent rich mixes with small aggregate. Bars two and four represent lean mixes with large aggregate.

FIGURE 7.2 Proportions of materials used in concrete.

Masonry cement includes the following ingredients:

- Portland cement or blended hydraulic cement.
- Plasticizing material: finely ground limestone, hydrated lime, certain clays, shales, or water-repelling agents.
- Pre-milled pigments of white masonry cement.

More than 80 percent of all mortar used today is made with masonry cement.

PORTLAND CEMENT PLASTER

Portland cement plaster is a combination of portland cement and fine aggregate mixed with water to form a plastic mass able to adhere to a surface and harden, preserving any form and texture it had before setting. Portland cement plaster and portland cement stucco are the same material. Typically, the term stucco describes cement plaster used for exterior surfaces; in some areas it refers only to a factory-prepared finish coat mixture. For ease of understanding, the term plaster is synonymous with plaster or stucco.

BASES

Portland cement plaster usually is applied in three coats over metal reinforcement, with or without solid backing, but two coats can be used over solid masonry or concrete. The successful application of plaster to a base coat depends on the compatibility of the plaster and the base material, the soundness of the base, and the application procedure. Plaster is very compatible with concrete masonry and new concrete and might be compatible with old concrete or masonry, depending on the degree of deterioration. When the plaster is not compatible with a base, the quality of the base must be upgraded to gain both chemical and mechanical bonding, or metal reinforcement must be furred over the surface. Figure 7.3 illustrates methods of applying plaster to a variety of bases.

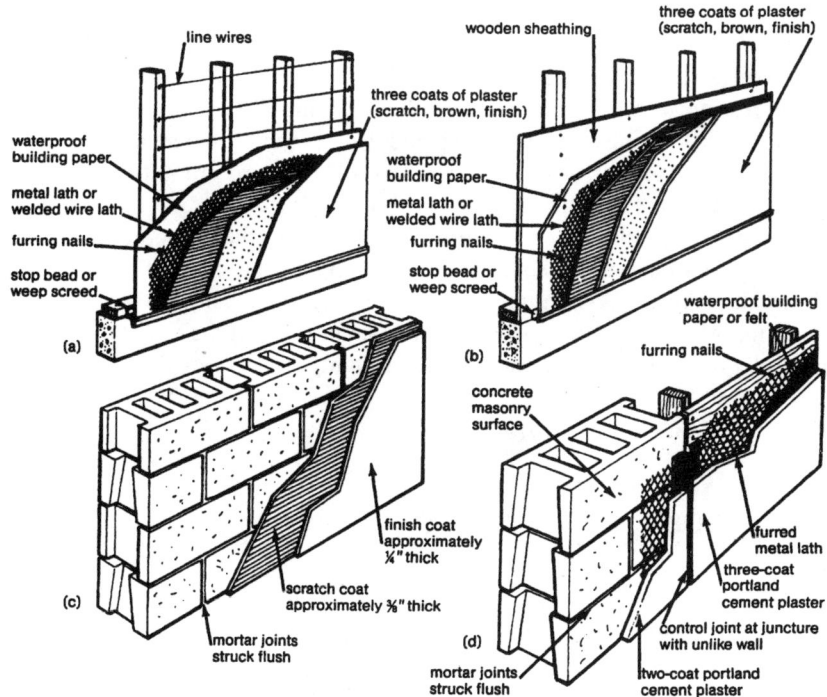

Stucco may be applied on a variety of bases: (a) open wooden framing, (b) sheathed wooden concrete masonry, and (d) unlike bases such as a combination of wood and masonry.

FIGURE 7.3 Stucco applied to a variety of surfaces.

CRACKING IN FLOORS

Although concrete is affordable, durable, and extremely versatile, it is not without flaws. One of the most troublesome characteristics of concrete is the fact that it cracks. Random cracks in floors are unsightly. They make floor cleaning difficult and, especially in installations such as food-processing plants, they can harbor bacteria. The use of forklifts, pallet jacks and other cargo-moving equipment is very demanding on warehouse floors. Joints and cracks tend to deteriorate in high-traffic areas, resulting in stress and wear on mobile equipment. In addition, spalling concrete becomes a grinding compound on the adjacent floor surface, causing further damage. Severe deterioration can cause structural failure. Maintenance and repair of joints and cracks is continuous and onerous. The busiest traffic areas are interrupted for days for repairs that are rarely satisfactory. All this can be avoided by installing joints at proper intervals during floor construction.

CAUSES OF FLOOR CRACKING

Two of the most common causes of floor cracking are drying shrinkage and contraction caused by cooling. Both of these cause a shortening tendency in slabs, which is restrained by friction between the base and the concrete resting on it. The result is that tensile stress builds up and, because the concrete is weak in tension, it cracks. Cracking also can occur when floor slabs are not free to move independently of the building elements with which they are in contact. Movement of the slabs is likely to be different from that of other structural members, and if they are joined rigidly to other structural members, they may be unable to accommodate this differential movement and, again, cracking is the result.

Curling of slabs on grades also causes cracking. When the top of a slab dries and shrinks more than the bottom, the edges curl up. The weight of the unsupported edges, plus any other load on the floor, creates tensile stresses in the top of the slab and cracks occur [5].

HOW VARIOUS JOINTS WORK

Joints permit concrete to move slightly, creating planes of weakness where cracks can form. The idea is that if concrete cracks or separates at a joint, it is less likely to crack at other locations. Joints are neater in appearance than random cracks and are easier to keep sealed and clean. In addition, the edges of a properly maintained joint are less likely to chip or spall than the edges of random cracks. Three types of joints are used in concrete floor construction. Other chapters in this book will discuss joints and design in greater detail.

Isolation Joints

Sometimes called expansion joints, isolation joints allow both vertical and horizontal movement. The joints separate or isolate concrete slabs from columns, walls, footings, and other points of restraint, such as machine foundations and stairwells. No connection should be made across an isolation joint by either keyways or bond reinforcement.

Control Joints

Frequently called construction joints, control joints allow movement only in the plane of the floor and control cracking caused by the restrained forces resulting from drying and cooling. Load is transferred across control joints by the interlocking of aggregates in crack faces formed at the joint, or by contact of the tongue and groove in a keyed joint, as illustrated in Figure 7.4. Dowel bars also are used to transfer loads that must support heavy-wheeled traffic. When wire mesh is used in the slab, it should not cross a contraction joint. If it does, the joint will not open properly and cracking can occur at that location.

Construction Joints

These joints are stopping places for a day's work. They can be made to function as isolation joints by lining the bulkhead with a preformed sheet material or as control joints by using keyed bulkheads. If construction joints are located where no movement is wanted, tie bars or welded wire fabric can be used to hold adjacent slabs together.

HOW JOINTS ARE FORMED

Isolation Joints

These permit slabs to move slightly up or down relative to walls or columns. The vertical surfaces of the joints are covered with joint material to eliminate concrete-to-concrete contact. Asphalt-impregnated sheets or other expansion joint fillers must be thick enough to permit the required movement. Joints with rough concrete surfaces require thicker joint fillers. At columns, isolation joints are boxed out with square or circular shapes, as shown in Figure 7.5. When square shapes are used, the wood form is placed so that its corners point at the control joints along the column lines. Unless this is done, cracking can occur. When circular forms are used, they generally are made of plastic or fiberboard.

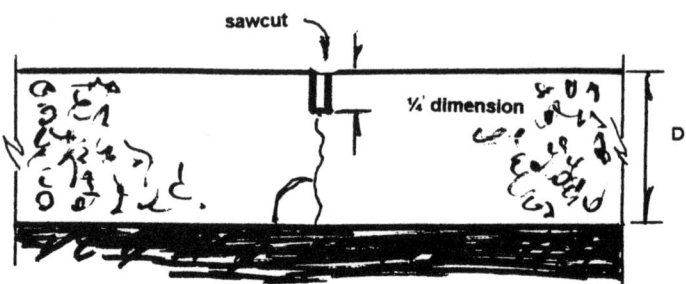

SAWED CONTROL JOINT

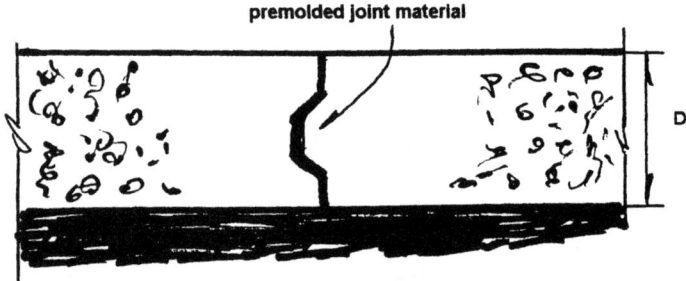

TONGUE AND GROOVE CONTROL JOINT

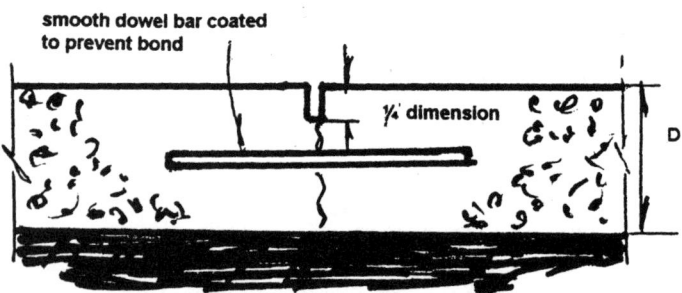

CONTROL JOINTS WITH DOWEL

**Control joints may provide load transfer
through aggregate interlock, keyways or dowel bars**

FIGURE 7.4 Control joints may provide load transfer.

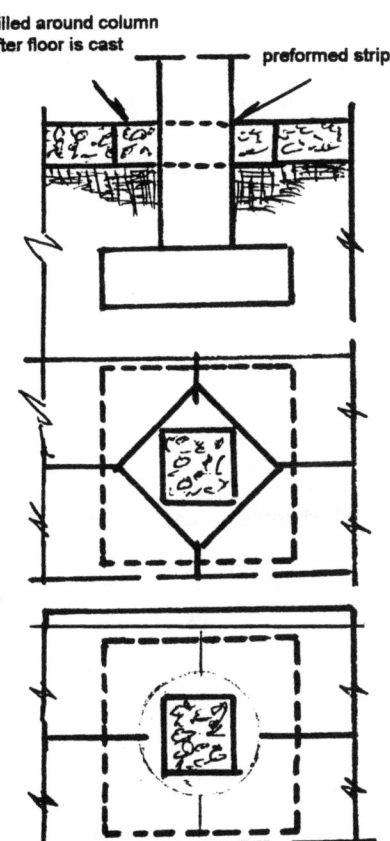

FIGURE 7.5 Isolation joints separate concrete slabs from columns.

Control Joints

These joints can be produced in several ways: by saw-cutting, tooling, or inserting plastic strips. Cutting a groove with a power saw is the most common method of making a control joint. A variety of diamond-tipped blades is available to cut concrete containing hard or soft aggregates and blade suppliers can recommend the best choice. Lower-cost, reinforced abrasive (silicon carbide) blades are suitable when the concrete is made with soft, free-cutting aggregates. One way of getting longer life from saw blades is to precut the joint with a pointing trowel after the fresh concrete has been bull-floated. This moves the aggregate particles aside so that when the joint is sawed later, only the mortar is cut. A straight board can be used as a guide for the trowel cutting. Further floating and troweling follow, making it important to mark exactly where the precut was made so the saw operator will know where to cut the joint.

Saw cutting must be done as soon as possible after the concrete hardens; otherwise, random cracks can form in the slab. Saw operators should make trial cuts, starting a few hours after the concrete hardens, then observe the results. If aggregate particles come loose, it is too soon to begin sawing. When the blade doesn't tear or damage the concrete, it is hard enough. Joints should be sawed to a depth of about one-quarter the thickness of the slab. Shallower cuts are unlikely to weaken the concrete sufficiently at the joint, and this can cause cracking between joints.

Tooling is another method of forming control joints. Tooling is usually less efficient and less economical than sawing, but it can be done on smaller jobs. Tooled joints are formed with a jointer having a bit deep enough to cut one-quarter the thickness of the slab. When making the groove, a straight board can be used as a guide to ensure that the joint is straight.

Preformed plastic ribbon also can be used to make control joints. Placed on edge just beneath the surface of a freshly poured concrete slab, a plastic ribbon creates a vertical plane of weakness in the slab. As the concrete hardens and the slab shrinks, the slab will crack along this plastic ribbon.

To be certain the ribbon is in a straight line, a screed or board can be used as a guide for the float pan, or a string line can be struck on the concrete slab and used as a guideline. The ribbon usually is installed immediately after the concrete has been screed, but it also can be installed when the concrete is just starting to set and is able to support the weight of a worker. The ribbon is placed in the concrete with a tool. The float pan of the tool is pushed across the slab with a wooden handle or pulled across by a string attached to the front of the float pan. As the float pan is moved across the slab, the vertical steel blade underneath the float pan cuts a groove in the slab and the ribbon is fed into this groove through a slender opening in the steel blade. Concrete then is finished over the top of the ribbon. The ribbon is left in place, even after the slab cracks. Any of three ribbon widths up to 1½ inches can be placed by the same tool with only minor alterations. Regular surveyor's tape 1-5/16 inches wide can be purchased from most building materials suppliers. By this method, a joint 25 feet long can be installed in about 30 seconds; the cost to install joints is less than 3 cents a foot. The tool used to place the ribbon costs approximately $100.

If specifications require that joints are to be sealed, the manufacturer should recommend a polyurethane sealant. Because the crack is so narrow, less sealant is needed than would be required in other types of joints.

Construction Joints

Butt-type construction joints, as described in Figure 7.6, are formed by a bulkhead and are satisfactory for thin floors that aren't heavily loaded. Dowels can be added at butt-type joints to provide load transfer in floors carrying heavier loads. The side forms or bulkheads for slab-on-grade construction also can be keyed so that the slab will have a tongue-and-groove construction joint after the concrete has been cast on both sides. These keyed joints act as a control joint, permitting slight horizontal movement, but no vertical movement. On the surface, a keyed joint often looks like a ragged line unless both sides of the joint are edged or a sawcut is made, as shown in Figure 7.7.

SPACING AND LAYOUT OF JOINTS

Figure 7.8 shows a typical layout of joints in a floor on grade. It is common practice to construct floors in long lanes and to locate control joints, whether sawed or keyed, along column lines. The contractor can start at one end and place each successive lane after the previous one has hardened. Another approach is to construct alternate lanes, but this may require more side forms. Joint spacing also should be chosen so that concrete sections are approximately square. If sections have length-to-width ratios greater than 1½ to 1, they are likely to crack between joints.

Often, locating joints along column lines doesn't provide close enough spacing to prevent cracking between joints. Table 7.3 suggests spacing of control joints for various thicknesses of slabs. A rule of thumb for plain slabs is that joint spacing, in feet, shouldn't exceed two and one-half times the slab thickness, in inches. However, other factors influence the required spacing. These include the slump of the concrete and the maximum size of the coarse aggregate.

CRACKING IN CONCRETE REPAIRS—
A COMPLICATED PROCESS

Repair of concrete structures is a million-dollar business, and a good portion of every concrete repair dollar goes towards specialty mortars [6]. Selecting a material to patch concrete appears to be simple—why not use concrete? Sometimes

7.16 HANDBOOK OF ADHESIVES AND SEALANTS IN CONSTRUCTION

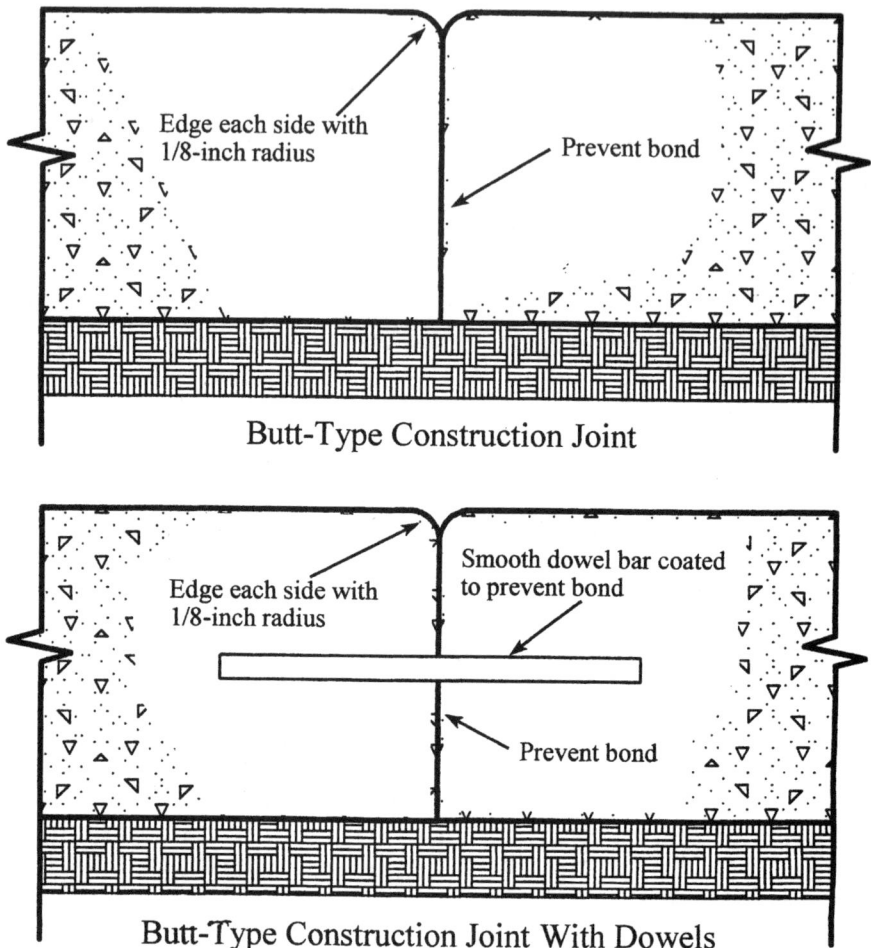

Butt-Type Construction joints may be formed with or without provision for load transfer. Bonded construction joints are occasionally needed and can be built using tie bars instead of dowels.

FIGURE 7.6 Butt-type construction joints may be formed with or without provision for load transfer.

by adding certain chemicals, sand, stone, and water recipes, manufacturers claim to make the repair foolproof; and many times they do. The addition of chemicals to enhance one property, such as set time, however, can cause reductions in other properties, such as strength. This is one of the reasons a material supplier has so many products to offer—a balance of good and bad must be matched to the individual repair constraints. While this provides specifiers with great flexibility, it also requires them to have extensive knowledge of the products.

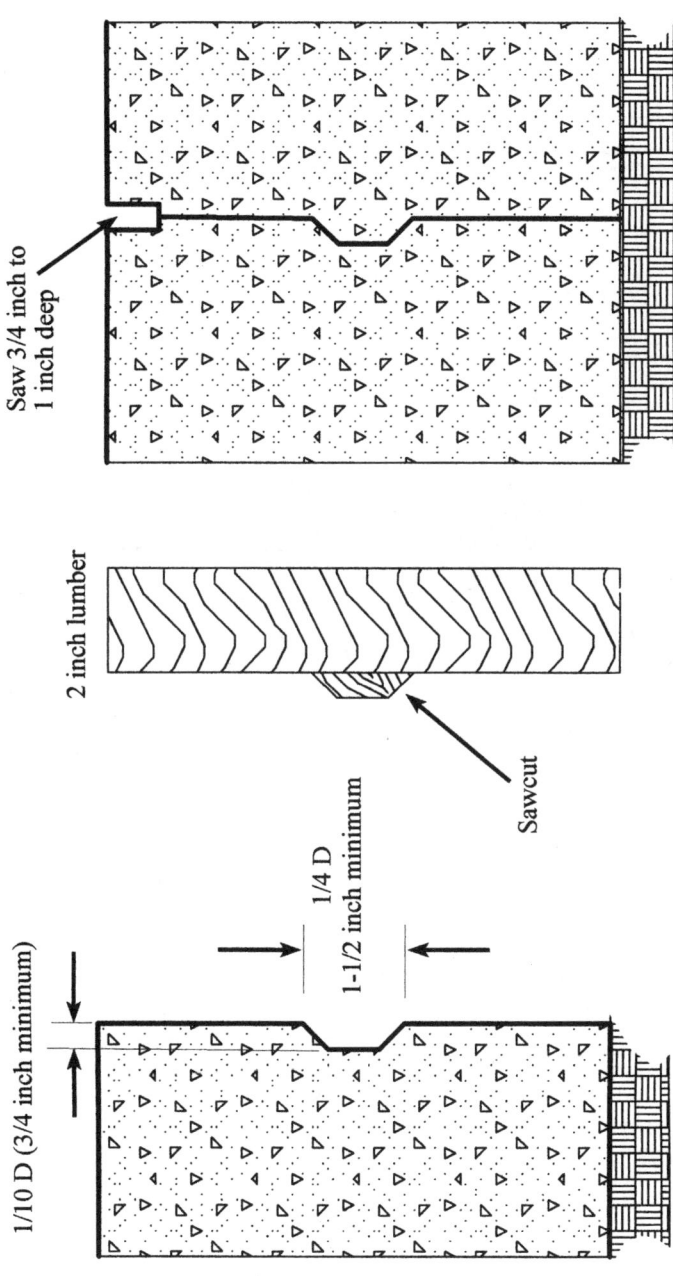

FIGURE 7.7 Tongue and groove or keyed joints.

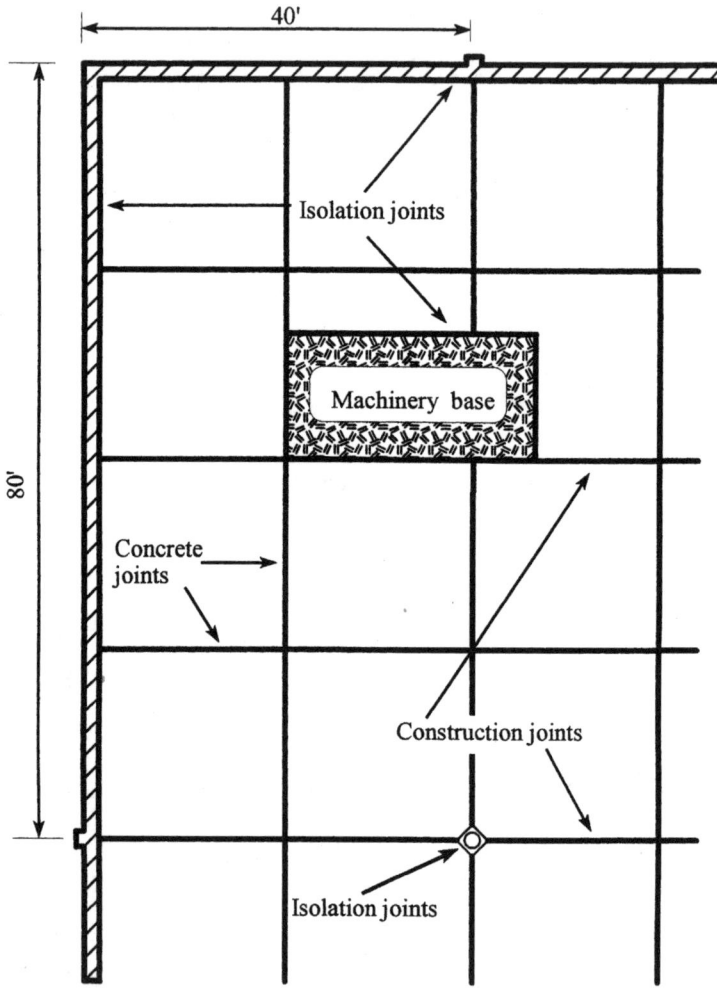

Joints should be located along column lines and intermediated points so that panels are as nearly square as is practical. Construction joints in this layout, function as control joints. Note also the isolation joints around walls, columns, and machinery bases.

FIGURE 7.8 Joints should be located along column lines.

TABLE 7.3 Suggested Spacing of Control Joints

Slab thickness, inches	Less than ¾-inch aggregate: Spacing, feet	Larger than ¾-inch aggregate: Spacing, feet	Slump less than 4 inches: Spacing, feet
4	8	10	12
5	10	13	15
6	12	15	18
7	14	18	21
8	16	20	24
9	18	23	27
10	20	25	33

Spacings also apply to the distance from the control joint to parallel isolation joints or to parallel construction joints. Spacings greater than 15 feet show a marked loss of effectiveness of aggregate interlock to provide load transfer across the joint.

Specifiers must consider:

- Placement method: hand applied, form and pour, form and pump, or shot applied.
- In-place environmental exposure: weather, chemicals, de-icing salts, and abrasion.
- In-place load exposure: compressive, tensile, and shear loads.
- Patch geometry: thickness, width, and length.
- Interaction of the patch material with the substrate: bond, strength gain with time, shrinkage, and resistance to cracking during curing.

Most manufacturers' brochures offer expert advice about the placement method. Special formulations are available to make moldable hand-applied mortars, flowable material for placement in tight forms, and less rebound to shot-applied materials. Most of the specialty products also resist chemical and environmental attack, because the polymers that usually are added make the material much less permeable than normal concrete. It usually appears easy to pick a material strong enough for in-place load exposure, since compressive-strength testing is so generic and well understood by people in the concrete industry. In addition, concrete patches occur where compressive strength is sufficient, since the patches usually are concentrated near corroding reinforcement in areas under tension, not compression.

On the other hand, compatibility with the modulus of elasticity (a measure of the material's stiffness necessary to convert strain to stress) between the substrate and patch material can be critical. As an example, in column repairs with compression loading, a too soft material (low modulus of elasticity) will leave the substrate overloaded, while a too-hard material (high modulus of elasticity) can attract more than its share of the load, causing the repair to spall. Both problems can occur regardless of the compressive strength of the patch material.

Producers of patching materials usually offer adequate instructions about which materials work with certain patch geometries. Special materials are required for very thin patches, and some materials will not work in very large patches. It is necessary to take precautions with any applications that are covered too specifically in the data sheets supplied.

UNDERSTANDING CRACKING

Despite apparent comprehensive assistance from the industry, experienced concrete-repair specifiers often are shocked to find that their well-researched, newly installed repair material is cracked several days after placement. Over the past several years, many concrete experts have blamed lack of crack resistance on the drying-shrinkage properties of repair materials. The material manufacturers have been chastised for their apparent inability to supply nonshrink materials and have asked the material suppliers to produce a repair material that has little or no shrinkage compared to normal concrete, and have insisted the shrinkage be reported accurately in their data sheets so a comparison can be made with competitive patching systems.

On first analysis, it makes sense that an introduction of a shrinkage-prone patch material into an inert patch cavity will lead to cracking upon drying. In addition, there is a general inconsistency in the industry about how to measure shrinkage. One often needs to read between the lines of data to determine if the shrinkage is an air-dried exposure (the real world) or a wet-bath exposure (no shrinkage will occur because drying has been prevented). Even with real-world data, the great variability of material properties and test methods for the reported properties can be confusing. An example of the confusion is shown in Table 7.4, which compares properties of several hand-applied patching mortars. The resulting emphasis on drying-shrinkage alone oversimplifies a complex interaction between several time-dependent properties: drying time, tensile strength, tensile stiffness, and tensile creep. None (except shrinkage) is usually reported. Careful consideration must be taken when specifying the correct patching system.

Figure 7.9 shows a plot of the volume-change stresses in a patch superimposed over a plot of the tensile strength of the patch, both against time. The volume-change stress in the restrained environment of the patch (usually the patch is bonded to a relatively immovable substrate on several planes) is a function of unrestrained volume-change strain (change in volume caused by drying shrinkage and thermal cooling from peak temperatures associated with hydration of the cement in the patch material); tensile creep (a measure of relaxation in the

TABLE 7.4 Comparison of Hand-Applied vs. Dry Pack Materials

Material	Wet density (lbs/ft³)	Pot life (minutes)	Workable time (minutes)	Coefficient of thermal expansion (per °C)	24-hour compressive strength (psi)	7-day compressive strength (psi)	28-day compressive strength (psi)
Product 1	104	30	Not listed	7.30E-06	Not listed unsure by telephone	2860 unsure by telephone	5200-5800 ASTM C 109
Product 1A	107	15 @75	30 @75F/50RH	7.30E-06	1650 ASTM C 109	3500 ASTM C 109	5100 ASTM C 109
Product 2	Not listed	15	20-60	Not listed	3500 ASTM C 109	6500 ASTM C 109	8000 ASTM C 109
Product 3	130		120 after initial set	Not listed	2690 ASTM C 109	6825 ASTM C 109	7050 ASTM C 109
Product 4	Not listed	Not listed	25-60	Not listed	2500 ASTM C 109	5000 ASTM C 109	6500 ASTM C 109
4000 psi Substrate	145	60	60-90	9.90E-06	500-1000	3000	4000-5000

(Continued)

TABLE 7.4 Comparison of Hand-Applied vs. Dry Pack Materials *(Continued)*

Material	28-day flexural strength (psi)	28-day tensile strength (psi)	28-day splitting tensile strength (psi)	28-day slant shear bond strength (psi)	28-day shrinkage (micro in./in.)	Modulus of elasticity (psi)
Product 1	925 7-day test by telephone	Not listed	590 unsure by telephone	1600 ASTM C 882 mod	1120 ASTM C 157 by telephone	2.00E+06 unsure by telephone
Product 1A	790 7-day test by telephone	Not listed	400 unsure by telephone	700 AASHTO T-237	1103 ASTM C 157 by telephone	1.90E+06 unsure by telephone
Product 2	2000 ASTM C 293	Not listed	900 ASTM C 496	2200 ASTM C 882 mod	900 ASTM C 157 by telephone	4.70E+06 ASTM C 293 (10 day) by telephone
Product 3	930 ASTM C 348	600 ASTM C 190	Not listed	1075-1605 (14 days) ASTM C 1042	700 ASTM C 157	Not listed
Product 4	1500 ASTM C 78	Not listed	750 ASTM C 496	2200 ASTM C 882 mod	750 ASTM C 157	3.20E+06 ASTM C 496
4000 psi Substrate	475				400-800	3.60E+06

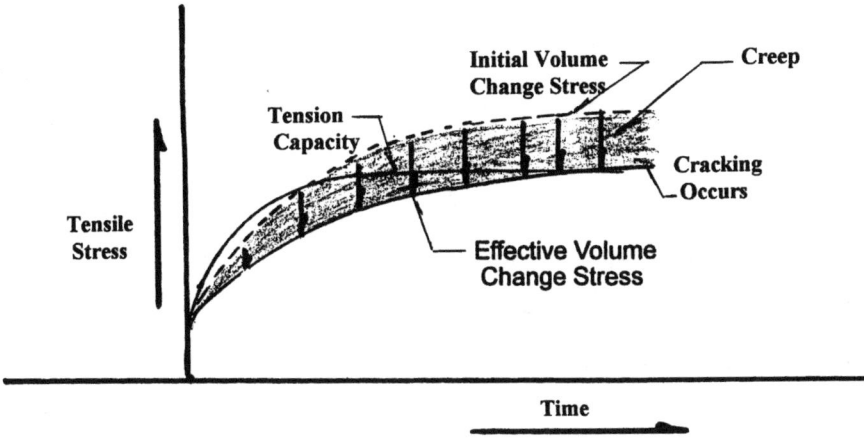

FIGURE 7.9 Volume change effects.

unstrained material under load generally reduces the effective strain); and the tensile modulus of elasticity. These properties and, therefore, the effective, in-place, volume-change stress, change rapidly in the first three to seven days after placement, usually stabilizing by 28 days. As the volume-change stress changes, so does the tensile strength or resistance to the volume-change stress. If the stress exceeds the strength at any time during this period, cracking will occur.

VISUALIZING STRENGTH

Fosroc, Inc., a repair-material manufacturer, has asked Simpson, Gumpertz and Heger, Inc., to apply the color-graphics capabilities of modern finite element computer analysis (FEM) to develop simulations of how time-dependent materials, patch geometry, extended loading, and curing conditions affect a patch material's performance, especially its crack-resistance in service. The objective of the FEM study is to develop a tool that explains to material suppliers the need to evaluate why some materials that might have higher drying-shrinkage properties than other products perform well in service.

Based on drying-shrinkage data alone, a 1,200-micro-strain ($1 \times 10\text{-}6$ in./in.) product does not compare well with a competitor's 800-micro-strain product. But if the 1,200-micro-strain product has a modulus of elasticity that is one-half that of the 800-micro-strain product, there will be less strain and fewer tendencies to crack, associated with restraint of the higher shrinkage product. Similarly, a material with high creep (ability to relax under load) will generate less effective strain and stress.

The efforts have been directed at development of generic two- and three-dimensional simulations of a concrete repair. (MSC-NASTRAN finite element analysis software was used in the study.) The model in Figure 7.10, simulates a patch in the rib of a two-way waffle slab. The rib is 5 in. (125 mm) wide and reinforced with two #5 bars. The patch extends over the bottom 3 in. (75 mm) of the full rib width and is 16 in. (405 mm) long. The patch and the concrete are considered symmetrical about the longitudinal vertical plane through the depth of the rib and about the midlength of the patch. The substrate concrete is assumed to be restrained except for the elastic material deformation adjacent to the patch. With boundary conditions that account for symmetry, the full patch can be simulated with a smaller model (one-fourth of the full patch), greatly reducing computer time. (Additional data can be obtained from the Fosroc, Inc. in Georgetown, KY.)

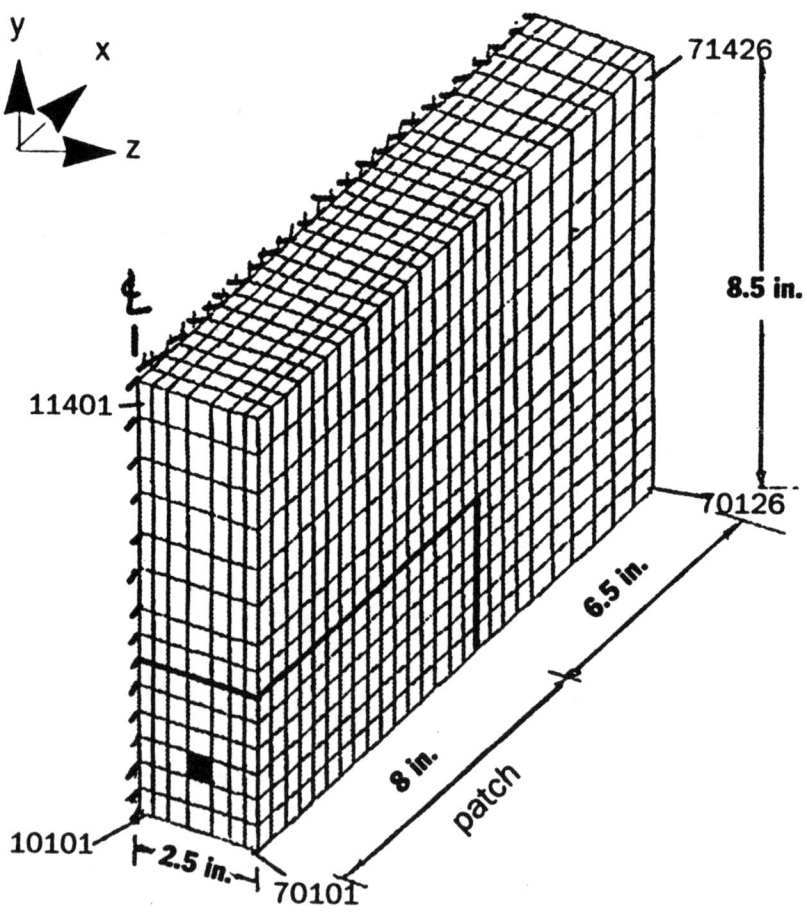

FIGURE 7.10 Finite element model.

SHRINKAGE [7]

Concrete is basically a mixture of variously sized aggregates, cement, and water. The aggregates are sized so that the smaller particles fill the voids between the larger pieces, and the cement binds the two together. Chemical reactions occur when half of the water is added to the mix. On the job site, the remaining half of the water usually is added so that the concrete can be worked. Sometimes even more water is needed to make the mix wet enough for the contractor to handle. It must pour smoothly out of the truck and spread easily within the form work.

Concrete shrinks as it dries and more water requires more drying and produces more shrinkage and movement. The concrete is still weak when most of this movement occurs. Shrinkage-compensating concrete reduces cracking by minimizing or delaying shrinkage until the concrete is stronger.

KEEPING CRACKS TO A MINIMUM

Shrinkage-compensating concrete has successfully limited cracks in millions of square feet of concrete floors in all types of service structures. This type of concrete also eliminates the substantial first coat for steel armoring of additional joints in conventional floor designs.

Basically, an agent is introduced into the concrete that promotes the growth of crystals, which occupy space and creates expansive forces. Although the usual hydration and drying processes still occur, the concrete actually expands slightly (because of the crystal growth) instead of experiencing the normal shrinkage. The reinforcing system restrains the expansion, putting the steel reinforcing bars in tension. As the slab continues to dry and shrinkage begins to exceed expansion, the tension in the reinforcing is relieved. These opposing forces keep the concrete from cracking as it shrinks. Shrinkage-compensating concrete's unique characteristics enable contractors to pour slabs of 8000 ft2 to 12,000 ft2 (745 m2 to 1115 m2) without a single joint. As shown in Figure 7.11, a 40,000 ft2 (3715 m2) floor could be built with only two 400-ft (120 lineal meters) joints. The same floor with conventional contraction and construction joints would have 3600 ft (1095 lineal meters) of joints, assuming 20-ft (6 m) spacing. Figure 7.12 depicts typical length change characteristics of shrinkage-compensating and portland cement concretes.

HAVING THE RIGHT CONCRETE

ASTM defines three types of expansive cements, K, M, and S, but recognizes that other formulations also can be used. Much of the information about shrinkage-compensating concrete can be confusing (and sometimes can appear contradictory). The following issues deserve particular attention:

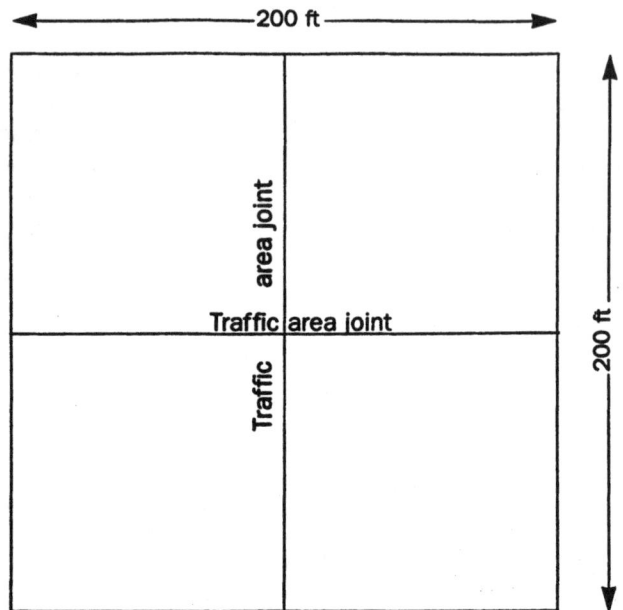

FIGURE 7.11 Traffic area joint.

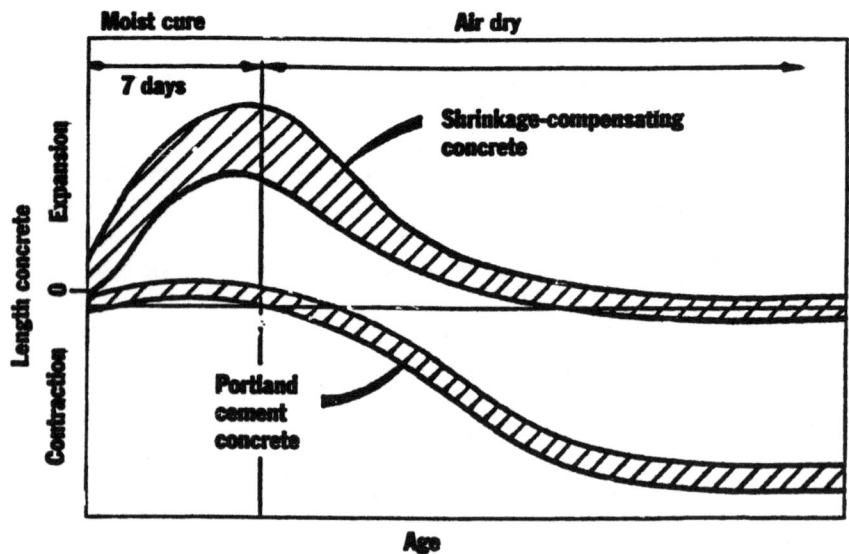

FIGURE 7.12 Length change characteristics.

Expansion Tests

The ASTM C 806 test for expansive cement mortar and the ASTM C 878 concrete expansion test determine how much expansion a specified mix design obtains. Local aggregates can have a significant effect on the amount of expansion produced. The C 878 test determines the amount of expansive concrete or component needed in the mix design to achieve the desired amount of expansion. The laboratory test is very sensitive to variables such as temperature and operator error. Differences frequently are found between lab results and actual field performance.

Expansion Rate

Some expansive cements do a better job than others in matching the opposing forces and retaining slab movement. Timing of the expansion can be at least as important as the total amount of expansion achieved. With the C 878 test, there is a tendency to specify expansions of 0.05 percent or greater, when satisfactory field performance can be obtained with expansions of 0.03 percent. This a case in which more is not necessarily better. Since overspecifying expansion can result in an inferior slab, it is important to choose the right specification for the product being used rather than using arbitrary standards for the C 878 test.

Interground Versus Expansive Components

Many specifiers refer to Type K cement (as specified by ASTM C 845) when they want shrinkage-compensating concrete, without realizing that they might be excluding some viable alternatives. Type K expansive cements are produced by intergrinding the expansive clinker with the portland-cement clinker and gypsum. This process has some benefits and some drawbacks.

The principal benefit is that the materials are blended well under controlled conditions. A drawback is that the expansive clinker generally is softer than the portland-cement clinker. The resulting expansive material is finely ground, resulting in excessive water demand and rapid slump loss. To compensate by undergrinding and premixing the expansive material with the portland cement means that to vary the amount of expansion achieved (by varying the amount of expansive material), one also must vary the amount of portland cement. This obviously reduces control of the mix design. The alternative is to obtain the preferred expansive material (separately ground) and have it added to the concrete mix at the batch plant.

Experience

The characteristics of shrinkage compensation will change, depending on such variables as the aggregates used and the temperature and humidity at the construction site. A contractor who has worked with the specific type of shrinkage-compensating concrete under similar conditions can cope with these and other variables. The concrete floor will arrive at the job site as a raw material. Although quality materials are important, they will not make up for a contractor who lacks the skill or integrity required to do the job right.

JOINT SOLUTIONS

Combining efficient design, skilled construction crews and new processes and materials, it now is possible to produce industrial and commercial concrete slabs on grade that are flatter and more durable than ever. Despite these improvements, however, joints continue to be a source of owner complaints.

BASIC JOINT TYPES

Construction Joints

Generally speaking, a joint is any deliberate break in the continuity of the concrete slab. Joints often occur from the practical limitations of how much concrete the contractor can place and finish in one day. A day's work typically is bounded by forms that will be removed before adjacent concrete is placed. The remaining joints are called construction joints (Figure 7.13).

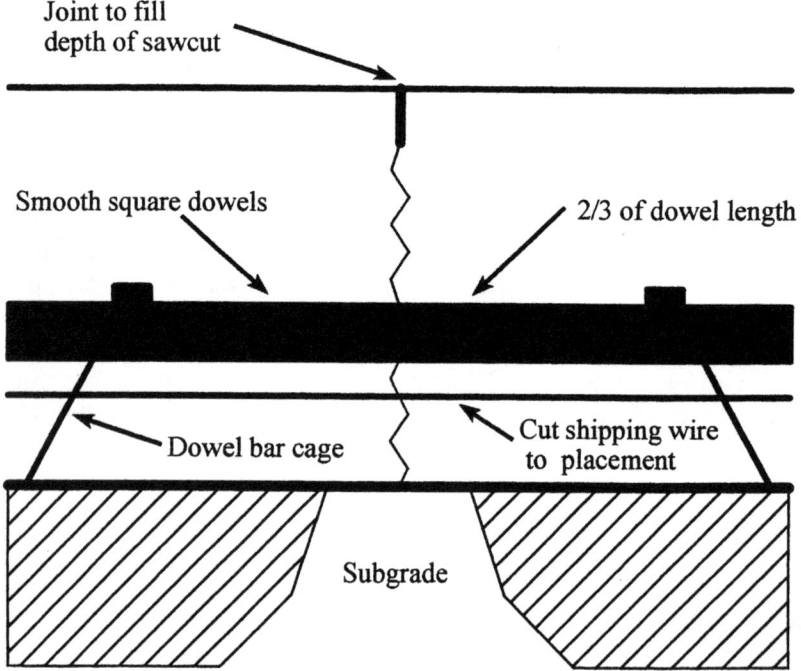

FIGURE 7.13 Continuous joint with dowel basket.

Isolation Joints

Similar to construction joints, these slabs are formed separately to allow each one to react to temperature, loads, or other factors without affecting an adjacent area.

Shrinkage-Contraction Joints

Cut shortly after placing the concrete, these joints provide a weakened plane to relieve stresses produced as the concrete shrinks during curing. Without them, such stresses would cause numerous uncontrolled shrinkage cracks. Unlike construction joints that penetrate the entire depth of the slab, contraction joints are cut only into the top third. Typically only about ⅛ inch (3 mm) wide when initially cut, they tend to widen as the slab shrinks.

TYPICAL PROBLEMS

Joints represent several potential problems for the long-term building owner or occupant. Depending on how the building is to be used, sanitation can be an issue. The need to keep the facility clean can be severely complicated by joints that trap dirt and debris. A more severe (and expensive) problem is spalling, the gradual wearing away of the joint edge as wheeled traffic repeatedly moves across it. A wheel rolling smoothly across the surface of the floor strikes the edge of the adjacent slab, causing rapid deterioration. Since conventional construction processes actually encourage joint spalling, deterioration at the construction joint begins even before the structure is occupied. Most of the slab will be finished using mechanical troweling equipment; however, the slab edge will be finished by hand. The mechanical troweling process forces the aggregate firmly together to create a denser surface than is practical in hand finishing. To facilitate hand finishing, the finishers tend to pull the cream, a wet mixture high in cement and fine aggregates, to the edge of the slab. The material finishes well, but is weaker than the rest of the slab because of the high water-to-cement ratio and the absence of larger aggregate. Spalling also occurs after wood forms are removed, taking concrete with them from the joint edge. The resulting ragged edge will deteriorate more rapidly than a smoothly finished edge.

Many attempts—some more successful than others—have been made to overcome the inherent weaknesses of floor joints. While not a complex review of all available options, the following information might be useful when comparing the strengths and weaknesses of various design choices.

FILLING THE JOINT

Filling the void area with a suitable material would eliminate the difficulties associated with joints. The challenge is to find a material that is strong enough to support wheeled traffic but still sufficiently elastic to permit continued slab movement.

Rigid Epoxy Filler

This was an early solution to joint spalling. Unfortunately, concrete slabs continue to shrink significantly for two years after installation, and thermal expansion and contraction occur for the life of the building. Rigid epoxy doesn't stretch or compress enough to allow for this movement. It either debonds from the concrete or maintains its bond so that cracks develop elsewhere to relieve stress caused by the slab's movement as shown in Figure 7.14

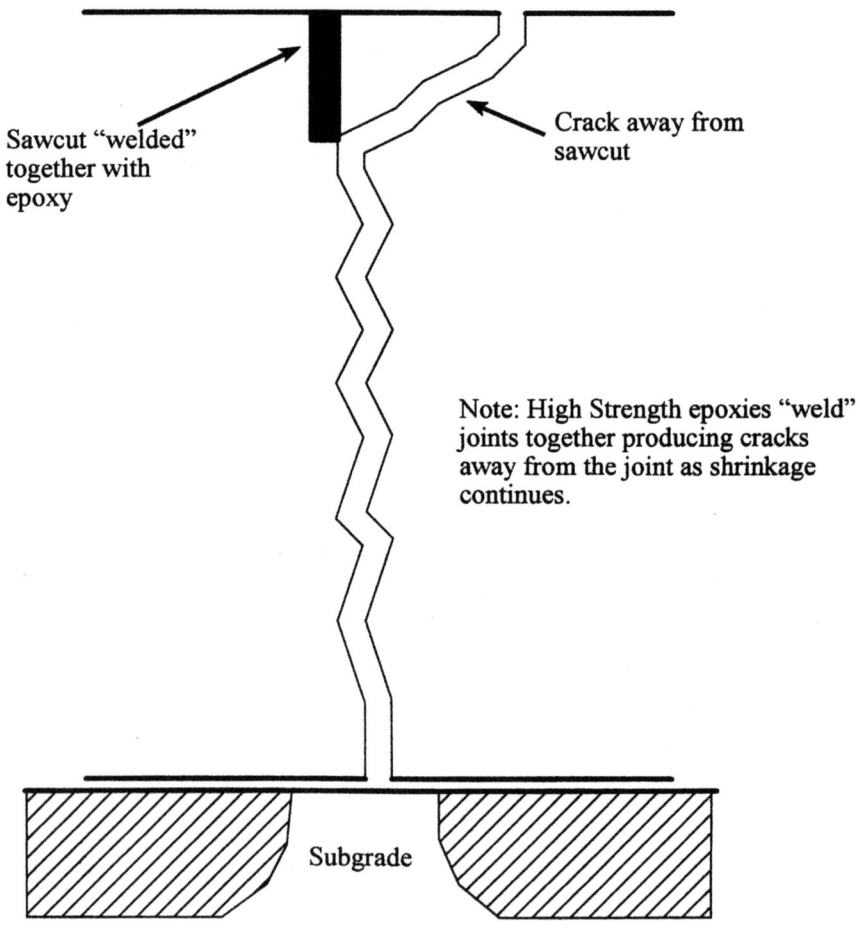

FIGURE 7.14 Rigid epoxy filler.

Semi-Rigid Epoxy Filler

This will stretch and will de-bond before it can damage a shrinking slab, making it relatively safe to use. Wait at least 30 days after placing concrete before installing most semi-rigid epoxy products—60 days to 90 days is recommended. Since significant slab movement will occur after the material is installed, frequent replacement or re-filling is required to keep the joint protected. The initial cost of the semi-rigid epoxy is not prohibitive, but re-filling changes the equation. Figure 7-15 is a description of a semi-rigid epoxy.

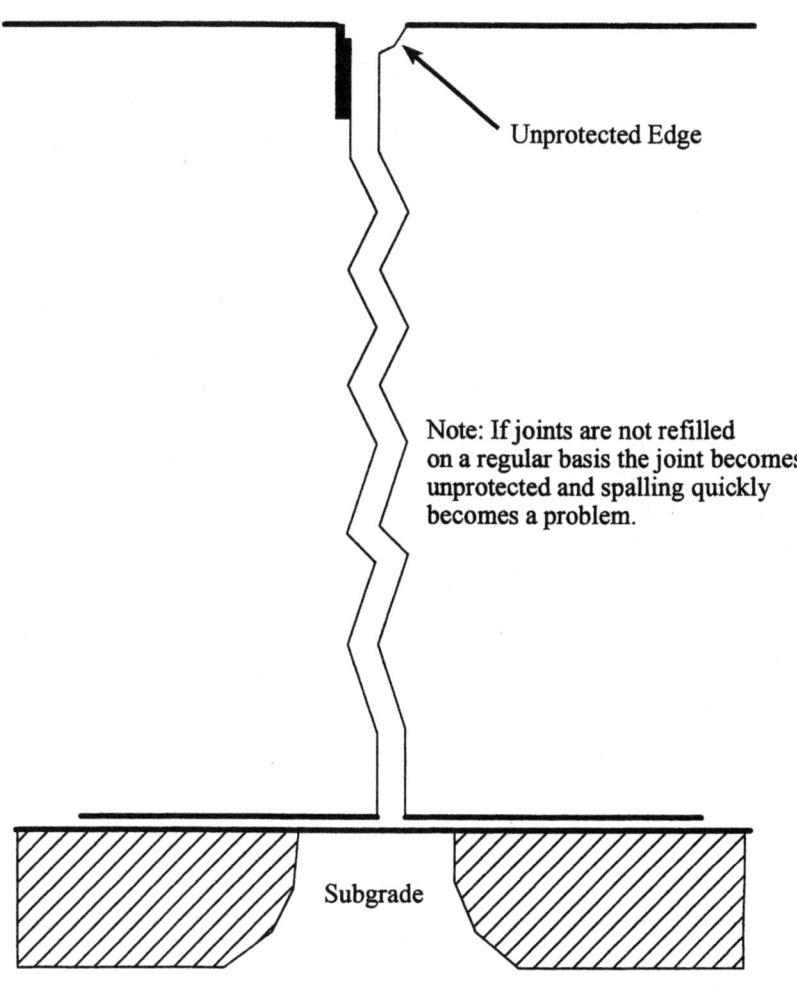

FIGURE 7.15 Semi-rigid epoxy.

Elastomeric (Soft) Fillers

Elastomeric (soft) fillers stretch and compress more readily than semi-rigid epoxy products. Unfortunately, that compressibility makes the flexible fillers relatively ineffective in preventing joint spalling. When wheeled traffic strikes the edge of the concrete slab the filler gives way under the load, as illustrated in Figure 7.16. Flexible fillers are useful in filling joints for cleanliness and sanitation purposes after the joints have been otherwise protected from damage.

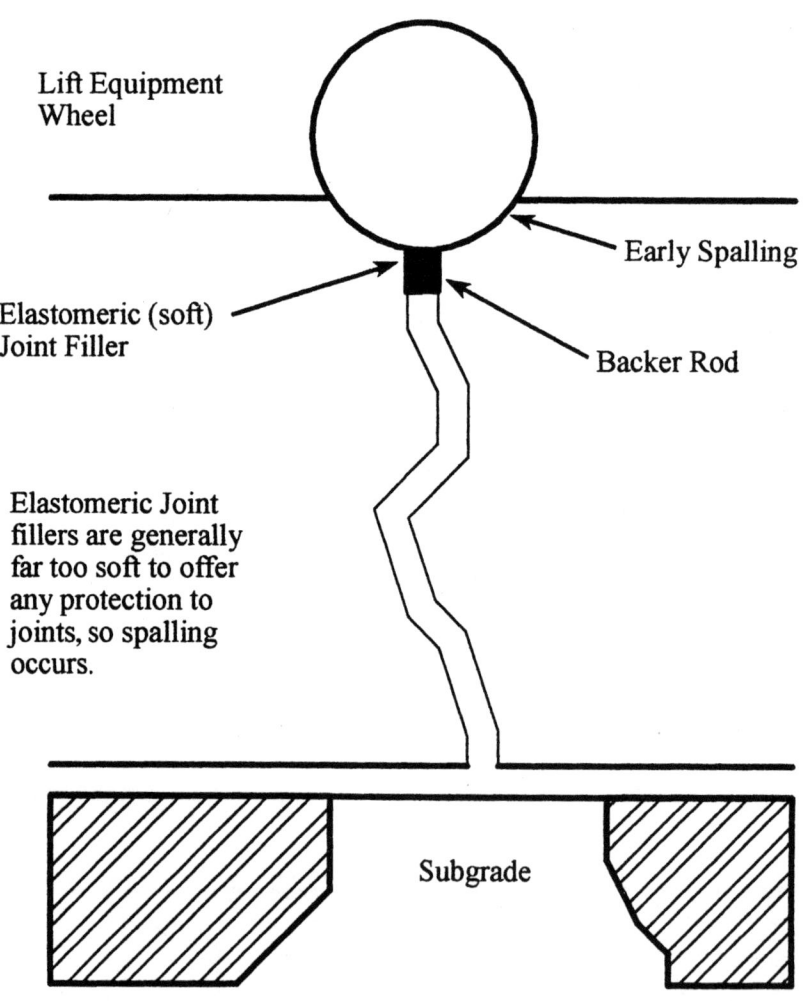

FIGURE 7.16 Elastomeric (soft) filler.

LEAVE-IN-PLACE FORMS

Leave-in-place forms left at construction joints avoid the problems associated with weak concrete at the edge, since the form constitutes the edge. Manufactured concrete forms simplify the process of forming since they become a permanent part of the floor. This eliminates the cost of stripping and cleaning forms as well as the damage caused by their removal. Since they are factory-made, the forms are not subjected to the distortion and warpage so common with wood forms. Manufactured concrete forms are ideal for applications that demand flat floors.

STRENGTHENING THE JOINT EDGE

Considering the difficulty associated with finding a suitable joint filler, it is not surprising that attention has shifted to developing ways to strengthen the slab edge.

Rigid Epoxy Filler

This filler, with an enbonded center, is an ingenious solution to the problem associated with rigid epoxy. The concrete at the joint is cut away to make a wider joint, which is filled with rigid epoxy. A narrow plastic strip is inserted to separate the epoxy-filled joint down its center so that the slab can shrink freely. This approach works well when installed properly, but it is extremely expensive because of the amount of labor and epoxy required.

Hard Aggregate

This material loaded at the joint is another option. Aggregate can withstand the impact of wheeled traffic better than ordinary concrete, but the joint is only as strong as the cement paste that holds the aggregate together. Hard aggregate can simply break away, leaving the joint to spall.

Conventional Steel Protected Joints

These have been used for years to prevent spalling. The concept is sound, but several persistent problems can be traced back to field fabrication. In many cases, steel angles are aligned improperly, resulting in uneven floor surfaces at the joint. Welding studs to the angles (to tie them to the floor slab) also causes problems; the heat of the weld and placement of the stud can cause the angle to distort.

Inadequate compaction of concrete under the steel angle is one of the most serious, but least obvious problems. Properly compacted concrete supports the steel angle and helps it distribute energy from the impact of wheeled traffic. The steel angle distorts when it lacks that support and can produce stress points that damage the adjacent floor.

The amount of welding, grinding, and other field work required to fabricate steel joint protection makes this an expensive solution.

Factory-Produced Steel Angle Joints

This type differs from field-fabricated counterparts in several ways. Figure 7.17 illustrates the makeup of a steel angle joint on grade. Most differences stem from the quality control and production efficiencies possible in a factory environment. Specially designed jigs are used to place and hold the stud in the optimal location for strength and bonding to the surrounding concrete. Heat sink systems and high-voltage welding processes quickly fuse the metals together with very low heat dispersion

The compaction of concrete under the steel angle is increased by inverting the steel angle and casting the low moisture, high-strength concrete above the angle. Vibration and gravity do the rest. Properly aligned dowel holes also are cast into the concrete at this point. Precise alignment is maintained because the steel angle is bolted together at the factory for easy, accurate, job site installation. Because of the production efficiencies gained in the manufacturing environment, factory-produced steel angle joints are frequently less expensive than their field-fabricated alternative.

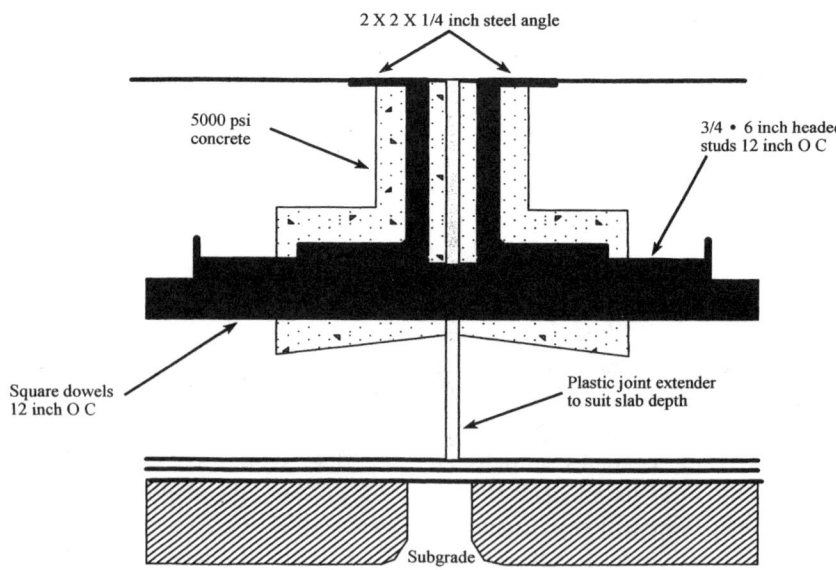

FIGURE 7.17 Steel angle joint on grade.

GYPSUM CONCRETE [9]

Gypsum concrete (ASTM C 317) is made by using a mixture of plaster of Paris (calcium sulfate hemihydrate), sand, water, and perhaps portland cement. It is flowable, self-leveling, inexpensive, and offers good fire retardance and sound transmission resistance. It can provide excellent flooring and sub-flooring.

PLASTER AND WATER

Plaster of Paris and water are the major raw materials in gypsum floors and gypsum plaster walls and ceilings. The chemical reaction is:

$$CaSO_4 \cdot \tfrac{1}{2}H_2O \text{ (plaster)} + 1\tfrac{1}{2} H_2O \text{ (Water)} = CaSO_4 \cdot 2H_2O \text{ (gypsum)}$$

The parts of water used by weight per 100 parts of plaster is termed consistency. A consistency of about 19 provides all the water needed for a chemical reaction. Figure 7.18 shows that additional water causes a precipitous decrease in the product's strength, and ordinary plaster requires a consistency of about 60 to 70 to make a workable mix.

Using special manufacturing techniques, plaster can be produced with particle sizes and shapes that allow workable mixes with much less water. This alpha plaster (or alpha gypsum) is more expensive than ordinary plaster, but it allows a consistency as low as 30 to 40 to produce a workable mix. For either type of plaster, compressive strength is essentially the same at the same consistency.

AGGREGATE AND PORTLAND CEMENT

Sand-sized or larger aggregates are incorporated in a concrete mix to produce an economical gypsum floor. The resulting product's strength falls with increasing aggregate content in a manner similar to what occurs with increasing amounts of water, as shown in Figure 7.19.

Many manufacturers of flooring patching and grouting materials add portland cement to their proprietary products. Literature and technical data sheets for these products usually produce no clear reason why gypsum plaster and portland cement are combined. Perhaps such systems can be best evaluated by considering them from opposite approaches:

- Adding plaster to portland cement allows quick setting and hardening.
- Adding portland cement to plaster can allow greater water resistance, ultimate strength and decreased corrosion.

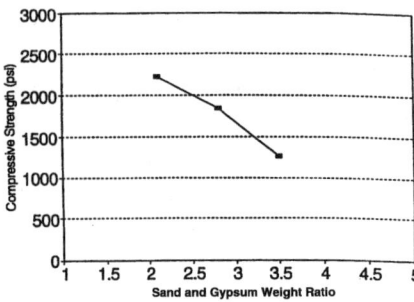 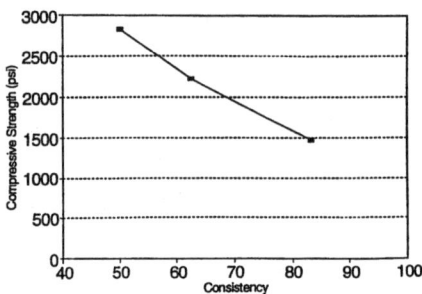

FIGURE 7.18 Sand and gypsum ratio versus strength.

FIGURE 7.19 Consistency versus compression strength.

Gypsum systems soften or dissolve when exposed to water, leading to decreased performance or failure. Further, as members of the chemical class called salt, they promote corrosion of embedded or contacting metals. Although adding portland cement to a gypsum system might decrease or eliminate its corrosivity due to the alkalinity of the cement, such a mixture presents its own serious problem.

Portland cement (specifically, calcium aluminate component) reacts with calcium sulfate systems and water to produce ettringite, a 125-atom molecule whose formation is an expansive process. The reaction is:

$$3\ CaO \cdot Al_2O_3\ (\text{tricalcium aluminate}) + 3\ CaSO_4 \cdot 2\ H_2O\ 9\text{gypsum}) + 26\ H_2O =$$

$$3\ CaO \cdot Al_2O_3\ 3\ CaSO_4 \cdot 32\ H_2O\ (\text{ettringite})$$

Ettringite formation can occur very early, producing a shrinkage-compensating or expansive-cement product that tightly fits the space it is placed in, or it can occur after the system becomes hard, leading to cracking or complete disruption of the product.

QUALITY CONTROL

Since the strength of gypsum concrete depends on water and aggregate content, quality control of the parameters is required. Unfortunately, sand frequently is measured by the shovel and water by hose. Complicating the situation, contractors often are unaware of the importance of the ratio of the amount of water in the aggregate component. A dry aggregate might differ from the more usual damp or wet one by only enough water to produce an enormous change in consistency and strength if mix water content is varied to compensate for aggregate water content.

The aggregate's moisture content seldom is monitored, nor is such control required in many operations. Quality control can be achieved easily in the field by using standardized measurements.

IN-SERVICE ASPECTS

The use of gypsum concrete in kitchens, bathrooms, or on exterior porches and balconies is not advisable unless tile or coatings can be used to protect the concrete from moisture. Such protective systems must be well-maintained. Also, gypsum concrete is usually of substantially lower strength than portland cement, and it might not resist point loads such as those produced by couch or piano legs. Using tile or carpet to distribute the load helps to minimize such distress.

CONSTRUCTION AND INSTALLATION

When used as flooring, gypsum concrete frequently is placed on plywood. Because the concrete is normally only ⅝ in. to 1 in. (16 mm to 25 mm) thick, it will have less flexural strength. The joists supporting the plywood must not be so far apart that they allow significant flexure of the plywood, and thus of the concrete. Manufacturers of such concrete specify that joist spacing be based on the thickness of the plywood. Many proprietary flooring material producers specify a glue-like coating for the plywood to prevent the gypsum from moving during service and to protect the plywood from the wet, fresh concrete. This coating should be tacky when the concrete is poured.

FORENSIC STUDY OF FAILURES

Common causes of gypsum product failures include:

- Low strength caused by excessive water or aggregate content
- Cracking caused by reaction between portland cement and gypsum, or from plywood flexure, often compounded by failure to use adhesive
- Softening caused by moisture exposure
- Corrosion of embedded or contacting metals
- Cracking of surrounding structure because of rust buildup during corrosion of embedded metals and ettringite production in the gypsum concrete
- Deterioration of the plywood floor support caused by exposure to water in the gypsum mix, often compounded by failure to use a sealer-adhesive

Petrographic examination, chemical analysis, and physical testing may be required to determine the cause of a gypsum concrete failure. Erlin, Himes and Associates' investigations of gypsum concrete distress have led to the development of several procedures for such failure studies. The principles include:

- Petrographic study to identify components and reveal distress manifestations
- Sulfate analysis to determine gypsum content, with correction for sulfate in any portland cement
- Sand content determined as acid insoluble residue if a petrographic study discloses it to be siliceous. (For calcareous aggregates, calcium analysis also might be required, correcting for calcium in gypsum and portland cement.)
- Acid-soluble silica analysis to determine portland cement.
- Infrared spectroscopy to identify plywood sealant and adhesive.
- Kerosene adsorption, after calibration with knowns, in favorable cases to determine consistency. (Systems containing large amounts of portland cement might not be amenable to analysis.)
- Cutting cubes of the largest possible dimension and testing in compression between cardboard and neoprene sheets to determine compressive strength Calibration using knowns similarly cut might be necessary. Based on the Erlin-Himes study of some gypsum systems, such sawed cubes are about two-thirds of the strength of cubes prepared and tested in accordance with ASTM C 472, Physical Testing of Gypsum, Gypsum Plasters, and Gypsum Concrete.

GROUT SPECIFICATIONS [8]

Is it possible to improve your nonshrink grout specification and lower your in-place cost? ASTM C 1107, Standard Specification for Packaged Dry Hydraulic Cement (nonshrink) can eliminate potential problems on projects. ASTM C 1107 consolidates two specifications under one umbrella. The first specification is ASTM C 827 Standard Test Method for Change in Height at Early Ages of Cylindrical Specimens from Cementitious Mixtures. The second is the U.S. Corps of Engineers CRD C 621 Post Hardening Volume Adjusting specifications is the recognized specifications for cement-based nonshrink grouts.

ASTM C 1107 was introduced to the public in 1989 and has undergone a few revisions, the latest in 1991, and continues to be improved. ASTM C 1107 classifies grouts in three categories:

- Class A—volume control of grout caused by expansion before hardening occurs
- Grade B—volume control caused by expansion after grout has hardened
- Grade C—volume control caused by a combination of both processes

In addition to height changes, critical elements of working time (place ability), and compressive strength are measures at varied temperature ranges and at varied consistencies or water content. The grout is tested at the manufacturer's maximum allowed water-to-dry solids ratio and is rated as plastic, flowable, or fluid. Grout consistency is tested both after initial mixing and at the end of the grout's stated working time at the manufacturer's allowed temperature ranges. Compressive strength tests are run at maximum and minimum temperatures both initially and at the end of the working time. For compliance, a grout must achieve a conservative 28-day strength of 5000 psi (34,500 kPa).

Although C 1107 addresses grout limitations with respect to temperature, consistency, and working time, it fails to set limits for these parameters, allowing manufacturers to do so. A grout with very narrow parameters could result in failure. For example, a product that is non-shrink at 70° F (20° C) may not be at 90° F (32° C). C 1107 does not address bleeding. A grout that bleeds free water caused by consistency or temperature results in a void under the base plate (or surface) after the water has evaporated.

HEIGHT CHANGE

Although all grouts are tested for height change of hardened grout at 1, 3, 14, and 28 days in accordance with ASTM C 1090, Standard Test Method for Measuring Changes in Height of Cylindrical Specimens from Hydraulic Cement Grout, Grade A grouts do not have to meet the hardened maximum expansion required for B and C grouts. The higher expansion allowed for Grade A grouts can result in lower compressive strengths and might not compensate for long-term drying shrinkage. Grade C grouts will perform as long as they have adequate post-hardening expansion to counteract drying shrinkage over a wide temperature range. Only B and C Grade grouts address post-hardening shrinkage.

SHORT-FORM SPECIFICATIONS

Specifying a grout material to comply with the stringent requirements for fluid consistency, working time, and temperature range reduces potential problems on projects that experience those extremes. The following is a sort-form specification for nonshrink grouts:

1. Grout shall be prepackaged, nonmetallic, and nongaseous. It shall be non-shrink when tested in accordance with ASTM C 1107 Grade B or C at a fluid consistency (flow cone) of 20 to 30 seconds. Thirty-minute-old grout shall flow through flow cone after slight agitation in temperatures of 40° F to 90° F (4° C to 32° C).

2. Grout shall be bleed-free and attain 7500 psi (51,700 kPa) compressive strength in 28 days at fluid consistency.
3. Certified independent test data shall be required.

SUMMARY

In a study of sealants and adhesives, it is important to be familiar with the basics of concrete, cement, and other cementitious materials used in concrete. It is necessary to have an understanding of the formulations and types of cements available, and to know how, when, and where to use them.

REFERENCES

1. Neville, Adam M. 1999. *Cement and Concrete.* New York: Grolier Electronic Publishing, Inc.
2. Wilkes, Joseph A. 1988. *Encyclopedia of Architecture, Design, Engineering, and Construction.* New York: John Wiley & Sons.
3. Wilkes, Joseph A. 1988. *Encyclopedia of Architecture, Design, Engineering & Construction.* New York: John Wiley & Sons.
4. The Portland Cement Association. 1988. *The Homeowner's Guide to Building with Concrete, Brick, & Stone.* Rodale Press.
5. Use Joints to Control Floor Cracks. June 1984, *Concrete Construction.*
6. Kelley, Paul L, Brainerd, Michael L, Liepens, A. Atis, and Todd Bergstrom. July 1995. Understanding Cracking in Concrete Repairs. *The Construction Specifier.*
7. King, Stephen M. July 1995. Shrinkage—Compensating, Concrete Floors, *The Construction Specifier.*
8. Schwietz, Mark B. July 1995. L&M Construction Chemicals, Inc. *The Construction Specifier.*
9. Hime, William G., Erlin, Hime Associates. July 1995. *The Construction Specifier.* Northbrook, IL

FURTHER READING

1. *Concrete Handbook.* Popular Mechanics Press.
2. Love, T.W. 1986. *Construction Manual: Concrete & Formwork.* Craftsman Book Co.
3. Wynne, George B. 1981. *Reinforced Concrete Structures*, Reston Publishing Co., Inc.

CHAPTER 8
BUTYL SEALANTS

INTRODUCTION

Butyl rubber plays a major role in the construction industry, and is a significant part of the insulating glass industry, as well as an ingredient for preformed tapes and gaskets. (See Chapters 16 and 17.) Both curable and noncurable butyl contain sealants that are used in high-rise construction. The combined volume sold of these sealants is approximately 20 million pounds. The sale of butyl caulks in the over-the-counter market, which includes ropes, gaskets, and other repair compounds around the house, is considerably greater. Excellent weather resistance and low cost makes butyl rubber-based compounds very popular in all phases of construction and in the automotive and insulating glass industry.

Butyl rubber refers to a common synthetic elastomer prepared from isobutylene and isoprene. The isoprene is present in much lower quantities (1.5 percent to 4.5 percent) than the isobutylene. The unsaturation introduced by the isoprene is sufficient to enable vulcanization by a classical sulfur cure. Butyl sealants described here are in the form of solvent release sealants, deformable preformed tapes, hot applied sealants, and polyisobutylene caulks.

Butyl rubber was developed in the 1930s in the laboratories of the Standard Oil Company. During World War II the elastomer was produced in plants operated by the U.S. government and was issued for inner tubes on military vehicles.

A number of variations in the diene portion of the new raw material have been studied, including butadiene, dimethylbutadiene, and piperylene (1, 3-pentadiene). None, however, has demonstrated the wide suitability of properties of the isoprene type.

BASIC CHEMISTRY [1]

$$CH_3$$
$$|$$
$$H_2C=C-C=CH_2$$
$$|$$

Isoprene

$$H_3C \quad CH_3$$
$$| \quad\quad |$$
$$HC = CH$$

Butylene

$$CH_3$$
$$|$$
$$H_2C = C$$
$$|$$
$$CH_3$$

Isobutylene

$$\begin{array}{c} CH_3 \\ | \\ -CH_2-C- \\ | \\ CH_3 \end{array} \begin{array}{c} CH_3 \\ | \\ CH_2-C \\ \end{array}_{50} = CH-CH_2- \begin{array}{c} CH_3 \\ | \\ CH_2-C- \\ | \\ CH_3 \end{array}$$

Butyl Rubber

$$-CH_2-CH_2-CH\ CH_2-\overset{CH_3}{\underset{|}{\underset{CH_3}{C}}}\ CH_2\ CH-$$
$$\quad\quad\quad\quad |\quad\quad\quad\quad\quad\quad\quad |$$
$$\quad\quad\quad\quad CH_3\quad\quad\quad\quad\quad\quad CH_2$$
$$\quad\quad\quad\quad\quad\quad\quad\quad\quad\quad\quad\quad\quad |$$
$$\quad\quad\quad\quad\quad\quad\quad\quad\quad\quad\quad\quad\quad CH_3$$

Polybutene

$$-CH_2-\overset{CH_3}{\underset{|}{\underset{CH_3}{C}-}} \quad -CH_2-\overset{CH_3}{\underset{|}{\underset{CH_3}{C}}} \quad -CH_2-\overset{CH_3}{\underset{|}{\underset{CH_3}{C}-}}$$

Polyisobutylene

The preparation of butyl rubber is unique in that it is carried out at 140° to 150° F (95° to 100° C) with an aluminum chloride catalyst and it is prepared as a crumb stock directly, without going through the latex stage. The unsaturation level is lower than natural rubber and, consequently, vulcanization systems must be highly reactive to use the reduced number of sites. The corresponding reduced unsaturation in the main polymer backbone has a significant effect in several properties including ozone resistance.

Carbon black reinforcement of butyl rubber, unlike natural rubber, does not result in any increase in tensile strength but it does improve tear resistance. Dispersion of the carbon blacks in butyl rubber is important because poor dispersion reduces tensile strength. Several grades of butyl are marketed with different ranges of molecular weight and different levels of unsaturation. Blending with natural rubber is not recommended; neoprene, however, has been blended successfully with butyl. Butyl rubber has wide use in electrical applications, construction sealants, and caulks because of its good resistance to heat, ozone, and weathering. The good chemical resistance of butyl prompts its use in linings for ware and chemical applications.

Because most butyl rubber caulks are either non- or slow-curing, they are not recommended for large joint movement applications. The maximum joint movement capability is plus or minus 7.5 percent. Butyl caulks do not harden or crack and are quite stable against the elements. They can be compounded as solvent-based using tough rubber as a starting point, or plasticized with polybutene oils that permanently soften the caulk and do not volatilize. The field is wide open to formulators. Although there are only a few manufacturers of butyl rubber, such as Exxon, Cities Service, and Polysar, dozens of formulators purchase the raw material and sell the finished sealant, adhesive, or caulk as specialty items (i.e., gutter sealants, driveway crack sealers, and foundation crack sealers). Skylights, roofs, and gutters offer common service problems because the majority of roofing leaks occurs in flashings, parapet walls, copings, and gutters. Applying butyl caulk can significantly reduce maintenance costs. Table 8.1 lists the various properties of skinning and nonskinning butyls to assist the formulator in choosing the correct base material.

Sealants based on the above chemistry, or combinations of it, usually are sold as single-package, ready-to-use products that require no additional mixing or preparation [2]. Two-part systems, introduced a few years ago by Enjay Chemical Corp., are based on butyl LM for low molecular weight [3]. The butyl—semiliquid that can cure at room temperature—is produced in either nonchlorinated or chlorinated form. Butyl rubber sealants are limited in that they are usually noncuring. Table 8.2 lists the advantages and limitations of a compounded butyl product.

TYPES OF BUTYL SEALANTS AND CAULKS

Butyl and polyisobutylene polymers are used in a variety of mastics, caulks, and sealants. A high viscosity mixing phase prepares the butyl sealants and mastics in one of several ways.

TABLE 8.1 Properties of Sealing Compounds

Property	Butyl Skinning type	Butyl Nonskinning type
Ingredients	Butyl polymers, inert reinforcing pigments, nonvolatile plasticizers, and polymerizable dryers	Butyl polymers, inert reinforcing pigments, nonvolatilizing and nondrying plasticizers
Primer required	None	None
Curing process	Solvent release, oxidation	No curing; remains tacky
Tack-free time, hours	24	Remains tacky indefinitely
Cure time, days	Continuing	N/A
Elongation, %	40%	N/A
Joint movement, %	± 7.5%	N/A
Joint width, inches	0.75	N/A
Resiliency	Low	Low
Resistance to compression	Moderate	Low
Resistance to extension	Low	Low
Service temp. range °F (°C)	−20° to 180° F (−4° to 82.22° C)	−20° to 180° F (−4 to 82.22° C)
Application temp. range °F (°C)	40° to 120° F (4.44° to 48.89° C)	40° to 120°F (4.44° to 48.89° C)
Weather resistance	Fair	Fair
UV resistance, direct	Good	Good
Cut, tear, abrasion resistance	N/A	N/A
Life expectancy, years	10	10
Hardness, Shore A	20–40	N/A

TABLE 8.2 Advantages and Limitations of Butyl Sealant Compounds

Advantages	Limitations
1. Good glazing compound	1. Noncuring
2. Little odor	2. Permanently tacky
3. Good package stability	3. High shrinkage rate
4. Low cost	4. Poor color selection
5. Primer not required	5. High compression set
6. One- and two-component systems	6. Limited to ± 7.5% movement
7. Good water resistance	7. Not recommended for expansion joints
8. Good adhesion	
9. Readily available	
10. Good color stability	
11. Meets TT-S-001657	
12. Meets ASTM-C-1085	
13. Meets Canadian spec 19GP14 for ± 5.0%	

1. A heavy-duty kneader is used to make a butyl rubber solution, normally with mineral spirits. The solution is made by charging the rubber into the kneader and working the solvent into the rubber. The solvent must be added very slowly at first, and each increment of solvent must be incorporated into the batch before more is added. If the solvent is added too fast, the polymer will be dispersed in the form of small pellets that can be dissolved by extended subsequent mixing. Adding the solvent too quickly results in either greatly increased batch cycle time to obtain uniformity or in polymer particles in the final solution. The finished solution is transferred into a smaller, less powerful mixer for further compounding with plasticizer pigments, resins, and other ingredients.
2. The butyl is mixed in a heavy-duty kneader with all the nonvolatile ingredients and then solvated. The same precautions apply to the mixing and solvation steps as outlined above; that is, each increment must be well dispersed before subsequent additions are made. Two different orders of addition are used in this technique:
   a. The rubber is added first to the kneader and the dry ingredients then are dispersed slowly.
   b. The upside-down technique, has been found to result in better dispersion in less time, especially in the case of mastics based on partially crosslinked butyl. In this technique, the fillers and a small amount of plasticizer are blended into a smooth, stiff paste in the kneader, and the rubber is added last.
3. A masterbatch of polymer and filler can be prepared in a mill or a Banbury and then solvated in another mixer. For very high viscosities, such as blends of rubber with filler and/or plasticizers, mills and Banburys are the best choice.

Butyl rubber and some polyisobutylene grades are used as the base or binder in a wide range of caulking or sealing compounds [4]. The inherent properties of polymers of the polyisobutylene family, particularly the chemical inertness, age and heat resistance, long-lasting tack, flexibility at low temperatures, and favorable FDA position on selected grades, make these products commercially attractive. They are used in a variety of ways as pressure-sensitive compounds and as other adhesive and sealant compounds in architectural applications and as coatings. An added dimension of the polyisobutylene polymers is their ability to achieve specific properties by easily blending with each other and with other adhesive/sealant polymers, such as natural rubber, styrene-butadiene rubber, EVA, low molecular weight polyethylene and amorphous polypropylene. For example, they can be blended to enhance age and chemical resistance. A description of the polyisobutylene polymer family in adhesive and sealant applications follows [5]:

- Pipe wrap and electrical tape applications exploit age resistance, low water absorption and permeability, inherent tack, and good electrical properties. Figure 8.2 shows a formula for this adhesive.

- Sealants commercially compounded from butyl rubber and sometimes certain polyisobutylene grades are used as binders or the base: Figures 8.3 and 8.4 are typical starting points for caulking compounds based on different types of butyl rubber. Caulks and sealants are available in three forms: *bulk sealants* (usually one-component, plasticized systems with inert fillers and solvents designed for standard application through a hand caulking gun); *preformed tapes* (described in Chapter 1); and newer formulas of 100 percent *solid hot-melt combinations* (described in Chapter 5).

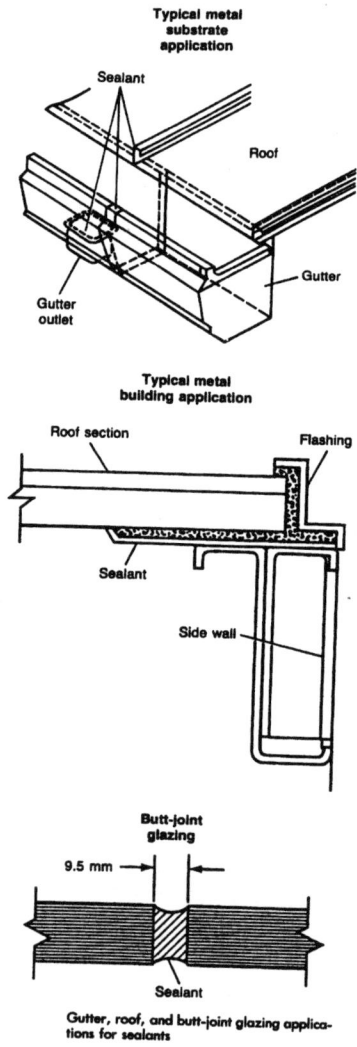

Gutter, roof, and butt-joint glazing applications for sealants

FIGURE 8.1 Properties of sealing compounds.

Raw materials	Parts by weight
Exxon butyl 065	100
N550 (FEF) carbon black	100
Mistron vapor talc	200
Parapol 950 polybutene	100
Escorez 1304 resin	75
Amorphous polypropylene, A-Fax 600	50
Paraffin process oil	5

FIGURE 8.2 Butyl mastic for pipe tape wrap.

Raw materials	Parts by weight
Exxon butyl 065, 50% in mineral spirits	200
Vistanex LN-MS	20
Isostearic acid	5
Fibrous talc	300
Atomic whiting	200
Titanium dioxide, rutile grade	25
Schenectady SP-553 resin	35
Parapol 1300 polybutene	100
Blown soy oil, Z₃	15
Cabalt naphthenate drier, 6%	0.5
Cab-O-Sil M5	20
Mineral spirits	55

Physical properties	Values
Solids, %	84
Density	1.37
Weight per gallon, lbs	11.4

FIGURE 8.3 Typical formula for butyl based caulking compound.

- Insulated window sealants used in the fabrication of sealed insulating glass units (double pane) must not only have excellent weathering, aging, and durability characteristics but, must also have a low moisture vapor transmission rate and be nonfogging. The butyl rubber polymers fulfill all these requirements and have grown extensively for several years in this application. Figure 8.5 offers an extruded mastic for insulating glass. This type of formula, using Vistanex LM, has more than 20 years of field service experience. Other insulating glass formulas are covered in Chapter 5 in the analysis of hot melts.

A few other typical formulations—Figures 8.3, 8.4, 8.5—will assist the chemist in developing caulks, sealants, and coatings for construction applications. Figure 8.6 is a general purpose architectural sealant; Figure 8.7 is a unique sealant for aquariums; Figure 8.8 is a black butyl sealant; and Figure 8.9 is a light-colored roof coating formulation.

Raw materials	Parts by weight
Polysar Butyl 100	100
Mineral spirits	100
Atomite	300
International 3x talc	100
Titanium dioxide RA 50	10
Irganox 1010	1
Piccolyte S-125	12
Bentone 38	10
Methanol: water (95:5)	5
Indopol H-100	70
Sunpar 110	12
Thixatrol GST	10
Mineral spirits	15
Physical properties	*Values*
Shrinkage, %	20.7
Slump, mm	2.5
Extrudability, sec/ml	4.3
Stain index	1.0
Tack-free time, hrs	None
Bond cohesion, cm^2	
Aluminum	None
Glass	None
Mortar	None

FIGURE 8.4 Solvent release butyl caulk.

Raw materials	Parts by weight
Vistanex LM-MH	100
Mistron vapor talc	48
N 990 (MT) carbon black	2

FIGURE 8.5 Insulating glass extrude mastic.

Raw materials	Parts by weight	
	1	2
Enjay Butyl LM 430	50	50
Calcene TM	30	—
HiSil 215	—	20
Titanium dioxide rutile grade	—	10
Zinc oxide	—	5
Indopol H-100	60	—
Silane A-187	4	—
GMF	3	—
Lead peroxide 0.5μ	—	7
Toluene	5	25
Stearic acid	—	1

This formulation has sufficient resistance to cyclic deformation to pass the cyclic durability test of Federal Specification TT-S-00227E.

FIGURE 8.6 General purpose architectural sealant.

Raw materials	Parts by weight	
Enjay Butyl LM 430	50	50
OMYA BLH	60	—
HiSil 215	—	20
Indopol H-100	—	30
Silane A-187	4	—
GMF	3	—
Lead peroxide 0.33μ	—	8
Toluene	12	19

This product is nonsag and can be used for the construction and repair of aquariums. This formulation has performed well between layers of marine plywood in submerged test samples.

FIGURE 8.7 Aquarium sealant.

Raw materials	Parts by weight	
	1	2
Enjay Butyl LM 340	90	10
MT black	150	—
Titanium dioxide rutile grade	—	2
Irganox 1010	2	—
Silane A-187	6	—
Epon 872	5	—
GMF	3.5	—
2-pyrrolidine	2	—
Magnesium dioxide	—	10
Neodecanoic acid	—	0.5
Toluene (dry)	80	4.5

The high loading of black results in a thixotropic consistency, often very desirable.

Physical properties	Values
Work life	< 1 hour
Cure time	~ ½ day

FIGURE 8.8 General purpose black butyl sealant.

Raw materials Part A	Parts by weight	
	Dry	Wet
Enjay Butyl LM 430 (75% NV)	100	133.8
T 674 D Phthalo Green	3.3	3.2
Asbestos Fibre 7R4	16.2	16.2
Cabolite F-1	13.5	13.5
Albron 2240 aluminum paste (68% NV)	20.2	31.0
Para-quinione dioxime	3.5	3.5
Hexane	—	26.9
Mineral spirits	—	49.1
Part B curing agent		
Lead peroxide 0.5μ	7.5	7.5
Stearic acid	0.75	0.75
Dibutyl phthlate	6.75	6.75

Physical properties	Value
Solids, wt	58.7

FIGURE 8.9 Light-colored roof coating.

SUMMARY

Butyl caulks and sealants are single- and two-component, although the single-component is more versatile, setting to a rubber-like consistency. Solids content is generally about 75 percent to 84 percent and is accompanied by shrinkage. These are low modulus sealants with low recovery. The industrial grades give good performance when used properly; the grades for the homeowner might be satisfactory in nonmoving joints. Butyls do not compete with polysulfide, silicone, or polyurethane, except when less movement is required. Butyls might compete with acrylic latex caulks, some solvent release caulks, some solvent-based sealants, and oil and resin-based caulks in both the construction and consumer markets. Butyl caulks and sealants have the advantage of low cost, which can be the deciding factor when materials that perform equally well have been found.

Butyls are recognized for their limited performance capabilities. Their limitations, particularly in the construction market, and their relative market volume will remain static over the next several years.

REFERENCES

1. Panek, Julian R. and John P. Cook. 1991. *Construction Sealants, and Adhesives*, 3rd Edition. New York: John Wiley and Sons, Inc.
2. Courtesy of Du Pont de Nemours & Company. 1998. Technical literature.
3. Courtesy of Enjay Chemical Company. 1994. Technical Literature.
4. Courtesy of Bayer Corporation. 1997. Technical literature about polysar butyl.
5. Maslow, Philip, H. December 1985. Sealants, Joints, and Membranes for Concrete. *The Construction Specifier*.

FURTHER READING

Amstock, Joseph S. *Handbook of Glass Construction*. New York: McGraw-Hill.

CHAPTER 9
PERMAPOL® POLYMERS

INTRODUCTION

Products Research and Chemical Corp. developed two families of novel polymers trademarked Permapol P2 and Permapol P3. Both have been created in the early 1970s to meet the exacting demands required in structural materials for high performance sealants and adhesives that adhere to a variety of substrates used in the building trades [1]. These compositions also show desirable properties of high temperature, hydrocarbon fuel and chemical resistance, and bond durabilities after exposure to boiling water for 100 hours.

Diamond Alkalai Corp. attempted to introduce polymercaptans in the late 1960s and 1970s. These had a backbone of polyoxyalkylene and were sensitive to water immersion, obviously making them poor candidates for sealants in the building market. Patents for these systems later were sold to the Thiokol Chemical Corp and all promotion by Diamond Alkalai stopped.

In the mid-1970s, Hooker Chemical introduced a polymercaptan that had more sulfur (55 percent) than the Thiokol polysulfide polymer (37 percent). All the other properties were about the same but the adhesion was somewhat poorer. The principal complaint was a strong, obnoxious odor. Many compounds of the sealants were sampled, but little interest was generated and this project was also abandoned.

Permapol P2 [2]

Permapol P2 is a family of mercaptan-terminated polymers available in a range of molecular weights and functions. The structure of these Permapol P2 polymers allows a number of valuable characteristics to be incorporated into a sealant or sealant combinations. Because of the greater compatibility of the Permapol P2, base systems can tolerate large amounts of plasticizers and fillers, giving these systems economic advantages without sacrificing their physical properties.

Permapol P3

This family of liquid polymers contains about 30 percent polymer. They exhibit good resistance to hydrocarbon fuels, have good hydroelectric stability, and excellent elastic properties over a wide range of temperatures when cured. They also exhibit excellent high ultraviolet and thermal resistance. Unlike conventional polysulfide polymers, they do not contain any S-S or formal linkages (the weakest links) in the structure. In addition, the polythioether polymer is highly resistant to organic fluids.

CHEMISTRY

The chemical structure of the polythioether polymer is compared with conventional polysulfide in the following diagram:

$$\{(O-CH_2CH_2-S-CH_2CH_2)_x-(O-CH-CH_2-S-CH_2CH_2)_y\}_n$$
$$|$$
$$CH_3$$

Polythioether structure

$$\{CH_2CH_2-O-CH_2-O-CH_2-CH_2-S-S\}_n$$
$$..$$
$$..$$
$$..$$
$$S\ S$$

Where x = 2, y = 1

Conventional polysulfide polymer structure

In addition to the common polysulfide terminal of mercaptan, the options include hydroxy, silyl, olefin, or non-reactive end groups. The unique ability to terminate this polymer in a variety of ways allows the formulator access to several curing mechanisms. Table 9.1 lists a few advantages and limitations for sealants based on the polythioether polymers.

TYPICAL PHYSICAL PROPERTIES

Since this is a proprietary product, basic formulations from the manufacturer are not known. However, it is possible to submit a series of physical data for both single- and two-component sealants. Tables 9.2 and 9.3 outline the properties of a Permapol® polymer.

TABLE 9.1 Advantages and Limitations of Polythioether Systems

Advantages	Limitations
1. Money savings because of low-density system	1. Strong odor
2. Good package stability	2. Insufficient field data
3. Quick cure	
4. Low weight loss after cure	
5. Good application properties	

TABLE 9.2 Typical Properties for a Single-Component Polythioether Based Sealant

Physical properties	Values
Rheology, slump	
Vertical @ 122° F (48.89° C), in.	$3/32$
Vertical @ 40° F (4.44° C)	$3/32$
Horizontal @ 122° F (48.89° C)	None
Horizontal @ 40° F (4.44° C)	None
Weight loss, %	
Chalking after heat aging	2.5
Tack-free time, hrs	12
Stain or color change	Conforms
Durability	
Glass	Conforms
Aluminum	Conforms
Mortar (primed)	Conforms
Peel adhesion, pli	
Glass	14 C
Aluminum	15 C
Mortar (primed)	12 C

The sealant meets or exceeds the requirements of ASTM C920-79, Grade NS, Class 25, use NT, A, and O.

TABLE 9.3 Typical Properties for a Two-Component Polymer P Sealant

Physical properties	Values

Specific gravity
  Base
  Catalyst
  Mixed
	
	1.3
	3.0
	1.49

Viscosity, base, Brookfield #7 spindle, poise
  @ 10 rpm
  @ 1 rpm
	
	10,000–11,000
	50,000–60,000

Solids, %	94
Slump, in.	$\frac{1}{8}$
Application life, 75° F (25° C), min	30–50

Cure rate at 75° F (25° C)
  after 6 hrs
  after 24 hrs
	
	15
	40

Color	Black
Hardness, Shore A	45
Tensile strength, psi	400 (28.2 kg/cm^2)
Elongation, %	600
Lap shear, psi	250 (17.6 kg/cm^2)
Peel strength, pli	65 (11.6 kg/cm)
Low temperature flexibility	−60° F (−52°C)

SUMMARY

Because of the limited use of this type of polymer, there are not many known applications. Interested parties should contact PRC DeSoto International, Glendale, CA, for additional data [3].

REFERENCES

1. Cagle, Charle, V. 1973. *Handbook of Adhesive Bonding*. New York: McGraw-Hill.
2. Amstock, J.S. 1973. *Handbook of Adhesive Bonding*, Chapter 7. New York: McGraw-Hill.
3. Courtesy of PRC DeSoto International. Glendale, CA.

CHAPTER 10
NEOPRENE, HYPALON,® AND NITRILE

INTRODUCTION

This chapter deals with adhesives and sealants made from neoprene [1], nitrile, and hypalon rubbers that the construction industry has used for such applications as bridge pads, coatings, membranes, adhesives, caulks, and sealants.

Neoprene sealants were among the first elastomeric sealants offered to the construction industry as single-component, gun grade sealants cured slowly to a Shore A hardness of 35 to 45. First produced in the United States in 1931, neoprene was the first synthetic rubber developed that had many properties of natural rubber.

In the early 1920s, Dr. Nieuwland of the University of Notre Dame discovered that acetylene could be polymerized by passing it through a catalyst based on cuprous chloride. The main product of this reaction was divinyl acetylene. In 1930, a team of DuPont chemists studying the polymerization of acetylene found that by varying the conditions of polymerization, mono-vinyl acetylene would be prepared in abundance while the divinyl acetylene produced was present only as an impurity. It was further found that mono-vinyl acetylene could be reacted with hydrochloric acid to form chloroprene, which could be polymerized under certain conditions to yield a rubberlike polymer. This material first was called *DuPrene* and later *neoprene*, the name by which it is known today. It became commercially available in 1936.

Before the introduction of neoprene, the only elastomer available for the manufacture of solvent adhesives and sealants was natural rubber, but rubber cements were limited in use because of several adverse properties:

- Solutions of unmilled or slightly milled rubber were very viscous even at low percentage solids, so it was difficult to apply an amount sufficient to give the satisfactory thickness of dried film.
- Rubber films had poor resistance to oils, weather, ozone, and oxidation — hence they would deteriorate after a comparatively short service life.
- The low cohesive strength of uncured natural rubber films could be increased by the use of room temperature accelerators. These accelerators, however, tended to overcome the rubber, resulting in early deterioration. This deterioration could be decreased if fewer accelerators were used.

In spite of these limitations, large amounts of rubber were produced before World War II for a variety of industrial applications. If a combining operation with natural rubber adhesive could not be accomplished, it had to be nailed, screwed, or sewn, or fastened by some other means.

At that time little thought was given to the use of neoprene as an adhesive base for the following reasons:

1. It was expensive, costing from three to five times as much as natural rubber.
2. Toluene, a comparatively expensive solvent, was required to dissolve neoprene while natural rubber was soluble in low-cost petroleum naphtha.
3. Considerably more neoprene than natural rubber was required to obtain a given viscosity.

This situation changed soon after the beginning of World War II when the use of natural rubber in adhesives was prohibited, and neoprene was the only alternative. The manufacturers were dependent on neoprene adhesives to fabricate products critical to the war effort.

NEOPRENE

The chemical equations listed on the following page illustrate the reactions required for the preparation of neoprene.

Any type of neoprene that can be dissolved can be used as a cement base. Adhesives and sealants made from the different types of neoprenes will vary in properties.

C (coal) + $CaCo_3$ (limestone) → CaC_2 (calcium carbonate)

$CaC_2 + H_2O$ → $CH≡CH$ (acetylene)

$CH≡CH$ + catalyst → $\underset{Cl}{C-}CH=CH_2$ (MVA)

$C-CH=CH_2 + HCl$ → $CH_2 =\overset{-}{C}-CH=CH_2$ (chloroprene)

$\underset{|}{\overset{Cl}{|}}$
$CH_2 = C-CH_2$ + polymerizer → $(-CH_2-\underset{|}{\overset{Cl}{\underset{|}{}}}C=C=CH_2-)_n$ (neoprene)

TYPES OF NEOPRENE SEALANTS AND ADHESIVES

Neoprene Type KNR

Noted for its ability to become more plasticized (chemically or by milling) than any other type used in adhesives, Neoprene Type KNR is not used extensively in adhesive formulations because of its low cohesive strength.

Neoprene AC

A general purpose elastomer, Neoprene AC has fast bond strength development, good stability, and cohesive strength.

Neoprene AD

A general purpose elastomer with fast bond strength development and good cohesive strength, Neoprene AD provides better *stability* and *viscosity* than AC.

Table 10.1 illustrates the cured physical properties of Type AD versus AC. Adding either NA-22 (a curing additive) or magnesia changes some of the physical properties.

TABLE 10.1 Cure Rate of Neoprene Type AD vs. AC

Raw materials	Parts by weight					
Compound No.	1	2	3	4	5	6
Neoprene AC	100	—	100	—	100	—
Neoprene AD	—	100	—	100	—	100
Magnesium oxide	4	4	2	2	4	4
Zinc oxide	5	5	5	5	5	5
NA-22	—	—	—	—	0.5	0.5
Properties	Values					
Mooney Scorch, min. to 5 pt. rise at 250° F (121.11° C)	28	43	25	42	15	10
Stress-strain (press-cured) 307° F (153° C)						
Tensile strength at break, psi						
5 min	400	350	2100	450	1700	3050
10 min	1750	740	2700	750	3200	3300
20 min	3300	1160	2850	1360	3400	3300
Elongation at break, %						
5 min	*	*	*	*	*	910
10 min	*	*	980	*	940	800
20 min	*	*	960	*	880	710
Hardness, shore A						
5 min	37	38	39	35	39	37
10 min	39	38	42	38	42	47
20 min	44	39	43	39	44	47

* No break at 1500% elongation

Neoprene AD-G

Neoprene AD-G is similar to AD but provides smoother, less stringy solutions and a longer pot life in isocyanate cured systems.

Neoprene AF

A very low crystallizing RT curing elastomer, the bond strength of Neoprene AF is superior to Neoprene AC and AD.

Neoprene AH

Neoprene AH is an acrylic copolymer of chloroprene that forms colloidal dispersions in hydrocarbon solvents. It has good application properties in high solids.

Neoprene FB

This version is a high viscosity fluid elastomer.

Neoprene Type GN and GNA

Neoprene Type GN and GNA crystallize slowly and are used where long tack retention is desired and high cohesive strength is not required. They are sulfur modified with a fast cure rate at room temperature.

Neoprene Type GRT

Similar to Type GN in most of its properties, Neoprene Type GRT differs because its slower rate of crystallization, which results in somewhat longer tack retention time and lower ultimate cohesive strength.

Neoprene Type W

A general-purpose elastomer, Neoprene Type W will crystalize or freeze more rapidly and to a greater extent than type GN. Therefore, it will have higher original and ultimate cohesive strength. Neoprene Type W cements are appreciably more stable than type GN in respect to both viscosity increase during curing and viscosity decrease during aging. Neoprene Type W films also can be cured at room temperature but not as rapidly as type GN or GNA films. Type W has lower cohesive strength than Neoprene AC and AD

Neoprene Type WRT

Neoprene Type WRT has similar properties to type W but differs in that it does not crystalize to any appreciable extent.

Neoprene WHV

WHV is similar to Type W but has a much higher plasticity value. Consequently, Type WHV cements are much higher in viscosity, and their films have considerably higher cohesive strength. Although useful in the preparation of adhesives to require high viscosity, it is particularly useful in blends with other polymers, such as those used for tile and wallboard laying.

Neoprene Type CG

Neoprene Type CG is the first of a series of quick-freezing polymers developed especially for use in solvent adhesives. It was used extensively during World War

II for seaming barrage balloons. After the war, it was used in the automotive and shoe industries. Even though Neoprene Type CG makes an excellent adhesive, it has several disadvantages that tend to limit its use.

1. The raw polymer discolors during aging and, in several months, changes from a light yellow, to green, and finally to black.
2. The plasticity value of the raw polymer decreases with aging; thus the ratio of solution viscosity to the solids content varies as the polymer ages.
3. Adhesives and cements made from Type CG become lower in viscosity as they age and have a limited storage life when they contain magnesium and zinc oxides.

In discussing the properties of polychloroprene sealants or adhesives, it should be pointed out that they contain a greater amount of solids per area than the natural rubber types, and have superior aging qualities. Their advantages and disadvantages are listed in Table 10.2.

TYPES OF NEOPRENE, SEALANTS, AND ADHESIVES

Neoprene adhesives and cements can be classified into two types: curing and noncuring.

Curing adhesives and sealants are employed where service is at a temperature high enough to soften an uncured film and cause bond failure. They can be cured at room temperature by the use of ultra-accelerators or by application of heat on less active ones. For service at somewhat elevated room temperatures, it is usually unnecessary to cure a neoprene film, since it has excellent uncured cohesive strength.

Neoprene solvent adhesives can be utilized as both construction and industrial adhesives and can be cured or uncured. Uncured adhesives are used to adhere porous material to wood or metal, such as in curtain-wall construction panels.

TABLE 10.2 Advantages and Limitations of Neoprene Sealants

Advantages	Limitations
1. Compatible with bitumens with good adhesion	1. Up to 40% shrinkage
2. Compatible with asphalt with good adhesion	2. Only dark colors available
3. Compatible with neoprene gaskets	3. Very slow cure
4. Good water resistance	4. No dynamic movement
5. Available in single component	5. No specifications
6. Good adhesion to metals	6. Stains stone and wood
7. ± 12.55 movement	

The various types of neoprene can be used in formulations to adjust shear strengths, viscosity, or other properties required by the intended application.

The following are typical formulas for some construction applications. Figure 10.1 illustrates three suggested starting formulations for a mastic adhesive, based on neoprene latex. Figure 10.2 is a typical neoprene-based sealant, and Figure 10.3 suggests formulas for compounding an anionic neoprene latex contact adhesive.

HYPALON®

Chlorosulfonated polyethlene [2], the chemical name for Hypalon®, (a DuPont trademark), is a synthetic rubber that can be used as a base for a resilient sealant or adhesive as a one-part material in the construction market. It can be compounded to give excellent mechanical properties, such as high tensile strength and abrasion resistance. The product first was developed in 1940 and first was sold under the Hypalon® trademark in 1952. Chemically formulated, it can be represented as:

$$RH^{**} + Cl \rightarrow R + HCl$$

$$R^{\cdot} + SO_2 \rightarrow RSO^{\cdot}_2$$

$$RSO^{\cdot}_2 + Cl \rightarrow RSO_2CL$$

$$R^{\cdot} + Cl \rightarrow RCl$$

The systems used for curing a chlorosulfonated polyethylene sealant are:

- PbO-litharge
- PbO-MgO2. litharge-magnesia
- Organic materials such as epoxies, and organic sulfates

TYPES OF HYPALON® SEALANTS AND ADHESIVES

Several grades and types of Hypalon® are available for a variety of end-use requirements. A general description of each of these products with its advantages and limitations is shown in Table 10.3. All these systems can be processed and used in the usual manner for solid elastomers. Several types are also of value in unvulcanized applications and solution coatings applications. The total market for these types of adhesives and sealants is small, since they are competing with solvent-based acrylics that are now supplied by many manufacturers. Table 10.4 lists a general outline of compounding ingredients and their functions. A typical formulation for Hypalon® caulk is listed in Figure 10.4.

Table 10.5 lists the advantages and limitations of a Hypalon® sealant.

Raw materials	Parts by weight		
	1	2	3
Neoprene Latex 750	100	—	—
AquaStik™ 2900	—	75	—
Neoprene Latex 735A	—	25	—
Neoprene Latex 617A	—	—	100
Darvan WAQ	2	2	1
Nonionic stabilizer	0.5	0.5	—
Sodium silicate	0.25	0.25	0.25
Dispersion masterbatch			
Zinc oxide	10	10	15
Nonstaining antioxidant	2	2	2
Thiocarbanilide	1	1	1
Sulphur	0.5	0.5	0.5
Terpene phenolic resin	60	85	—
Tall oil pitch	—	50	—
Pate ester gum resin	—	—	100
Water floated kaolin	50	50	—
Ammonium caseinate	2	2	—
Freeze-thaw stabilizer			
Methanol	—	—	20
Siponic 218	—	—	1
Dimethylamine caseinate	—	—	2
Thickener masterbatch			
Light process oil	20	—	—
Emulsifier	0.5	—	—
Methocel 4000	4	3	3
Kerosene	—	6	6

Physical properties	Values		
	Neoprene Latex 750	671	AQS 2900
Oulutac 90D* phr	25	50	25
Tack life, hr	< 6	24	< 1
Peel strength, kN/m^3	3.6	2.7	2.4
Creep at 74° F (23° C) under 3 kg/2.5 cm/30m, mm	2.0	7.0	17
Creep at 158° F (70° C) under 500 g/2.5 cm/30m, mm	—	34	26

*A pentaerythritol ester of tall oil rosin.

FIGURE 10.1 Mastic adhesive based on neoprene latex.

Raw materials Component A	Parts by weight
Neoprene W	100
Magnesium oxide	4
Antioxidant	2
Calcium carbonate	150
Petroleum process oil	30
Amine-modified bentonite	3
Zinc oxide	5
Total, Component A	294
Component B	
Butyl phenol formaldehyde, resin (heat reactive)	45
Xylol	115
Zinc oxide	7
Total, Component B	167

Physical properties	Values
Pot life, min	15
Weight, lbs/gal	11.5
Hardness, Shore A	
(initial, 1 day @ RT)	30
(7 days @ RT)	70
Nonvolatile (blended), %	80
Application temperature, °F (°C)	40°–122° F (5°–50° C)
Tensile strength, psi	900
Elongation, %	400

FIGURE 10.2 Neoprene based sealant, two-component.

NITRILE

Nitrile rubbers are copolymers of diene and a vinyl unsaturated nitrile [3]. They were developed in the 1930s as Buna-N adhesives suitable for fabrication of aircraft structures and now have found some use in curtain-wall construction. They have uses similar to general construction adhesives and sealants and are often referred to as construction adhesives. While nitrile rubbers can be prepared in low viscosity, most of the products for the construction industry today probably would fall into the heavier mastic classification. Blended with phenolic resins, they are superior to the lower-priced SBR construction sealants in heat resistance, oil resistance, and strength.

$$\left(H_2-CH=CH-CH_2\right)_x \left(CH_2-CH\right)_y$$
$$|$$
$$C\equiv N$$

Raw materials	Parts by weight					
	1	2	3	4	5	6
Neoprene Latex 671	100	100	100	100	100	—
Neoprene Latex 750	—	—	—	—	—	100
Zinc oxide	5	5	5	5	5	5
Antioxidant	2	2	2	2	2	2
Ferquatac® 100-D	30	—	—	—	—	—
Dresinol® 215	—	—	9	—	—	—
Aquatac® 90-D	—	30	20	—	—	—
SP-560	—	—	—	30	—	—
Dymerex® resin	—	—	—	10	—	—
Pentalyn® K	—	—	—	—	24	—
Sylvatac 295	—	—	—	—	6	—
Sylvatac 140	—	—	—	—	6	—
Durez 12603	—	—	—	—	—	50–85
Picolyte S-25	—	—	—	—	—	25–45
Properties	Values
---	---
Oulutac 90-D, phr	25–50
Tack life, hr	24
Peel strength, kN/m³	2.7–4.1
Creep at 74° F (23° C) under 3 kg/2.5 cm/30m, mm	2–7
Creep at 158° F (70° C) under 500 g/2.5 cm/30m, mm	8–34

FIGURE 10.3 Anionic neoprene latex contact adhesive.

TYPES OF NITRILE SEALANTS AND ADHESIVES

These materials require high solvent levels to produce gunnable sealants, hence are little-used in the construction market. Reclaimed rubber goes into asphaltic sealants, and many reclaims are neoprene or nitrile. Nitrile rubbers can be used to make a solvent-based fluid-caulking compound to seal small cracks that are either discontinuities in metal sash or shrinkage cracks that are not moving. The sealant adheres to all substrates, remains flexible and is not affected by UV.

The clear sealant is used for very small cracks up to ⅛ inch and is pumped into the crack using a modified oil can or Plews gun. These sealants are recommended for use in sealing small openings in sash, masonry, and various narrow construction joints. This includes needle glazing around the perimeter of leaking channel-glazed sashes and skylights; miter and butt joints; cracks in brick, tile, and masonry joints, and porcelain and enamel construction. The sealant consists of 62 percent polymer and 38 percent solvent.

The pigmented sealant has essentially the same solids content as clear sealant, but has added fillers. It is used in the same kinds of applications as the clear

TABLE 10.3 Descriptions of Types of Hypalon®

	20	30	45	HPG-6525
Description				
Chlorine content, %	29	43	24	27
Sulfur content, %	1.4	1.1	1.0	1.0
Physical form	Chips	Chips	Chips	Chips
Color	White	White	White	White
Odor	None	None	None	None
Specific gravity	1.12	1.27	1.07	1.10
Mooney viscosity, ML 1+4 @ 100° C (212° F)	28	30	37	90
Storage stability	Excellent	Excellent	Excellent	Excellent
Distinguishing features	Readily soluble in common solvents. Good low-temperature flexibility.	Readily soluble in common solvents. Forms hard, glossy films.	High uncured strength. Good heat resistance. Good low-temperature flexibility.	High polymer viscosity. Good low-temperature and heat resistance. Good processing at high extensions.
Processing characteristics				
Extruding	Fair	Fair	Good	Excellent
Molding	Good	Fair	Excellent	Excellent
Calendaring	Fair	Fair	Excellent	Good
Solution properties				
Brookfield viscosity, MPa·s				
25 wt % polymer in toluene	1300	400	—	—
5 wt % polymer in xylene	9	4	60	45
Vulcanizate properties				
Hardness, Durometer A	45–95	60–95	65–98	40–95
Tensile strength, MPa				
Carbon black stocks	Up to 20.6	Up to 24.2	Up to 27.6	Up to 27
Gum stocks	Up to 8.2	Up to 17.2	Up to 27.6	Up to 27.6
Color stability	Excellent	Excellent	Excellent	Excellent
Low-temperature properties	Good	Poor	Excellent	Excellent
Tear strength	Fair	Fair	Good	Good
Resistance to				
Abrasion	Very good	Very good	Excellent	Excellent
Chemicals	Good	Excellent	Good	Very good
Compression set	Fair	Poor	Good	Very good
Flame	Fair	Very good	Fair	Fair
Heat-aging	Very good	Fair	Very good	Very good
Ozone	Excellent	Excellent	Excellent	Excellent
Petroleum oils	Fair	Excellent	Fair	Fair
Weathering	Excellent	Excellent	Excellent	Excellent

All values on this table are approximate. They are presented to describe the various products, and are not intended to serve as specifications.

TABLE 10.3 Descriptions of Types of Hypalon® (*Continued*)

	40S	40	4085	48
Description				
Chlorine content, %	35	35	36	43
Sulfur content, %	1.0	1.0	1.0	1.0
Physical form	Chips	Chips	Chips	Chips
Color		White	White	White
White				
Odor		None	None	None
None				
Specific gravity	1.18	1.18	1.19	1.27
Mooney viscosity, ML 1+4 @ 100°C (212°F)	46	56	94	78
Storage stability	Excellent	Excellent	Excellent	Excellent
Distinguishing features Viscosity. Improves processing	Low polymer viscosity. Improves processing of dry, stiff stocks.	Medium polymer viscosity. Versatile, suitable for many applications.	High polymer viscosity. Good green strength. Improves processing of soft or highly extended stocks.	High polymer viscosity. Excellent oil and fluids resistance. High uncured strength.
Processing characteristics				
Extruding	Excellent	Excellent	Excellent	Good
Molding	Excellent	Excellent	Excellent	Good
Calendaring	Excellent	Excellent	Excellent	Good
Solution properties				
Brookfield viscosity, MPa·s				
25 wt % polymer in toluene	—	—	—	—
5 wt % polymer in xylene	20	25	50	12
Vulcanizate properties				
Hardness, Durometer A	40–95	40–95	40–95	60–95
Tensile strength, MPa				
Carbon black stocks	Up to 27.6	Up to 27.6	Up to 27.6	Up to 27.6
Gum stocks	Up to 27.6	Up to 27.6	Up to 27.6	Up to 24.2
Color stability	Excellent	Excellent	Excellent	Excellent
Low-temperature properties	Good	Good	Good	Poor
Tear strength	Good	Good	Good	Good
Resistance to				
Abrasion	Excellent	Excellent	Excellent	Excellent
Chemicals	Excellent	Excellent	Excellent	Excellent
Compression set	Good	Good	Good	Fair-good
Flame	Good	Good	Good	Very good
Heat-aging	Very good	Very good	Very good	Good
Ozone	Excellent	Excellent	Excellent	Excellent
Petroleum oils	Good	Good	Good	Excellent
Weathering	Excellent	Excellent	Excellent	Excellent

All values on this table are approximate. They are presented to describe the various products, and are not intended to serve as specifications.

TABLE 10.4 Compounding Ingredients and Their Function

Function	Ingredients normally recommended
Acid acceptor	Litharge Magnesia (high activity) Epoxy resin of epoxidized oil Calcium hydroxide
Colorants	Carbon black Titanium dioxide Organic and inorganic pigments
Filler	Carbon black Mineral Cork Magnetic filler
Flame retardant	Antimony oxide Hydrated alumina Halogenated hydrocarbons

Plasticizer	Petroleum oils
	  Aromatic
	  Naphthenic
	Esters
	Chlorinated paraffins
	Polymeric
	Polyesters

Processing aids	Waxes Stearic acid Low mol. wt. polyethylene Polyethylene glycol Fatty acid amides Nordel® 2744 or *cis*-4 PBD Tackifiers
Sulfur bearing	TMTD (or TETD) Sulfur Tetrone® A MBTS NBC DOTG Sulfasan® R
Vulcanizing agent* Nonsulfur	Peroxide plus a coagent HVA-2® plus a coagent Pentaerythritol

*For some applications, no vulcanizing agent is used.

Abbreviation	Chemical name	Abbreviation	Chemical name
DOTG	Di-ortho-tolyl guanidine	TETD	Tetraethyl thiuram disulfide
MBT	2 mercaptobenzothiazyl disulfide	TMT	Tetramethyl thiuram disulfide
NBC	Nickeldibutyl dithiocarbamate		

Raw materials	Parts by weight
Hypalon 40	10
Hypalon 30	90
Chlorinated paraffin	100
Asbestos 7RF	15
Viscotrol A	15
Hi-Sil 233	30
Ti-Pure R610	80
Talc	50
Tri-Mal	40
Staybelite resin	1.5
MBTS	1
Thiuram M	0.5
Xylene	60
Tributyl phosphate	53
Fractol A	17
Isopropyl alcohol	10

Physical properties	Values
Weight per gallon, lbs	10.9
Solids by weight, %	88
Solids by volume, %	83
Sag, Boeing test jig	Neg.
Tensile strength, psi @75° F (23.5° C)	100
Elongation, % @75° F (23.5° C)	60
@ –40° F (–40° C)	60
Tensile adhesion, psi Al to Al	100
Steel to steel	60
Shear strength, psi Al to Al	40
Glass to glass	50
Wood to wood	150

FIGURE 10.4 Hypalon® caulk.

TABLE 10.5 Advantages and Limitations of Hypalon® Sealants

Advantages	Limitations
1. Impervious to water	1. Slow cure of up to four months
2. Remains flexible	2. Poor package stability
3. Fair recovery	3. Higher cost
4. Good chemical resistance	4. High shrinkage
5. Good UV resistance	5. Not for interior use
6. Good ozone resistance	6. Not for traffic bearing areas
7. Has ± 12.5% movement	7. Not for sidewalks
8. Single component	8. Difficult to extrude

sealant, but the filler gives the pigmented sealant some structure and makes it non-sag when used in slightly wider joints not exceeding ³⁄₁₆ in. The sealant dries to a rubberlike consistency and will take a small degree of movement, ± 7.5 percent.

The market for these sealants is small, and outlets are found mostly in the construction industry and small segments of over-the-counter retail business. They are available in screw-top pint cans and larger sizes by special order. The materials are expensive compared with conventional sealants but, nevertheless, are a necessary adjunct for any large-scale commercial sealant applicator. This is because, inevitably, the applicator will find a small crack that can be sealed only with a thin-viscosity small-joint filler. The advantages and limitations of nitrile rubber sealants are listed in Table 10.6. A typical nitrile-like structural adhesive is presented in Figure 10.5.

Raw materials	Parts by weight		
	1	2	3
Nitrile rubber	100	100	100
Phenolic resin	50	80	100
Zinc oxide	5	5	5
Sulfur	1.5	1.5	1.5
Benzothiazole disulfide	1.5	1.5	1.5
Stearic acid	1.5	1.5	1.5

Physical properties	Values
Viscosity	Paste
Typical cure, hr at 300°–400° F (150° –230° C)	1-2 at 300° F
Tensile lap-shear, ksi, Mpa	
@ –70°F (–55° C)	3.2–4.6 (22.1–32.7)
@ 75°F (25° C)	3.2–4.6 (22.1–32.7)
@ 300°F (150° C)	1.5–2.5 (10.3–17.2)
Heat resistance °F (° C)	500 (260)

FIGURE 10.5 Typical nitrile, structual anaysis.

TABLE 10.6 Advantages and Limitations of Nitrile Rubber Sealants

Advantages	Limitations
1. Very low viscosity	1. Up to 45% shrinkage
2. Gunnable from an oil can	2. Poor UV resistance
3. Good to metals and masonry	3. Very high cost
4. No primer required	4. Used only in cracks
5. Long life	5. Only in screw-top cans
6. Single component	6. No available specification
7. Cures to a rubber	7. ± 12.5% movement
8. Clear sealant used in cracks up to ⅛ in.	8. Two colors only, black/bronze and aluminum
9. Good weather and UV resistance	
10. Pigmented sealant used in cracks up to ³⁄₁₆ in.	

SUMMARY

Neoprene Hypalon® and nitrile rubber sealants and adhesives are the least-used in the construction markets. They are applied when high-volume usage is not required. Their importance, however, cannot be minimized. Each formulated compound fulfils a specific function when used by competent applicators.

REFERENCES

1. "Neoprene®, Polychloroprene." 1997. A Product of DuPont Dow Elastomers
2. "Hypalon®, Chlorosulfonated Polyethylene. Types, Properties and Uses of Hypalon®." 1997. A Product of DuPont Dow Elastomers.
3. Cagle, Charles V. 1973. Handbook of Adhesive Bonding. New York: McGraw-Hill.

CHAPTER 11
POLYSULFIDE AND LP™ POLYMERS

INTRODUCTION

The first synthetic rubber, a polysulfide, was manufactured in the United States in 1929. Its most interesting property was an unusual inertness to solvents and hydrocarbon fuels in contrast to the easy swelling of natural rubbers. In early 1942, work began that led to the invention by J.C. Patrick and H.R. Ferguson of a process of reductively cleaning, to a predetermined degree, a portion of polymeric polysulfide groups to a curable rubber (SH polymer chain terminals). These studies led subsequently to a wide range of liquid polymers, ranging from less than 1000 molecular weight to viscous liquids of about 7000 molecular weight. LP™ became the trademark for these liquid polymers. Polysulfides are polymers of bis (ethylene oxy) methane containing disulfide linkage. The reactive groups are mercaptans (general structure is HS).

MERCAPTAN-TERMINATED LIQUID POLYMERS [1]

Conventional mercaptan-terminated liquid polymers (HS-R-Sh) are available in a wide range of molecular weights with cross-linking densities ranging from 0.05 mol percent to 2.0 mol percent. Table 11.1 lists the physical properties of the liquid polysulfide polymers, which provides the formulator with a range of selection.

TABLE 11.1 Properties of LP Liquid Polysulfide Polymers

Typical properties	LP-3	LP-33	LP-977	LP-980	LP-2	LP-32	LP-12	LP-31
Color-MPQC-29-A	50 max.	30 max.	70 max.	50 max.	70 max.	50 max.	40 max.	100 max.
Viscosity-poises 77° F (25° C)	9.4-14.4	15-20	100-150	100-150	410-525	410-525	410-525	800-1450
Moisture content, %	0.1 max.	0.1 max.	0.3 max.	0.3 max.	0.3 max.	0.3 max.	0.3 max.	0.3
Mercaptan content, %	5.9-7.7	5.0-6.5	2.50-3.50	2.50-3.50	1.70-2.20	1.50-2.00	1.50-2.00	1.0-1.5
General properties								
Average molecular weight	1000	1000	2500	2500	4000	4000	4000	8000
Refractive index n/D	1.5649		1.568	1.566		1.5689		1.5728
Pour point, ° F (° C)	-15 (-26)	-10 (-23)	40 (4)	40 (4)	45(7)	45(7)	45(7)	50(10)
Flash point (PMCC); ° F (° C)	<350(177)	<350(177)	<350(177)	<350(177)	<350(177)	<350(177)	<350(177)	<350(171)
% Cross-linking agent	2.0	0.5	2.0	0.5	2.0	0.5	0.2	0.5
Sp Gr @ 77° F (25° C)	1.31	1.27	1.29	1.29	1.29	1.29	1.29	1.31
Avg. viscosity poises 40° F (4° C)	90	165	770	770	3800	3800	3800	7400
Avg. viscosity poises 150° F (65° C)	1.5	2.1	11	11	65	65	65	140
*Low temp. flex., $G_{10,000}$ ° F (° C) (708 kb/cm^2)	-65 (-54)	-65 (-54)	-65 (-54)	-65 (-54)	-65 (-54)	-65 (-54)	-65 (-54)	-65 (-54)

* Cured compound

LP™ POLYSULFIDE BASIC CHARACTERISTICS

Polymerization of liquid polysulfide polymers to a high molecular weight elastomer usually is accomplished by oxidizing the thio (SH terminals to disulfide -S-S bonds).

LP™ Polymer $HC(C_2H_4OCH_2H_4SS)_x C_2 H_4 OCH_2 OC_2H_4 SH$

Curing by oxidation $2\sim RSH + MnO_2 \rightarrow \sim RSH \sim\rightarrow MnO_2 \pm H_2O$

Curing with epoxy resin

$$2\ R\text{-}CH\text{-}CH_2 + HS\text{-}R'\text{-}SH \rightarrow R\text{-}CH\text{-}CH_2\text{-}S\text{-}R'\text{-}S\text{-}CH_2\text{-}CH\text{-}R$$
$$\begin{array}{ccc} \backslash\ / & | & | \\ O & OH & OH \end{array}$$

CURING MECHANISMS

The curing agents most commonly used are compounded into pastelike materials as oxygen-donating substances. They include:

MnO2, manganese dioxide, which provides improved heat resistance and lower toxicity. The amount required is 7.5 to 10 parts/100.

CaO2, calcium peroxide, which requires moisture for activation. It can be used to provide a light-colored, one-part moisture curing system. The desired amount is 10.0 to 12.0 parts /100.

PbO2, lead peroxide, (fast, medium, or slow grades) which provides an easily controllable LP polymer curing rate. The desired ratio is 7.5 to 10.0 parts/100.

Cumene hydroperoxide which provides a convenient liquid form for obtaining pourable sealants that have good compression set resistance. The required mixing ratio is 8.0 parts 100 ZnO2. Zinc peroxide reacts slowly with LP polymers and can be used to provide white formulations with moderate heat stability. Ratio is 10.0 parts/100.

NaBO3•H2O, sodium perborate monhydrate, which provides light-colored, nonstaining one- or two-part sealants able to be manufactured in a range of colors for the building industry. They are low modulus, highly elastic, and water-, UV-, and mold-resistant. The amount required is 4.0 parts to 100.

Quinone-dioxime and lower valence metallic oxides, other organic oxides, metallic paint dryers, and aldehydes also can function as a catalyst. The most widely used, however, are selected dichromates, (protected by U.S. Patents 2,787,608 and 2,964,503) manganese dioxide, and lead peroxide. The chromate-type catalyst is predominantly for integral fuel tank sealants in aircraft application. As with most formulations, selection of ingredients, such as fillers, pigments, and plasticizers is governed by the end use. This is particularly true of sealants used to seal insulating glass window units. Compared with other elastomeric sealants, polysulfide has some of the most extensive history as sealants. Polysulfide sealants and caulks were introduced as a commercial compounded product into the construction market in early 1950. They enjoyed increasing popularity for the next 10 years and began sharing the market with solvent-based acrylics and polyurethanes through the 1970s. (See Chapter 12 for additional information about polyurethanes.) Table 11.2 lists the advantages and limitations of both single- and two-component polysulfide sealants.

TABLE 11.2 Advantages and Limitations of One-Component Polysulfide

Advantages	Limitations
1. One-component sealant	1. Requires moderate temperature for faster cure
2. Broad color range	2. Requires high humidity for faster cure
3. Good durability	3. Slow cure at low temperatures
4. Good adhesion	4. Poorer recovery
5. Can meet TT-S-00230C	5. Limited package stability
6. Can meet ASTM C 920	6. Not recommended for pedestrian traffic areas
	7. Not recommended for sidewalks
	8. Slight odor

Advantages and limitations of two-component polysulfides	
Advantages	Disadvantages
1. Overall better physical characteristics; recovery, adhesion-in-peel, tensile-adhesion	1. Requires mixing, but easily mixed
2. Fast through cure	2. Slower cure below 40°F
3. Better UV resistance	3. Light colors a problem
4. Better water resistance	4. Limited pot life
5. Life expectancy over 20 years	5. Very short pot life at 100°F
6. Non-staining to masonry	6. Slight odor
7. Can meet TT-S-00227E	7. Poorer UV resistance compared to urethane and silicone sealants
8. Can meet ASTM C 920	8. Poorer recovery compared to urethane and silicone sealants
9. Cost slightly lower than one-component polysulfides since tubing and labor are expensive	9. Primers needed for porous substrates

TYPES AND USES OF POLYSULFIDE SEALANTS

When structures were built using traditional, empirical methods, sealants were called caulks, reflecting their limited use for filling cracks. With today's lighter-weight building components that have been engineered to even closer tolerances, we could not do without high-performance sealants, particularly in types of construction that require a barrier to water penetration. Yet selecting sealants can be like walking in a minefield. Sometimes the specifier needs to have the background of a chemical engineer and at the same time imagine what life is like on the business end of a caulking gun. As a comparison, Table 11.3 shows the differences among the major sealant backbones used in the construction industry. No definite standard exists to assist the designer in choosing among generic types of sealants, leading many architects to specify high-performance products in terms of movement range and life expectancy, when other criteria might be more important in the field.

Tables 11.4 and 11.5 cover the typical physical properties of one- and two-component sealants. Typical formulations of these types of sealants are found in Figures 11.11 and 11.12, respectively.

The construction industry generally used polysulfides in two-component kits, designated as 1 ½-gallon kits and in single component $1/10$ cartridges. The two-part packages are premeasured for mixing on the job site. A standard half-inch drill with a Jiffy-type mixing blade is required to thoroughly mix the two-component sealant. For sealants used in inside industrial applications, a meter-mix machine capable of handling a 55-gallon drum kit can be used.

Both one- and two-part polysulfides are recommended for all joints subjected to structural movement such as:
- Sealing metal curtain wall panels
- Window glazing and setting of the glass
- Caulking of marble and mosaic panels
- Caulking section joints in tilt-up construction
- Sealing concrete conduit
- Sealing of expansion and contraction joints

Polysulfide sealants have stood the test of time. Many buildings sealed 25 to 35 years ago still have the original sealant in place and are performing well. A case in point is the McKesson Plaza in San Francisco, California.

The polymers used in producing construction sealants are Thiokol LP™ polymers manufactured by Morton International, a Rohm and Haas company [2]. The compounded sealants are available in a wide range of colors to match most exterior building materials.

TABLE 11.3 Representative Properties of Sealants

Test method sealant base	Tensile strength, psi ASTM D412	Elongation, % ASTM D412	Shore "A" Durometer ASTM 676	Abrasion resistance	Operating temperature, °F	Shrinkage, %
Polysulfide	50-125	150-500	15-60	Fair to good	-60 to 250	0-3.0
Polymercaptan	50-125	150-500	15-50	Good to excellent	-65 to 250	0-3.0
Polyurethane	55-300	250-950	10-50	Excellent	-65 to 200	Nil
Silicone	400-600	250-1000+	10-55	Fair to good	-60 to 550	0-5.0
Neoprene	1000-1500	250-350	30-80	Excellent	-45 to 300	0-10.0
Epoxy-modified	1200-3500	10-20	40-60	Good to excellent	-50 to 300	0-3.0
			Shore "D"			
Acrylic	50-400	100-270	5-50	Fair to good	-10 to 300	5.0-15.0
Butyls-mastic type	—	5-150	5-70	Poor	0 to 200	15-40
Polybutene	—	5-20	20-40	Poor	-20 to 250	0-3.0

TABLE 11.4 One-Component Sealant

Properties	Gun grade, self-leveling	Pour grade, self-leveling
Tack-free time at 21° C (50% RH), hr	12-24	12-24
Cure time ($1/4$ in. × $1/4$ in., channel) days	5-10	4-8
Tensile strength, psi	300-400	250-500
Ultimate elongation, %	500-600	100-200
Modulus at 100% elongation, psi	40-60	60-80
Shore A hardness, initial	20-30	35-45
% recovery after 100% elongation	90	85
Continuous service range, ° C	-51 to +135	-43 to +135
Water-immersion properties	Good	Good
Dilute-acid resistance	Good	Good
Dilute-alkali resistance	Very good	Very good
Solvent resistance	Excellent	Excellent
Fire resistance	Excellent	Excellent
Electrical insulation	Good	Very good
Aging properties	Excellent	Excellent

TABLE 11.5 Two-Component Building Sealant, Metallic Oxide Cure

Nonvolatile content		95% min
Application life at 75° F, 50% RH		3 hr min
Tack-free time at 75° F, 50% RH		36 hr
Cure time at 75° F, 50% RH		72 hr max
Hardness, Rex (5-sec reading)		15-25
Shrinkage		Negligible
	Tensile adhesion, psi	Ultimate elongation, %
Adhesive strength in tension and ultimate elongation:*		
At 75° F	45-75	200-250
At 75° F after 9 days at 160° F	40-70	150-200
At 75° F after 21 days in water	25-40	150-200
At -20° F	70-90	200-250
Tensile strength and ultimate elongation:**		
At 75° F	90-160	400-700
At 75° F after 12 days at 160°F	100-190	300-650
At 75° F after conditioning in Atlas Twin-Arc weatherometer (ASTM E 42):		
For 500 hr	110-180	300-500
For 1100 hr	70-100	200-330
Temperature range -40° to 200°		

* Test specimens consisted of two small concrete blocks prime sealed with a seam of two-component polysulfide sealant $1/2$ in. × $1/2$ in. and cured for 14 days at 75° F

** Tested in accordance with ASTM D 412

DESIGN CHARACTERISTICS FOR LP™ POLYSULFIDE SEALANTS

LP™ polysulfide-based sealants are solid elastomers that adapt themselves to changes in joint dimension. As joints expand and contract or experience shear movement, the sealant's shape changes accordingly, but the volume of the sealant remains constant.

As the sealant shape changes, the magnitude and type of stress also changes because the sealant can be either in compression, expansion, or shear, as illustrated in Figure 11.1. Thus, in designing joints, it is vital that the proper width-to-depth ratio be specified so the width of the joint is consistent with the capability of the sealant to endure daily and seasonal cycles for prolonged periods. Figure 11.2 is a graph for anticipated joint movement in a joint dependent on the length and composition (the coefficient of linear expansion) of a panel section. The amount of movement that takes place for panels of various construction materials for a temperature gradient of 130° C (72° C) is graphically illustrated. Shown in Figure 11.3 (see dashed lines) are the recommended joint width for sealants capable of ± 25 percent joint movement. Wider joints are required for sealants with less movement capability; otherwise, the product will fail. Figure 11.4 shows the expected movement of materials and consequent changes in width. The joint widths are based on polysulfide based sealants with a ± 25 percent movement capability.

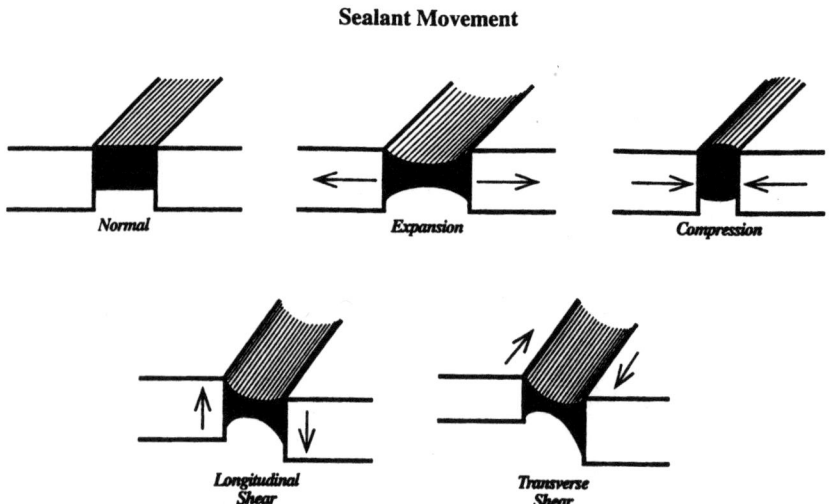

FIGURE 11.1 Sealant movement.

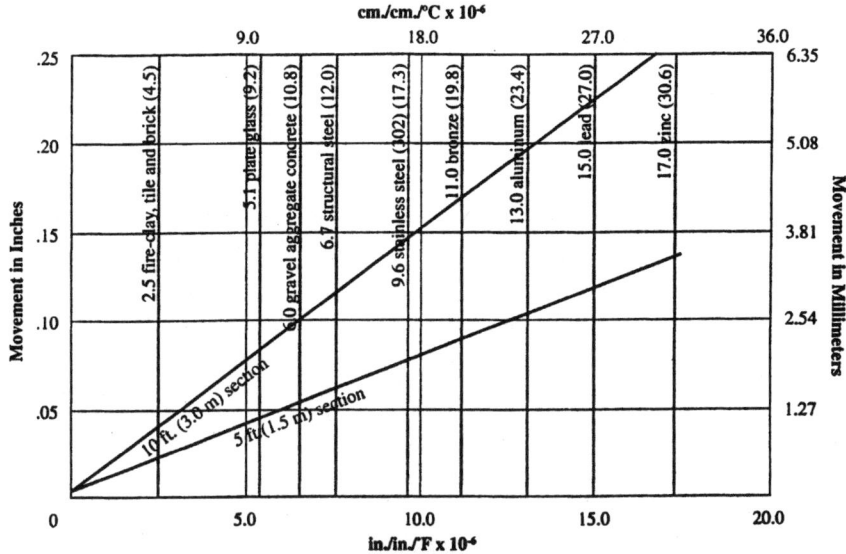

FIGURE 11.2 Joint movement for sections of various building materials at $\Delta T=130°$ F (72° C).

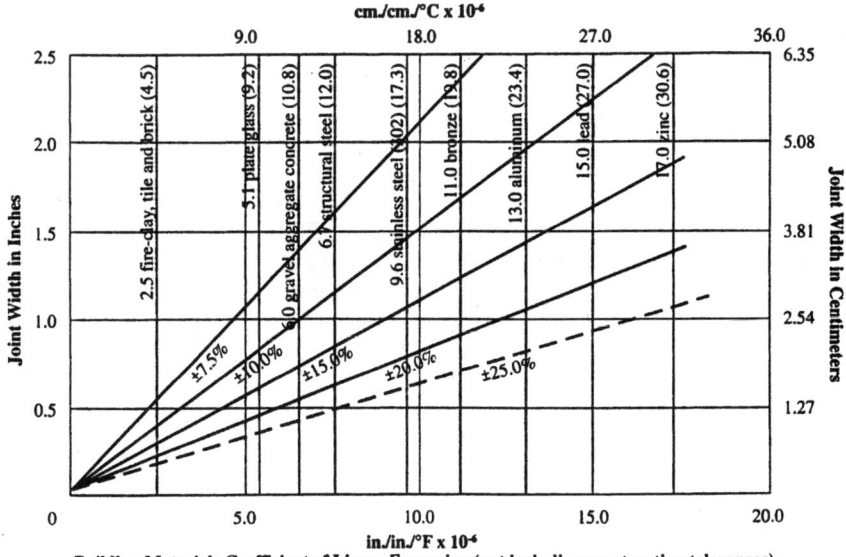

FIGURE 11.3 Joint with 10-foot sections for $\Delta T=130°$ F (72° C) with various building materials.

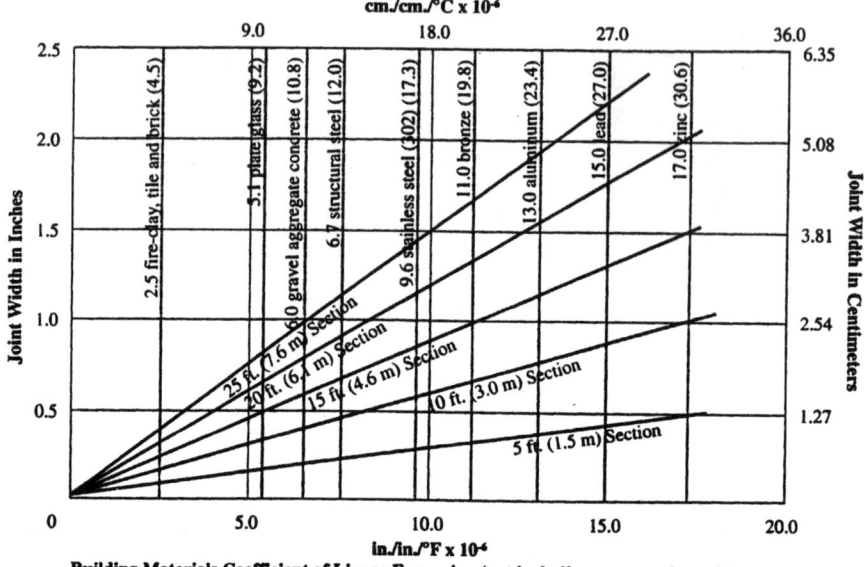

FIGURE 11.4 Joint movement for sections of various buildings materials at $\Delta T=130°$ F (72° C) for LP polysulfide-based sealant.

Figure 11.5 provides drawings of typical joints experienced in buildings and suggests the use of backup materials for joint design. Good joint design requires the proper selection and use of backup materials for optimum performance of sealants. Generally, the diameter of the backup should be about 25 percent greater than the joint width. A backup material performs the following functions:

- Controls depth of sealant in the joint as illustrated in Figure 11.6
- Provides support for tooling of sealant in joints
- Serves as a bond breaker to prevent sealant from bonding to the back of the joint
- Acts as a temporary joint filler in some cases
- Supports the sealant in horizontal deck joints subject to pedestrian or vehicular traffic
- Supports sealant in glazing applications and assist in cushioning and positioning glass sections

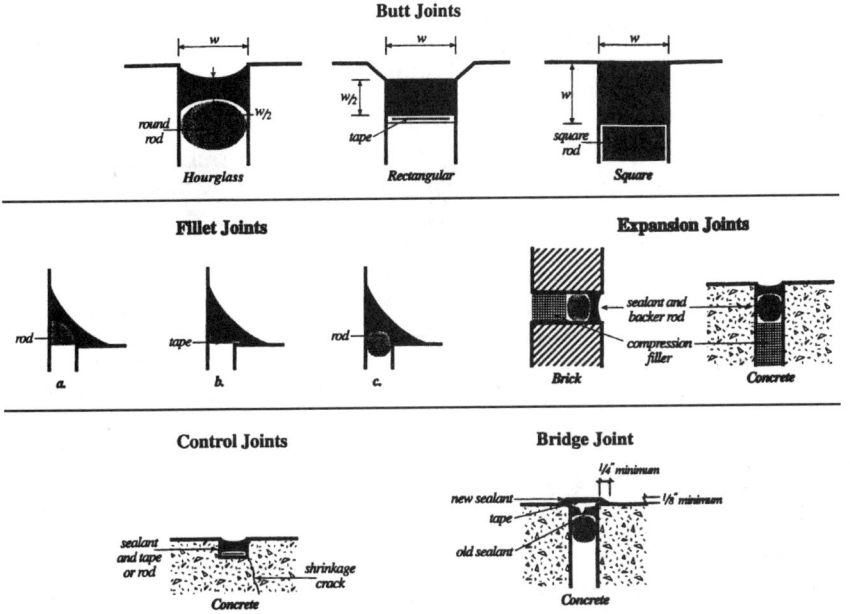

FIGURE 11.5 Typical joint types.

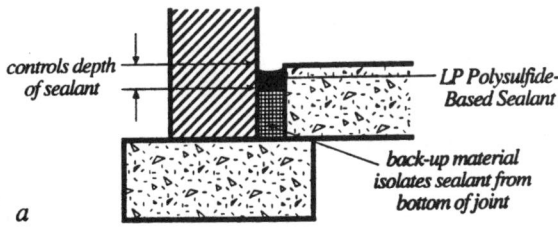

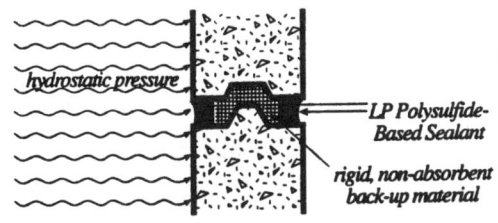

FIGURE 11.6 Back-up Materials.

Undesirable Effect With Sealant Bonded To Bottom of Joint

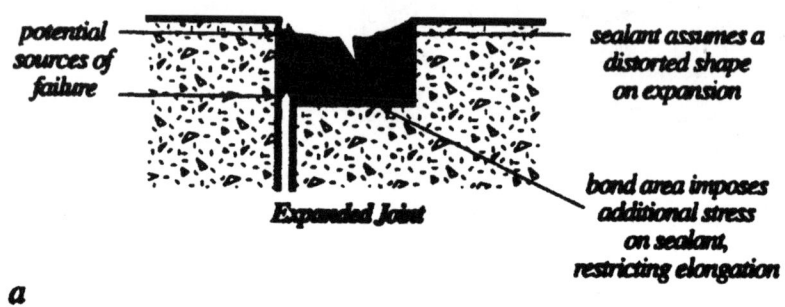

a

Use of Bond Breaker

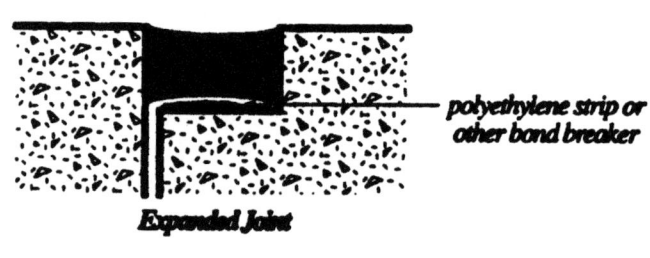

b

FIGURE 11.7 Bond breakers.

Some of the figures in this chapter also are applicable to Chapters 8, 10, 12, and 13. Bond breakers and their use are illustrated in Figure 11.7. In cases where the sealant has adhesive tendencies to the backup materials, a bond breaker must be used. The bond breaker film may be adhesive to the backup materials but must not exhibit any adhesion to the sealant. Materials impregnated with oil, bitumen, non-curing polymers, and similar materials are not to be used.

TYPICAL FORMULATIONS

The following formulas are typical starting points for the chemist or designer as he or she become familiar with initiating work on the development of these types of compounds:

Figure 11.8 is data on an adhesive for bonding to aluminum panels.

Figure 11.9 makes a comparison of different ingredients for a concrete adhesive.

Raw materials	Straight epoxy control	LP/epoxy adhesive
	Parts by weight	
Part A	1	2
Thiokol LP-3 polymer	—	100
Liquid epoxy resin, epoxide equivalent 175 to 210	—	—
Calcium carbonate	—	71
DMP 30	15	15
Part B		
Thiokol LP-3 polymer	—	—
Liquid epoxy resin, epoxide equivalent, 175 to 210	100	100
Calcium carbonate	100	179
DMP 30	—	—
Physical properties	Values	
A/B ratio	1/13.3	1/1.5
Cure rate	1 hr @ 250° F (123° C)	1 hr @ 250° F (123° C)
Tensile shear, psi, in./min	0.05	0.05
Peel strength, lbs/in.	8	18
Bend bond strength, lbs	100	140

FIGURE 11.8 LP/epoxy adhesive for aluminum.

Raw materials	Straight epoxy control	LP/EP concrete adhesive
	Parts by weight	
Part A	1	2
Thiokol LP polysulfide	—	100
White silica, HDS 100	—	80
DMP 30	7.5	20
Epoxy resin, epoxide equivalent	—	—
Part B		
Thiokol LP polysulfide	—	—
White silica, HDS 100	50	—
DMP 30	—	—
Epoxy resin, epoxide equivalent, 175 to 210	100	200
Physical properties	Values	
A/B ratio	1/20	1/1
Tensile adhesion, psi, 7 day cure	150	345
After water immersion, 7 days @ RT	150	335
Flexural strength, psi, 7 days @ RT	35	335
Shear strength, psi, 7 days @ 150° F (65.6° C)	400	4200

FIGURE 11.9 LP/epoxy for concrete.

Figure 11.10 suggests several formulas for two-component sealants. Figure 11.11 is a two-part, gun-grade gray sealant, and Figure 11.12 provides data for a single-component system using a calcium peroxide cure mechanism.

Raw materials	Parts by weight						
Part A	1	2	3	4	5	6	7
LP-2	100	100	100	100	100	—	100
LP-3	—	-30	—	—	—	100	—
SRF #3 black	30	1.0	—	—	—	10.0	—
Stearic acid	1.0	—	1.0	1.0	—	1.0	1.0
Durez 10694	5.0	—	5.0	—	—	—	—
Calcene TM	—	—	25.0	25.0	—	—	—
Titanox AMO	—	—	10.0	10.0	—	—	—
Lithopone	—	—	—	—	50	—	—
Sterling MT	—	—	—	—	—	10.0	—
Sulfur	—	—	—	—	—	0.15	—
Thermax	—	—	—	—	—	—	100
Santicizer E-15	—	—	—	—	—	—	50
Kenflex A	—	—	—	—	15	—	—
Part B							
C-5 catalyst	15	15	15	15	15	—	—
C-9 catalyst	—	—	—	—	—	13.8	—
Lead peroxide	—	—	—	—	—	—	25*
Dibutyl phthalate	—	—	—	—	—	—	—
Stearic acid	—	—	—	—	—	—	—

```
*C-9                    75
  Dibutyl phthalate     60
  Stearic acid           3
```

Color	Black HT	Black HT	Tan HT	Tan HT	Tan	Black SL
Suggested use	Adhesive	Sealant	Adhesive	Sealant	Electrical potting	Deck seal

FIGURE 11.10 Suggested starting formulas for LP polysulfide sealants, two part.

Raw materials	Parts by weight
Part A	
LP 32, polysulfide	100
Phthalate plasticizer	20
Calcium carbonate, precipitated	35
Titaniun dioxide, rutile grade	10
Carbon black, furnace	5
Stearic acid	1
Bentonite clay	3.0
Fumed silica	2.5
Phenolic resin	5
Gamma-aminopropyl-triethoxy silane	0.15
Sulfur	0.1
Toluene	2
Total	183.75
Part B	
Lead peroxide, technical grade	7.5
Stearic acid	0.75
Dibutyl phthalate	6.75
Total	10.0
Lbs/gal	13.5
Nonvolatile, %	99+

For physical properties, see table 11.5

FIGURE 11.11 Polysulfide sealant, two-component, gun grade, gray.

Raw materials	Parts by weight
LP-32 polysulfide	50
Epoxidized soya oil	4
Pyrogenic silica	2
Calcium carbonate	5
Titanium dioxide, rutile grade	22
Hydrated lime	2
Synthetic zeolite	2
Calcium peroxide	4
Phthalate plasticizer	2
Gamma-aminopropyl-triethoxy silane	1
Toluene	4-6
Total	100
Lbs/gal	12
Nonvolatile content, %	96

For physical properties, see table 11.4

FIGURE 11.12 Single component polysulfide-base, gun grade, white.

SUMMARY

The development of polysulfide materials for joint sealants has provided a means of sealing joints between concrete panels and other types of construction such as curtainwalls, floors, and decks in a manner superior to earlier materials. Failures have been extremely rare and joint maintenance has been reduced to a very low point.

Although performance records of these sealants have been highly satisfactory, the manufacturers continue to strive to improve the existing products and to develop new and better sealants. It might be too early to judge their performance, but we can look forward to even better sealants.

REFERENCES

1. Amstock, Joseph S. June 1998. "LP™ Polysulfides," *Glass Magazine*
2. Morton International Polymer Systems. 1997. Chicago, Ill.

CHAPTER 12
POLYURETHANE (URETHANE)

INTRODUCTION

In recent years, the use of polyurethane sealants and caulks in the construction market has shown a dramatic increase. Sealants are important in present-day construction technology, and a wide variety of materials is available, with highly different properties and applications. The unique versatility of urethane technology allows for the formulation of a wide range of products for the construction industry [1]. Urethane polymers are among the most widely used in both new and remedial construction markets. Sealing of joints is an important part of the modern building process. Polyurethane was developed in 1937 in Germany by Dr. Otto Bayer, who discovered the poly addition polymerization reaction of polyisocyanates that leads to the production of high molecular weight products. The rapid popularity it has attained throughout Europe and the United States was stimulated by shortages of several natural rubber materials during World War II.

It usually is necessary to seal joints to avoid penetration of air, water, vapor, odors, noise, insects, and so forth. Sealing against air, water, and vapor is considered most important although the requirements might not differ much. Successful sealing can be achieved by the selection of the right sealant and including sensible details in the design. The second point can be as important as the first one. The formulation data are illustrated later in this chapter.

All urethane prepolymers are manufactured by the reaction of a polyisocyanate with a polyol, in the presence of a catalyst, which results in the formation of stable chemical links or bonds creating a urethane polymer. There can be many variations to the reaction to achieve various degrees of polymer quality and performance success. The polymer then is compounded with various other raw materials to create the desired finished product.

The basic reaction on which polyurethane chemistry is based was discovered more than 140 years ago. In 1849, Wurtz reacted isocyanates with compounds containing hydroxyl groups into esters of carbamic acid, which were named urethanes [2].

$$R-N=C=O \; + \; R'-OH \; - \; R-NH-(C=O)-RH'$$
$$\text{Isocyanate} \qquad \text{Alcohol} \qquad \text{Urethane}$$

SIMPLE URETHANE REACTION—POLYOLS

$$2OCNRNCO + HOR'OH \rightarrow OCNRNH(CO)OR'O(CO)NHRNCO$$

In the above equation, the ratio of the NCO:OH was 2:1 but if it had been less than 1:1, some of the molecules would have been terminated with a hydroxyl group. In most prepolymers, the polyols will consist of both diols and triols—the latter to produce cross-linked in the cured polymer. The ratios of diols and triols (D/T) has a big effect on both the rheology and the cured properties of the prepolymer. The D/T on rheology is important.

Polyurethane-based sealants are available in both single- and multi-component formulations; and a variety of performance properties is available with each. Furthermore, these sealants or caulks can be designed to be self-leveling for horizontal joints or non-sag for vertical applications. Upon application, polyurethane sealants are chemically cured by various mechanisms, including direct moisture cure, moisture-triggered chemical cure and direct chemical cure.

Polyurethane-based sealants have existed in laboratories for a number of years, but have been very slow to be developed and come to market. The types now available and used in the construction industry have generally higher performing properties than polysulfides. Many of the brands on the market are superior in some respects in regard to puncture and abrasion resistance, which makes a urethane a highly desirable sealant for horizontal joints in traffic areas. Another major advantage is that primer treatment of the substrate often can be omitted.

Those developed for adhesive and sealant use are predominately elastomer compounds; that is, products having a high extension and an elastic, or pseudoelastic, characteristic similar to vulcanized natural rubber. They are generally thermoplastic; however, thermosetting materials are also available.

POLYURETHANE CHEMISTRY

We will discuss the two types of cure involved in sealants for the construction market.

1. The one-component, which cures by the absorption of moisture or oxygen from the air.
2. And the two-component, in which the second component contains an active hydrogen that cures the isocyanate containing the first component.

One-Component Sealants [3]

The advantage of a single-component polyurethane is obvious. The sealant or caulk is supplied to the end user in cartridges or sausage-Pak, or the can supplied to the applicator contains the final product. This eliminates the near-certainty that some worker in the field will find a way to omit one or the other components. Another advantage is that the moisture-curing reaction produces urea linkages, which are stronger and, hence, more tear-resistant than the urethane linkages produced in a two-part reaction.

The disadvantages, however, are many. From the point of the manufacturer, manufacturing a single component sealant is far more capital-intensive. The process requires a great deal of development work before it will work. For example, the easier-to-use thixotropes all have active hydrogens that tend to react with the free isocyanate of the prepolymers. Those components that can be used must be dry. Prepolymers and other liquids must be mixed with pigments, fillers, and thixotropes in an air-tight reactor with accurate temperature control, vacuum, and dual mixing ability. Once manufactured and applied, the sealant might develop bubbles because of CO_2 or moisture absorption, which is released by the curing reaction; thus, an unsightly-looking surface might result. The final, and perhaps most difficult problem, is the slower cure. Hence, building movement can cause failure while the sealant or caulk is in a tender, uncured state.

The free isocyanate of the prepolymer reacts with moisture from the air. This extends the chain, curing the sealant. If the functionality exceeds 2, the sealant will be cross-linked. The reaction takes place in two steps. the first forms an amine and releases CO_2. Since amines react very rapidly with isocyanates, the amine intermediate quickly reacts with available NCO, extending the chain. Now, the polyol groups are linked by urea bonds.

$$RNCO + H_2O \rightarrow [RNHCOOH] \rightarrow RNH_2 + CO_2$$

$$RNH_2 + RNCO \rightarrow_{fast} RNH(CO)NHR$$

For the overall reaction:

$$2RNCO + H_2O \rightarrow RNH(CO)NHR + CO_2$$

In many instances, the excess NCO groups will react with the active hydrogen of the ureas to form biuret crosslinks as in the above equation.

THE BASICS OF URETHANE AND POLYURETHANE REACTIONS

Urethane Reaction

Both urethane and polyurethane reactions have two components commonly called the A side and the B side.

The A side contains an isocyanate group, which is made up of nitrogen, carbon, and oxygen atoms. When there is more than one -N-C-O- group, it is referred to as a polyisocyanate.

The B side of a urethane reaction contains an alcohol group composed of an oxygen and hydrogen atom. This is called the hydroxyl group, or -OH unit. When there is more than one hydroxyl group on the B side, it is called a polyol. When an isocyanate and an alcohol react, a simple urethane is formed. When a polyisocyanate and a polyol react, the reaction repeats itself, causing the product to gel and form a solid material: a polyurethane. This reaction is called a polymerization reaction because the molecules of a monomer are linking together to form a polymer. See Figure 12.1 for a simple urethane reaction. Figure 2.2 is a polyurethane reaction.

Polyurethane Reaction

If the polyol in the polyurethane reaction has two reactive sites, it is considered difunctional and called a diol. If the polyol has three reactive sites, it is trifunctional and called a triol. The product of a polyurethane reaction with a diol on the B side will be a linear polymer. The product of a reaction with a triol on the B side will be a branched or cross-linked polymer.

The lower the functionality of a polyol, the more flexible the resulting polyurethane.

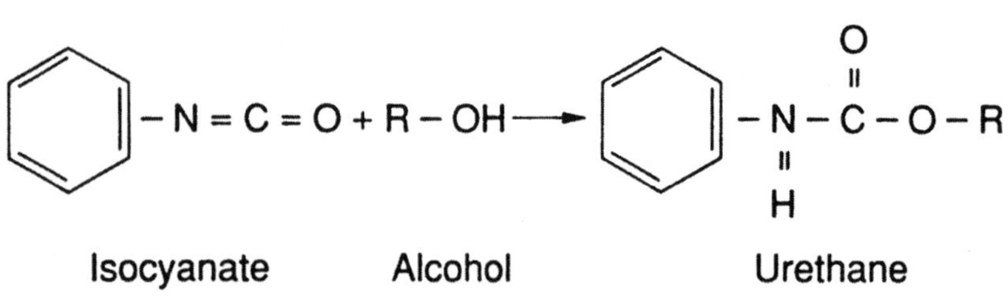

$$\text{Ph}-N=C=O + R-OH \longrightarrow \text{Ph}-\underset{H}{\overset{\|}{N}}-\overset{O}{\overset{\|}{C}}-O-R$$

    Isocyanate   Alcohol   Urethane

FIGURE 12.1 Simple urethane reaction.

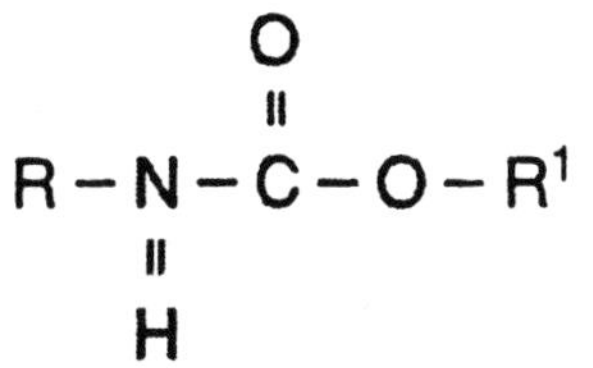

$$R-\underset{\underset{H}{\|}}{N}-\overset{\overset{O}{\|}}{C}-O-R^1$$

FIGURE 12.2 Polyurethane reaction.

FIGURE 12.3 Linear, branched and cross-linked polymers.

Isocyanates

The two types of isocyanates are aliphatic and aromatic. In an aliphatic isocyanate, the -N-C-O- group is linked to a molecule that does not contain a benzene ring. Aliphatic isocyanates are more expensive and less reactive than aromatic isocyanates. They often are used in weatherizing sealants and coatings. Aromatic isocyanates have -N-C-O- groups linked to molecules that contain benzene rings. The two most commonly used aromatic isocyanates are toluene diisocyanate (TDI) and methylene diphenyl diisocyanate (MDI), both of which are manufactured by the Dow Chemical Company. In fact, 95 percent of all polyurethanes are based on these two isocyanates. Figure 12.4 illustrates the production of isocyanate.

TDI. In the manufacture of TDI, benzene is converted to toluene, which then is converted to diamine. The diamine then is reacted with phosgene to yield TDI, a liquid used primarily in the production of flexible foams. TDI is the most frequently used diisocyanate. It is a liquid and requires very careful handling. Figure 12.5 illustrates TDI production.

MDI. Methylene diphenyl diisocyanate is produced when benzene is converted to aniline, which reacts with formaldehyde to form a polyamine. This reacts with

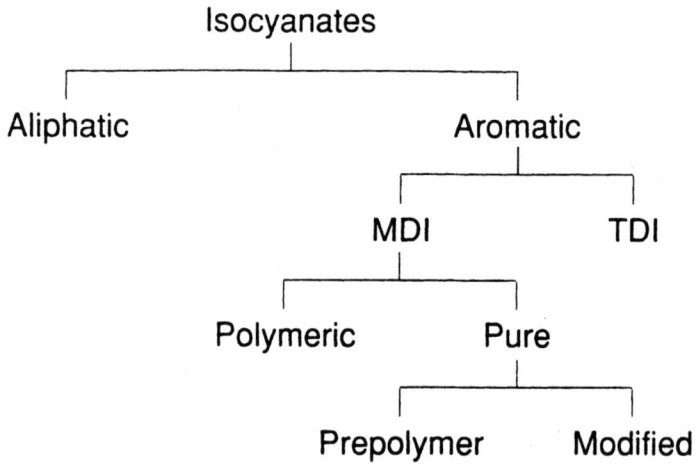

FIGURE 12.4 Isocyanate production.

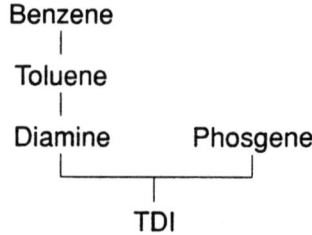

FIGURE 12.5 TDI production.

phosgene to create polymeric MDI or PMDI. If PMDI is further distilled, pure MDI results. If pure MDI is reacted with itself in the presence of a catalyst and heat, modified MDI results.

The functionality of an isocyanate, or the number of reactive sites on a molecule, is expressed as an average and dictates the amount of flexibility in the end product. The lower the functionality, the more flexible the product; the higher the functionality, the more rigid the finished product. Three types of modified MDI and one type of pure MDI are offered commercially. Pure MDI is a solid and is a 4,4 isomer. Pure MDI can react with itself if stored improperly, forming a separate molecule called a dimer. This is a solid and is considered an impurity, which can affect polymer properties during polyurethane production Figure 12.6 shows MDI production.

Prepolymers

To minimize storage and handling problems with pure MDI, prepolymers often are used. Prepolymers are isocyanates prereacted with some or all of the polyol needed in the formulation. The full reaction does not take place, so a complete polymer is not formed. Dimer is less likely to form in prepolymers than in pure or polymeric MDI. When forming prepolymers, not all of the isocyanate is used. Figure 12.7 shows the amount of free -N-C-O- determined by this reaction.

Polyester Polyols

There are two main types of polyols: polyether polyols and polytetramethylene ether glycols.

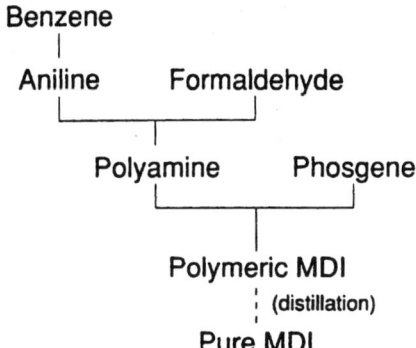

FIGURE 12.6 MDI production.

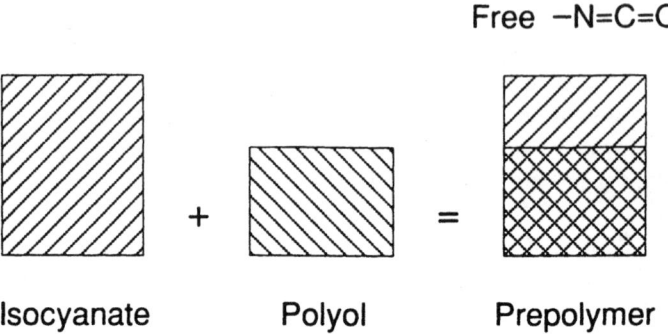

FIGURE 12.7 Prepolymer formation.

Polyether Polyols. Comprising about 90 percent of industry usage, polyether polyols are produced by combining a initiator with one or more oxides. Initiators such as glycerine, water, or amines are molecules that contain active hydrogens and affect the functionality of the polyol. Oxides, generally propylene oxide (PO) or ethylene oxide (EO), affect the reactivity of the polyol, which, in turn, determines the properties of the polyol.

At least one carbon-oxygen linkage, called an ether group, is formed during polymerization. Depending on the type of initiator and the ratio of the initiator to oxides, three types of polyester polyols are available:

1. All PO
2. PO-based, but EO-capped
3. Heteropolyols, which are a random mixture of both EO and PO. See Figure 12.8.

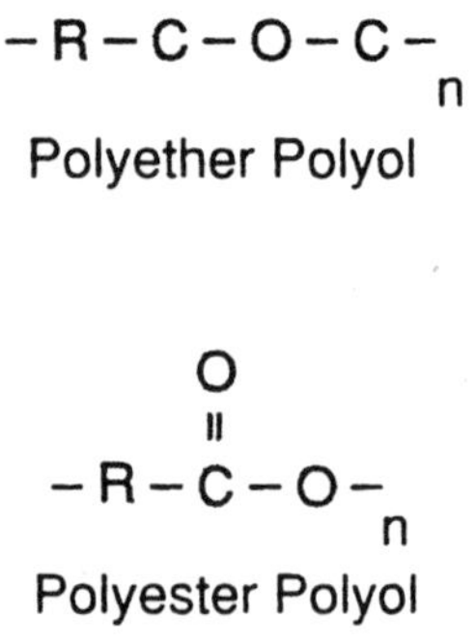

$$-R-C-O-C-_n$$
Polyether Polyol

$$-R-\overset{\overset{O}{\|}}{C}-O-_n$$
Polyester Polyol

FIGURE 12.8 Polyether and polyester polyols.

Polytetramethylene Ether Glycols. PTMEGs are another type of polyether polyol, but are comparatively expensive and used in elastomers. They incorporate at least two COOC groups, called an ester group. They are more expensive, but offer excellent cut, tear, and abrasion resistance. Polycaprolactone polyols are a type of polyester polyol.

Auxiliary Components

Besides the isocyanate and polyol, other chemicals, called additives or auxiliary components, are added to a polyurethane formulation. Their purpose is either to control the speed of the reaction (reactivity) or to modify the properties of the end product.

Blowing agents, such as water, chlorofluorocarbons (CFCs) and methylene chloride, are used to produce foam. They cause bubbles or cells to form during polymerization, expanding the polyurethane into a range of rigid or flexible foams. See Chapter 16 for foamed tapes.

Catalysts control the speed of the reaction. Surfactants, such as silicones, are used to maintain uniform cell size. Fillers are materials added to improve properties, such as product strength, or to decrease flammability or cost. Chapter 15 discusses firestops and flammability.

Pigments add color to the final product and increase sunlight resistance. Plasticizers add flexibility. Defoamers are low molecular weight silicones that eliminate bubbles in the cast elastomer and sealants.

ADVANTAGES AND LIMITATIONS OF URETHANE SEALANTS

As with any other type of construction materials, there are limitations in where and how they are used. Table 12.1 outlines the advantages and limitations.

SUGGESTED STARTING FORMULATIONS FOR CONSTRUCTION SEALANTS

The following series of tables of formulation is from a variety of sources and illustrate what typically is being used in the building industry today. Also the physical data offered is of a basic nature. The companies are noted in Chapter 23. Sources can be contacted for additional or more specific data concerning an end use or application.

TABLE 12.1 Advantages and Limitations of Polyurethane Sealants

Advantages	Limitations
1. Can be used in joints up to 6 inches	1. Light colors can discolor
2. ± 25% movement capability	2. Poor water immersion resistance
3. Non-staining	3. Limited elevated temperature properties
4. Excellent recovery	4. Not recommended for wet joints
5. Excellent UV resistance	5. May require priming
6. Excellent ozone resistance	6. Limited package stability for one-component
7. Fast cure for multi-component	7. One component requires more cure time
8. Adjustable work life for multi-components	8. Multi-component requires mixing
9. Negligible shrinkage	9. Single component not recommended for traffic areas
10. Excellent weather resistance	10. Not recommended for stopless or structural glazing
11. Fair to good chemical resistance	11. Possible reversion
12. Excellent durability (20 to 30 years)	12. Variety of formulations can cause wide differences in performance
13. Can meet ASTM C 920 for all systems	
14. Superior to polysulfides	
15. Paintable	

Polyurethane-Based Insulating Glass Sealant [4]

Table 12.2 gives a basic formula for a Poly bd resin-based, two-component, insulating glass sealant. This suggested formula and the physical and application properties are also listed. In addition, we offer another polyurethane-based sealant, differing from the example in Figure 12.2. Table 12.3 is a two-component IG sealant using different chemistry.

The relative advantages of a polyurethane insulating glass sealant, compared to a polysulfide benchmark, can be divided into three categories: *application*, *performance*, and *cost*.

Over the past few years, polysufides have been reformulated to improve wet application characteristics and the ability to go on a stack of insulating glass lights more easily and smoothly. Once applied, the sealant should not run out or sag, especially at the corners of large air space units. Urethane-based sealants are lower in viscosity and can be formulated without the addition of solvents with better flow properties and sag resistance. Polyurethane sealants are applied easily by gun or trowel with full penetration into the spacer profile, but with good resistance to sagging. Lower viscosity translates into easier application and better wetting of the glass spacer without stringing or pulling out of the sealant. The lack of stringing is especially useful in automated lines, because it minimizes or eliminates the cleanup of units after the robot gunning heads have applied the sealant.

TABLE 12.2 Two-Component, Insulating Glass Formula, Poly bd® Based Sealant

Raw materials	Base	Accelerator
Poly bd® Resin R45HT (OH value = 0.85 MEG/G)	100.00	
Insol 100 (equivalent weight = 104)	3.00	
Super-Pflex 200 (calcium carbonate)	85.00	
Ultra-Pflex (calcium carbonate)	40.00	
Sterling R (carbon black)	2.00	
Santicizer 278	75.00	
Cab-O-Sil N-70-TS	7.5	
Vanox NBC	2.0	
Organosilane ester A-187	2.0	
T-12 catalyst	0.045	

Part B		Accelerator
Isonate 143-L (equivalent weight=143.5)	(NCO/OH-0.95)	15.49
	(NCO/OH-0.925)	15.10

Elastomeric physical properties	NCO/OH ratio 0.95	NCO/OH ratio 0.925
Tensile strength, psi	237	256
Elongation, %	383	473
Tensile set, %	6	9
Tear strength, pli	57	52
Hardness, Shore A	43	39
Modulus, 100%, psi	131	105
300%, psi	211	179
Water vapor transmission, g/m^2/24 hr	8.29	–
Permeance, g/m^2/24 hr	0.29	–

Application properties at NCO/OH -0.95

The mixing ratio (weight 100 parts base/4.74 parts accelerator); the ratio is variable as part or all the Santicizer 278 plasticizer can be put on the isocyanate accelerator side.

Work life, adjustable*, min.	20-45
Tack-free time, adjustable*, hrs.	3-8
Sag, channel at 70°F (21°C), in.(a)	0.15
Peel adhesion, aluminum to glass, pli (b)	12-14
Lap shear, glass to glass, psi (c)	> 130 glass failure

* Can be adjusted by changing the T-12 content

Aluminum channel, $3/4$" wide, $3/4$" deep × 6 in long; $1/16$ in. wall, maximum

(a) Sigma Test Method A.3.B
(b) Sigma Test Method P.7.A
(c) Sigma Test Method P.6.A

TABLE 12.3 Polyurethane Insulating Glass Sealant

Raw materials	Pounds
Part A	
Liquiflex H	160.11
Santicizer 261	153.6
Abitol	4.8
Thorcat 535	0.7
Lorol C-18	0.6
Calofort S	298.3
Eskal 430	367.5
Purmol	9.0
Total	994.61
Part B	
Isonate M 143 L	353.6
Santicizer 261	237.7
X 30943 carbon black masterbatch	300.0
p-toluenesulfonyl isocyanate	1.0
Silane A 187	77.7
Aerosil R 972	30.0
Total	1000.0
Formulation properties	
Tensile strength, psi	237
Elongation, %	383
Tensile set, %	6
Tear strength, pli	57
Hardness, Shore A	43
Modulus, 100%, psi	131
300%, psi	311
Water vapor transmission, $g/m^2/24$ hr	8.29
Permeance, $g/m^2/24$ hr	0.20
Work life, adjustable min*	24-45
Tack-free time, adjustable, hr*	3-8
Sag, inch	0.15
Peel adhesion, al to glass, pli	12-14
Lap shear, glass to glass, psi glass failure	>130

*Sigma test methods

The single most important advantage of a polyurethane sealant is that it does not contain any solvent. The lack of solvent in the polyurethane-based sealant results in several improvements for the insulating glass manufacturer.

United States ASTM E 774 specification is a modification of the durability test of the Sealed Insulating Glass Manufacturers Association (SIGMA). But, instead of SIGMA's single level, it has three levels, two of them more difficult than SIGMA's single requirement.

One-Part Pourable Joint Sealer

The formulation shown in Table 12.4 is recommended as a starting point toward a high modulus joint sealer for traffic areas requiring exceptional toughness, resistance to penetration, water resistance, adhesion, and cure under adverse weather conditions. The ultimate cured properties are such that oil extension and other modifications can be employed without degrading performance to a marginal level. Table 12.5 shows a gun-grade, self-leveling, one-part joint sealant that can be used for horizontal joints in sidewalks or similar type joints in flat pavement and is capable of sealing joints in concrete walls, curtainwall panels, roof copings, etc. The ways of formulating gun-grade versions of polyurethane are virtually endless, depending on the choice of prepolymers and the variety of fillers, curing agents, and so on. The single components are generally moisture-cure type. All the materials are well-mixed and packaged in one-component cartridges, and premixing before application is not necessary. The shelf life of a properly packaged single-component sealant should exceed one year.

The isocyanate-terminated prepolymers are tailored in accordance with the requirements for rate of cure, hardness of cured product, and stability of the pigmented formulation.

OTHER FORMULATIONS

The following formulations will provide the end user with a variety of data to choose from, depending on his application

P-Toluene Sulfonyl Isocyanate (PTSI)

The formula in Table 12.6 uses a low-viscosity, reactive additive useful as a water scavenger of specialty urethane products, including sealants, adhesives, and coatings. The reaction of PTSI with water generates generous carbon dioxide and the corresponding toluenesulfonamide, which is generally inert toward further reaction with alkyl and aryl isocyanates. The sulfonamide is usually soluble in common coating solutions and presents no significant toxicity hazards.

TABLE 12.4 Two-Component Joint Sealant, High Traffic Type

Raw materials	Part A	Part B
1. Poly bd® R45HT	100	
2. Isonol 100	18	
3. CW 85-60	18	
4. Aromatic process oil	85	
5. Irganox 1076	2	
6. Admex 710	0.5	
7. Purecal O	110	
8. Plack Paste	1	
9. Foamkill 8D	0.05	
10. KOP Lime	5	
11. Thixcin R/xylene (3/1 by wt.)	9	
12. Lead octoate, 24%	0.10-0.15	
Part B		
13. Isonate 143L		39
14. DUP		13

Physical and application properties	
Mixing ratio-parts A/B	100/15 parts by weight
Work life at 70° F (21.11° C), hrs	1-1½
Tack-free time, hrs	12-24
Full cure	2 days
Slump, %	Up to 5% grade
Tensile strength, psi	500
Elongation, %	400
Hardness, Shore A	50
Tear strength, pli	100

PTSI [5] provides the formulator of specialty urethane products with an expedient and efficient alternative to physical methods of dehydration in common use. It is further recommended for the storage stabilization of purified diisocyanates against deterioration and discoloration.

The chemical formula for p, Toluenesulfonyl Isocyanate is as follows:

$$CH_3-C\underset{\underset{CH=CH}{\diagdown\;\;\diagup}}{\overset{\overset{CH-CH}{\diagup\;\;\diagdown}}{}}C-SO_2NCO$$

TABLE 12.5 Single Component, Gun Grade, Polyurethane Sealant

Raw materials	Formula 1	Formula 2
NCO prepolymer (a) equivalent weight 1272, slow curing	400	
NCO prepolymer (a) equivalent weight 1355, fast curing		450
Carbon black, p33 (b)	300	30
Titanium dioxide, R-900 (c)		20
Calcium carbonate, Atomite (d)		280
Silica (e)	20	50
Toluene	10	10

(a) Isocyanate-terminated prepolymer, Spencer Kellog Div., of Textron, Inc.
(b) R.T. Vanderbilt Co.
(c) E.I. DuPont de Nemours & Co.
(d) Thompson, Weiman & Co.
(e) "Santocel" FRC; trademark Monsanto Chemical Co.

Physical and application properties

Properties	Gun grade
Tack-free time @ 66° F (21° C), 50% RH, hrs.	12-24
Cure time ($1/4$ in. × $1/4$ in. channel), days	5-10
Durometer, Shore A hardness	20-30
Elongation, %	500-600
Tensile strength, psi	300-400
Modulus @ 100%, psi	40-60
Continuous service temperature range, °F	-58 to +275
Continuous service temperature range, °C	-50 to +135

Resistance to

Dilute acid	Good
Water	Good
Dilute alkali	Very good
Solvent	Excellent
Fire	Excellent
Electrical insulation	Good
Aging properties	Excellent

TABLE 12.6 Moisture Cure Urethane Zinc Primer

Raw materials	Pounds
Novacite L-207A	114.3
Zinc Dust #64	1370.1
Vertal 5	114.3
Acronal 700L (10% in A-100)	2.2
MPA-60X	5.6
DC-11 (1% in A-100)	4.1
Ircogel 905	9.9
A-100/PMA 1:1	265.6
p-toluenesulfonyl isocyanate	61.2
Desmodur E-type resin	204.0
Total	2151.3

Formulation properties	
Solids by weight %	87.0
Solids by volume %	63.4
Weight per gallon, lbs	21.5
Suggested film thickness, mils	3–5
Recoat time, min., 77° F (25° C) 50% RH, hours	4
PVC, %	53.6
P/B ratio	6.0
VOC, lbs/gal	2.8
Viscosity @77° F (25° C) Kreb units	73 ± 2

TABLE 12.7 Moisture Cure Urethane Coal Tar Formulation

Raw materials	Pounds
Desmodur E-type resin	278.4
KC-261, coal tar	199.7
p-toluenesulfonyl isocyanate	28.5
Bentone 34 Gel	135.9
Cyprufil 325	270.7
Xylene	93.7
DC-11 (1% in A-100)	3.5
Additive OF	9.3
Total	1019.7

Formulation properties	
Solids by weight, %	68.2
Solids by volume, %	55.0
VOC, lbs/gal	3.2
Weight per gallon, lbs	10.2
Viscosity @77° F (25° C) Kreb units	70.0

(Courtesy of VanDeMark Group, Lockport, N.Y.)

TABLE 12.8 Single Component, Polyurethane Based Sealant

Raw materials	Pounds
Plastisol Tan	1820.00
P-Toluene Sulfonyl Isocyanate	22.00
Caloxol DOP/V	53.00
Prepolymer Rus 21	1043.00
Irgastab 15MS	3.88
Silane A 187/Z6040	5.88
Aromatic 100	43.00
Total	2,999.76

Typical physical and application properties

Properties	Values	Test methods
Work-tool life	70 minutes	
Tack-free time	6 hours	TT-S-00230 C/ASTM C 679
Curing time @ 77°F (25°C)	3-6 days	Varies with RH
Flow or sag	0.1 inch	TT-S-00230 C/ASTM C 639
Staining	None	TT-S-00230 C/ASTM C 510
Hardness, Shore A	38	ASTM D 2240
Modulus @ 100% elongation	70 psi	ASTM D 412
Elongation	950%	ASTM D 412
Adhesion in peel	> 35 piw	TT-S-00230 C/ ASTM C 794
Stain and color change	None	TT-S-00230 C/ASTM C 510
Ozone resistance	Excellent	
Joint movement capability	± 25%	TT-S- 00230 C/ASTM C 719
UV resistance	Good	ASTM C 793

Poly bd-Based Resin Sealants

Poly bd R-45HT and R-45M resins are liquid, hydroxyl-terminated polymers of butadiene with number average molecular weights of approximately 2700. The Poly bd resin is a lower-viscosity, lower molecular weight version of the R-45HT. Poly bd resins' terminal hydroxyl groups are predominantly primary and of the allylic type. The formula below this structural combination results in high reactivity in both condensation polymerization and the preparation of derivatives. See Table 12.9.

Table 12.10 illustrates an industrial sprayed applied membrane; Table 12.11 shows a starting formula for a high-strength metal adhesive; and Table 12.12 illustrates an industrial membrane that can be applied with a trowel.

TABLE 12.9 Poly bd® Resins R45HT, R45M and R20LM

$$HO-\left[\left(CH_2\underset{CH=CH}{\diagup}CH_2\right)_{.2}\left(CH_2-CH\underset{|}{\overset{|}{\underset{CH=CH_2}{}}}\right)_{.2}\left(CH_2\underset{CH=CH}{\diagup}CH_2\right)_{.6}\right]_{n*}-OH$$

$n* \approx 50$ for R45M and R45HT
$n* \approx 25$ for R25JM

Predominant microstructure:
Cis	1,4 unsaturation	20%
Trans	1,4 unsaturation	60%
Vinyl	1,4 unsaturation	20%

TABLE 12.10 Industrial Membrane, Spray Applied

Raw materials	Pounds
Part A	
Poly bd® R451HT resin	100.00
Voranol 220-530	18.00
Regal 600 carbon black, dry	0.50
Thixin/xylene, 3/1	10.00
Cyanox 2246 (antioxidant)	1.00
Byk 070 (defoamer)	0.10
Part B	
Isonate 2143L	38.60
Total	168.20

Formulation properties		
Viscosity, cp @ 73.3°F (23°C) Brookfield RVT #7		Mpa.sec.
	.5 rpm	3,320,000
	1 rpm	1,334,000
	2.5 rpm	700,000
	5 rpm	438,000
	10 rpm	295,000
	20 rpm	198,000
Thixotropic index	.5/5 rpm	7.6
	1/10 rpm	4.5
Hardness, Shore D		26
Hardness, Shore A		83
Tensile strength, psi		840
	(Mpa)	(5.8)
Elongation, %		550
Tear strength, pli		173
	(N/mm)	(30.5)
Modulus, 100%, psi		410

TABLE 12.11 Polyurethane Metal Adhesive

Raw materials	Pounds
Part A	
Poly bd® R45HT resin	100.0
Voranol 220-530	17.15
Dixie clay	100.00
Silquest A 187™	1.086
Foamkill 8D	0.1
Part B	
Isonate2l43L	37.2
NCO/OH ratio	1.05
Equivalents Voranol 220-530:Poly bd® R45HT	2:1
Formulation properties	
Tensile strength, psi	1460
Elongation, %	400
Tear strength, pli	190
Adhesion, psi lap shear initial	1330
H$_2$O boil, 2 hr soak	1400
Heat aged, 2 weeks @ 194°F (90°C)	1470
H$_2$O immersion 28 days @ 77°F (25°C)	1300
Lap shear on oily galvanized steel initial	1520
Lap shear on aluminum, initial	1440

TABLE 12.12 Trowelable, Industrial Membrane

Raw materials	Pounds	
Part A		
Poly bd® R45HT resin	100.0	
2-ethyl 1-3-hexanediol	20.0	
AC 5 asphalt	100.0	
Atomite CaCO$_3$, dry	100.0	
Foamkill 8D	1.0	
Cyanox 2246	1.0	
Aerosil 202, dry	5.0	
Part B		
Isonate 2l43L	54.6	
Formulation properties		
Viscosity @ 73.4° F (23° C) Brookfield RVT #7 spindle (Mpa.sec)		
2.5 rpm		156,800
5 rpm		118,000
10 rpm		88,000
20 rpm		67,000
30 rpm		49,000
Thixotropic index, 5/50		2.4
Hardness, Shore A		87
Hardness, Shore D		28
Tensile strength, psi		600
(Mpa)		4.1
Elongation, %		260
Tear strength, pli		107
(N/mm)		18.8

SUMMARY

Polyurethane chemistry undoubtedly has provided the adhesives and sealants industry with one of the most versatile raw materials classes, making it possible to meet the stringent requirements of almost any application in the construction industry. It will always be necessary to find ways to make different materials adhere in any building structure where there is a variety of substrates.

In the past, solvent systems have been popular and, therefore, present today a mature segment of polyurethane adhesive and sealant technology. The future challenge will be to meet environmental regulations and requirements with *non-solvent* systems, which include reaction systems, aqueous dispersions, and, possibly, radiation-curable polyurethanes and hot melts.

REFERENCES

1. Dormish, J. "Polyurethane Adhesives," *Technology and Development Trends*. Bayer Corp. Akron OH 44236. Distributed with permission of ASC.
2. *Sealants: The Professionals' Guide*. 1995. Sealant, Waterproofing & Restoration Institute.
3. Evans, R.M. 1993. *Polyurethane Sealants, Technology and Applications*. Lancaster, PA 17604. Technomic Publishing Co.
4. Technical literature. Elf-Atochem North America, Inc. Philadelphia, PA 19103.
5. Technical literature. Vanchem, Inc. Lockport, NY 14094.
6. Technical literature. OSI Specialties, Inc. Endicott, NY 13761.

CHAPTER 13
SILICONES

INTRODUCTION

With the debut of neutral cross-linking silicones, many applications for insulating glass and construction applications once considered particularly difficult, have become quite feasible. Before understanding the use of silicone sealants, it is important to remember what has been learned about other sealant backbones and how they are used in various construction joint applications [1].

Most silicone sealants belong to a family of products called RTVs (room temperature vulcanizing), versions of which have been in use since the early 1960s. When the two components of the rubber are mixed, the resulting mixture vulcanizes (cures) without heat. The use of silicone sealants in fabricating sealed, insulating. glass window units began in the mid-1970s. These are single-component systems, requiring no mixing, and they vulcanize by means of moisture from air. The adhesives and sealants used in construction are of this type.

Cured silicone rubbers have a non-stick surface, a property that makes them especially effective in another type of application, as mold linings of polyurethane, epoxy, and other plastics capable of strong adhesion to other surfaces.

Silicones are silicon-containing polymer materials that have found wide use in industry because of their great stability. They are available as fluids, sealant-adhesives, moldable resins, and rubbers. We will discuss only the sealant-adhesives in this chapter.

When the first silicone oil was made in the 1870s, its insensitivity to both high and low temperatures was noted, but the first silicone rubbers were not invented until 1943. In the 1950s, silicones were developed commercially for the aerospace and electronics industries, but rapidly found applications in many fields, especially in construction.

The desirable properties of silicones are derived from their combination of inorganic and organic chemical groups in a polymer chain. The chain consists of a pair of silicon and oxygen atoms, with two organic groups. Usually both are methyl (CHO), but sometimes other groups are attached to each silicon atom. Without organic side groups, the material would be as refractory and brittle as silicon oxide, such as quartz. The organic groups impart flexibility. The silicon-oxygen chain is unaffected by ultraviolet radiation, oxygen, or high temperature—agents that degrade many standard polymers that have a straight carbon chain. In addition, the chain makes the silicones non-burning. Silicones are clear and colorless but can be pigmented with any color. Silicon rubbers are useful through an unusually wide temperature range and are inert to agents that degrade other rubbers and sealants, although their qualities of elongation, tensile strength, and tear resistance are not as high as those of standard rubbers.

Silicone sealants, followed by urethane, are now preferred for use in high-rise structures because of their non-staining performance on masonry. The use of silicones as structural sealants for glass is a good contribution to this industry. Silicones now have greater than a 25 percent market share in the insulating-glass sector with high growth potential because of their superior performance. They play an extremely large role in highway and bridge joints because of their extensive movement capability.

HISTORY AND CHEMISTRY

The earliest work on silicones probably dates to the synthesis of silicone tetrachloride by Swedish chemist Johan Berzelius in the early 1800s. Boot provides some early history on silicone chemistry in various parts of the world. In the early 1930s, the General Electric Company and the Corning Glass Works began to develop high temperature electrical insulations. Corning Glass and the Dow Chemical Company formed the Dow-Corning Corporation to carry out the work and used the Grinard process for their early production of silicones. GE looked for less expensive methods and developed a process by which silicone metal is reacted with an alkyl halide, such as methyl chloride, to yield chlorosilane intermediates. These products can then be reacted with water to form basic polymers, which can be convert to reactive polymers. The typical reactions are shown in the equation on page 13.3.

Intermediate A is dimethyl polysiloxane and either can be formulated or reacted to produce polymers that will cure at room temperature. Reactant A is tri-acetoxy methyl silane and introduces very active acetoxy terminal groups, which, in the presence of moisture, co-react to yield cured polymer and acetic acid. This is representative of the earlier silicone sealants, which give off acetic acid upon curing. Acetic acid has a vinegar-like odor.

These sealants had movement a capability of only ± 25 percent. Other reactants can be substituted for Reactant A to give terminals such as amine, amide, phenol,

$$Cl-\underset{\underset{CH_3}{|}}{\overset{\overset{CH_3}{|}}{Si}}-Cl + H_2O \rightarrow HO-\underset{\underset{CH_3}{|}}{\overset{\overset{CH_3}{|}}{Si}}-\left[O-\underset{\underset{CH_3}{|}}{\overset{\overset{CH_3}{|}}{Si}}-\right]_n OH$$

Intermediate A

$$CH_3-Si(O\overset{\overset{O}{||}}{C}CH_3)_3 + \text{Intermediate A} \rightarrow$$

$$\left[HO-\underset{\underset{CH_3}{|}}{\overset{\overset{CH_3}{|}}{Si}}-\right]\left[O-\underset{\underset{CH_3}{|}}{\overset{\overset{CH_3}{|}}{Si}}-\right]_n\left[O-\underset{}{\overset{\overset{CH_3}{|}}{Si}}-(O\overset{\overset{O}{||}}{C}CVH_3)_2\right] + \text{acetic acid}$$

Intermediate B

alkoxy, oximes, and alcohols, producing a wide range of polymers that have various activities and properties. Upon reaction, the by-products include amines, amides, alcohols, phenols, and oximes, some of which are reactive with metals and masonry. It is necessary to use a neutral cure when sealing concrete joints.

Intermediate B is the base polymer that reacts with moisture in the air to yield the cured polymer and acetic acid. Other base polymers can be made by combining suitable reactants with Intermediate A.

If polyurethane represents a second generation in quality, then in some aspects of performance, the silicones represent a *third* generation. The general class of silicone sealants and adhesives includes a variety of performance characteristics, although some properties are common to all true silicone sealants. A true silicone sealant contains only silicone polymers, generally is filled with minerals or other inorganic fillers, has special additives for special purposes, and is a functional silane or siloxane cross-linker.

There are now five major manufacturers of silicone sealants: General Electric, Dow Corning, Wacker, Rhone Poulenc, Shin-Etsu Silicones, and Tremco. These firms have stabilized the industry because their finished sealants give high-quality performance. Dow and Union Carbide manufacture the silanes, and Mobay and Perennator-North America sell reactive silicone intermediates to licensed manufacturers.

TYPES OF SILICONE SEALANTS AND APPLICATIONS

The complex chemistry of silicones has resulted in the formation of various groups of sealants having widely different movement capabilities, ranging from ± 25 percent, ± 50 percent to +100 percent to −50 percent. The silicone sealants have advantages and limitation as do other generic sealants as shown in Table 13.1.

When introduced, high-modulus sealants had acetoxy terminals. Although they are still being used to some extent, they have been gradually replaced with medium- and low-modulus compounds that do not corrode surfaces, have greater movement, and achieve adhesion to many more surfaces, both with and without primers. Both high- and medium-modulus sealants can be used as structural sealants, and the high-modulus types are used for insulating glass in structural glazing applications. The low-modulus sealants are used in problem areas, where joints are too narrow for conventional sealants, and in highway and bridge joint applications. Figure 13.1 shows the general properties for four sealants of different hardness, based on several polymers. The most popular silicone sealants have alcohol or oxime terminals with ± 50 percent movement. Table 13.2 is a table of typical cured properties of RTV silicone adhesives/sealants.

TABLE 13.1 Advantages and Limitations of Silicone Sealants

Advantages	Limitations
1. One- or two-component sealant	1. Slightly more expensive
2. Available in most colors	2. Limited colors in single components due to batch size
3. High temperature resistance	3. Dirt pickup
4. Low temperature application qualities	4. Poor tear resistance
5. Excellent UV resistance	5. Some odor problems with acetoxy-type sealant
6. High recovery	6. Short tooling time
7. No shrinkage	7. Short skin time
8. Color stability is good	8. Primer required with some sealants
9. Low heat-aged weight loss	9. Critical surface preparation; do not apply to wet or damp surfaces or during inclement weather
10. No hardness increase with time and temperature	10. Aluminum surfaces a problem with some sealants
11. Improved tear resistance with medium and high modulus	11. Not recommended for prolonged water immersion
12. Various moduli available	
13. Movement capability excellent	
14. Excellent fire resistance	
15. Long durability of over 20 years	

TABLE 13.2 Typical Cured Properties of RTV Silicone Adhesive/Sealants

Category	Hardness Shore A	Tensile strength psi	Elongation %	Tear strength lb/in.	Peel strength lb/in.	Max. continuous °F	Dielectric strength V/mil	Dielectric constant @ 60 Hz	Dissipation factor @ 60 Hz
Acetoxy cure									
General-purpose paste	27–30	320–400	400–450	up to 45	up to 40	400	500	2.8	0.001
General-purpose liquid	25	325	325	—	15	400	400	2.8	0.001
High Strn	25	750	800	100	100	400–500	500	2.8	0.001
High temp paste	30	375	400	40	40	500	500	2.8	0.001
High temp liquid	20	350	350	—	25	500	400	2.8	0.001
Alkoxy cure									
General-purpose paste	35	475	400	—	50	400	450	2.8	0.001
General-purpose liquid	25	275	275	—	10	400	500	2.8	0.001
High Strn	37	800	650	—	80	400	600	2.8	0.001
Flame-retardant	45	650	250	50	—	400	500	2.8	0.001
Neutral cure									
General-purpose paste	23	220	400	35	40	400	500	2.8	0.001

Raw materials	Parts by weight			
	1 IG for structural use	2 High modulus	3 Med. modulus	4 Low modulus
Silicone polymers, %	80-85	80-85	80-60	45-55
Fumed silica	10-15	6-10	2-5	—
Plasticizer	—	—	5-20	—
Calcium carbonate	—	—	20-30	45-50
Silanes	5-7	5-7	5-7	5-7
Miscellaneous	3-5	3-5	3-5	3-5
Physical properties	Values			
Hardness, Shore A	45-55	35-45	25-35	10-15
Tensile strength, psi	140	120	100	30
Elongation, %	85	140	250	1000+
25% modulus, psi	60	40	30	10
50% modulus	90	60	40	11
75% modulus	130	80	60	12
100% modulus	—	100	70	13
Tear strength, pli	50-70	50-70	20-40	20-25
Movement, ±	25	25	50	+100 to -50

FIGURE 13.1 Sealant formulations for various modulus silicones.

THE AGE OF CHEMICAL FASTENERS [2]

Until chemical fasteners came upon the scene, all products were assembled mechanically. Screws, bolts, nuts, rivets, welding, and press fits were the principle components and methods used. Chemical fastening, however, came into its own as adhesive development produced stronger, more reliable compounds.

There are a number of other reasons why adhesives have gained in importance, particularly in conjunction with the increased use of plastics in industrial, consumer, and construction products. This use might lead to premature part failure in plastics, which are weaker than metals. Adhesives often yield better results because they are applied over a wider area, which reduces stress levels and spreads contact loads. Adhesives also look better than mechanical fasteners. The exterior surface of an adhesively bonded assembly is smooth, uninterrupted, and cosmetically attractive, because deformations that could be caused by point fastening are eliminated.

Another reason for the growth of chemical fastening is that adhesives usually perform two or more functions. For instance, they can seal or insulate as well as adhere. Electrical insulation between two disparate metal substrates can prevent galvanic corrosion. In addition, elastomeric adhesives can compensate for differentials in thermal expansion/contraction between surfaces. Seven specific types of adhesives and sealants are used as chemical fasteners: epoxies, cyanoacrylates, anaerobics, polyurethanes, hot melts, acrylics and silicones. The aforementioned materials are covered in different chapters in this book. Each requires a different method of application or curing to produce peak performance

Structural Glazing Applications [3]

This is one of the fastest-growing and most innovative forms of curtainwall construction the commercial façade market has seen in many years. Structural glazing has changed the face of architectural design by giving architects and designers the freedom to create dramatic structures. The effective and economical technology allows architects to make glass, ceramic, metal, stone, and/or composite panels adhere to a building frame by using silicone adhesives/sealants. Silicone structural glazing technology has been gaining rapid popularity throughout the world as increasing numbers of architects realize the wide range of creative possibilities that this technique offers. Aesthetically appealing designs are important to the building owner, but it is also critical that the structure be structurally sound.

When used in structural silicone glazing, the sealant serves a dual function. It seals the building from the elements and it supports the panel itself, transmitting wind loads to the frame structure. This system withstands flexure, tension, compression, thermal shear stresses, and continuous movement when properly designed, specified, and fabricated. It also withstands weathering, extreme temperature changes, and chemical corrosion. The sealant forms watertight bonds, adheres through weather extremes, withstands designer-specified joint movements, and remains flexible.

In addition to giving greater design freedom and performance, this technique offers economic advantages. Structural silicone glazing technology promotes optimum sound and heat insulation. Moreover, the elimination of thermal breaks and unnecessary metal can further reduce building costs. Properly installed, four-sided structural glazing systems provide leak-free performance because silicone sealant, used as an adhesive by design, keeps out air and water, and, with certain design systems, can eliminate weep-hole systems.

Although initially more expensive to buy than other sealants, silicone sealants cost much less in the long run because they demonstrate longer service life, which translates into reduced energy losses, reduced maintenance, and the prevention of building damage. The suppliers of silicone structural sealants strongly recommend complete cooperation with the architects, designers, consultants, structural engineers, curtainwall fabricators and erectors, and glass and component suppliers throughout all phases of planning and construction.

The materials used in structural glazing must be matched for adhesive properties, compatibility, and proper dimensions. Pre-job testing is a must. Written approval from the sealant manufacturer must be required for each structural application. This approval process is structured to help ensure that the steps leading to a successful project are taken.

Two basic forms of structural glazing are possible: two-sided or four-sided. Both types can be fabricated in the shop or directly on the job site (field-glazed). In general, factory-glazed elements can be fabricated under more easily controlled conditions, eliminating exposure to airborne dust and debris, and thereby assuring maximum cleanliness, sealant cure, and development of adhesive properties. These elements then are moved to the job site and mechanically attached to the curtainwall structure.

Two-sided structural glazing provides mechanical support for the window head and sill, as shown in Figure 13.2. Only the vertical joints are structurally adhered to the supporting structure. The dead load of the glass weight is supported mechanically. The live load is carried on two sides by structural silicone sealant and at the head and sill by mechanical fastening.

Four-sided structural glazing provides no mechanical support for the façade fronting materials as shown in Figure 13.3. Silicone sealant is used for glazing on all four sides.

Dead loads are supported either mechanically by a horizontal fin or by the sealant, depending upon the design. Four-sided systems are less likely to leak because the adhesive-sealant blocks water and air from entering the building interior. This design can incorporate a split mullion that acts as an expansion joint, thereby simplifying the building design. Structural glazing also can incorporate insulating-glass units as illustrated in Figure 13.4. A horizontal fin is used to support the weight (dead load) of the insulating-glass unit. These IG units also must be sealed with a silicone-edge seal. See Figure 13.4 for details. Each project carries its own set of specifications.

Highway, Bridge and Airfield Sealants [4]

These sealants are one-part silicone formulations that can be installed over a wide temperature range. They cure on exposure to atmospheric moisture to form a flexible, low-modulus, high-elongation silicone rubber seal. The joint sealant material is designed for use in highway joints that experience a high degree of movement, such as transverse pavement expansion, and contraction joints. Because of its modulus characteristics and good extension/compression recovery (+100 percent to −50 percent of original joint width), the silicone highway joint sealant performance outstandingly in highway joints in which extreme movement occurs. It is also ideal for use on concrete airport aprons, runways, parking ramps, and bridges.

Highway concrete contraction/expansion joints generally are sealed to prevent concrete "slab growth" and/or "blow-up," erosion of the sub-base, or corrosion of metal tie bars embedded in the concrete, which occurs when

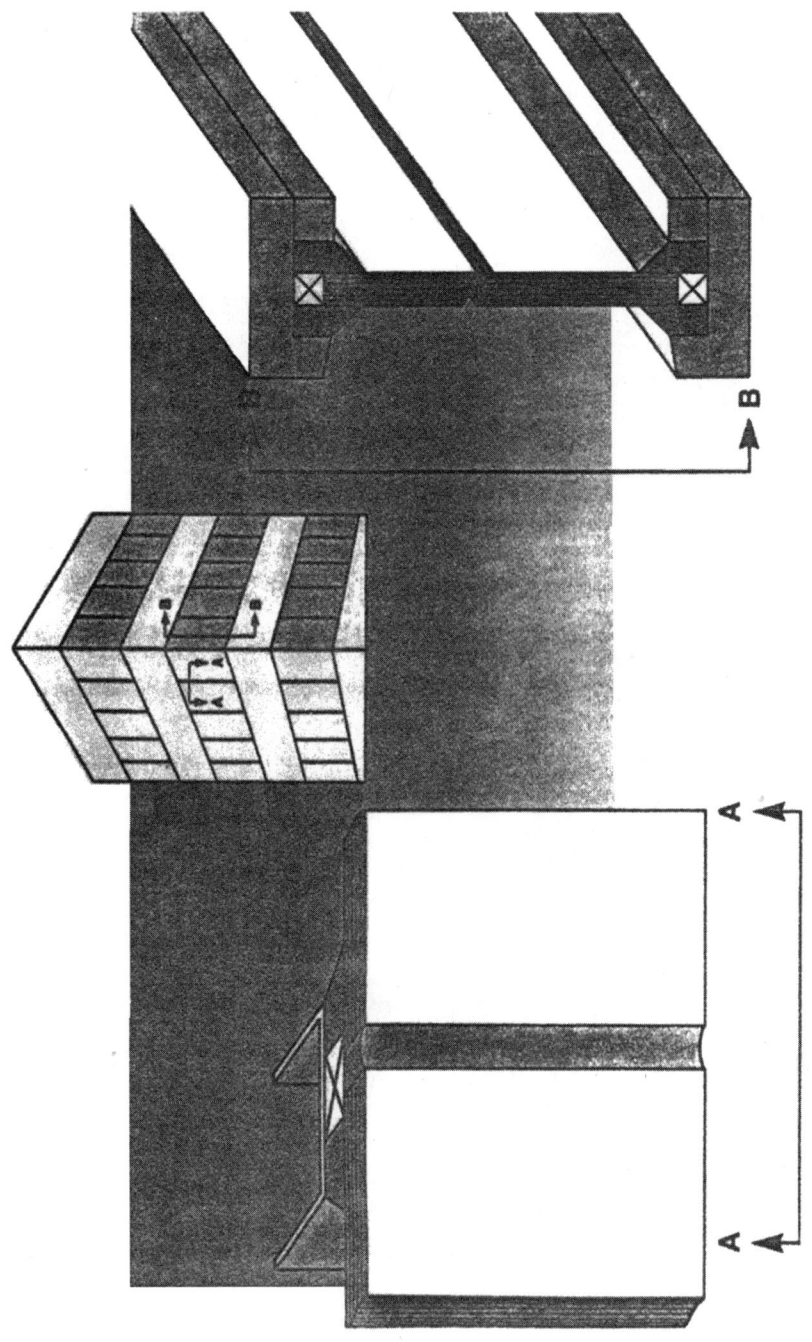

FIGURE 13.2 Two-sided structural silicone glazing.

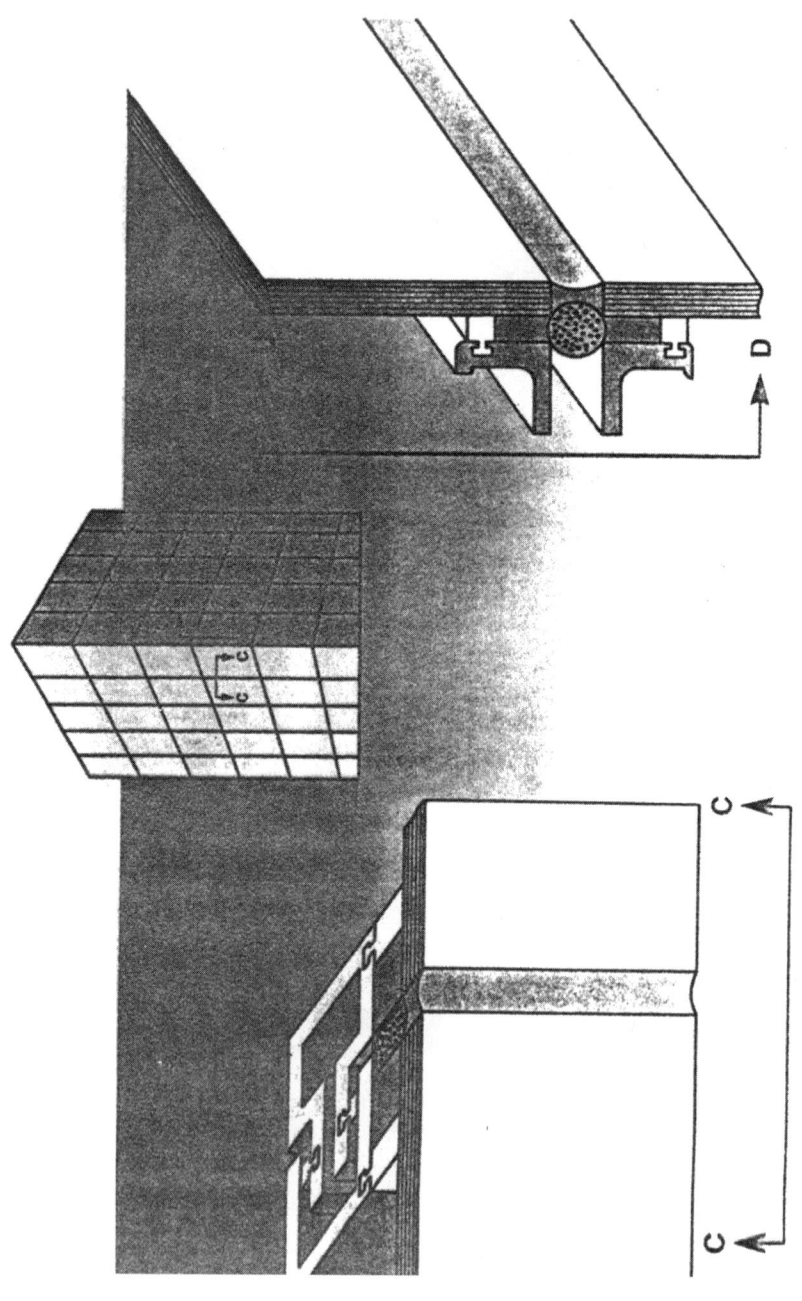

FIGURE 13.3 Four-sided structural silicone glazing.

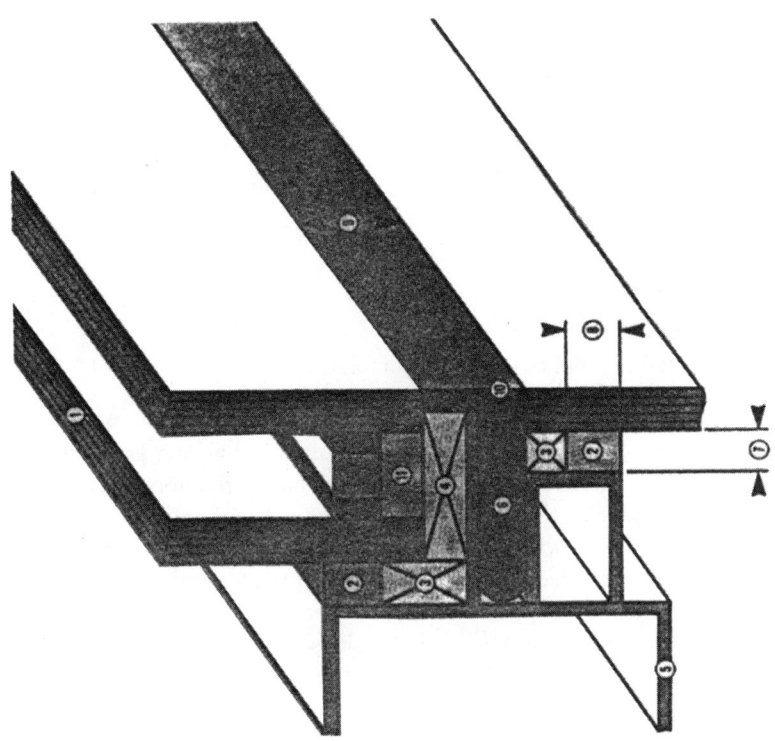

1 Insulating glass unit
2 Structural silicone glazing sealant
3 Silicone rubber spacer
4 Silicone rubber setting blocks
5 Aluminum support
6 Backer rod
7 Width of SSG joint
8 Depth of SSG joint
9 Width of weatherseal
10 Silicone weatherseal
11 Silicone insulating glass sealant

FIGURE 13.4 Design system for insulating glass units.

incompressibles such as dirt, stones, water, and deicing chemicals enter the joints at the joint surface. Some of the characteristics are as follows:

- Ease of application and all-temperature gunnability
- Unprimed adhesion
- Seals irregular surfaces
- Low modulus
- Fully elastic and resilient
- Good weatherability
- High movement capability
- Fast cure and long life reliability
- Conforms to federal and commercial standards

Generally, a primer is required and makes it possible to apply a silicone sealant that will be subjected to excessive moisture. The primer prevents premature sealant replacement in such applications as canals, irrigation channels, dams, spillways, and roadways that experience flooding and/or have little or no drainage. Figure 13.5 shows good joint designs where the joint is a new cut; a shallow cut is recommended where the backer rod is placed on the shelf or bottom of the joint. This design provides a firm support for the sealant tooling, making the sealant easier to install and provides more insurance for a good sealant-concrete contact.

In repair work where previous sealing materials have been of a joint filling sealing type or the joint is not broadened by sawing, the standard joint design is suggested where the backer rod is slightly above the shelf. Enough space between the backer rod and the shelf should be provided for possible pumping of the old

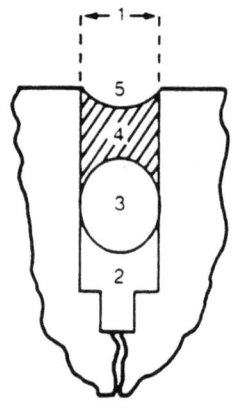

STANDARD JOINT

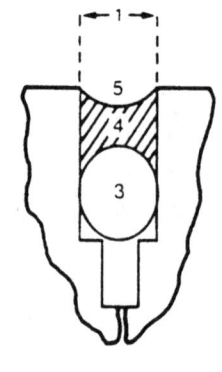

SHALLOW CUT JOINT FOR NEW CONSTRUCTION

FIGURE 13.5 Good joint design.

joint sealant—the filling from the bottom of the joint. It is recommended that care be given to selection of the proper oversized backer, so that a firm tooling support is obtained (generally about ¼ inch larger than the joint works quite well). The silicone highway joint sealant is part of a system that must include the proper backer rod and proper installation procedures. The backer rod must be expanded closed-cell polyethylene foam. Several backup materials (paper rope and open-cell foam) are available but have proven unacceptable. There are several manufacturers of backer rod materials. Another application worth noting is in a butt-glazing design as shown in Figure 13.6. Another illustration, Figure 13.7, for a similar application, is structural glazing of monolithic glass in a continuous span, also known as ribbon glazing It is important that a clear silicone sealant be used in glazing the lights of glass.

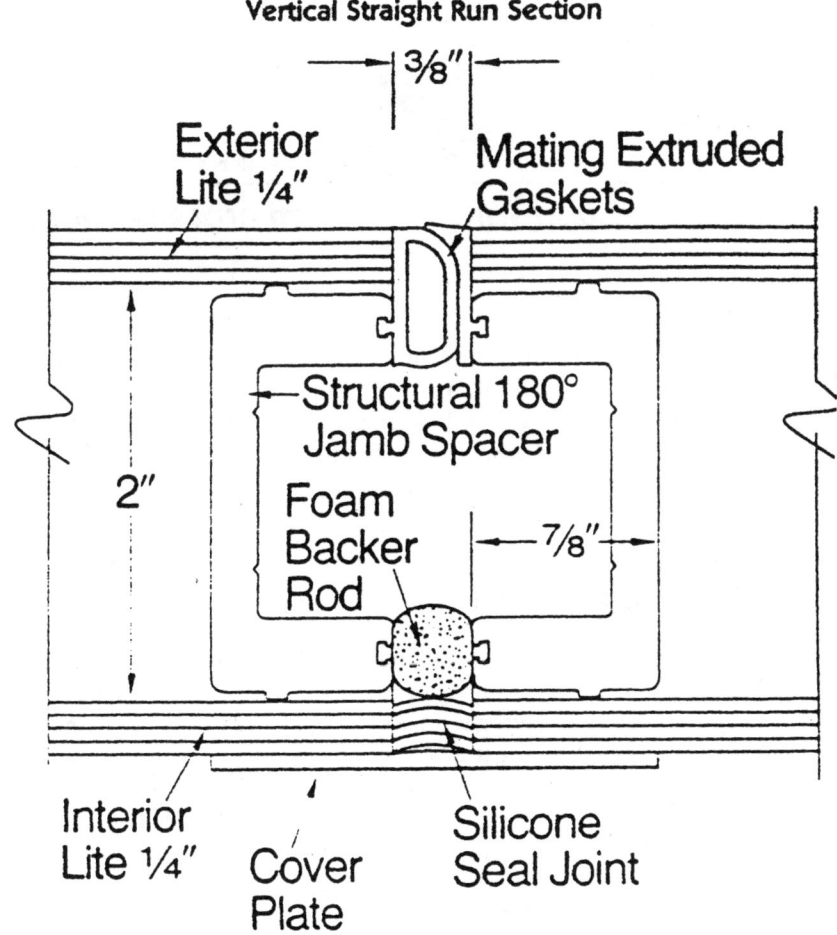

FIGURE 13.6 Butt glazed.

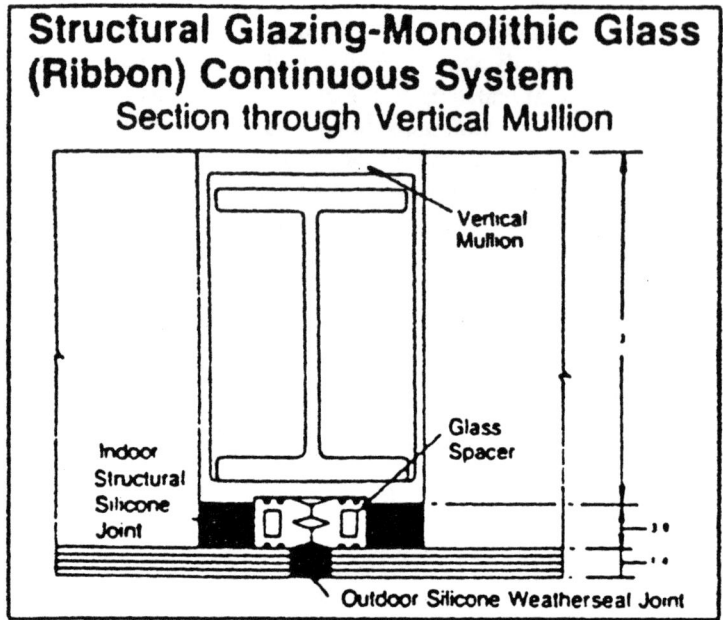

shows a ribbon window system cross-section.

FIGURE 13.7 Ribbon window system.

Figure 13.8 is a typical extrusion pump used to install the sealant. After all contaminants, such as dirt and oil, are removed from the joints, they can be sand-blasted or mechanically cleaned. All loose dust and other debris must be blown from the joint and surfaces must be frost-free. Install the backer as previously stated. See Table 13.3 for recommended backer rod installation (shallow cut).

Preformed Silicone Profiles [5]

Joint sealant failures are an all-too-common occurrence in the envelope of a building. The reasons for such failures are numerous and varied, and any one of them might lead to a complete failure of the envelope's waterproofing properties. Repairing these failures can be extremely costly and time-consuming, particularly in exterior insulation finish systems (EIFS) where affected areas of the system have been cut out, restructured, and resealed. Sil-Span extruded profiles offer an innovative and reliable approach to solving these and other problems.

Sil-Span is a specially formulated, high-molecular weight, low-modulus silicone compound extruded through a process that is new to the industry. It is designed to repair joints without removing the failed sealant while maintaining the original aesthetics of the substrate. To accomplish this, custom colors and

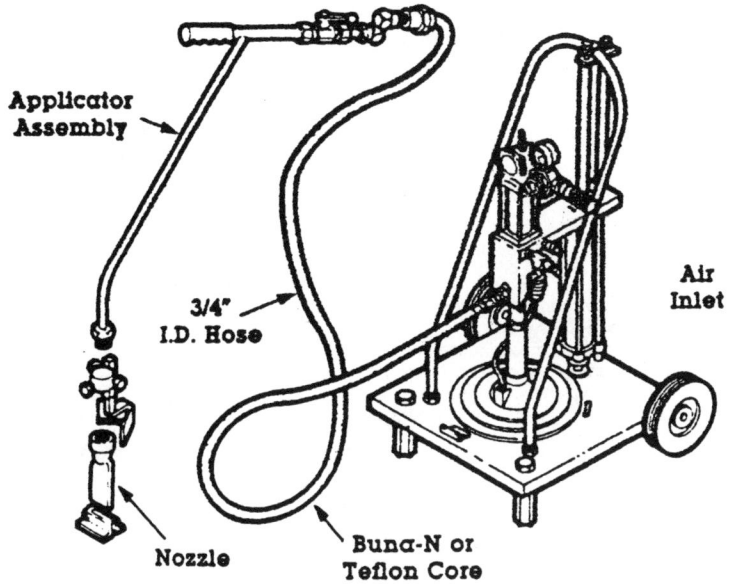

FIGURE 13.8 Extrusion pump.

TABLE 13.3 Recommended backer rod installation (shallow-cut)*

Joint width	$1/4$ inch	$3/8$ inch	$1/2$ inch	$3/4$ inch	1 inch
Recessed below surface	$1/4$ in.	$1/4$ in.	$1/4$ in.	$1/4$ in.	$1/2$ in.
Sealant thickness	$1/4$ in.	$1/4$ in.	$1/4$ in.	$3/8$ in.	$1/2$ in.
Backer rod diameter	$3/8$ in.	$1/2$ in.	$5/8$ in.	$7/8$ in.	$1 1/4$ in.
Total joint depth	$7/8$ in-1 in.	1 in.-$1 1/8$ in.	$1 1/8$ in.-$1 1/4$ in.	$1 1/2$ in.-$1 5/8$ in.	$2 1/4$-$2 3/8$ in.

* On road surfaces where grinding is planned at a later date, the sealant and backer rod should be installed so that they are approximately $1/4$ inch below the road surface after grinding is complete.

textures of the profile can be provided using a proprietary factory cold-vulcanizing process.

Sil-Span also offers a cost-effective method of connecting failures in metal curtainwalls, stucco, skylights, roofing details, reglets, coping stones, and flashing details. These can be areas in the envelope that are difficult or almost impossible to reseal or replace through conventional methods. Benefits of this silicone system are:

- 200 percent movement capability
- Resistance to UV rays and weathering
- Wide temperature performance range

- Minimal dirt pickup
- No need to remove old sealant or gasket materials before application
- Custom color and texture matching available

Figure 13.9 offers a typical application of Sil-Span, effectively bridging gaps found in an EFIS system.

Sil-Span's design properties enable it to be used in a variety of construction applications with expansion joints in EIFS systems being the most common. Sil-Span is also well suited for use in glass-to-glass and glass-to-metal glazing details where the bond between these two materials has failed. Sil-Span effectively bridges the gap in these applications and creates a clean, even sight line. The details below illustrate a few of the uses.

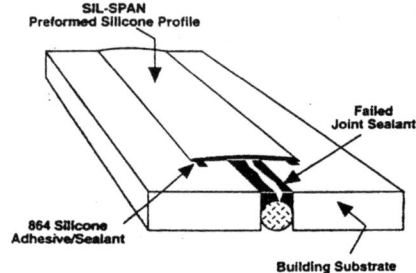

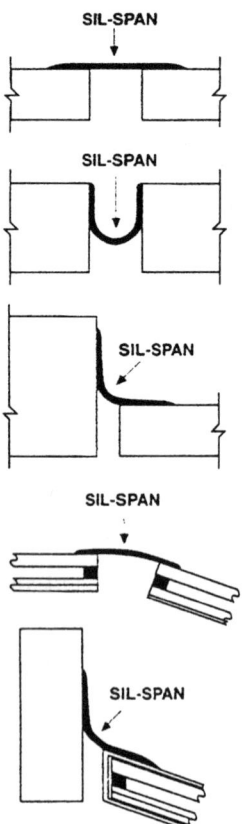

FIGURE 13.9 Sil-Span preformed silicone profile.

Remedial Caulking with Silicone Sealants

Because of the increasing cost of new construction, interest in preserving the character of the original building, and emphasis on energy savings, building owners are more frequently asking the question: "How do I reseal my old, leaking building?" This is the challenge to the architect, the sealant supplier, and the caulking contractor because of the wide variety of problems that can be encountered in such a project.

The object of this section is to provide a series of checkpoints to assist with the successful completion of such projects. The following points will be covered:

- Locating points of water or air infiltration
- Determining causes of sealant failure
- Criteria for sealant selection
- Removal of old caulking
- Application considerations for resealing or glazing

We are concerned about two types of penetration—water and air. The following procedures will aid in locating penetrations of each type.

Water. If exact water leakage points are not obvious, the easiest way to find the entry points is with a simple water hose test. Starting outside at the bottom of the structure, water is sprayed on the suspect leakage areas. Then the areas are checked for moisture on the building interior. This process is repeated floor by floor, up to and including the roof and flashings. This testing must be done from the bottom to the top because, if reversed, water entering from upper floors could run down structural members or conduits to lower floors and give the appearance of a leak in that area. Be sure to allow enough time for water to get to the low points on the building interior and seep through porous materials before proceeding to the next floor.

Air. Air leakage points might be more difficult to detect than water leakage points. In cold weather, the interior entry areas are identified easily, but the origin of the outside air might not be so obvious. Careful visual examination might reveal exterior air leakage points, but air seals are hidden in many window units and curtainwalls. If detail drawings of the building are available, careful study by a qualified architect, engineer, contractor, or consultant can pinpoint suspect areas. Disassembly of window and curtainwall façades might be necessary to confirm the leakage point and make necessary repairs.

Common Problem Areas. When hunting for leaks, look for common failure locations, such as caulking and open joints around window frames, joinery in metal curtainwalls, expansion joints, coping stones, parapet flashings, vent collars, air conditioners, ducts, electrical conduits, and cracks in the structure . Of course, thorough inspection of the roofing materials is essential.

In glazing assemblies, look for brittle and cracked glazing compounds or caulking tapes, sealants that have been pumped out of their original position, and openings or adhesion failures in cap, toe, and heel beads.

Failure Analysis. Once the leakage areas are found, the next step is to determine the proper course of remedial action. Visual inspection of failed caulking usually reveals the cause of the failure. The old sealant might have become brittle and dry because of weathering or cohesively cracked because of excessive joint movement. On the other hand, the sealant might show loss of adhesion to the joint surfaces because of poor surface preparation, weathering, movement, surface incompatibility, or poor sealant performance. In glazing assemblies, many sealants lose adhesion to glass because of degradation of the sealant at the bond line caused by ultraviolet radiation.

This inspection and analysis helps the architect, designer contractor, or sealant supplier decide what type of sealant will perform the best in a particular application. If failures are cohesive, one needs a sealant with good extensibility and resilience. If the old sealant is dry and brittle, it is obvious that a high-quality material with good weathering resistance is required. If most problems result from adhesion loss, a sealant with known good adhesion to that particular substrate should be chosen. In glazing applications, the sealant should be able to maintain adhesion after prolonged ultraviolet exposure. Often sealants fail as a result of a combination of factors, so, naturally, a sealant with high-performance characteristics should be the preferred choice.

When Not to Seal. If areas of air and water penetration were previously unsealed, a judgment must be made about the consequences of sealing these areas. Many curtainwall systems are designed to allow water entry, which is channeled back to the exterior of the building through interior gutters or weep holes. Sealing these gutters or weep holes would cause water to back up and run to the interior, causing serious damage to the structure and its contents. Glazing channels where insulating glass is installed requires that a weep system be used. Sealing weep holes in these assemblies can cause failures and void any insulating glass warranties because of the possibility that standing water might degrade the edge seals in the IG units or freezing water might cause the glass to break.

Many buildings are designed to allow the entry of air for pressure equalization or venting of moisture vapor. In your zeal to seal, be sure you do not interfere with the design function of the building. Review of drawings by a qualified architect or engineer can determine which areas must remain unsealed.

Sealant Selection. What will give the best performance in a given area is complicated by the variety of sealants on the market, all of which are designed for a specific function and application.

High-quality architectural-grade caulks and sealants are categorized by generic types, use areas, and movement capability.

Interior/exterior—Little or No Movement. As a general rule, paintable silicone caulks are used in interior or exterior applications where little or no movement is expected. These sealants normally have fair- to -good weather resistance and adhesion, can be used on a variety of substrates, and will accept and hold paint. Various specialty sealants are recommended for use in bathrooms, showers, and other sanitary facilities. These sealants should contain additives that retard the growth of fungus or mildew, commonly found in high-moisture interior areas.

Exterior—Moderate to Extreme Movement. In exterior applications where moderate to extreme movement, weather exposure, ozone and ultraviolet light (sunlight) are factors, high-performance silicone sealants should be considered. As can be seen in Table 13.4 the polysulfide and polyurethane sealants are recommended for movements from ± 12.5 percent to ± 25 percent. Because of their resistance to weathering and flexibility and resilience at temperature extremes, silicone sealants are recommended for movement as high as ± 50 percent. It is important that the movement capability of the sealant can accommodate the expected movement for a given joint width. For example, if the expected joint movement in butt joints is known to be ± 1/16 in. and the width is only 1/8 in., only a ± 50 percent movement capability sealant will perform in that joint.

Because of different stress factors, elastomeric sealants will take more movement in shear than in tension. Figure 13.10 shows the manufacturer's recommendations for joint width vs. expected movement for sealants with various movement capability. In some structures, movement may be cumulative and

TABLE 13.4 Movement Capabilities of Currently Available Sealants

Sealants	Movement capability
Gun-grade butyls	± 5% to ± 10%
Acrylic latex	± 5% to ± 10%
Solvent-release acrylics	± 12.5%
One-part urethane	± 12.5% to ± 25%
Two-part urethane	± 25%
Two-part epoxy, modified urethane	+ 40% to − 25%
One-part polysulfide	± 12.5% to ± 25%
Two-part polysulfide	± 25%
One-part silicone, acetoxy cure	± 25%
Two-part silicone, low modulus	± 50%

Movement capability of sealant	Suggested joint widths (D) XN	
	Movement in shear	Movement in tension and compression
	Dimension "D" must be at least	Dimension "D" must be at least
± 12.5% urethanes/ polysulfides	3x's expected movement	6x's expected movement
± 12.5% urethanes/ polysulfides		
Acetoxy cure silicone sealants	2x's expected movement	4x's expected movement
± 50% low modulus silicones	Equal expected movement	2x's expected movement

FIGURE 13.10 Suggested joint widths in the application of currently available sealants.

impossible to predict. In such cases, sealants with high movement capability should be used to provide maximum safety factor.

Job Site and Weather Conditions Affecting Sealant Selection. The proper choice of sealant also depends upon job site and weather conditions at the time of application. If mixing a two-part sealant on the job site will cause problems because of untrained personnel, logistics, added costs, temperature extremes, and so forth, a one-part sealant is the logical choice. If sealing must be accomplished at low or extremely high temperatures, the manufacturer must recommend the sealant for application under these conditions. At low temperatures, most sealants, except silicones, become viscous, hard to gun, and might not wet the surfaces properly for best adhesion. Conversely, the same sealants might exhibit a tendency to sag and flow in the joint if application or substrate temperatures are too high.

Application Temperature Consideration. In addition to application problems, low temperatures retard the cure of many sealants. Some organic sealants that cure through in days at 70° F (21° C) and 50 percent RH might take several weeks to cure through at temperatures below 40° F (4.5° C). Consequently, surface cracking, deformation, or splitting through the sealant might occur because of movement during the cure. Sealing at temperature extremes affects the width of the butt joints to be sealed. A sealant that is acceptable for a joint installed at 70° F (21° C) might not be acceptable for use in the same joint at expected temperature extremes. Figure 13.11 shows why this is true. If a sealant is installed when the joint is at its minimum or maximum width, the total movement from that period of time is, in one direction, subjecting the sealant to

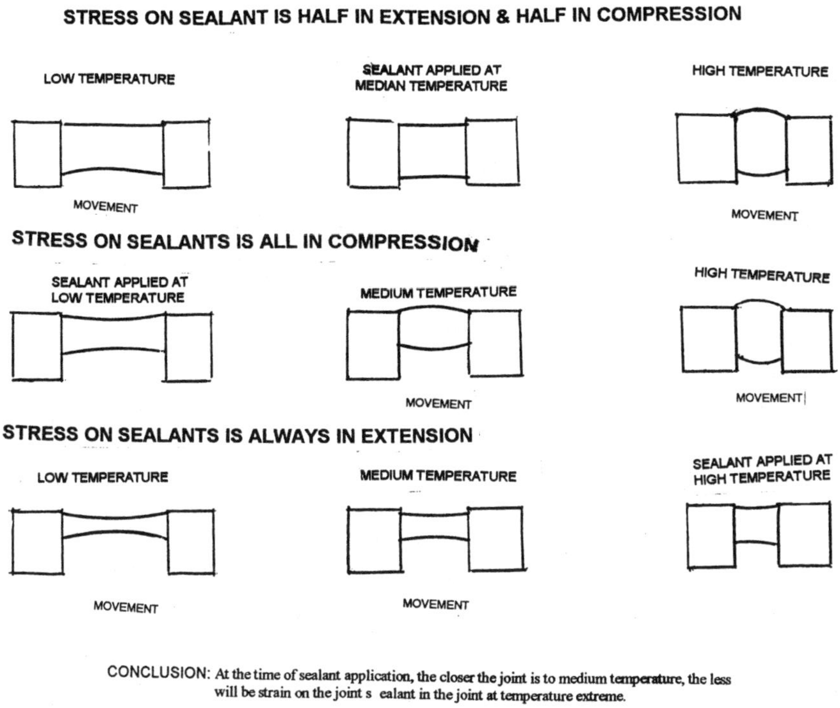

STRESS ON SEALANT IS HALF IN EXTENSION & HALF IN COMPRESSION

LOW TEMPERATURE   SEALANT APPLIED AT MEDIAN TEMPERATURE   HIGH TEMPERATURE

MOVEMENT   MOVEMENT

STRESS ON SEALANTS IS ALL IN COMPRESSION

SEALANT APPLIED AT LOW TEMPERATURE   MEDIUM TEMPERATURE   HIGH TEMPERATURE

MOVEMENT   MOVEMENT

STRESS ON SEALANTS IS ALWAYS IN EXTENSION

LOW TEMPERATURE   MEDIUM TEMPERATURE   SEALANT APPLIED AT HIGH TEMPERATURE

MOVEMENT   MOVEMENT

CONCLUSION: At the time of sealant application, the closer the joint is to medium temperature, the less will be strain on the joints sealant in the joint at temperature extreme.

FIGURE 13.11 Stress on sealant during extension/compression cycle.

extreme stress either in compression or tension. In lap shear joints, the relative stress is always in tension, but in different directions, depending on the temperature at time of installation.

Sealant Manufacturer's Recommendations. Sealant manufacturers make specific recommendations for use and how to treat various construction surfaces for best adhesion and compatibility. Manufacturers' data sheets should be consulted to be sure the sealants chosen are recommended for the surface in question.

Removal of Old Caulking. In remedial caulking applications, all sealant manufacturers recommend that all old sealant be removed, especially if the new sealant is of a different type. Removal of the old sealant can be very difficult and usually requires cutting to the bond line, then using solvent or mechanical means to remove all traces of the sealant film from the joint surface. After this is accomplished, test applications with the new sealant are recommended to determine whether proper adhesion of the new sealant will be obtained. Be sure to allow adequate time for the sealant to cure before testing for adhesion. The time required for optimum adhesion varies with each sealant.

It is unlikely that silicone sealants will fail, even over extended periods. If a silicone sealant is mechanically damaged, however, it can be repaired simply by resealing over the damaged areas with fresh silicone caulk. Care must be taken to make certain that the original silicone is well-adhered and clean.

If Old Caulking Cannot be Removed.   If the old caulking cannot be removed, a technique sometimes referred to by experienced caulkers as the "Band Aid" can be used. This method involves the placement of a polyethylene bond breaker tape over the old sealant and application of the new sealant over and beyond the edges of the tape, forming a new joint. The section on Sil-Span referred to earlier in this chapter is a more sophisticated method than this approach. The technique might not be aesthetically acceptable for all applications, but, where possible, it can result in considerable time and labor savings. Always contact the sealant manufacturer to make certain there are no compatibility problems between the old and the new sealant. A number of variations of this technique also can be used to spread movement in joints that are not wide enough to accommodate the proper sealant width for expected movement. See Figure 13.12 for a remedial approach to repair.

Remedial Glazing Applications.   There is an increasing tendency to replace old single-glazed units with insulating glass units for energy conservation. In these cases, the sealant must be compatible with the insulating glass units, especially if it is used as a heel bead or toe bead in direct contact with the edge seals. If old frames or glazing assemblies are reused, it is important that the surfaces to receive the new sealant completely free of old sealant residues and contaminants. A competent glazing contractor should install new units according to good practice, and the sealant manufacturer's recommendations must be followed. The sealant or insulating glass supplier should be consulted to determine if any compatibility problems exist.

Application of the New Sealant.   A number of factors influence the performance of a sealant, and manufacturers' literature should be consulted for the following data;

- Sealant limitations
- Application temperatures
- Surface preparation
- Primer requirements
- Joint width and depth requirements
- Backup materials
- Sealant placement
- Cleanup
- Safety

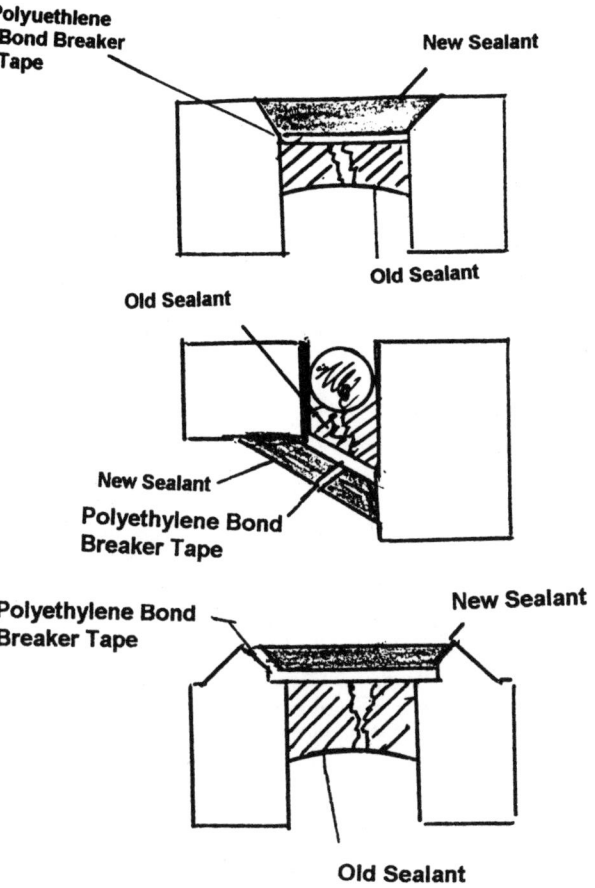

**POLYETHYLENE BOND BREAKER TAPE
MAY BE USED TO ISOLATE SEALANT OR
SPREAD MOVEMENT OVER A WIDER AREA**

FIGURE 13.12 Isolating failed sealant with bond breaker tape.

SUGGESTED STARTING FORMULATIONS OF SILICONE

In addition to Figure 13.1, which lists four suggested formulas (an IG formula and three that vary in modulus for use in construction applications), the four sealants shown in Figure 13.13 use different cure mechanisms resulting in different degrees of modulus.

Raw materials	Percentage			
	High modulus acetoxy	Medium modulus oxime	Medium modulus oxime	Low modulus aminoxy siloxane
Silanol polymer	80-85%	80-85%	60-80%	46.0%
Fumed silica (treated and/ or untreated)	6-10%	5-10%	2-6%	—
Acetoxy cross-linker	5-7%	—	—	—
Tin catalyst	0.05-0.1%	0.05-0.1%	0.05-0.1%	—
Oxime cross-linker	—	5-7%	5-7%	—
Calcium-carbonate	—	—	20-30%	50.3%
N-methylacetamide chain extender	—	—	—	3.0%
Amino siloxane cross-linker	—	—	—	—
Silicone plasticizer	—	—	5-20%	—
Aminoxy siloxane cross-linker	—	—	—	0.7%
Physical properties		Values		
Skin time, min.	5-7	10-20	20-30	50-60
Tack-free time, min.	10-20	20-30	30-60	10-15
Durometer, Shore A	25-35	25-35	20-30	15
Tensile strength, psi	175-300	175-300	125-200	85-120
Elongation, %	200-400	200-400	400-700	>1000
Modulus, 100%, psi	75-125	75-125	50-75	20-25
Tear strength, pli	35-70	35-70	20-40	20-25

FIGURE 13.13 Basic silicone formulations of varying modulus.

SUMMARY

The building owner has many incentives to maintain the integrity of building sealing and to prevent penetration of air and water. Elimination of exterior leaks prevents costly interior water damage; air and water leaks waste energy and increase costs of heating and cooling; and water leaks in masonry can result in freeze-thaw damage. Careful inspection is required to locate leaks and to determine the cause of sealant failure.

The study of silicone sealants and caulks is essential so that the architect, designer, or contractor can make an intelligent decision in choosing the correct most economical material.

REFERENCES

1. Grolier Interactive, Inc. 1999. Danbury, CT 06816.
2. *The Age of Chemical Fastening*. 1999. General Electric Co. Waterford, NY
3. *Structural Silicone Glazing*. Dow Corning Corp. Midland, MI.
4. *Information about Silicone Sealants*. Dow Corning Corp. Midland, MI.
5. *Sil-Span Extruded Profiles*. 1999. Pecora Corporation. Harleysville, PA .

CHAPTER 14
NEW POLYMERIC SYSTEMS

INTRODUCTION

Within the past few years, sealant makers and/or manufacturers of raw materials have introduced several new polymeric systems. It is not the intent of the author to discuss any of these products in detail, but to an introduction to their background and possible end uses.

POLYMERIC SYSTEMS

Polytetrafluoroethylene (PTFE) and *polychlorotrifluoroethylene* (PCTFE) are two fluorocarbons that are used predominantly as sealant materials. They are used in static applications, in a cut-and-place form, and as a solid form in place material.

There are three types of polytetrafluoroethylene:

- Granular: a free-flowing powder or granulated powder used for gaskets packing and seals and, in some cases, tapes.
- Fine powder: an agglomerated powder used for seals and gaskets
- Dispersion: aqueous dispersion, impregnation coating, and packing

The major application area is gaskets and industrial seals.
A few companies manufacture these products or offshoots of the polymer:

- Asahi Glass Company Ltd makes Alfas.
- DuPont makes Viton and Kalrez.
- 3M makes Fluorel.

These products are used principally in automotive and aerospace industries; however, a small amount is used in construction gasketing.

Polyether silicone sealants, adhesives, and caulks are important architectural and construction materials. Typical applications involve waterproofing, weatherproofing, sealing, and bonding. Polyether silicone is a common name applied to silicon- or silane-modified-polyethers in their cured state; they contain Si-O-Si bonds.

Polyether silicones are moisture-cured and form an elastomeric product similar to RTV silicone sealants. They have outstanding adhesion to a variety of surfaces encountered in the construction industry. The finished products are usually supplied to a high viscosity with solids exceeding 90 percent.

Figure 14.1 is a typical formula suitable for construction applications. Unlike the two-part systems, this formula is kept in a completely anhydrous condition to prevent curing from taking place within the package. It is a stable product until it is applied and exposed to the atmosphere.

Polyether silicones oriented to the construction market are supplied in the United States by:

- BASF Corporation, Coatings and Color Division.
- Essex Specialty Chemicals
- Union Carbide Corporation

Ethylene-acrylic ionomer has been commercialized by DuPont and is an ionomer based on a copolymer of ethylene, methyl acrylate, and a carboxylic acid under the trade name Vamac VMR 5245. Its intended use is as a base for a hot melt sealant for the insulating glass market. It is supplied as a partially compounded base containing 30 percent by weight of carbon black. It exhibits good elevated temperature properties and good weatherability. It can be formulated into a preformed tape, mastic, and solvent-based adhesive.

1,2-Syndiotactic Polybutadiene [1] is a thermoplastic elastomer based on 1,2-syndiotactic polybutadiene with more than 90 percent vinyl content and is manufactured by Japan Synthetic Rubber Co., Ltd. It can be obtained in the United States from JSR America, Inc., New York, New York and is supplied in pellet form. The polymer has a melting point in the range of 167° to 194° F (75° to 90° C). It is compatible with all the typical plasticizers and tackifiers, exhibiting good adhesion and weatherability. It has been evaluated for adhesive and sealant applications.

Raw materials	Parts by weight
Silmod 20A	600
Silmod 300	400
Plasticizers	500
Calcium carbonate	1200
Titaniun dioxide	200
Thixitropic agent, Cab-O-Sil	30
Antioxidant	10
UV absorber	10
Dehydrating agent	20
Adhesion promoter	30
Dibutyl dilurate	20
Lauryl amine	5.0
Physical properties	Values
Hardness, Shore A	50-65
Tensile strength, ksi (Mpa)	1.0-1.5 (6.9-10.3)
Elongation, %	200-300
Tear strength, lbs/in. (N/mm)	200-300 (35-53)

FIGURE 14.1 Typical sealant formulation based on polyether silicone.

Fluoroalkoxy-polyphosphazene elastomers are marketed by Firestone Tire and Rubber Company under the trade name PNF. Early work was conducted by Horizons, Inc., Cleveland, Ohio, conducted early work under a U.S. government contract and the exclusive worldwide rights were obtained by Firestone in the early 1970s. The polymer has good high- and low-temperature properties, and its cured compositions provide excellent oil resistance below $-40°$ F ($-40°$ C). They are tougher than fluorosilicones and can be used for oil seals, gaskets, and specialty coatings.

Unsaturated polyether is a joint sealant marketed as Radseal by Radiation Technology, of Rockaway, N.J. It is a two-component system with a density of 9.5 lbs/gallon. Its principal use is for horizontal and vertical joints in the concrete slabs in cold storage and freezer warehouses. Its main advantage is that it can cure quickly at $-40°$ F ($-40°$ C). It is not a high sales item because of its intended end use.

Epoxy tape markets a ribbon form under the trade name Kneadatite. This product, which is used in plumbing and home repairs, cures at room temperature and is an epoxy-polyamide material in two parts. The base and curing agents are extruded side by side on a release paper liner. The composition can be readily kneaded together by hand or machine. It is formulated with high molecular weight resins and is covered by U.S. Patents 3,708,379; 3,837,981; and 4,160,064. Applications include plumbing and home repairs. Polymeric Systems, Inc., Pheonixville, Pa., has licensed Loctite to market this system to over-the-counter outlets.

SUMMARY

Proper adhesives, sealants, tapes, and gaskets are essential to the new technologies being encountered in the contemporary architectural and construction industry. We, as educators, architects, designers, contractors, and building owners, are grateful that the raw material suppliers and manufacturers of the systems are constantly working to provide the industry with new and proven products.

REFERENCES

1. Skiest Inc., Whippany, N.J.

CHAPTER 15
FIRESTOPS AND FLAMEPROOFING

INTRODUCTION

A fire is one of the most devastating disasters anyone could experience. Fortunately, today's building codes address the need for fire safety on several levels, including fire-rated walls, floors, ceilings, and joints.

In the early stage of the industry's development, firestop protection was not regulated and was done with few, if any, guidelines. Firestopping usually was performed by filling the opening to be protected with ceramic fire insulation, mineral wool, rock wool, or concrete.

The forces unleashed during a building fire are deadly. Temperatures can rise higher than 2000° F (1093.3° C), and smoke and toxic gases are released. Tiny openings in walls, floors, and ceilings can become blowtorch nozzles, spitting fire into adjacent rooms and jets of toxic smoke that can fill the room in minutes. As fire disasters in public buildings across the country have demonstrated, lives often are lost because of smoke inhalation, even though the victims are not directly threatened by fire. Building penetrations to be firestopped include steel, cast iron, and plastic pipes; electrical wiring and conduits; telephone and video cables, HVAC ductwork, beams, and structural elements. Other construction openings include expansion joints, curtainwall gaps, and interfaces at wall/floor and wall/ceiling points [1].

Fire-resistant walls, floors, and ceilings are designed to compartmentalize and contain fire within a small area in larger buildings. Fire-rated assemblies, however, can be rendered ineffective, however, when openings are created to accommodate pipes, tubes, conduits, vents, wires, electrical cables, and other penetrants. These openings can lead to the spread of smoke, toxic gases, and flames at astounding speed if any of the active fire-prevention systems, such as alarms, sprinklers, and smoke detectors, fail [2].

To overcome this hazard, fire codes now require the use of passive firestop systems designed to prevent the spread of smoke and fire through penetrations. The system maintains assembly integrity during a fire. It doesn't take much to inhibit the spread of fire before it starts. A firestop system can be installed for about one-one thousandth of the cost of constructing a new building [3]. Today, several large firms offer the most complete range of reliable firestop technologies in the world. These products have a history of reliability that spans more than 25 years, long before the need for passive fire-prevention systems was recognized.

CURRENT MODEL BUILDING CODES

The current model building codes, such as the ICBO Uniform Building Code, the SBCCI Standard Building Code, the BOCA Basic/National Building Code, and the National Fire Protection Association Standard NFPA 101, include detailed provisions governing firestopping. These up-to-date specifications and standards implement our improved understanding about the spread of structural fires that threaten life, safety, and property.

For specification data about firestop-related materials see Chapter 18.

ELEMENTS OF FIRESTOP TECHNOLOGY

Firestops are designed to maintain the rated protection of fire barriers for a wide range of conditions, including the size of the opening, the number and size of the penetrants, the annular space (the space between the penetrant and the opening), and the thermal and mechanical properties of both the penetrant and the rated assembly.

Although the wide range of design variables might suggest to design professionals that the selection and specification of the firestop systems is a complex and confusing process, firestop technology is very straightforward after several concepts are understood.

Several reference guides are available, including Nelson Application Details, Firestop Products (Table 15.1) [3]; FlameSafe® Firestop System Selector Guide (Table 15.2) [2]; and the Specified Technologies, Inc. firestop selector guide (Table 15.3). These guides give the architect and design engineer an overview of firestop technology and help one ask the right questions required to select the appropriate firestop systems for virtually any building condition. The first step in the firestop selection process is to answer these two questions:

1. Are the openings in the rated assembly large or small?
2. Are the penetrants combustible or non-combustible?

TABLE 15.1 Nelson Application Details, Firestop Products

Penetrating Item	Assembly Thickness	Ratings Hours	Product	UL System	Nelson FS Detail
Gypsum Wallboard Walls, Steel Pipe	5	2	CLK	WL 1031	FS-0098
Gypsum Wallboard Walls, Steel Pipe	5	2	FSP	WL1005	FS-0039
Gypsum Wallboard Walls, Copper Pipe	5	2	LBS	WL1094	FS-0154
Gypsum Wallboard Walls, Cable Tray	5	2	PLW	WL4003	FS-0097
Gypsum Wallboard Walls, Bus Duct	5	2	FSP	WL6003	FS-0132
Concrete /Masonry Walls, Steel Pipe, 1½ in. multiple	4½	3	LBS	CAJ8062	FS-0164
Concrete/ Masonry Walls, Steel Pipe, 4 in . Sch 5	4½	2	FSP	CAJ1054	FS-0092
Concrete/Masonry Walls, Insulated Steel Pipe, 1½ in. AB/PVC	6	2	FSP	CAJ5054	FS-0107
Concrete/Masonry Walls, Insulated Steel Pipe, 2 in. Sch 10, ¾ in fiberglass	4½	2	LBS	CAJ5104	FS-0195
Concrete/Masonry Walls, Control Cable, 2 AWG, 15% fill	4½	3	FSP	CAJ3003	FS-0003
Concrete/Masonry Walls, 12 in. x 24 in., HVAC Duct	6	2	FSP	CAJ7010	FS-0128
Concrete/Masonry Walls, Communications Cable, 25 pair	5	2	CLK	CAJ3010	FS-0101
Concrete/Masonry Walls, Duct, 3000A Copper	4½	2	CLK	CAJ6004	FS-0102
Fire and Thermal Protection, Grease Duct, 24 in. x 48in.			FSB	CAJ7024	FS-0211
Concrete to Concrete, 4in. Floor	5½	4	CLK	FFS1011	FS-0119
Wood Joist Floors, Steel Pipe, 4in., Sch 5	10 solid	2	CLK	FC1012	FS-0112
Wall Joist Floors, Non-metallic Pipe, 2in., Sch 40 PVC	10 Solid	1	LBS	FC2097	FS-0206

Opening Size

As the exposed area of an opening in a rated assembly increases, so do the challenges to thermal, mechanical, and structural properties of the firestop system. Small openings create technical difficulties because minimal annular space can prevent the installation of enough firestop material to provide effective protection from smoke, flames, and combustible gases.

TABLE 15.2 FlameSafe® Firestop Selector Guide

Category	Penetrating item	Wall-floor type	Rating (hours)	Product	UL system	Page
Dynamic & static joints	4" joint maximum	Dynamic floor to floor	2	FS 2900	FFD1005	5
	4" joint maximum	Dynamic floor to wall	2	FS 2900	FWD1004	5
	Gypsum board to Deck	Futed deck	2	C 700S / FS 1900	PSV1094 (WH) / PSV1095 (WH)	5 / 4
Metallic penetrants	24" steel or 6" copper maximum	Concrete wall/floor	2 and 3	FS 900 or FS 1900 / FS 2900	CAJ1235 / CAJ1236	4 / 5
	24" steel or 6" copper maximum	Gypsum wall	1 and 2	FS 1900	WL1089	4
	4" vent pipe, 3" copper, EMT, steel pipes maximum	Wood joist Floor/ceiling	1 and 2	FS 1900	FC1020 / FC1021	4
	4" vent pipe, 3" copper, EMT, steel pipes maximum	Chase wall	1 and 2	FS 1900	FC1022 / FC1023	4
Non-metallic penetrants	4" PVC, cc-PVC, FRPP maximum	Concrete floor	3	FlameSafe Sleeve FS 900	CAJ2172	7 / 4
	2" PP, FRPP maximum	Concrete floor	2 and 3	FlameSafe Sleeve FS 1900	CAJ2175	7 / 4
	2" Non-metal PVC, CPVC maximum	Wood joist floor/ceiling	2	FS 1900	FC2042	4
	4" PVC, CPVC, cc-PVC maximum	Chase wall	1	FS 1900	FC2082	4
	2" Non-metal PVC, CPVC rigid non-metallic	Gypsum wall	1 and 2	FS 1900 / FSP 1100	WL2038	4 / 5
Cables	Common cables	Concrete wall/floor	2	FSP 1000	CAJ3016	5
	SER, Romex, Alarm 100 pr #24	Wood joist floor/ceiling	1 and 2	FS 1900	FC3019 / FC3018	4
	SER, Romex, Alarm 100 pr #24	Chase wall	1 and 2	FS 1900	FC3021 / FC3020	4
	Common cables 40%	Gypsum wall	1 and 2	FS 1900	WL3061	4

TABLE 15.2 FlameSafe® Firestop Selector Guide *(Continued)*

	Penetrating item	Wall-floor type	Rating (hours)	Product	UL system	Page
Cable tray	36" × 5" tray	Concrete wall/floor	3	FlameSafe Bag	CAJ4038	6
Insulated pipe	8" metallic and 2" insulated maximum	Concrete wall/floor	3	FS 1900	CBJ5008	4
	1½" & 1" insulated pipe (FG) 1½" copper, steel & cast iron pipe	Wood joist floor	1 and 2	FS 1900	FC5010 FC5011	4
	1½" & 1" insulated pipe	Chase wall	1 and 2	FS 1900	FC5012	4
HVAC ducts	12" × 12" duct (no angles)	Gypsum wall	1 and 2	FS 1900	WL7006	4
	12" × 12" duct (no angles)	Block wall	2	FS 1900	WJ7002	4
Blank openings	Blank opening 1280 sq. in. maximum	Concrete wall/floor	4	KBS Sealbag	CBJ0009	6

TABLE 15.3 Specified Technologies, Inc, Firestop Selector Guide

This matrix provides a quick index of the most common firestopping applications. It includes only a small percentage of STI's classified firestop systems. Be sure to consult the complete drawing index as well as the most recent *UL Fire Resistance Directory* for additional designs.

Penetration	Concrete or masonry		Gypsum board/frame		Product
	Floors UL Sys.	Walls UL Sys.	Floors UL Sys.	Walls UL Sys.	
Electrical					
Metallic conduits	CAJ1079	WJ1055	FC1010	WL1049	SSS100 Sealant
	CAJ1213	CAJ1213		WL1088	LC150 Sealant
	CAJ1198	CAJ1198		WL1062	PEN300 Silicone
PVC conduits ≤ 2"	CAJ2031	WJ2018	FC2032	WL2093	SSS100 Sealant
Cables	CAJ3133	WJ3022	FC3015,16,22	WL3076	SSS100 Sealant
Cable trays	CAJ4029	CAJ4029		WL4008	Firestop pillows
	CAJ4020	CAJ4020			Firestop mortar
		WJ4009		WL4005	SSS100 Sealant
Bus duct	CAJ6009	CAJ6009			Firestop mortar
	CAJ6008	CAJ6008		WL6001	SSS100 Sealant
Mechanical					
Steel pipes	CAJ1079	WJ1055	FC1010	WL1049	SSS100 Sealant
	CAJ1213	CAJ1213		WL1088	LC150 Sealant
	CAJ1198	CAJ1198		WL1062	PEN300 Silicone
≤ 2" CPVC sprinkler pipes	CAJ2031	WJ2018	FC2032	WL2093	SSS100 Sealant
Fiberglass insulation	CAJ5087	CAJ5087	FC5014,29	WL5014	SSS100 Sealant
	CAJ5058	WJ5006		WL5033	Wrap strips
AB/PVC insulation	CAJ5079	WJ5005	FC5014,29	WL5054	SSS100 Sealant
Fan coil	CAJ8053	CAJ8053			SSS100 Sealant
	CAJ8052	CAJ8052			Firestop pillows
Air cond. line set	CAJ8054	WJ8006	FC8011	WL8011	SSS100 Sealant
			FC8011		Wrap strips
Small ducts	CAJ7023	WJ7005	FC7002	WL7019	SSS100 Sealant
Larger air ducts	CAJ7027	CAJ7027		WL7009	SSS100 Sealant
Plumbing					
Copper pipes/tubes	CAJ1079	WJ1055	FC1010	WL1049	SSS100 Sealant
	CAJ1213	CAJ1213		WL1088	LC150 Sealant
	CAJ1198	CAJ1198		WL1062	PEN300 Silicone
Plastic supply lines	CAJ2031	WJ2018,21	FC2032	WL2093	SSS100 Sealant
≤ 2" PVC tuck-in	CAJ2064	CAJ2064	FC2019	WL2048	Wrap strips
≤ 4" most plastics	CAJ2038,45	CAJ2038,45	FC2033	WL2029	Factory collar
≤ 4" most plastics	CAJ2124	CAJ2124	FC2033	WL2059	Wrap strips
Plastic sanitary tee			FC2034		Wrap strips
Plastic closet flange			FC2037		Wrap strips
Plastic tub drain			FC2036		Wrap strips

Penetrant Combustibility

Combustible penetrants, such as plastic pipe and conduits, insulated piping, or telecommunication and electrical cables, are consumed by fire. Most firestop systems designed for combustible penetrants, therefore, consist of intumescent materials that expand when subjected to heat to fill the void left after the combustible penetrant decomposes. A wide range of intumescent materials is available, with expansion capability from 400 percent to more than 1500 percent, to meet the requirements of different penetrants and annular space configurations. Intumescent materials expand in proportion to temperature, starting at about 250° F (121.11° C) and reaching full expansion at about 1000° F (537.78° C).

Non-combustible penetrants, such as steel, copper, and cast-iron pipes, remain in place during a fire and produce heat transfer paths for flames. Consequently, firestop systems designed for non-combustible penetrants consist of endothermic materials that contain chemically bonded water to absorb the heat. After the required opening size and level of penetrant combustibility is established, other conditions, such as the fire assembly material, the required fire rating of the firestop system, and the need for retrofit capability, will determine the specification of the proper firestop. The data in Tables 15.1, 15.2, and 15.9 will establish which firestop systems are applicable to your building conditions. A long history of high-quality service is available, backed by expert technical engineering support and a commitment to quality customer service.

A FIRESTOP—BASICS TO A BETTER UNDERSTANDING [4]

The active/passive fire protection chain consists of four essential links: evacuation, extinguishment, detection, and compartmentalization. If any of these links is broken during a fire, the probability exists for great loss of property or even human life.

The first three links—evacuation, extinguishment, and direction—are considered active fire protection. Passive fire protection systems function by virtue of their inherent attributes, while active fire protection systems require outside and/or mechanical power, such as firefighters and sprinkler systems. Active and passive systems can provide protection from smoke, flames, and water. Firestop systems promote compartmentalization.

Approved firestop products are used in areas where fire-rated assemblies are penetrated. The products have been tested to resist the passage of smoke, flames, and water or hose stream. The most widely recognized testing standards are ASTM E 84 for surface-burning characteristics of building materials, ASTM E 814 for through-wall penetrations, and ASTM E 119 for fire-resistant joint treatments. It is the manufacturer's responsibility to submit firestop materials for testing in accordance with these standards at recognized testing laboratories such as Underwriter's Laboratories (UL).

Firestopping must include both penetration and expansion or control joints in order to be effective. Unless properly addressed, openings and holes in construction form major obstacles to optimum compartmentalization of smoke, flame, heat, water, and gas. Certain considerations should be taken into account: material performance, ease of application, specifier requirements, and manufacturer's commitment.

MATERIAL PERFORMANCE

Materials must adhere well to the substrate with minimal preparation. Poor adhesion will allow for "blow by" from heat, gas, smoke, or water. In determining the correct firestop material for the application, consider the following:

1. Does the material require special surface treatment?
2. Does it compromise adhesion to substrates, such as PVC pipe, by reacting negatively with them?
3. Will it contact with all surfaces to prevent the spread of flames, smoke, and water?
4. Does it maintain good adhesion during a fire?
5. When exposed to flames, does it turn into a coal-like char?
6. Does it cycle back and forth and still maintain adhesion?

Silicones, for example, are medium- and low-modulus sealants designed to absorb movement with durable adhesion characteristics. The only limiting factor, however, might be when they are used in combustible penetrations, such as plastic pipe. In these instances, an intumescent material will probably be required. Self-leveling products are useful in terms of durable adhesion and installation.

The firestop material should exhibit little or no shrinkage during the cure process. This eliminates stress and failure to bond-lines or the need to reapply sealant. Many products exhibit shrinkage rates up to 40 percent. The percentage of solvent or water is often a better indicator of shrinkage potential than published shrinkage data.

An inorganic composition must not give off gas or smoke, and it should not support combustion before or after the material has cured. Ratings, in accordance with ASTM E 84 and close to zero in all categories, are the optimum. In addition to performance criteria in a fire, another concern is toxicity during installation. Be conscious of sealants that release an inordinate amount of solvent into the atmosphere. Check labels and material safety data sheets (MSDS).

EASE OF APPLICATION

It is far too easy to make firestop products complicated. Many products require varying depths of application for specific ratings. Look for consistency of application depth. Some manufacturers test at a half-inch depth or more. The more consistently a material is applied, the more likely it will be installed correctly. The minimal criteria for specifying a firestop product lie with the architect's specification and third-party testing. Testing should reflect on-site conditions, if possible. Volume is not always an indicator of a product's flexibility.

MANUFACTURER'S COMMITMENT

A firestop manufacturer should offer technically sound materials that are in keeping with industry standards and offer technical assistance to the architect, designer, and end-user.

Since no firestop product can do everything in every application nor be tested in every condition, the manufacturer should be willing to make logical and educated engineering judgments or test the specific site conditions as needed.

The manufacturer should also offer field assistance with qualified representatives who are readily available to building officials, specifiers, installers, and distributor representatives. Some general guidelines to minimize problems in the installation of firestops are:

1. Use self-leveling products (horizontal) where possible. This will ensure the likelihood of an effective water and smoke seal.
2. Beware of shrinkage potential. The effects of significant shrinkage can be catastrophic.
3. Always put the onus on the manufacturer to justify the material's use in the correct application.

FIRESTOP SYSTEMS—AN INTRODUCTION

Firestops for floor-ceiling and partition assemblies generally consist of five components: the penetration, the penetrant, the annular space, the forming material, and the sealant [5].

- The penetration is a hole cut out or formed through the entire floor/ceiling or wall assembly.
- The penetrant typically is a pipe, conduit, duct, or cable bundle, passing through the penetration.
- The annular space is the difference between the size of the penetration and the size of the penetrant.
- The forming material (when required) is used as a backing or a dam for sealants. In the assemblies described in this chapter, the forming material is a special sealant that is flexible, yet effectively bonds to the hole perimeter to keep the firestop in place, and becomes an integral part of the system.
- The sealant is a material troweled, poured, gunned, wrapped, or caulked over the forming material around the penetrant(s) to form a complete seal.
- Firestop: A specific construction using special materials designed to fill the annular space. The purpose of the construction is to prevent the passage of fire through a fire-resistant partition or floor-ceiling assembly.

- ASTM E 814 or UL 1479 are the published test procedures that determine the effectiveness of a through-penetration firestop. Other ASTM or UL approved systems are also applicable.

FIRESTOPPING SEALANTS

Mortars, Caulks, Intumescents [3]

Mortar-type materials are applied wet over the forming materials (where applicable) and then set and dry to form a tough durable seal. For many firestopping products, large complex openings pose equally large problems. To solve large penetration problems, simply form the opening, mix the mortar with water, and then pour, pump, or trowel the mortar in place. Mortar is a cost-effective choice for medium to large openings through masonry walls or floors. When combined with intumescent products, wraps, or collars, it becomes the perfect choice for large, complex openings, including single or multiple combustible penetrants.

Generally, mortars are tested and approved for a variety of penetrants: steel and aluminum cable trays, steel, iron and copper pipes, fiberglass and foam plastic insulated pipes. Mortars have been engineered to be lightweight (as in a mix with water) and low in density (when dried). Damming is easier because forming materials require less support. Low-density materials make retrofit easier. Mortars are UL-Classified and Factory Mutual Systems-approved, generally with F ratings up to and including three hours. They are also low in cost and available in powder or ready-mix form.

Caulks, the most common materials used in the firestopping industry, are applied from a caulking tube or pail, and then cured to form a flexible seal. Because of their ease of application and labor savings, these sealants tend to be the product of choice among applicators, architects, and design engineers. The most obvious keys to product selection are the ease of use and economy. The less obvious, but equally important, factor is range of application.

Because of environmental concerns, the industry is switching to water-based technologies. For many materials, the switch from solvent systems to water systems has meant a sacrifice in some aspects of performance. Some water-based sealants, however, make no concession in this area. They have expansion ratios that are up to three times greater than that of other water-based sealants. Their active ingredients are insoluble in water and assure performance that is unaffected by humidity or moisture. These sealants adhere tenaciously to all common construction materials and penetrants and should not shrink while curing. An added benefit from water-based sealants is extended shelf life.

With heat-activated expansion of up to 10 times, they are equally well-suited for non-combustible penetrants or PVC pipes up of to 2 inches trade size. From simple pipe penetrations to complex combinations of pipes and cables in a single opening, these sealants cover a broad range of applications better than any other single product [6]. These sealants exhibit exceptional shear-thinning, can be troweled for vertical applications, and also can be bulk-loaded into caulking guns or caulked from standard 10.3-fl. oz. cartridges into openings in floors or overhead.

Today a great majority of sealants are water-based, latex, and single-component. These sealants apply easily and demonstrate excellent adherence to almost any common construction material. They have no noxious or potentially toxic solvent odors. They are safer for contact with penetrants and common construction materials. Since solvents are not used, the sealants do not soften plastic pipes or cable-jacketed materials. These sealants have been tested and approved for direct contact by manufacturers' plastic pipe resins. And with a neutral pH, the sealant has no corrosive effect on metallic pipe.

The single-component sealants systems save money. With about two-thirds of the cost of a project paying for labor, any product that can minimize labor costs should be given serious consideration. In many applications, a sealant can be used by itself where other systems might require collars, wrap strips, or composite boards. As little as ½ inch of sealant will provide up to three hours rating. Imagine being able to firestop insulated pipe with up to 2 inches fiberglass with sealant only! Cleanup is easy: simply use soap and water. These compounds have been fully tested to rigorous industry standards. UL-Classified and Factory Mutual Systems Approved, the sealants have been tested to the time-temperature requirements of ASTM E 119, as well as to ASTM E 814 (UL 1479).

Intumescent-type materials greatly expand when exposed to high temperatures. These materials are necessary only when high temperatures compromise the integrity of the penetrants, such as PVC (plastic) pipe or some insulated pipes. In the few cases where intumescent material is required, some companies wrap strips in conjunction with the intumescent compounds to achieve more economical UL-rated assemblies. Intumescent materials are UL-classified, expensive, and available in wrap or strip form.

FIRESTOPS—LIFE SAFETY PROTECTION

The first priority in building safety is containing both the smoke and flames in the area of origin. The firestop compound creates a system that combines both economy and performance, providing protection from fire and smoke. The firestop system should seal out smoke and toxic gases, the major causes of building fire fatalities, and seal out water, the major cause of equipment damage. It also stops dust infiltration and sound leakage.

FIRESTOP MATERIALS AND COMPOUNDS

Underwriters Laboratories, Inc., Factory Mutual, and Warnock Hersey, Inc. recognize more than 1200 firestop systems. The following is a list of a few of the commercial products available:

FlameSafe® Firestop Sealants

- Intumescent/elastomeric sealants
- Dynamic movements approved
- Endothermic sealants
- UL and ULC approved/Factory Mutual Certified/Intertek (Warnock Hersey) Listed
- 1-, 2-, and 3-hour rated systems
- Various construction applications

Nelson CLK™ Firestop Sealants

- Available in non-sag and self-leveling systems
- Extremely good in high-vibration areas
- UL and FM approved

SpecSeal® Series ES Elastomeric Sealant

- Control/expansion joints (floor and wall)
- Slab edge conditions
- Head of wall—concrete slab to masonry wall
- Head of wall—concrete fluted deck to gypsum wall

Nelson FSC™ Fire Protective Coatings

- Easily installed by brush or spray
- Highly intumescent
- Flame spread index is low
- Suitable for electrical and control cables

FlameSafe® Firestop Coatings

- Intumescent/elastomeric coating
- Dynamic movement approved
- UL- and ULC-approved
- 1-, 2-, and 3-hour rated system
- Construction joints, top of wall, and metallic pipe application

Pensil® 300 Sealant

- Expansion and seismic joints
- Curtainwall and head of wall joints
- Piping—non-combustible steel and iron
- Jacketed cable
- Bus ducts with wrap strip and cover plate

A number of other certified applications covering putties, intumescent sealants, collars, putty pads, wrap strips, foams, and mortars. Figures 15.1 through 15.5 are some typical drawings and installation guides to give the reader an idea of the use of these systems. Figures 15.1a and 15.1b illustrate four applications of an elastomeric sealant, the physical properties of which are listed in Table 15.4. Figures 15.2a and 15.2b, firestop for metallic piping through concrete floors is a 3-4 hour rated firestop. It has the following physical properties as listed in Table 15.5. Figures 15.3a and 15.3b are for Pen 300 Silicone Sealant. Its physical properties are listed in Table 15.6.

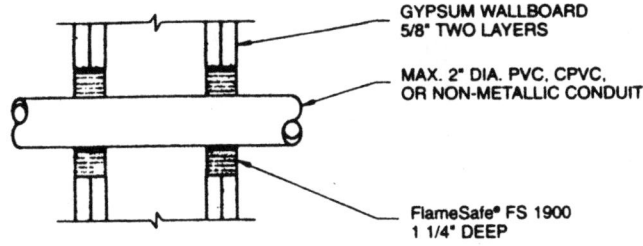

CPVC & PVC System
UL System WL 2038 – 1 & 2 Hour

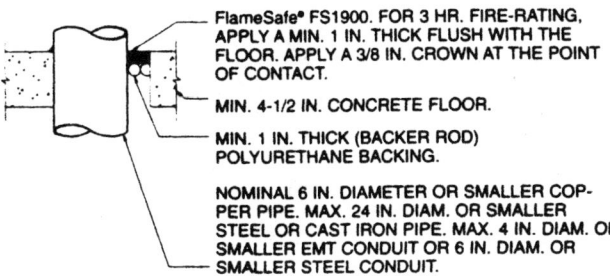

Metallic Pipe Penetration
UL System CAJ1235 – 3 Hour

FIGURE 15.1a FlameSafe® FS 1900 high performance elastomeric sealant.

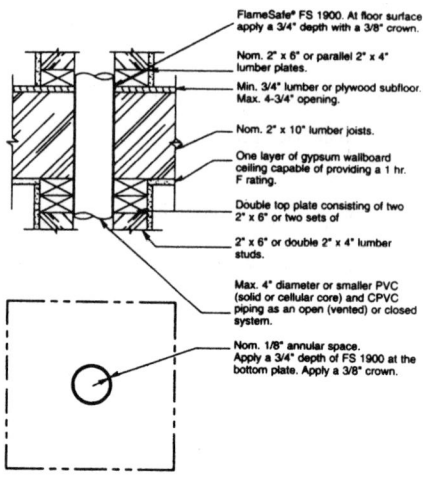

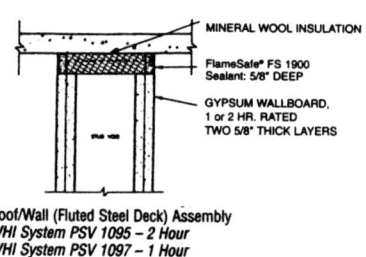

FIGURE 15.1b FlameSafe® FS 1900 high performance elastomeric sealant.

TABLE 15.4 Physical properties of an elastomeric sealant

Color	Red
Odor	Faint latex
Vehicle	Water-based
Weight	10.0 lbs/gal (1.2 kg/L)
pH	10-11
Volume expansion	700 percent
Installation temp.	40° to 90° F (4° to 32° C)
Storage temp.	40° to 90° F (4° to 32° C)
ASTM C719	Passed
Flame spread ASTM E84	0
Smoke developed ASTM E 84	5

Metallic Pipe Penetrations — Concrete/Masonry Floors & Walls

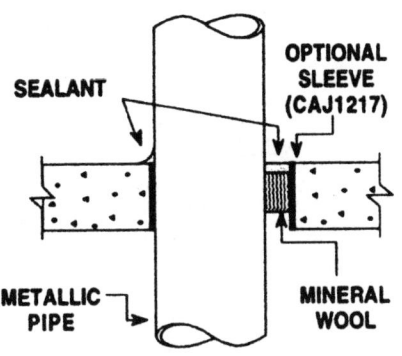

UL SYSTEM NO. CAJ1079
F RATING: 3 HR. T RATING: 0
Steel or Iron Pipe: ≤ 24", Copper Pipe ≤ 4"
Annulus: Point Contact to 2-1/4"
Sealant Depth: 1/2"
Forming Material: Nom. 4 p.c.f. Mineral Wool
Thickness: 1-1/2" for ≤ 6" Steel or Iron Pipe
3" for ≤ 4" Copper or ≥ 6" Iron or Steel Pipe

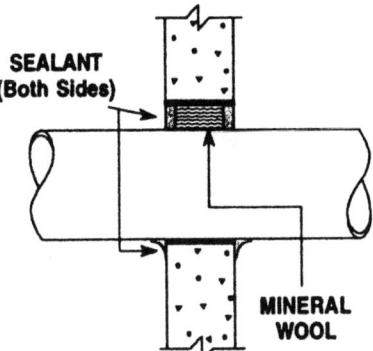

UL SYSTEM NO. CAJ1217
F RATING: 3 HR. T RATING: 0
Steel or Iron Pipe: ≤ 8", Copper Pipe ≤ 4"
Annulus: Point Contact to 1-1/4"
Sealant Depth: 1/2"
Forming Material: Nom. 4 p.c.f. Mineral Wool
Tightly Packed to a 3" Depth.

Insulated Metallic Pipe Penetrations — Concrete/Masonry Floors & Walls

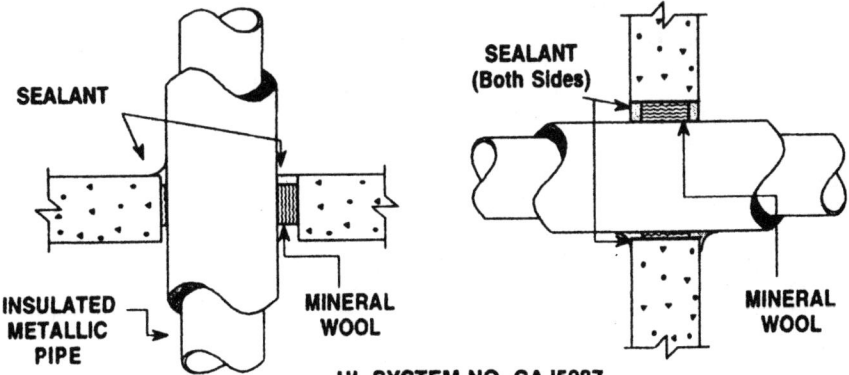

UL SYSTEM NO. CAJ5087
F RATING: 2 HR. T RATING: 1
Steel or Iron Pipe: ≤ 12", Insulated with ≤ 2"
Thick
Fiber Glass or Mineral Wool Pipe Insulation
Annulus: 1/2" to 1-1/2"'
Sealant Depth: 1/2"
Forming Material: Nom. 4 p.c.f. Mineral Wool
Tightly Packed to a 3" Depth.

FIGURE 15.2a Metallic pipe and electrical penetrations.

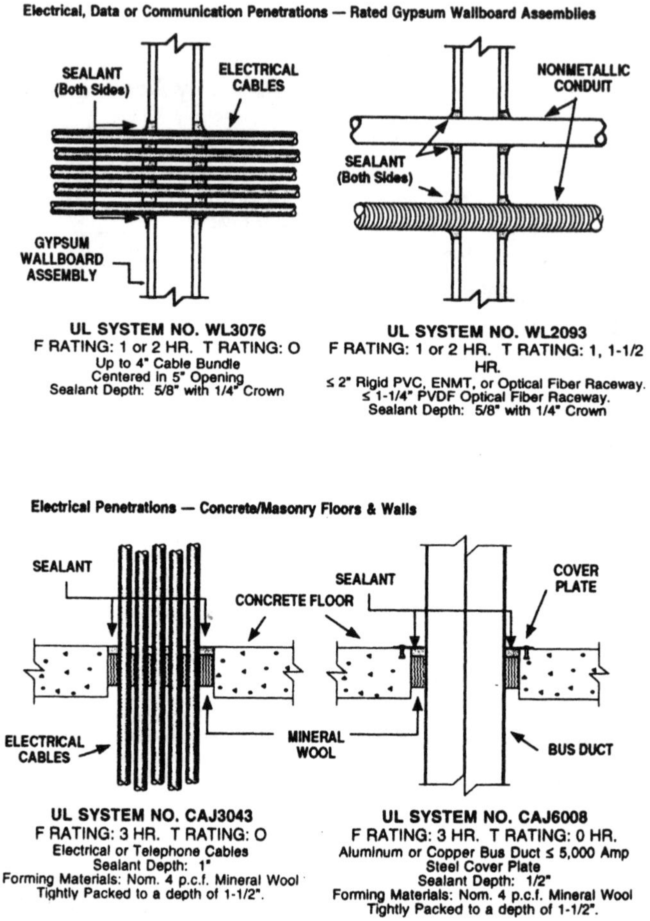

FIGURE 15.2b Metallic pipe and electrical penetrations.

TABLE 15.5 Physical Properties of a Firestop for Metal Piping Through

Color	Red
Odor	Mild latex
Density	9.4 lbs /gal.
Solids	80 percent ±2 percent
pH	8.3
Expansion begins	230° F (110° C) 1st stage • 350° F (177° C) 2nd stage
Expansion range	230° F to > 1,000° C • 110° F to > 538° C
Volume expansion	> 500% free expansion
In-service temp.	≤ 130° F
Flame spread	0
Smoke development	10

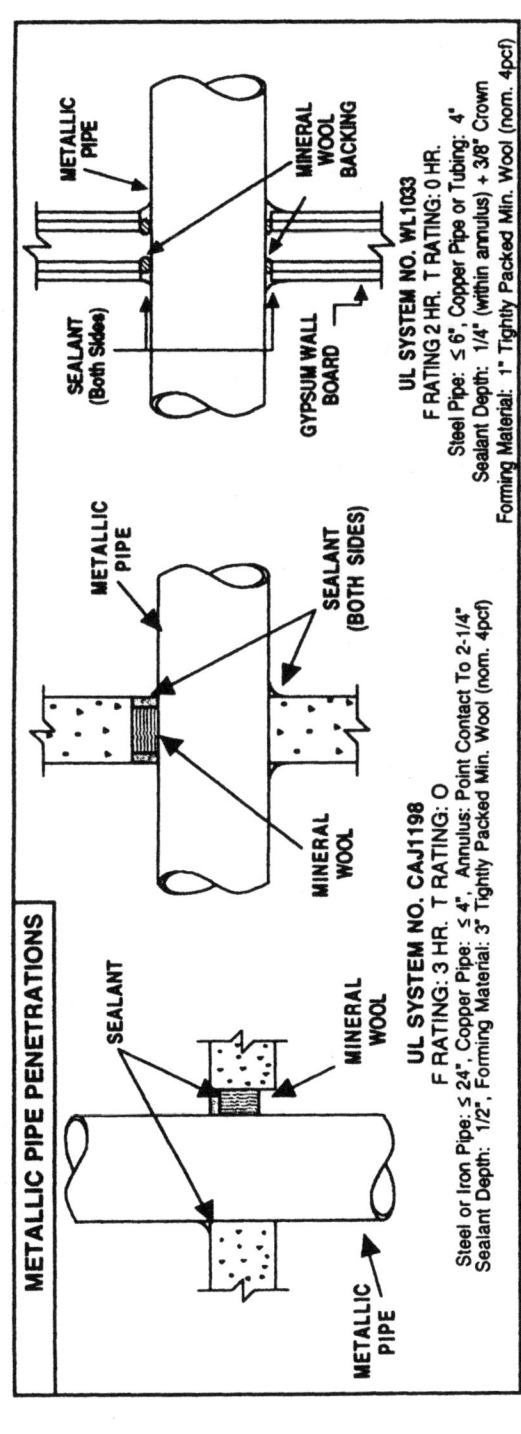

FIGURE 15.3a Pen 300 silicone sealant.

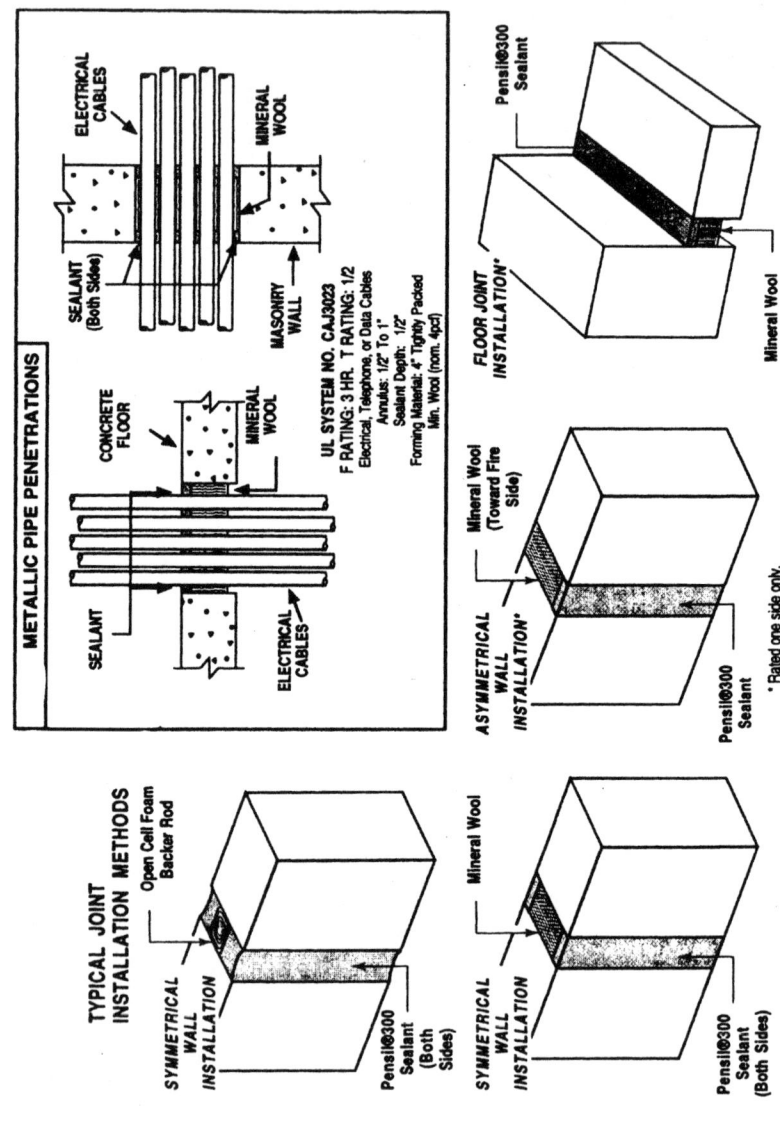

FIGURE 15.3b Pen 300 silicone sealant.

TABLE 15.6 Physical Properties of Pen 300 Silicone Sealant

Color	Concrete
Flame spread	5
Smoke development*	45
Hardness, Shore A, ASTM D2240	25
Tensile Strength, psi, ASTM D412	270
Peel Strength, piw, ASTM C794-80	55
Movement Capability, ASTM C719	±50%
Stress @ 50% extension, ½ × ½ in.	35 lbs/in.
Tooling time, minutes	30
Tack-free time, hours	5-9

* ASTM E 84 (UL 723) @ 14% coverage

Figures 15.4a, 15.4b, and 15.4c are illustrations of a FlameSafe® FS 2900 Firestop Coating. The physical properties of this coating are listed in Table 15.7.

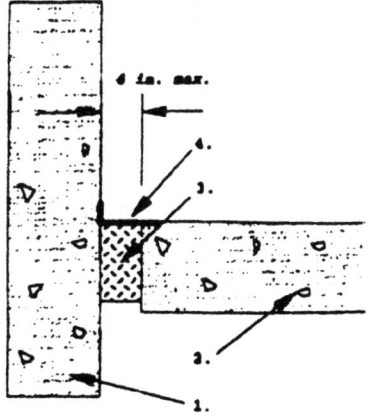

**Tested to ASTM E1399 (500 cycles @ 11 cycles/min.)
Max. 20% compression or elongation**

1. Min. 4-1/2" in lightweight or normal weight concrete or masonry wall with 2 hour fire rating.
2. Min. 4-1/2" in lightweight or normal weight concrete floor with 2 hour fire rating. Max. separation of wall and floor to be 4".
3. Min. 4" thickness of mineral wool (4 lb. density or heavier) compressed 60% into the wall/floor joint.
4. FlameSafe FS 2900 Coating. Min. 1/8" thickness (wet) sprayed or brushed onto the joint on top of the mineral wool. Provide a minimum 1/2" overlap onto the surface of the floor and wall.

Wall to Floor Construction Joint – 2 Hour

FIGURE 15.4a FlameSafe® FS 2900 firestop coating.

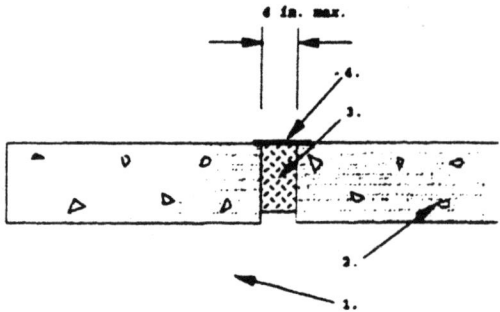

**Tested to ASTM E1399 (500 cycles @ 11 cycles/min.)
Max. 20% compression or elongation**

1. Min. 4-1/2" in lightweight or normal weight concrete or masonry wall with 2 hour fire rating.
2. Min. 4-1/2" in lightweight or normal weight concrete floor with 2 hour fire rating. Max. separation of floor to floor to be 4".
3. Min. 4" thickness of mineral wool (4 lb. density or heavier) compressed 60% into wall/floor joint.
4. FlameSafe FS 2900 Coating. Min. 1/8" thickness (wet) sprayed or brushed onto the joint on top of the mineral wool. Provide a minimum 1/2" overlap onto each floor surface.

Floor to Floor Construction Joint – 2 Hour

FIGURE 15.4b FlameSafe® FS 2900 firestop coating.

TABLE 15.7 Physical properties of FlameSafe® FS 2900 firestop coating

Color	Red
Odor	Mild latex
Vehicle	Water-based latex
Weight	8.0-9.0 lbs/gal (1.05 kg/L)
pH	10-11
Volume expansion	300% min
Installation temperature	40° to 90° F (4° to 32° C)
Storage temperature	40° to 90° F (4° to 32° C)
Solids, percent	62
Viscosity, cps	240,000 min.
Elongation/compression	Approved at 20% (2 hr) and 30% (1 hr)

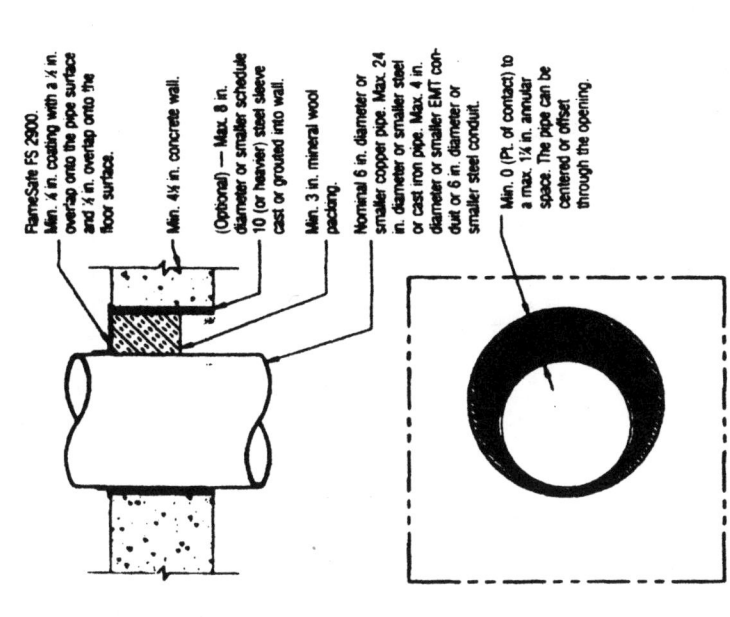

FIGURE 15.4c FlameSafe® FS 2900 firestop coating.

Figures 15.5a, 15.5b, and 15.5c illustrate a Firestop Putty and Putty Pad system with the following physical properties as listed in Table 15.8.

Metallic Pipe Penetrations — Concrete/Masonry Floor

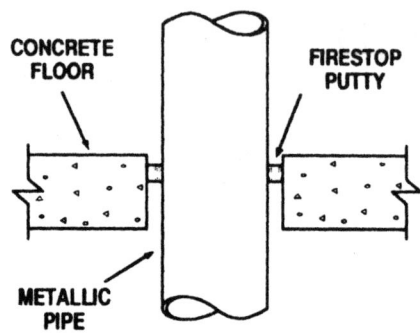

UL SYSTEM NO. CAJ1079
F RATING: 2 HR. T RATING: O
Steel or Iron Pipe: ≤ 6", EMT ≤ 4"
Annulus: Nominal 11/16"
Putty Depth: 1"
Forming Material: Optional

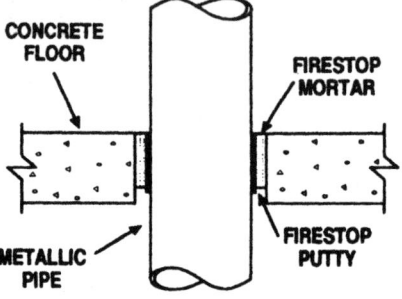

UL SYSTEM NO. CAJ8055
F RATING: 2 HR. T RATING: O
Steel or Iron Pipe: ≤ 6", EMT ≤ 4"
Annulus: 1" to 6 1/2"
SpecSeal Mortar Depth: 3 1/2'"
SpecSeal Putty Pad: 1 Layer Encircling Penetrant

Metallic Pipe Penetrations — Walls

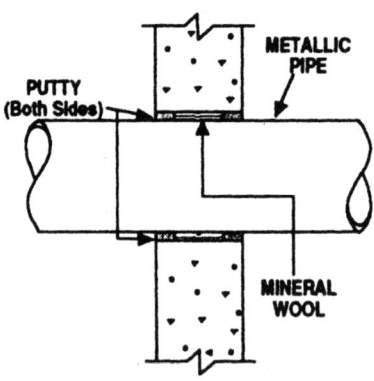

UL SYSTEM NO. CAJ1079
F RATING: 2 HR. T RATING: O
Steel or Iron Pipe: ≤ 6", EMT ≤ 4"
Annulus: Nominal 11/16"
Putty Depth: 1"
Forming Material: Optional

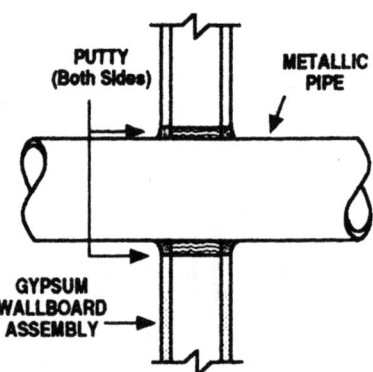

UL SYSTEM NO. WL1029
F RATING: 1 & 2 HR. T RATING: O
Steel, Iron Pipe, or EMT ≤ 4"
Max Diameter of Opening: 6"
Putty Depth: 5/8" Plus 1/4" Crown
Forming Material: Fiberglass Insulation

FIGURE 15.5a Firestop putty and putty bags.

Cable Penetrations — Concrete/Masonry Floors & Walls

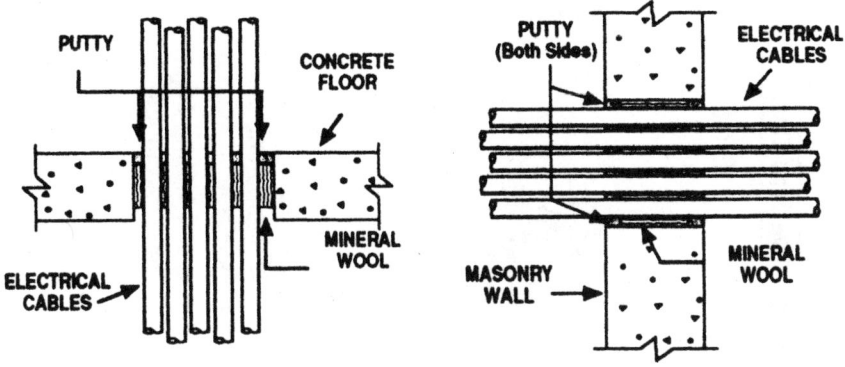

UL SYSTEM NO. WL3042
F RATING: 3 HR.
T RATING: 0 HR
Max 20% Fill Telephone, Romex Cable
Max Diameter of Opening: 5"
Putty Depth: 1"
Forming Material: 1-1/2" 4 pcf Mineral Wool

Cable Penetrations — Gypsum Board Walls

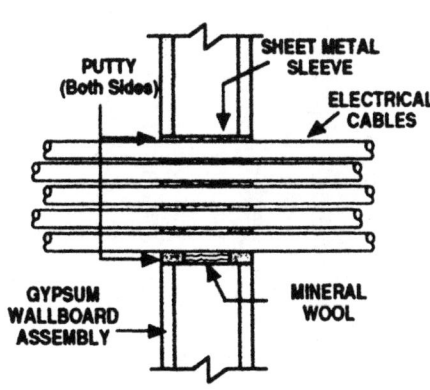 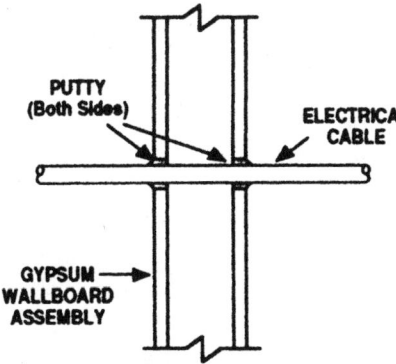

UL SYSTEM NO. WL3025
F RATING: 1 & 2 HR.
T RATING: 0 & 1/2 HR
Max 18% Fill Telephone, Romex Cable
Max Diameter of Opening: 5"
Putty Depth: 3/4" + 1/4" Crown
Forming Material: 4 pcf Mineral Wool

UL SYSTEM NO. WL3024
F RATING: 1 & 2 HR.
T RATING: 1/2, 1, & 2 HR
Telephone, Romex, SER, or MC Cable
Max Diameter of Opening: 2-1/2"
Putty Depth: 5/8" + 3/8" Crown
Forming Material: None Required

FIGURE 15.5b Firestop putty and putty bags.

INSTALLATION OF PUTTY PADS ON ELECTRICAL BOXES (Protective Wall Opening Material)

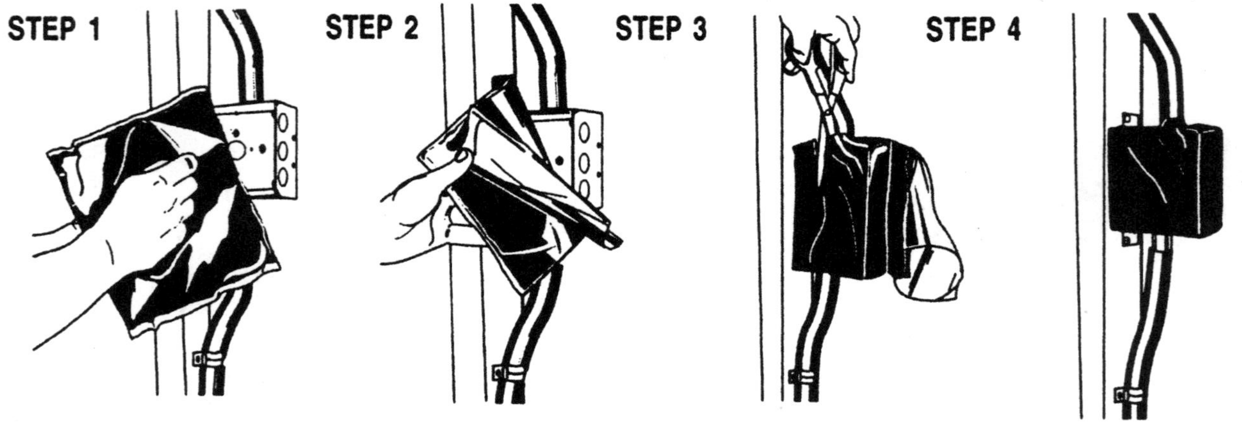

STEP 1 **STEP 2** **STEP 3** **STEP 4**

Remove poly liner from one side of pad (Step 1). Align pad to the side of box partially overlapping the stud and adhere. Working to the opposite side of the box to the edges (Step 2). If wall membrane is in place, pack putty into gaps between box and gypsum board slightly overlapping inner wallboard surface. If membrane is to be installed after pad installation, overlap front edge of box so that putty will be compressed around edges of box as wallboard is installed. Cut slits in pad to fit around conduits or cables. (Step 3). Press pad to surface of top, bottom, and sides of box (Step 4). Trim excess at corners and apply to conduit fittings connected to the box. Remove exposed poly liner. Optionally, putty may be packed into inside of conduit fittings to prevent passage of smoke.

FIGURE 15.5c Firestop putty bags.

TABLE 15.8 Physical properties of firestop putty and putty pad system

Color	Black
Odor	None
Density	1.45
Solids	100%
Expansion begins	230° F (110° C)
Volume expansion	>500% (free expansion)
Inservice temperature	± 130° F (55° C)

Figure 15.6 is an illustration of a firestop mortar and the typical physical properties are listed in Table 15.9.

In order to illustrate all the approved systems, it would take another several hundred pages. Therefore, we will let the approved sources—Specified Technologies, Inc., Dow Corning, Inc., 3M Company, International Protective Coatings Corp. and Nelson Firestop Products—supply what the architect, building owner, and contractor require.

For data relative to the amounts of materials required to seal joints and other openings, the reader can refer to the appendix.

TABLE 15.9 Physical properties of firestop mortar

Color	Red
Flame spread	0
Smoke developed	< 5
Odor	No perceptible odor
Density (as shipped)	Bulk approx. .45-.55
Density mixed, wet	0.80 ± 0.03 (6.4- 6.9 lbs/gal)
Density (dry mortar)	0.65 (5.4 lbs/gal) (40.5 lbs./cu ft.)
Water/mortar ratio	14.3/22 to 17.6/22
Yield	1250-1425 in^3
Application temp.	32° to 100° F
Open time	30 m

Metallic Pipe Penetrations — Concrete/Masonry Floor

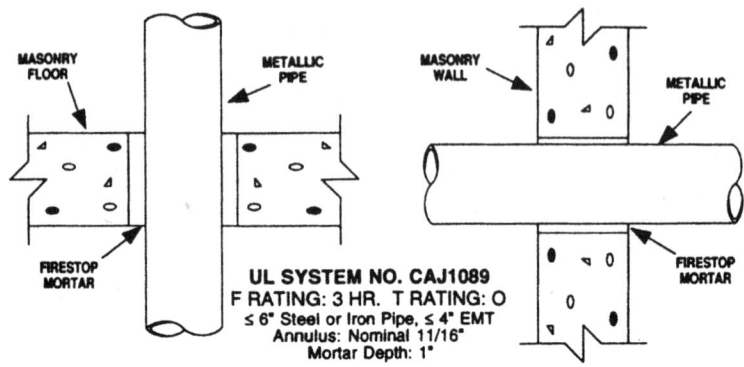

UL SYSTEM NO. CAJ1089
F RATING: 3 HR. T RATING: 0
≤ 6" Steel or Iron Pipe, ≤ 4" EMT
Annulus: Nominal 11/16"
Mortar Depth: 1"

Multiple Penetrations — Walls

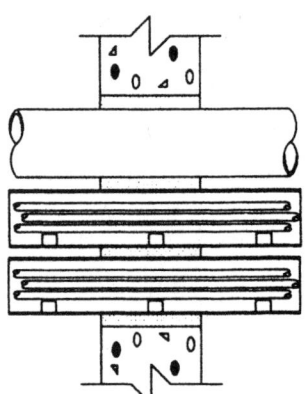

UL SYSTEM NO. CAJ0015, 16
F RATING: 3 HR. T RATING: 3
Blank Opening
Tested Opening Size: 576 in^2

UL SYSTEM NO. CAJ8016
F RATING: 3 HR. T RATING: 0
Steel or Aluminum Cable Trays
Steel, Iron, copper Pipe, Rigid Conduit, EMT
Tested Opening Size: 576 in^2

UL SYSTEM NO. CAJ8035
F RATING: 3 HR. T RATING: 0
Copper or Aluminum Cable Trays
Tested Opening Size: 576 in^2

Sealing Complex Multiple Penetrants

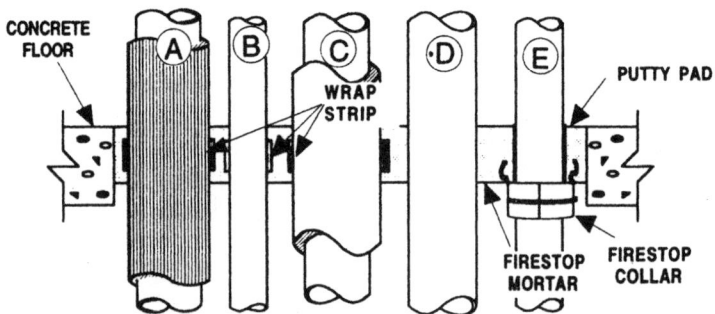

UL SYSTEM NO. CAJ8055
F RATING: 2 HR. T RATING: 0 To 2 HR (depending on penetrant type)
Penetrants: A. AB/PVC Insulated Metallic Pipe. B. Max 2" PVC Pipe.
C. Fiberglass Insulated Metallic Pipe. D. Bare Metallic Pipe. E. Max 4" PVC Pipe
Max Dimension of Opening: 24".   Mortar Depth: 3-1/2"
See UL *Fire Resistance Directory* for complete details.

FIGURE 15.6 Firestop mortar.

SPECIFICATIONS

The following is a typical specification that can be used when developing the requirements for firestopping.

Part 1—General

   1.01 Related documents

      A. The requirements of the GENERAL CONDITIONS, SUPPLEMENTARY CONDITIONS, AND DIVISION 1, GENERAL REQUIREMENTS, apply to the Work of this SECTION.

      B. Coordinate work of this section with the work of the following sections to properly execute the work in order to maintain the hourly ratings of the walls and floors.

         1. Section 03300 - Concrete Work
         2. Section 04200 - Masonry Work
         3. Section 07900 - Joint Sealers
         4. Section 0925 - Drywall
         5. Division 15 - Mechanical, Plumbing
         6. Division 16 - Electrical Work.

   1.02 References

      A. American Society for Testing and Materials Standards (ASTM):

         1. ASTM E 814-88: Standard Test method for fire *Tests Through-Penetration Firestops*

      B. Underwriters Laboratories Inc.

         1. UL 1479 *Fire Tests of Through-Penetration Firestops* (Consult UL Fire Resistance Directory)

   1.03 Description

      A. This SECTION describes the requirements for furnishing and installing firestopping for fire-rated construction. This includes:

         1. All openings in fire-rated floors and wall assemblies, both blank (empty) and those accommodating penetrating items, such as cables, conduits, pipes ducts, etc.
         2. Gaps (openings) between exterior curtainwalls and the outer perimeter edge of the structural floor.
         3. Openings at each floor level in shafts or stairwells.

   1.04 Quality Assurance

      A. Firestopping systems (materials and designs shall conform to both Flame (Fl) and Temperature (T1) ratings as required by local building code and as tested by nationally accepted test agencies per ASTM E 814 or UL 1479 fire tests in a configuration that is representative of field conditions. The F rating must be a minimum of one (1) hour but not less than the fire resistance rating of the assembly being penetrated. The T rating when required by code authority shall be based on measurement of the temperature rise on penetrating item(s). The fire test shall be conducted with a minimum positive pressure differential of 0.01 inches of water column.

B. Firestopping materials and systems must be capable of closing or filling through-openings created by 1) the burning or melting of combustible pipes, cable jacketing, or pipe insulation materials, or 2) deflection of sheet metal because of thermal expansion (electrical and mechanical ductwork).

C. Firestopping material shall be asbestos-free and shall not incorporate nor require the use of hazardous solvents.

D. Firestopping materials shall not shrink upon drying, as evidenced by cracking or pulling back from contact surfaces.

E. Do not use any firestop products that, after curing, dissolve in water. All firestopping materials shall be manufactured by one manufacturer (to the maximum extent possible).

F. Installation of firestopping systems shall be performed by a contractor (or contractors) trained or approved by the firestop manufacturer.

G. Equipment used shall be in accordance with the manufacturer's written installation instructions.

1.05 Submittals

A. Submit manufacturer's product literature for each type of firestop material to be installed. Literature shall indicate product characteristics, typical uses, performance and limitation criteria, and test data.

B. Material Safety Data Sheets (MSDS): submit MSDS for each firestop product.

C. Shop drawings: Show typical installation details for the methods of installation. Indicate which firestop materials will be used and where, as well as thickness for different hourly ratings

D. Submit manufacturer's installation procedures for each type of product.

E. Installer documentation: submit document from firestop manufacturer wherein manufacturer recognizes, i.e., approves, installer for said manufacturer's firestop products.

F. Upon completion, installer shall provide written certification that materials were installed in accordance with the manufacturer's installation instructions and details.

1.06 Product Delivery, Storage, and Handling

- A. Deliver material in the manufacturer's original, unopened containers or packages with the manufacturer's name, product identification, lot number, UL label, and mixing and installation instructions as applicable.
- B. Store materials in the original, unopened containers or packages, and under conditions recommended by the manufacturer.
- C. All firestop materials shall be installed before expiration of shelf life.

1.07 Product Conditions

- A. Conform to manufacturer's printed instructions for installation and, when applicable, curing in accordance with temperature and humidity. Conform to ventilation and safety requirements.

1.08 Sequencing

- A. Coordinate this work as required with the work of other trades.
- B. Firestopping shall precede gypsum board finishing.

1.09 Protection

- A. Where firestopping is installed at locations that will remain exposed in the completed work, provide protection as necessary to prevent damage to adjacent surfaces and finishes, and protect as necessary against damage from other construction activities.

Part 2—Products

2.01 General

- A. Firestopping materials and systems shall meet the requirements specified herein.
- B. Architect must approve in writing any alternatives to the materials and systems specified herein.
- C. All firestop products and systems shall be designed and installed so that the basic sealing system will allow the full restoration of the thermal and fire resistance properties of the barrier being penetrated with minimal repair if penetrants subsequently are removed.

2.02 Acceptable Manufacturers

- A. Dow Corning Corp., Midland, MI
- B. 3M Fire Protection Products, St. Paul, MN
- C. International Protective Coatings Corp., Oakhurst, NJ (800-334-8796)
- D. Nelson Firestop Products, Tulsa, OK, 07101, (900-331-5530)
- E. Specified Technologies Inc. (STI), Somerville, NJ 08876, (800-921-1180)

2.03 Materials
- A. Firestop Mortar
    1. Specified Technologies, Inc., SpecSeal Mortar®
    2. International Protective Coatings Corp., FlameSafe Mortar (IPC)
- B. Firestop Sealants and Caulks
    1. Nelson CLK Firestop Sealant
    2. STI SpecSeal® Sealant
    3. Dow Corning Firestop Sealant No. 2000
    4. 3M CP25WB Æ Caulk
    5. IPC FlameSafe®
- C. Firestop Putty
    1. Nelson FSP Firestop Putty
    2. STI SpecSeal® Firestop Putty Bars & Pads
    3. 3M MPS-2 Moldable Putty Stix & Putty Pads
    4. IPC FlameSafe® FSP 1000 and FSP1077
- D. Firestop Collars
    1. Nelson PCS Pipe Choke System
    2. STI SpecSeal® Firestop Collars
    3. 3M PPD Collars
- E. Wrap Strips
    1. Nelson WRS Firestop Wrap Strip
    2. SpecSeal® Wrap Strip
    3. 3M FS-195 Wrap Strip
    4. IPC FlameSafe® Intumescent Wrap Strip
- F. Accessories
    1. Forming/Damming Materials: Mineral fiberboard or other type recommended by manufacturer.

Part 3—Execution

3.01 Inspection
- A. Examine the areas and conditions where firestops are to be installed and notify the architect of conditions detrimental to the proper and timely completion of the work. Do not proceed with work until unsatisfactory conditions have been corrected by the contractor in a manner acceptable to the architect.
- B. Verify that environmental conditions are safe and suitable for installation of firestop products.

3.02 Conditions Requiring Firestopping
   A. General
      1. All through-penetrations, construction gaps, joints, and through-openings occurring in, adjacent to or between fire-rated floors, ceilings, and walls shall be firestopped as per the requirements of this specification.
      2. Insulation types specified in other SECTIONS shall not be installed instead of firestopping material specified herein.
      3. All combustible penetrants (i.e., non-metallic pipes or insulated metallic pipes) shall be firestopped using products and systems tested in a configuration representative of the field condition.

3.03 Installation
   A. General
      1. Installation of firestops shall be preformed by an applicator/installer qualified and trained by the manufacturer. Installation shall be preformed in strict accordance with manufacturer's detailed installation procedures.
      2. Apply firestops in accordance with fire test reports, fire-resistance requirements, acceptable sample installations, and manufacturer's recommendations.
      3. Coordinate with plumbing, mechanical, electrical, and other trades to assure that all pipe, conduit, cable, and other items that penetrate fire-rated construction have been installed permanently before installation of firestops. Schedule and sequence the work to assure that partitions and other construction that would conceal penetrations are not erected before the installation of firestops.
      4. Unless specified and approved, all insulations used in conjunction with through-penetrants shall remain intact and undamaged and may not be removed.
   B. Dam Construction
      1. When required to properly contain firestopping materials within openings, damming or packing materials may be used. Combustible damming material must be removed after appropriate curing. Noncombustible damming materials may be left as a permanent component of the firestop system.
   C. Field Quality Control
      1. Prepare and install firestopping systems in accordance with manufacturer's printed instructions and recommendations.
      2. Follow safety procedures recommended in the material safety data sheets.

3. Finish surfaces of firestopping that are to remain exposed in the completed work to a uniform and level condition.
4. All areas of work must be accessible until inspection by the applicable code authorities.
5. Correct unacceptable firestops and provide additional inspection to verify compliance with this specification at no additional cost.

3.04 Cleaning

A. Remove spilled and excess materials adjacent to firestopping without damaging adjacent surfaces.
B. Leave finished work in neat, clean condition with no evidence of spillovers or damage to adjacent surfaces.

3.05 Systems and Application Schedule

TYPICAL APPLICATION

Today, preventive medicine—the act of taking care of potential problems before they cause major damage—is practiced routinely. Prevention makes sense, and at the new Connecticut Children's Medical Center in Hartford, Conn., prevention started from the ground up. This new 138-bed hospital reflects significant trends in children's health-care facilities and meets the stringent demands of the building codes for health-care facilities. A very necessary preventive medicine—firestopping—has been practiced here. Firestopping maintains the integrity of a fire-rated wall, floor, or ceiling (substrate) altered by the insertion of pipes, cables, and/or conduits, or the creation of an empty space.

SUMMARY

Companies manufacture a complete line of water-based firestop materials—sealants, putties, collars, mortar, and bags. In general, sealants and putties are the most efficient, versatile, and effective firestops for plumbing and mechanical through-penetrations. These products are easy to work with, enable the penetration to be retrofit easily, and with putties, even the firestop material itself can be used. Also, elastomeric products, the latest technology, provide the required give so that the penetrations can move within the firestop material, and the intumescent materials take the worry out of compatibility with plastic or insulating piping and cables. In some products, both properties are present. As you have seen, there are many reasons why these particular firestops do best in these applications. Codes are in place and effective fire-protection methods are available. The key challenge to specifiers, contractors, designers, and inspectors is to become informed and to take full advantage of tested and proven fire protection technologies. The integrity of any of the firestop materials discussed in this chapter depends on the integrity of those involved in its manufacture, installation, and inspection. Some manufacturers feel strongly enough about proper application that they invest a great deal of time in educating contractors, professionals, and building officials. Factory-trained applicators ensure that firestop systems will be installed according to the listed design.

Additional consideration in the planning stage will be repaid many times over. Follow the guidelines, and years from now people who maintain fireproofing and other fire-prevention programs will be saluting the genius who designed the system. Plan! Plan! Plan!

REFERENCES

1. "Fire stopping protects both life and property," New York Construction News. November 1994.
2. IPC.FS.005 5, January 1997. International Protective Coatings Corp., Oakhurst, NJ.
3. Nelson Firestop Products. Tulsa, OK, 1999
4. Tyler, S. May 1991." The Basics to Better Understanding a Firestop." Walls and Ceilings. Tremco, Beachwood, OH.
5. SA727/1-97. United States Gypsum Co., Chicago, IL.
6. STI-1023 Product Catalog, 992. Specified Technologies Inc. Somerville, NJ.1

CHAPTER 16
GASKETS, FOAMED AND SOLID TAPES

INTRODUCTION

In a pure sense, the term sealant should be applied only to products that are applied as high-viscosity liquids to permanently seal or fill a void between two solid materials. Therefore, only mastics, including anaerobics, intumescents, and hot melts would qualify. Foam weatherstripping, semi-rigid and rigid PVC, Hypalon®, and neoprene products could just as well be grouped with gaskets. Preformed butyl tapes logically could be included in the tape market. Nevertheless, the primary purpose is to fill a void between two solid materials. The products are preformed like gaskets or tapes, but are not designed for filling joints of non-specific geometric shapes, such as window openings or glass edgings. This differs from what usually is perceived as a gasket—faucet washers, O-rings, and adhesive mending or bonding tapes.

TYPES OF TAPES, GASKETS, ROPES, AND FOAMS

Preformed flexible sealants are produced as ropes, tapes, foams, and gaskets. Ropes can be precured or partially cured extruded coils. Or they can be caulking cord, elastomeric ribbons of sealant (on separating release paper or film), and coils of fully cured plastic. Or they can be open rubber or closed cell foam serving typically as flexible weatherstripping or joint backing or filler material. These materials, along with foam sealants, are the second-largest product category in the sealant market.

It is estimated that the annual sales value is upwards of $100 million for preformed tapes and foam sealants. The foam and tape sealants can be grouped in two ways:

- As to their resiliency or ability to recover after being compressed or elongated
- As to whether they are solid in consistency or are bubble/cell-filled foam. Many different properties can be obtained within each class by varying the amounts of polybutene, butyl, and chlorobutyl rubber and polyisobutylene.

The three grades of resiliency are:

- Non-resilient
- Resilient
- Semi-resilient.

Table 16.1 outlines the physical properties of these types of systems.

Non-resilient tapes are based on polybutenes and are marketed as rope caulks, as shown in Figure 16.1. Its physical properties, such as durability and cure can be altered with the addition of butyl rubber. These products are used mostly for such noncritical areas as glazing and weatherstripping. Substitution for a lower molecular weight polybutene results in a knife-grade mastic.

TABLE 16.1 Preformed Tapes (100% Solids Sealing Material)*

Types	Non-resilient	Semi-resilient	Resilient
Chief ingredients	Polybutenes and selected inert fillers	Polybutenes, polysobutylenes and selected fillers	Polybutene or polysobutylene and fillers
Curing process	None	None	None
Staining of masonry	Yes	May occur	May occur
Primer	Consult manufacturer for requirements		
Ultraviolet resistance	Fair	Fair to good	Fair to good
Hardness, Shore A @ 75° F	Putty-like	5-25	5-30
Resistance to extension	Very low	Low to medium	Medium to high
Resistance to compression	Very low	Low to medium	Medium to high
Resiliency	None	Low to medium	Medium to high
Advantages: All Types	No special equipment normally required for installation. Forms functional seal immediately. Initial adhesion even at low temperature. No shrinkage. Good to excellent weatherability.		
Precautions: All Types	Failures of seals may result when sealants are: 1) Applied to moisture-containing materials such as concrete 2) Subjected to long periods of immersion in water 3) In contact with other sealants with which they are incompatible *Manufacturers should be consulted on these matters.*		

* Consult manufacturer for compatibility with sealed insulating glass and glass manufacturer for their recommendation.

Resilient tapes are based on either partially cured rubber or a combination of partially cured butyl and chlorobutyl rubber. A variety of plasticizers, pigments, and fillers is added to provide the desired properties. Partial curing of the rubber base imparts a high shear strength and weatherability. Chlorobutyl rubber provides a faster and more thorough vulcanization. The formulas in Figure 16.2 represent tapes that exhibit excellent recovery from compression and extension, which is essential when there is wide latitude of movement.

Semi-resilient tape formulations, such as in Figure 16.3, are based on butyl rubber for greater elasticity. Higher molecular weight polybutenes are added as plasticizers and tackifiers. The key to a successful formula depends on a balance of elasticity and compressibility.

The consistency types are:

- Unsupported solid tape, rope or ribbon
- Solid tape, supported or faced
- Foam, expanded or sponge.

Raw materials	Parts by weight
Polybutene, 1300-1500 MW	27.4
Asbestos fiber*	34.4
Calcium carbonate	32.6
Petrolatum	3.2
Tall oil fatty acid	0.6
Titanium dioxide, rutile grade	1.8

* It should be noted that asbestos fiber may be prohibited in many locales and substitute materials are available to impart body and thixotropy.

FIGURE 16.1 Typical polybutene base rope caulk.

Raw materials	Parts by weight	
Pre-cured butyl rubber	21.75	—
Exxon butyl 065	—	16.30
Polybutene MW 1300	30.45	31.25
Chlorobutyl rubber	—	10.90
Calcium carbonate	—	13.60
HAF LS carbon black	30.45	27.20
Silica	10.65	—
Resin	6.50	—
Zinc oxide	—	0.55
Magnesium oxide	—	0.10
Stearic acid	—	0.10

FIGURE 16.2 Resilient pre-formed butyl tape sealant.

Raw materials	Parts by weight
Exxon Butyl 268	21.20
Calcium carbonate	42.35
Polybutene MW=1300	23.30
Zinc oxide	2.10
Carbon black	0.05
Silica	10.60
Antioxidant	0.40

FIGURE 16.3 Typical semi-resilient pre-formed butyl tape.

Tapes are usually ribbons or bands of flexible plastic or rubber-like materials and are the largest and most varied form of butyl sealant. They consist of 100 percent solid materials and are supplied in a variety of widths and thicknesses, typically in the form of a roll on a core, with the layers separated by release paper. These tapes often contain a string or a cured rubber or metal insert core to prevent stretching or *drawdown* in the course of placement. In many cases, this core serves as a built-in spacer. The tape form allows clean, quick, and accurate installation.

Gaskets are used to prevent the entrance or exit of any gas or liquid, and can be termed a sealant. A wide range of products is available with highly different properties.

The principal use for all C_4 olefin-based sealants relates to glazing applications. Secondary uses are in lap joint and weatherstripping. Polybutene sealants produced as mastics are used in curtainwalls, hidden joints, acoustical sealants, and gap filling.

As noted earlier, tapes constitute the dominant usage form of pressure-sensitive adhesives. Some of the important areas of application involve air conditioner duct sealing, holding, bonding, or attachments.

Preformed foam sealants include foam strips cut to various sizes and coated with pressure-sensitive adhesives or impregnated with asphalt or polybutene. These materials possess the sealing characteristics of closed-cell foams but not the compression set or compression restrictions of closed-cell foams. They have the good compression/recovery characteristics of sponge, but will not absorb water more than 5 percent and will they last for more than 20 years. The tapes expand to seal between surfaces and stop air, vapor moisture, dust, sound, and vibration. Qualities of some common sealing materials include:

- PVC foam-closed cell—Rapid loss of recovery force. Hard to compress. Limited to approximately 50 percent compression. Takes absolute compression set with limited life.
- Rubber and PVC + rubber foams-closed cell—Dense, hard to compress. Not very moldable and takes limited compression set. Table 16.2 lists the properties of cellular preformed tapes.
- Polyurethane foam open-cell—Compresses easily with complete recovery. Does not provide an effective seal because cells are open. UV and ozone cause rapid deterioration.
- Saturated polyurethane foam open-cell—Important identifying features such as the type of saturant and the amount are important, along with the dispersion of the saturant.

Double-coated foam tapes are based on polyurethane, polyethylene, neoprene, PVC, and natural rubber foams with a pressure-sensitive adhesive on two sides. Their usage in the construction market is limited to adhering nameplates, mirror tiles, and the like.

Single-coated foam tapes are based on polyurethane, PVC, neoprene, or other foam, coated with a pressure-sensitive adhesive. Applications include gasketing in mobile homes, weatherstripping, sound-deadening, and vibration ab-

TABLE 16.2 Cellular Preformed Tapes*

Types	Closed or non-interconnecting cellular, compression seals, and does not to include impregnated open cell products.
Chief ingredients	Polyvinyl chloride, vinyl nitrile, neoprene, polyurethane
Adhesive systems	Pressure sensitive on contact
Curing process	Solvent dry only with contact adhesives
Percent compression required	Consult manufacturer
Resiliency	Medium to high
Ultraviolet resistance	Consult manufacturer
Resistance to extension	Consult manufacturer
Advantages	Forms functional seal immediately; no waiting for cure except with contact adhesives. Permanent, quick, accurate installation. No shrinkage. Good to excellent weatherability. No special equipment normally needed for installation.
Limitations	Must be held under compression. Tolerances must be close to ensure continuous compression. Joint movement should be limited depending upon tape formulation (consult manufacturer).
Precautions: All Types	Failures of seals may result when sealants are: 1) Applied to moisture containing materials such as concrete 2) Subjected to long periods of immersion in water 3) In contact with other sealants with which they are incompatible *Manufacturers should be consulted on these matters.*
Typical applications	Peripheral joints, singularly and in combination with other sealants

* Consult glass manufacturer for glazing recommendation with its product.

sorption. In a gasketing application, they find usage in dampening of noise and vibration around roof nails, heater ducts, and quarter panels. The mobile-home industry uses single-coated tapes principally around window frames and doors. In this type of application, they compete with felt strips, plastic and metal tension strips, and tubular gaskets.

Polyurethane foam is the only type of foam that is not a closed-cell structure. PVC foams prevent the passage of water, dust, or gases at a minimum compression set of 25 percent. Polyurethane foams require a much higher compression and have poor weatherability characteristics. Prices of single-coated tapes vary depending on size and types. A ½ in. × ¼ in. PVC or polyurethane tape costs about 5 to 8 cents per linear foot. Foam tapes based on EPDM rubber are the most expensive.

Impregnated foam tapes are polyester polyurethane foam impregnated with asphaltic bitumen. The latter constitutes 55 percent to 90 percent by weight of the material and forms a coating on the thin walls of the foam structure, leaving the cells unfilled. The density averages 10 to 11 lbs/ft3. When the tape is under compression in a joint, the asphalt-coated cells close and provide excellent watertight joint interfaces. To make a watertight seal, compression down to 25 percent of the original dimension is required. Several products on the market, such as Polytite, are patented. Table 16.3 outlines some of the properties of an impregnated foam tape.

TABLE 16.3 Tested Physical Properties of Impregnated Foam Tape

Properties	Values
Density, lbs/ft^3 (kg/cm^3)	14-16 (240-260)
Thermal conductivity, K cal/N° C	0.05
Temperature stability range, ° F (° C)	-40° to 212° F (-40° to 100° C)
Resistance to compression set, ASTM D 1564	Max. 2%
Peel strength, ASTM D 1000	At 1.57 rad. 0.25 revolutions/min. 12N/25mm
Softening point, ASTM D 816	Surpasses 50° C
Shear strength, N/cm^2	8 min.
Mildew resistance	Excellent
Staining	None
Flammability	Self-extinguishing per UL 94 HF 1
Flash point	590° F (310° C)
Outdoor exposure	Excellent resistance to UV
Accelerated aging, ASTM G 5377	825 hrs; minor surface degradation

APPLICATION TECHNIQUES [1]

A simple procedure used when glazing or sealing with tapes as shown in Figures 16.4a and 16.4b would be:

- Wipe clean with solvent the surface of the window frame or substrate to which the tape will be applied.
- Apply tape by unrolling it onto cleaned surface or applying precut strips with release liner still on one side.
- Use hand pressure to adhere most tapes to the frame member.
- Butt together and splice adjacent ends of the tape with hand pressure sufficient to form a continuous seal.
- Peel release liner from the tape and set the glass or panel in on setting blocks.
- Apply tape used to seal the interior of the window in a similar manner.
- Once tapes are in place, compress them against the window or frame when the interior sash or molding is snapped or screwed into place.

The techniques used to install sealing tapes will vary somewhat among particular applications, but the broad principles are the same. The glazing applications require tapes that are soft and compressible enough to be installed under any condition. Squeezeout or flow can occur because of glass deflection under wind pressure or temperature differential. A well-formulated, well-installed tape should have a service life of at least 20 years.

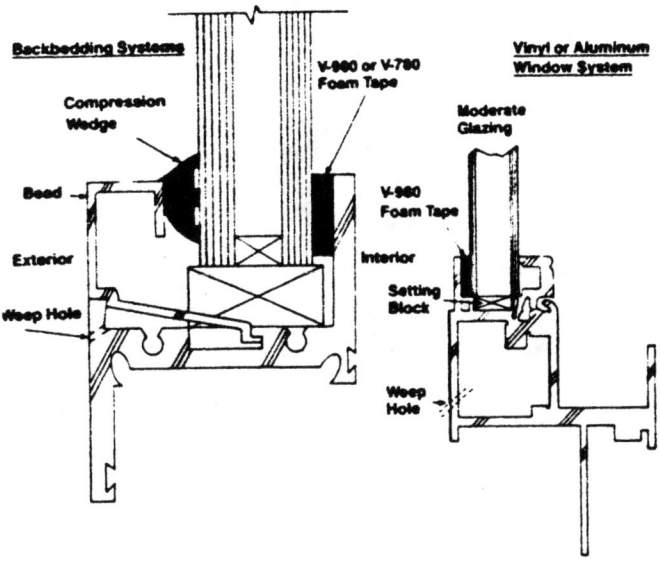

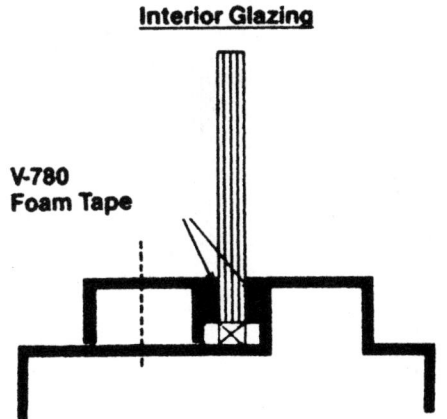

FIGURE 16.4a Typical installation of glazing and foam tapes.

When tapes are used in prefabricated construction, an immediate leakproof seal can be made as soon as the assembly is completed. This capability makes tapes ideal as sealants for factory assembly and prefabricated construction.

Finally, industrial buildings, farm structures, and mobile homes are all examples of tape sealant applications.

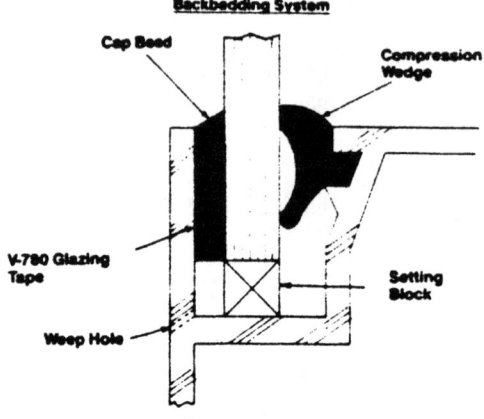

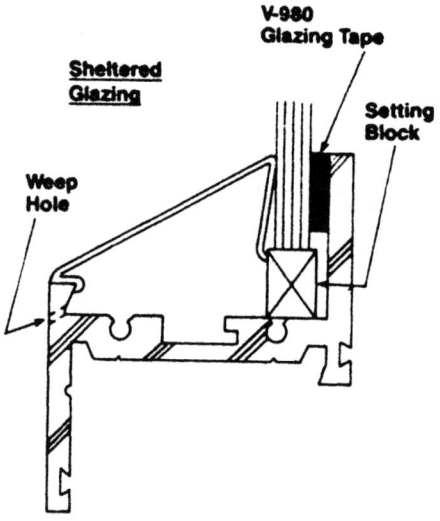

FIGURE 16.4b Typical installation of glazing and foam tapes.

GASKETS

Gaskets are used to prevent the entrance or exit of moisture and gases and are important in modern building technology. A wide range of products, with highly different properties and applications, is available. Classification of the products is difficult, but a general classification of gaskets has been developed and used with good results [2]. Different types of gaskets have different possible applications. A general classification system for joining materials in building construction is listed in Table 16.4. Building gaskets are quite resilient and are supplied preformed and ready to use. They base their sealing properties on an elastic defor-

TABLE 16.4 Gasket Classifications

Type of gasket	Description of material
Metal gaskets	Metallic strips; metallic strips combined with fiber or sponge material.
Fibrous gaskets	Woven organic material, sometimes combined with a core of open cell sponge. Impregnated wool felt. Pile of wool, nylon, etc.
Open cell sponge gaskets	Open cell sponge, with or without adhesive on one side. Open cell sponge with adhesive on both surfaces. Open cell sponge impregnated with asphalt, etc.
Closed cell sponge gaskets	Closed cell sponge with or without adhesive on one side. Open cell sponge with adhesive on both surfaces.
Resilient gaskets of solid rubber, plastics, etc.	Compression gaskets of various cross-sections (tubes, etc.) Sliding gaskets of various cross sections, zipper variety.

mation of the gasket material. To perform their intended function over time, gaskets must be kept under a certain pressure continuously and consequently must exercise a certain counter pressure. Figure 16.5 illustrates a variety of typical building gaskets.

Gaskets can be used as air barriers or rain screens and although it is possible to design double gaskets, which serve as both. In such cases, each part of the gasket performs one function. An alternate solution, with one single and simple gasket with air and water barrier combined at the same point, has been tried, but always with less than satisfactory results. The most important use for gaskets is in buildings as an air barrier between the sash and window frame and doors. Other important applications are air barriers in joints in large panel wall systems; as mounting strips in glazing; and as a rain screen in joints with a two-stage seal.

PROPERTIES AND EVALUATION

Since the primary function of a gasket is to seal, it must seal well, and it must do so for a long time. To serve the primary function first, a gasket must fulfill a number of secondary requirements derived from the primary. Because of the large selection of shapes and materials available, every single gasket has to be considered separately and tested accordingly. The following is a list of properties that should be used to evaluate a gasket:

- Shape—A building gasket is always supplied as a factory-made product with a certain shape and size. The cross-section of shapes is of vital importance for the gasket's ability to perform well.

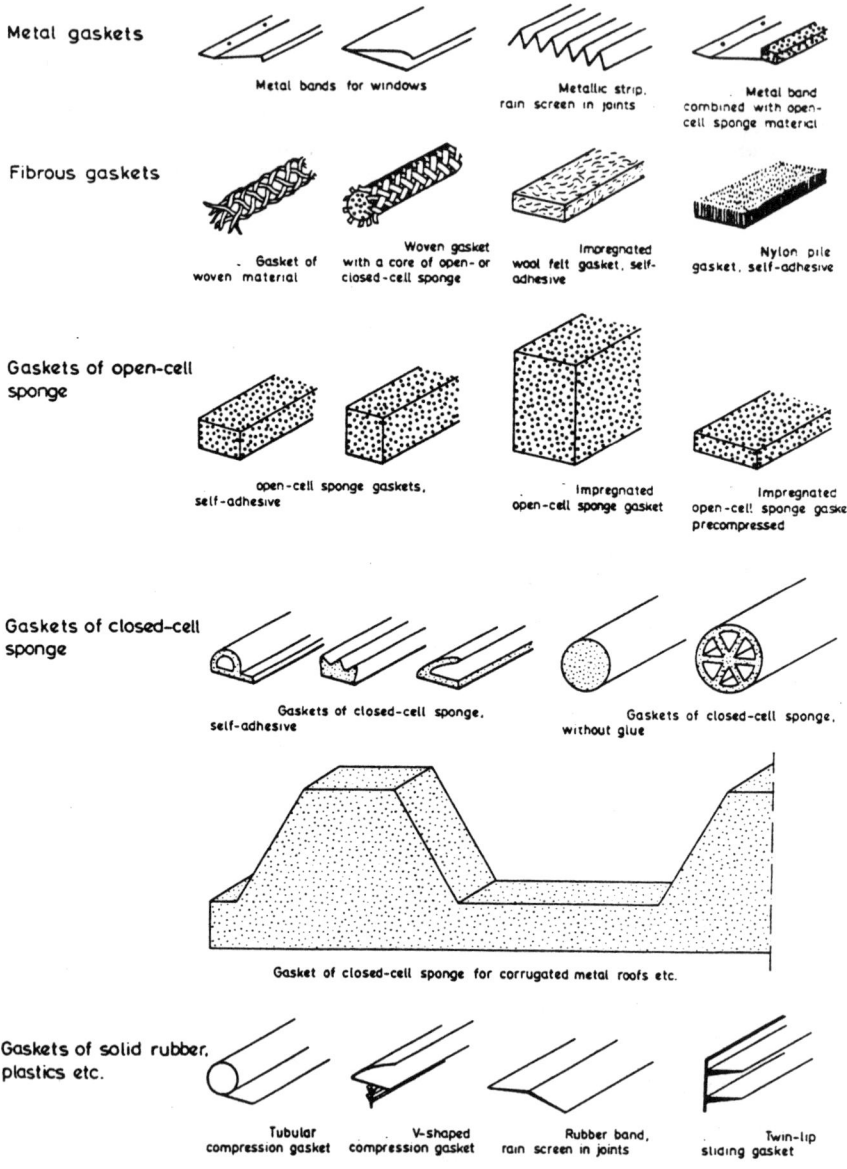

FIGURE 16.5 Typical building gaskets.

- Volume—Gaskets are in general lightweight. It is important that the manufacturer specifies the volume and package size.
- Weight—The density of the gasket material usually is not an important property, but the mass and weight per running foot can be useful to know in certain cases.

- Color—Gaskets can be found in black, white, gray, and brown. The color can influence the quality of the products. Black is the most common color.
- Toxicity—Building gaskets are supplied as vulcanized and cured products. However, they might give off certain smells when kept or used indoors. Gasket smells tend to disappear in time.
- Deformation resistance—Gaskets must be able to be compressed or deformed with reasonably heavy loads.

Building gaskets are always supplied as ready-to-use products in a certain size and shape. All that is required is to cut the individual lengths the correct way. The other necessary steps, such as pretreatment of the joint surfaces, fixing the gaskets, and joining, etc., have to be done properly.

- Pretreatment of joint surfaces—A smooth joint surface is one of the basic requirements in the use of building gaskets for joint-sealing purposes. Uneven surfaces rarely result in acceptable tightness. Damaged joints must be repaired to accept the gasket. Self-adhesive gaskets have approximately the same requirements for the condition of the joint surfaces. The substrates must be free from contaminants, such as oil, grease, and wax, and sufficiently clean and dry—otherwise the adhesive will not adhere well to the substrate.
- Fixing the gasket—In all normal applications, the gasket must be secured properly in the joints. If not, there will be an impending risk of gasket displacement. In some cases, the gaskets might even work themselves out of the joint because of the forces to which they are subjected when the joints are moving. For the current selection of building gaskets, five different mounting and anchoring methods can be used, as shown in Figures 16.6a through 16.6e. The anchoring methods are:
  - Adhering with self-adhesive gaskets
  - Anchoring with staples or tacks
  - Mounting in grooves
  - Mounting in joints
  - Mounting in building elements.

STRUCTURAL GASKETS

To meet the challenge of new construction methods, we have seen the introduction of structural gaskets in recent years. This concept was recognized as in the early 1940s with the sealing of the General Motors Technical Center. The concept caught on very quickly and spread. A technique called the zipper gasket was devised for those materials. Those gaskets eventually became known as lock-strip gaskets. A trial-and-error period took place at first to determine the proper selection of designs, some of which were successful, others that did not perform as well. Figure 16.7 shows several examples of this early gasket concept illustrating how the lock strip functions.

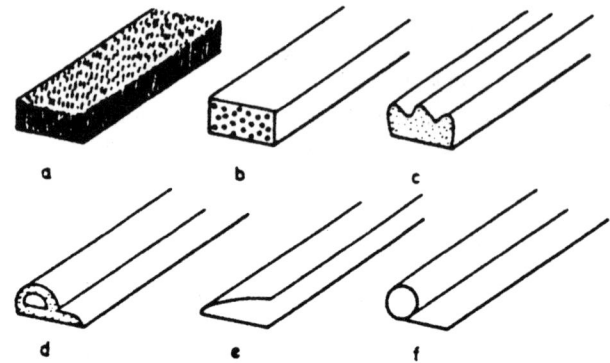

16-6a Self-adhesive gaskets. a) nylon pile gasket, b–d) gaskets of rubber or plastics sponge material, e–f) resilient gaskets of solid rubber or plastics

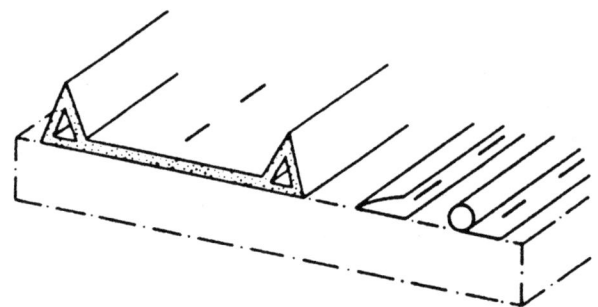

16-6b Gaskets fixed by staples or tacks

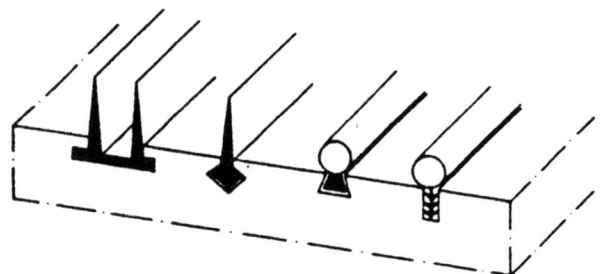

16-6c Gaskets mounted in grooves

FIGURE 16.6a Anchoring the gaskets.

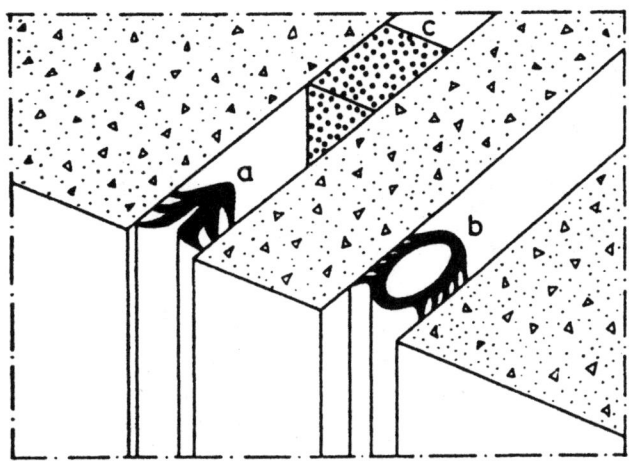

16-6d Gaskets mounted into joints. a) gasket forced into the joint, b) vacuum tube, c) gasket of precompressed open-cell sponge

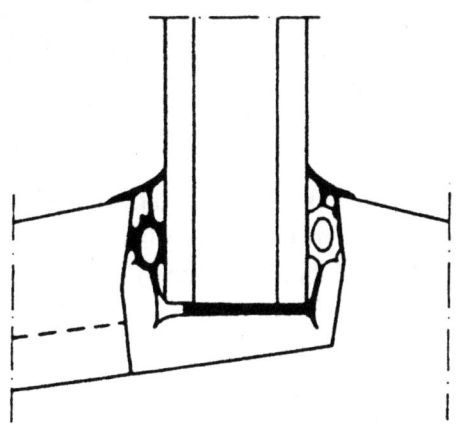

16-6e U-shaped gasket mounted on a sealed glazing unit

FIGURE 16.6b Anchoring the gaskets.

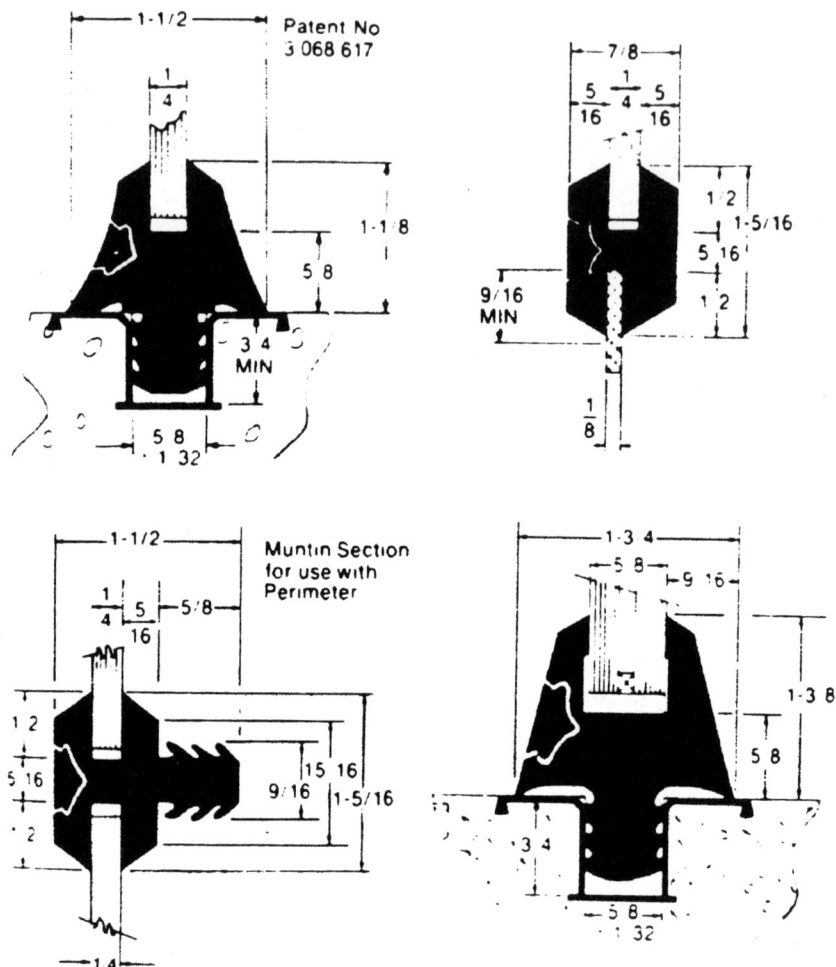

Structural neoprene glazing gaskets. These gaskets can be modified for various thicknesses of insulating glass from ¼ inch to 1 inch. (Courtesy of Maloney Rubber Products)

FIGURE 16.7 Structural glazing gaskets.

In addition, Table 16.5 outlines the requirements and test methods for lockstrip gaskets. Figure 16.8 shows that certain amounts of movement take place with a zipper-type gasket. Design configurations resulted in basic requirements such as lip pressure, which should be a minimum of 4 pounds per linear inch.

TABLE 16.5 Lock-Strips Gaskets, Requirements

Properties	Requirements
Tensile strength, psi, ASTM D 412	2000
Elongation %, ASTM D 412	175
Tear strength, pli, ASTM D 624 Die C	120
Hardness, Shore A, ASTM D 2240	75 ± 5
Compression set, 22 hours at 212° F, %	35
Brittleness temp. min, ASTM D 746	-40° F
Ozone resistance, ppm, ASTM D 1149 100 hours at 100° F at 22% elongation	no crack
Heat aging, 70 hours at 212° F, ASTM D 573 Change in hardness, max Loss in tensile strength, max, % Loss in elongation, max, %	+ 10-0 duro points 15 10
Flame propagation	none
Lip seal pressure	4 lbs/inch

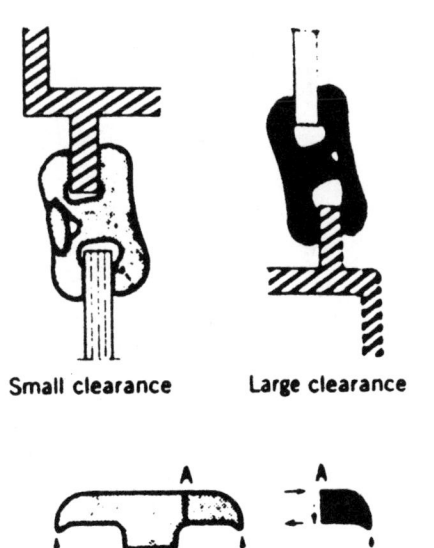

Action of structural gasket under load.
(top) Action under wind loads. (bottom) Internal stresses.

FIGURE 16.8 Action of a structural gasket under laod.

SUMMARY

The 100 percent solid preformed sealant tapes, because of their form and ease of handling, provide economies in installation often unattainable with other sealants. Tape rolls are available in a variety of sizes and shapes to accommodate the architects' designs. Most tapes have a degree of pressure sensitivity and are sufficiently soft to allow for ease of compression during installation.

A gasket's primary function is to seal out dirt, dust, or water in order to reduce discomfort in a structure. They can also serve a holding function as adhesives. Formed-in-place gaskets are a final example of adhesive sealants. They are available in a variety of shapes and are manufactured from backboned materials such as polyurethane, polyisobutylene, silicones, and polyesters. Proper selection of the material and design of the lock-strip gasket is extremely important, in addition to proper installation.

REFERENCES

1. "Glazing Products." 1990. Norton Performance Plastics Corp. A Saint-Gobain Co.
2. Gjelsvik, T., Berg, S.B., Johansen, T.S. 1988. "Building Gaskets." Norwegian Building Research Institute, Olso/Trondheim, Norway.

CHAPTER 17
FORMULARY FOR SEALANTS AND ADHESIVES

INTRODUCTION

Thousands of uses for adhesives, caulks, tapes, gaskets, coatings, membranes, and sealants can be found in the construction market. Many of the uses are similar to the industrial and do-it-yourself markets.

THE MARKETS

Many of these applications areas for the building construction market and the uses in each can be found in Table 17.1, Building construction market; Table 17.2, Glass and glazing; Table 17.3, Maintenance, commercial or industrial markets; Table 17.4, Insulating glass manufacturing; Table 17.5, Heavy construction; Table 17.6, Mobile homes; Table 17.7, Residential construction; Table 17.8, Do-it-yourself market; and Table 17.9, Contractors' uses.

TABLE 17.1 Building Construction Market

Architectural metal curtain wall joints
Window wall joints
Precast concrete walls, slabs, or sections (butt joints)
Floor joints, expansion/contraction
Mullion to wall panels
Mullion joints
Perimeter window sealing
Perimeter door sealing
Perimeter ventilation and air conditioning vent frame sealing
Decks, plazas, and sidewalk joints
Flashings, counter, and reglet
Masonry tuckpointing
Building sole plate sealing
Metal roof panels, butt and lap joints
Static or working joints on exterior of buildings
Bedding of metal panes, sheets, rails, or moldings
Damming or sealing exterior electrical or plumbing access holes
Acoustical sealing in drywall and other building spaces and partitions
Canopies and metal awning joints
General weatherstripping

TABLE 17.2 Glass and Glazing

Wet or dry, small- or large-pane glazing
Factory glazed windows at manufacturers plant
Back bedding
Heel bedding
Foam tape with or without shims

TABLE 17.3 Maintenance, Commercial, or Industrial

Glazing repair and maintenance
Joint repair and maintenance
Roof flashing repair
Gutters and downspouts
Sidewalk, plazas, platforms, and driveway cracks and joints
Wall and floor cracks
Masonry joints equipment and machinery (anaerobic) sealing

TABLE 17.4 Insulating Glass Manufacturing

Single and dual edge sealing
Hot melt sealant application
Heat Mirror® sealing
Warm edge technology sealing

TABLE 17.5 Heavy Construction

Airfield runways, aprons, highway and bridge joints
Bridge abutments
Concrete lined canals, traverse and longitudinal joints
Multilevel parking lot joints
Underground tunnel construction joints and sections
Utilities and processing plants

TABLE 17.6 Mobile Homes

Perimeter caulking of window and door frames
Window glazing
Vent and electrical access frame sealing
Roof-to-body joint and seam sealing
Roof line rain seam sealing
Shipping containers, joints and seams

TABLE 17.7 Residential Construction

Sealing and caulking around windows and doors
Sole plate sealing
Exterior chimney sealing and caulking
Sealing of through vents
Weatherstripping of doors
Sealing of cracks in patios, driveways, sidewalks, and deck sections
Joint sealing between house and attached patios and porches
Sealing between slab foundation and house proper

TABLE 17.8 Do-It-Yourself

Caulking and sealing around windows and doors
Weatherstripping
Roof flashings
Air conditioners and vent sealing
Tubs, sinks, and shower caulking and sealing
Re-glazing of windows
Minor roof patching
Swimming pool cracks, joints, and copings

TABLE 17.9 Contractors' Uses

Roofers: Flashings; coping joints; skylights, perimeter and flashings; valley cracks repair; patch work and minor roof repair

Heating, ventilation and air conditioning: Pressurized duct sealing; exhaust and intake vents; interior ducts, vent and mechanical rooms, etc.; seam and joint sealing for acoustical noise reduction; sealing joints and perimeters where ducts pass through walls and partitions; fire stops and flame proofing

Plumbers: Waste and drain pipe joints; toilet tank and bowl sealing; sealing of pass-through pipes and vent systems; waterproofing around tubs, sinks, and showers

Tile contractors: Sealing and waterproofing around tubs and other fixtures

Siding contractors: Pool and deck seams; flange sealing; joints, copings, and crack sealing

THE FORMULARY [1–10]

Sealant formulations cover a full series of sealant and caulks ranging from acrylics, butyls, polybutenes, and polyisobutylenes to polyurethanes and silicone materials, in addition to concrete curing systems and tile adhesives.

Tile and contact cement formulations also are included for study, using Figures 17.1 through 17.21.

Refer to Chapters 2, 3, 5, 8, 11, 12, 13, and 16 for additional formulas of a variety of basic generic sealants. Figure 17.1 offers three formulations based on solution acrylics for a clear sealant and two white materials.

Raw materials	SAC Clear	Parts by weight Sac White #1	SAC White #2
SM AR01, acrylic polymer (68% NVM in toluene)	80.0	48.70	—
SM AR01, acrylic polymer (83% NVM in xylene)	—	—	42.78
Drikalite, calcium carbonate	—	47.0	—
Duramite, calcium carbonate	—	—	46.40
Aerosil 200	6.30	—	—
Thixatrol ST	—	1.90	4.26
Toluene	13.60	2.30	—
Xylene	—	—	6.46
Memo, silane	—	—	0.10
Glymo, silane	0.10	0.10	—

Physical properties		Values			
% NVM by weight	60.7		82.1		86.3
Density, lbs/gal, ASTM D 1475	8.64		12.05		11.85
Pigment/polymer ratio	0.11/1		1.42/1		1.43/1
Slump, inches, ASTM D 2202	> 0.1		> 0.1		> 0.1
Extrusion rate, (sec/g) 60 psi, 0.104 inch orifice	<1		1		51
Tack-free time, min., ASTM D2377	30		30		180
Tensile strength, psi, ASTM D 412	20		30		80
Elongation, %	> 1000		> 1000		> 500
Recovery, %	80		80		50
Shore S hardness	20		35		50
Peel strength, pli, ASTM C 794	wet	dry	wet	dry	wet	dry
Glass	9/100	0/AF	10/100	0/AF	6/100	6/100
Anodized aluminum	9/100	8/100	9/100	7/100	6/100	6/100
Wood	9/100	4/AF	9/100	4/100	6/100	4/100

FIGURE 17.1 Solvent release systems, acrylic base. *(Courtesy of Schnee Morehead, Inc., Irving, Texas)*

In Figure 17.2, formulations are presented at different percentages of solids and compared to a control sample for an acrylic caulk. The intent of these formulations is to offer a comparison of costs while still maintaining good physical properties.

FORMULARY FOR SEALANTS AND ADHESIVES   17.5

Figure 17.3 is a suggested starting formula for a contact cement, solvent type. Another type of construction adhesive, based on an SBR raw material, is shown in Figure 17.4.

Raw materials	Parts by weight				
	CA-12 @ 40% solids		CA-12 @ 48% solids		CA-12 control
	1	2	3	4	5
Rhoplex CA-12 3-5546	75.5	75.5	90.6	90.6	100.0
Water	24.5	24.5	9.4	9.4	—
Vinol 540	2.4	3.9	0.95	2.2	—
Water	7.2	11.7	2.85	6.7	—
Totals	109.5	115.6	103.8	108.9	100.0
Physical properties			Values		
Solids, %	38.7	38.0	47.2	46.1	53.0
pH, initial	5.8	5.8	5.7	5.7	5.7
30 days	4.8	4.9	4.6	4.7	4.4
Viscosity, cP #4/60 rpm	1190	2850	1110	2900	180
Lap shear, psi	187	232	185	203	193

FIGURE 17.2 Contact cement, low cost, non-flammable. *(Courtesy of Rohm & Haas Co., Philadelphia, PA)*

Raw materials	Parts by weight
Methyl ethyl ketone	551.22
Hycar 1041 polymer	100.00
Durez 11078 resin	66.99
Diisobutyl ketone	35.18
Hi-Sil 233	15.00
Hexamethylenetetramine (Durez26309)	6.60
Zinc oxide	5.00
Benzothiazyl disulfide	1.50
Sulfur	1.50
Solids	70%

FIGURE 17.3 Contact adhesive, solvent type.

Raw materials	Parts by weight
SBR 1013	60.0
SBR 1009	40.0
Super Sta-Tac 80	75.0
Nirex V-2040 HM	55.0
Cyanox	2.0
Dixie clay	130.0
Atomite	130.0
Toluene	85.0
Hexane	126.0
Solids	70%

FIGURE 17.4 Construction adhesive, SBR base.

17.6 HANDBOOK OF ADHESIVES AND SEALANTS IN CONSTRUCTION

Figures 17.5 through 17.7 all are related to the sealing of insulating glass units. They are a butyl-based, a silicone-based and a butyl mastic.

In Figure 17.8, we offer a high-strength cement, based on a chlorobutyl rubber, which can be used as an adhesive in the metal building industry.

Raw materials	Parts by weight
Exxon Butyl 065	170.0
Vestoplast x3632	66.5
Black Masterbatch PZ700 BL-35	2.5
Calofort S	55.0
Calmote AD	167.0
Elvax 210	74.0
Stabelite Ester 10	200.0
Hyvis 2000	66.5
Silane A1120	11.0
Physical properties	*Values*
Appearance	Smooth, lump free, free of air and foreign matter
Softening point, ball and ring	258.8-320° F (126-160° C)
Viscosity, Atkinson-Nancarrow $^3/_{32}$ in. × 1 in. nozzle, $^1/_2$ in./min at 302°F (150°C)	6-10 lbs.
Tensile shear, psi, aluminum/aluminum	> 35 CF
Peel strength, pli	> 8 CF
Shelf life	> one year
Specific gravity	1.15-1.19

FIGURE 17.5 Hot melt butyl insulating glass sealant. *(Courtesy of Exxon Chemical Co., Houston, Texas)*

Raw materials	Parts by weight	
Butyl elastomer, Polysar XL 6810	156.0	166.0
Carbon black HAF, United N-330	235.0	255.0
Phenolic resin, Super Beckacite 2000	295.0	256.0
Polybutene H-900, Amoco	235.0	175.0
Alpha methyl styrene polymers, Resin 18, Amoco	—	59.0
Ethylene propylene rubber, Polysar EPM 306	79.0	89.0
Physical properties	*Values*	
Lap shear, psi, (kPa)	59 (407)	62 (427)
Hardness, Shore A	54	64
Penetration, mm ASTM C 782	12	7
Bead load, hr to failure at 158° F (70° C), 500 gm load	4.2	8+

FIGURE 17.5A Hot melt butyl insulating glass sealant. *(Courtesy of Amoco Chemical Co., Chicago, Ill.)*

Raw materials	Parts by weight
Silicone Resin E 50, hydroxy terminated polydimethylsiloxane	41.94
Silicone fluid M-1000, polydimethylsiloxane	4.01
Masterbatch, silicone	6.75
Crosslinker AC 3187	2.48
Cab-O- Sil LM 130	2.89
Omya BLR 3, calcium carbonate	41.85
Aromatic 100, solvent blend	0.07
Cotin 200, dibutyl tin dilaurate	0.03
Masterbatch Silicone Fluid M 1000 Elf Tex 12 *Crosslinker* Ethyltriacetoxysilane Methyltriacetoxysilane	5.69 1.04 1 1

FIGURE 17.6 Silicone insulating glass sealant. *(Courtesy of Dow Chemical Co., Midland, New Jersey)*

Raw materials	Parts by weight
Vistanex LM-HM	1000
Mistron vapor talc	480
Carbon black, N990 (MT)	20

FIGURE 17.7 Extruded mastic insulating glass sealant. *(Courtesy of Exxon Chemical Co., Houston, Texas)*

Raw materials	Parts by weight
Chlorobutyl 1066	100
Ditridecyl phtlate (DTDP)	5
Diethyl thiourea (DETU)	5
Toluene	360
Isopropyl alcohol	40

Mixing procedure

1. Preblend a paste of DETU in warm DTDP
2. Banbury or mill mix the compound for 5 minutes at 240°-280° F (115.56°-137.78° C)
3. The temperature must be kept below 300° F (148.89° C)
4. Solvate the base stock.

Note: High shear mixing at moderate temperature will result in the formation of a thiouronium salt, which, through ionic bonding, will increase the green strength of the gum polymer while still maintaining full solubility. This base is a good high-strength cement in its own right. To obtain a room temperature curing cement, add just before use:

Zinc resinate	10 parts
Toluene	10 parts

FIGURE 17.8 High-strength cement.

Figures 17.9 and 17.10 both are based on butyls and are of the solvent-containing type. The proportion of the solids varies only slightly, but these are proven efficient caulk materials for interior as well as exterior applications where movement is low.

As in many construction applications, tapes play an important role; therefore, Figures 17.11 and 17.12 are shown as tapes used for glazing purposes. One is a non-drying tape, while the other is semi-curing tape, offering three suggested formulas. In addition, Figure 17.13 shows a semi-resilient, butyl, sealing tape. Figure 17.14 is a white, butyl-based system of good quality.

Raw materials	Parts by weight
Bucar 5214	86.0
Super Sta-Tac 80	17.0
Indopol H 100	108.0
Camel Carb, calcium carbonate	432.0
IT-3X	216.0
Titanium dioxide, rutile grade	10.0
VM and PNaphtha	129.0
Solids	87.0%

Mixing procedure

1. Swell rubber in solvent
2. Charge into a Sigma blade mixer or suitable mixing equipment
3. Charge all resins talc, TiO$_2$, $^1/_2$ polybutene and $^1/_2$ calcium carbonate
4. Add remaining calcium carbonate
5. Add remaining polybutene
6. Mix until batch is homogeneous and free of gritty material

FIGURE 17.9 Butyl caulk, solvent. *(Courtesy of Reichhold Chemicals, Inc. Pensacola, Fla.)*

Raw materials	Parts by weight
Butyl 268	57.0
Vistanex L-100	106.0
Super Sta-Tac 100	82.0
Ondopol H-100	163.0
Atomite	82.0
IT-X	360.0
Mineral spirits	149.0
Solids	85.0%

Mixing procedure

1. Charge butyl rubber polyisobutylene to a Sigma blade mixer with solvent. Mix.
2. Add resin, $^1/_2$ polybutene, talc, $^1/_2$ calcium carbonate. Mix.
3. Add remaining calcium carbonate. Mix.
4. Add remaining polybutene. Mix.
5. Mix until system is homogeneous and free of grit.

FIGURE 17.10 Butyl-based sealant. *(Courtesy of Reichhold Chemicals, Inc., Pensacola, Fla.)*

Raw materials	Parts by weight		
	1	2	3
Polybutene H-300, Amoco	240.1	240.1	240.1
Amporphous polypropylene	50.50	50.50	50.50
Homopolymer, Eastman MSK	elim lines		
Butyl rubber, Kalar 5214	17.2	17.2	17.2
Clay, Sno-brite	163.7	163.7	163.7
Calcium carbonate, Atomite	440.5	520.5	520.5
Diatomaceous earth, Celite281	40.0	—	—
Cotton fiber, Filfloc S 160-900	48.0	—	—
Aramid fiber, Kevlar IF 302	—	8.0	—
Ceramic fiber, Fiberfrax HSA	—	—	8.0

Mixing procedure

1. Premix amorphous polypropylene with twice its weight polybutene under low heat to promote complete dispersion.
2. Charge the mixer with calcium carbonate and other ingredients in the listed order.
3. Mix entire mass for one hour after last addition
4. Extrude mass onto release paper.

Note: An AMH sealant mixer-extruder should be used.

Physical Properties		Values	
Formula	1	2	3
Basic visual screening, no cracks or bleeding	Pass	Pass	Pass
Yield strength, psi, > 6	17.9	15.6	12.3
Sag, inch,			
RT, 0 in max	0	0	0
after UV, $1/16$ in max.	0	0	0
heat aged, $1/16$ in max	0	0	0
Vehicle migration, inch			
RT, 0 in max	0	0	0
after UV, $1/8$ in max	0	0	0
heat aged, $1/8$ in. max.	$1/16$	$1/8$	$1/8$
Low temperature flexibility	Pass	Pass	Pass
Penetration after heat aging, > 50	75	69	106

FIGURE 17.11 Polybutene-based tapes, non-drying. *(Courtesy of Amoco Chemical Co., Chicago, Ill.)*

Raw materials	Parts by weight
Butyl XL 30102	100.0
Atomite, calcium carbonate	347.0
Camel Carb, calcium carbonate	25.0
Zinc oxide	15.0
Aluminum stearate	2.5
Irganox 1010, antioxidant	0.5
Monomix, talc	10.0
Indopol H-100, polybutene	75.0
Indopol H-1900, polybutene	145.0
Hi Sil 422, silica	90.0

Physical properties	Values
Tensile adhesion, psi	17.5
Elongation, %	650
Flow, 3/16 inch bead, 24 hrs. @ 158° F (70° C)	None
Cone penetration, 300gm/5 sec, units = 1/10 mm	
0° F	48.0
77° F	78.0
120° F	120.0

FIGURE 17.12 Architectural glazing tape. *(Courtesy of Polysar Chemical Co.)*

Raw materials	Parts by weight
Exxon Butyl 268	100
Silene 732D silica	50
Titanium dioxide	50
Atomite whiting	300
Mistron vapor talc	150
Parapol 1300 polybutene	200
AgeRite White antioxidant	2

FIGURE 17.13 Semi-resilient, butyl-based sealing tape. *(Courtesy of Exxon Chemical Company, Houston, Texas)*

Raw materials	Parts by weight
Polybutene H-300, Amoco	320
Butyl rubber, Katar 5214	80
Calcium carbonate, Atomite ECC	350
Attapulgus clay, Attagel 50	160
Diatomaceous earth, Celite 281	50
Titanium dioxide, Ti pure R 900	40

Physical properties	Values
Impact strength, ASTM C 766	Pass
Weight loss, %, ASTM C 771	1.78
Stain index, ASTM C 722	1.10
Slump, ASTM D 2376	Pass
Basic visual screening, no cracks bleed, etc.	Pass
Yield strength, > 6	11.4
Low temperature flexibility, no cracks, bond loss, etc.	Pass
Penetrating after heating, > 50	98.0

FIGURE 17.14 Butyl-based tape.

As in many construction glazing applications, there is a need for a clear sealant system, Figure 17.15 illustrates a possible choice.

For low-cost applications where gunning efficiencies are important, a gun-grade caulk is shown in Figure 17.16.

In Figures 17.17 through 17.22, suggestions are offered for materials of a cementitious nature, principally used in various flooring applications. They deal with a thin bed terrazzo, sand-filled flooring, trowel coating, seamless flooring, cement topping, and a thin bed mortar for setting tile.

Raw materials	Parts by weight
Thermoplastic elastomer (high cohesion), Kraton G 1652	148.0
Thermoplastic elastomer (high adhesion), Kraton G 1701	73.0
Tackifying resin, Escorez 5380	257.0
Reinforcing resin, Kristalex 1120	105.0
Alpha methyl styrene polymers, Resin 18	44.0
Adhesion promoter, Silane 189	5.0
Antioxidant, Irganox 1010	4.0
UV absorber, 1/1 Tinuvin 327/770	3.0
Xylene	251.0
Polybutene H-1500, Amoco	110.0

Physical properties	Values
Vertical slump, inch ≤ $3/16$ inch	0
Horizontal slump, none	None
Extrusion rate, sec. ≤ 45	31
Hardness, Shore A-2 >15 < 50	15
Tack-free time	Pass
Stain or color change	Pass
Adhesion in peel, ≤ 5 pli	
Glass	8
Aluminum	15
Concrete	12

FIGURE 17.15 Clear sealant, solvent-based. *(Amoco Chemical Co., Chicago, Ill.)*

Raw materials	Parts by weight
Polybutene, Amoco	200.0
Vinyltoluene/vegetable drying oil copolymer, Keltrol 1001	4.39
Tall oil fatty acid, Acintol EPG	4.9
Calcium carbonate, Atomite ECC	508.0
Platelet talc, Windsor C 85	58.0
Diatomaceous earth, Celite 281	32.0
Clay, Sno-brite	117.0
Mineral spirits	33.3
Cobalt drier, 6%	2.9
Physical properties	Values
Shrinkage, %, ≤ 20	13.2
Tenacity, 6-180° folds	Pass
Bond loss, %, ≤ 10	1.3
Slump, in (mm), 0.15 in. (≤ 3.8)	0.0
Stain index ≤ 6.0	3.2
Tack-free time, 72 hrs.	Pass
Extrudability, s/ml ≤ 9	3.9
Specific gravity	1.68
Density, lbs/gal	14.02

FIGURE 17.16 Polybutene caulk, gun grade. *(Courtesy of Shell Chemical Corp., Houston, Texas)*

Raw materials	Parts by weight	
	1	2
Epoxy resin, EEW 188	100	100
Jeffamine D-230	25	44
Nonyl phenol	16	16
N-aminoethylpiprazine (AEP)	4	4
# sand/silica flour 3/1 pbw	1160	1312
Physical properties	Values	
Brookfield viscosity, cPs	1250	1060
Gel time, minutes, 200 gm mass	42.6	88
Peak exotherm, °F (°C)	384.8° F (196° C)	233.6° F (112° C)
Time to peak temperature, min.	50.4	114.7
Compressive strength, psi	>12,000	11,000

FIGURE 17.17 Sand-filled flooring. *(Courtesy of Texaco Chemical Co., Austin, Texas)*

Raw material	Parts by weight
Component 1	
Epoxy resin, EEW 188	110
Clay, calcined kaolin type	80
Component 2	
Jeffamine D-400	105
Cab-O-Sil	17
Component 3	
Accelerator 399	20
Physical properties	*Values*
Izod impact strength, ft lbs/in.	0.31
Hardness, Shore A	82-76
Elongation, %	90
Flexural strength, psi	3700
Flexural modulus, psi	207,000
Compressive strength, psi	42,500

FIGURE 17.18 Trowel coating. *(Courtesy of Texaco Chemical Co., Austin, Texas)*

Raw materials	Parts by weight
Portland cement	150.0
Sand, fine water washed	300.0
Vinol 523C	2.0
Vinac RP251	8.0
Igepon AC 78, dry surfactant	0.5
Colloid 677 DD, dry defoamer	1.0

Note: The ratios of the ingredients may be altered depending upon the fineness of the sand.

FIGURE 17.19 Cement topping mix. *(Courtesy of Air products and Chemicals, Inc., Allentown, Pa.)*

Raw materials	Parts by weight
Component 1 Pigmented base coat	
Epoxy resin, EEW 188	800
Polyglycol diepoxide Dow, DER 732	200
Titanium dioxide	200
Component 2 Curative	
Jeffamine D-230	200
AEP	65
Nonyl phenol	150
Physical properties	*Values*
Gel time, minutes, 200 gm mass	36.2
Peak exotherm °F (°C)	140° F (51.9° C)
Time to peak temperature, min	42.5

FIGURE 17.20 Seamless flooring. *(Courtesy of Texaco Chemical Co., Austin, Texas)*

Raw materials	Parts by weight
White portland cement	940
Terrazzo chips color as selected	500
Vinol 540S	22
Surfynol TG	1.1
Colloid A77 DD	1.1
Water	As required
Yield	100 gallons

FIGURE 17.21 Thin bed terrazzo.
(Courtesy of Air Products and Chemicals, Inc., Allentown, Pa.)

Raw materials	Parts by weight
Gray cement	100
Silica sand 00	100
Vinol 540S	2.5
Lime	50

FIGURE 17.22 Thin bed tile mortar.
(Courtesy of Air Products and Chemicals, Inc., Allentown, Pa.)

SUMMARY

As in most construction projects, cost efficiencies, labor savings, and ease of application are paramount. These materials are offered to the reader as possible choices to suit application and satisfy the contract. In earlier chapters, formulas with specific or generic ingredients were presented for the architect, engineer, or formulator to study.

REFERENCES:

1. *Amoco Polybutene: Caulks & Sealants Formulary*, Amoco Chemical Co., Chicago, IL.
2. Air Products and Chemical, Inc. Allentown, PA.
3. Texaco Chemical Co., Austin, TX
4. Polysar Ltd, Canada.
5. Reichhold Chemicals, Research Triangle Park, NC.
6. Schnee Morehead Co. Irving TX.
7. Rohm & Haas Co. Philadelphia, PA.
8. Shell Chemical Co., Houston TX.
9. Exxon Chemical Co. Houston, TX.
10. Dow Chemical Co. Midland MI.

CHAPTER 18
SPECIFICATIONS, TESTING, AND QUALITY ASSURANCE

INTRODUCTION

For construction purposes, architects communicate their concepts to others by a set of documents containing both words and graphics. Although words and graphics are combined in the drawings, words predominate in bidding requirements, conditions of the contract, and specifications. Of these three, the one most closely related to the drawings are the specifications.

Whether you are an architect, design engineer, general contractor, building owner, subcontractor, material supplier, attorney, or building department official—whatever your role is in the building process—you will require specifications [1]. All the parties in the construction process need them for different reasons, and each can profit by an understanding of them.

Specifications must be written in a clear, concise, and precise manner if they are to give the proper directions to those who must be guided by them [2]. Some sealant specifics that must be included are: the type of sealant or adhesive to be used (i.e., acrylic, epoxy, silicone, or urethane, etc.), the types of joints to be sealed or adhered to, the required joint preparation, and method and conditions for application. It is important to avoid statements like "... cracks, crevices, and joints in construction to be caulked whenever necessary to provide a watertight seal." These kinds of statements are an open invitation to confusion and dispute as to the actual scope of work that is desired. It is important to list *specifically* every location where sealants or adhesives are to be applied. Whenever possible, name each sealant or adhesive by brand name and manufacturer to establish the level of quality desired. Be sure to spell out application conditions, including those conditions under which application should not take place. Remember, the added time spent writing a clear, concise specification will more than pay for itself in time saved on the project, improved project quality, and fewer misunderstandings with the installation contractor.

SELECTING SEALANTS, MANUFACTURERS AND APPLICATORS

The proper selection of a sealant manufacturer and applicator can make the difference between the success and failure of the project. There is, however, a step-by-step path to follow in the selection process. To assist you:

1. Establish sealant function and performance criteria. To do this, you must consider and answer the following questions:
   - What service life is expected?
   - What special conditions exist, such as water immersion, extreme temperature differentials, direct intense exposure to sunlight, chemical exposure, excess movement, or shear forces, abnormal fungal presence, or excessive abrasion?
   - What is the calculated range of movement expected of the joint to be sealed?
   - What chemical or special environmental attack will the sealant encounter?
   - Will special color or other aesthetic considerations be critical?
   - Does an inspection of a similar installation reveal any other specific or abnormal performance requirements of the sealant to be selected? (Sometimes discussing the application with the building owner may be beneficial.)
2. Select the sealant with the aid of the criteria listed from the literature supplied by the manufacturer's data sheets. Standard CSI form specification data sheets provide the most complete information guide. It is possible to select the generic type or types of sealants that perform according to the needs established. You also might find a SWRI sealant applicator helpful here.
3. Select a specific manufacturer of the acceptable type or types of sealant. Characteristics to seek in a sealant manufacturer are: consistency in quality, dependability in delivery, competitive pricing structure, established performance history, and competent technical service group. Quite often, the recommendations about specific manufacturers can be obtained from architects or specification writers, reputable sealants applicators, and astute interviews with the manufacturer's technical representatives.
4. Select the applicator using the following major criteria: reputation in the industry, number of years in the business, current technical expertise of its

management staff, and its financial stability. Consultation with architects or specification writers and material manufacturers and astute personal interviews with representatives of the sealant applicator also might be helpful in the selection process. Visiting other application sites where the prospective applicator has used the selected sealant is recommended, if time allows.

Every architectural and engineering team faces the important task of determining how to integrate specification writing into their practice. Decisions must be made about where to obtain the necessary professional talent, what resources and training to provide, and how to organize project responsibilities. Among the variables, however, are three primary options: assigning specification writing to an in-house specialist, relying on the project team members to prepare specifications along with other construction documents, or retaining the services of a specification consultant [3, 4]. The following profiles illustrate the three approaches.

DESCRIPTION OF TEAM MEMBERS

Specification Writer

Although some firms view specifications as a necessary evil or as a backroom function, others approach the discipline as a profit center. The first step in developing a good set of specifications is to hire a dedicated specification writer. You need a person who lives and breathes in a specification environment. This procedure generates income to support the specifications studio.

In an office with manual drafting, a specifier can look over the draftsman's shoulder to see what is on the boards. However, changing drawings in computer files may not be so easy. Designers redline and update drawings, but they do not plot progress sets, as they often used to. Drawings could be called up on a CAD terminal.

The alternative for the specification writer facing such a dilemma is to rely on effective communication with other project team members. Checklists are used for this purpose. When design changes occur or decisions are made, the specifier is notified in writing. This effort, combined with increased verbal communication, helps keep the workload and schedule on track in a busy, growing organization.

During the initial service phase of a new job, the architect must interview clients intensively, questioning them about what works and what does not. Process manuals are designed specifically for the client. The manual documents the client's decision-making process to acquire and develop sites, reach agreements, develop the design, produce drawings, and put a facility in place. A set of documents called *paper tools* is then created. These are used to record and communicate all decisions for each site adaptation. Each document is custom designed for each client, because no two clients do business exactly the same way. Too many loopholes can leave bidders wondering what the architect needs. Now, they feel that the price they quoted is the one at which they are going to build and make a profit, because they have a very buildable set of construction documents. A teamwork approach has contributed to the success of many building programs. You have a set of specifications, but each project must be tailored to provide a building for a specific climate, site, and program, and certain goals and code divisions. The computer sorts this to pick out appropriate items and generate a shopping list of specs by *zone*. It is that simple. The difficult thing is including the latest trends while making sure everyone knows about modifications.

Many problems can be experienced. Keeping the specifications coordinated with the drawings is a case in point. Computer-aided drafting has to be coordinated directly with the specifications, when certain items are pulled in the basic building that are also reflected in the specs. The process eliminates the need for double procedures.

Contract drawings and specifications should be revised and should approach specifications and drawings as a prototypical situation. Work out a master specification. Every spec must be custom designed to fit the specific needs of each client. Some owners do not understand specs and do not use them as well as they should. As a result, the difficult thing is to make sure that the decisions are made up front, before production work is issued based on the specification. The specification team should meet with the client to make every decision before they edit the spec.

Computers are the future of architecture and specifications. Everything that is accomplished, from marketing and word processing, to specifications and CAD drawings, is computer-related.

Training Architects as Specifiers

Project architects and engineers at many design firms are responsible for writing their own specifications. One reason for this approach is that the project architect and engineer probably know the job better than anyone else. This procedure allows for developing the specification simultaneously with design decisions and drawings, reducing the time required for the specifier to become familiar with a project.

The drawback to this approach is that project architects and engineers are not always trained and experienced in specification writing, and without a full-time specifier to establish office specification standards, quality control can be affected.

For that reason, teaching project architects and engineers to write their own specifications can be most important. In developing the firm's master specifications, coordinating production, and providing the technical expertise that only an experienced specifier can offer, techniques and options become more diverse. The same is becoming true for the task of writing construction specifications.

In small architectural firms, it is not often feasible to staff a dedicated specification writer. Yet, the principal and the more experienced senior members frequently find themselves managing the office and developing new business, in addition to completing actual design work. In an effort to keep a small firm's jobs on schedule, it can be advantageous to hire a consulting specification writer on an as-needed basis.

Similarly, large firms, even those with a dedicated spec writer on staff, may find the workload so pressing that retaining an outside consultant is the best means of meeting short-term specification needs.

DESCRIPTION OF TEAM MEMBERS

Owner

Without definitions of what constitutes *well enough*, the owner is pretty much forced to take *whatever he gets*. It is important that the owner do a little homework to benefit from any discussions with members of the design team. This person could be the actual owner or the owner's representative. The owner has the right and responsibility to request and approve contracts connected with the work, require contract bonds from the contractor, approve the proposed surety, inspect work as it proceeds, and approve and make payments to the contractor for accepted work performed.

The owner can approve and direct changes to the work in the absence of a project manager, architect, or engineer. However, the owner must not intrude on the direction and control of the work. By doing so, the owner may assume responsibility for the accomplished work and become liable for any negligent acts committed by the contractor during the course of construction.

The owner usually has the right to stop work, or to terminate the contract and carry out the work himself, depending on the signed contractual requirements. It is critical that the owner understands his rights in a contract document.

Architect-Engineer (A-E)

The architect uses the specs to define those things that cannot be illustrated in the drawings. The architect also needs a place for the information he develops to assist the owner in obtaining prices for the work, executing the contract, and administrating the project as it is built. The location of the bidding and contract documents is in the front of the specifications or, if the project is very large, in a separate volume of the project manual.

The architect or engineer designs construction projects. Building construction is primarily of an architectural nature with support services provided by structural, electrical, and mechanical engineers as the projects warrants.

The A-E team can provide a host of services for the owner, including value engineering, building analysis, consulting work, project design, preparation of specifications and building documents, drawings, and forms of agreement.

During a project, the A-E team can act as the owner's representative and as a point of control and liaison between the owner and the contractor. The architect can consult and advise the owner as well as inspect and approve the contractor's work. He can also help solve project disputes or interpret conflicts in contract documents.

It is important that the duties of the A-E are defined clearly before execution of a contract with the owner. It is also important that the duties and role of the A-E are defined clearly with respect to the contractor. These conditions should be understood prior to the execution of an agreement between the owner and the A-E.

Contractor

The contractor is that person or organization that brings the diverse elements and inputs of the construction process into a single and coordinated effort. This effort is performed within the terms of a written and enforceable contract. In its simplest form the contractor executes an agreement with an owner to provide goods and services for which he received payment. Depending on the scope of the project, the contractor's responsibilities and duties will vary. In general, the contractor usually provides project supervision and management, labor and/or materials, and equipment to complete the work as defined in the contract. The contractor is bound by the terms of the contract and project specifications. Straying from these puts the contractor in jeopardy for retrieval of payment as well as the assumption of additional liabilities. It is important that the contractor is aware of his rights and has a clear understanding of the specifications before entering into a contract with the owner.

Specifications can be extremely useful to the contractor. The specifications can help the contractor define the scope of the subcontracts. They are not written to mandate subcontracting work, and specifiers do not and should not assume the responsibility for doing so, but specifications divide the project into logical units of work. Specifications should include everything indicated on the drawings which assists the contractor in determining which of the subcontractors is obligated to perform certain parts of the work.

The contractor uses specifications when managing the project. Requests for substitutions, methods of submittal, procedures for preparation, and submissions of payment requests are a few of the administrative processes described in detail in the specifications. By understanding and following the procedures as described, the contractor greatly improves his chances of obtaining the desired results, obtaining permission to effect a substitution, making acceptable submittals, and receiving prompt payment.

Specialty Contractor

The specialty contractor is a person or organization that performs a specific type of work within the construction industry on a regular basis. The focused area of performance, in combination with current available technologies, quality materials, and knowledge of the particular industry, can create an organization that can be a great asset to an owner. Specialty contractors can operate in the role of a prime contractor with an owner or as a subcontractor to a general contractor. This usually is determined by the scope or variety of work covered by the contract and the financial stability of the contractor.

Specialty contractors also can provide consulting and cost-estimating services within their area of expertise for the owner, general contractor, or the architect/engineer.

Estimators

Specifications are essential for the contractor's estimator. Estimators use the 16 Division system of the Master format extensively in their work. Most estimating systems are based on Master format, allowing the estimators to easily compare a list of spec sections for their work.

Manufacturers Require Specifications

For the manufacturer and material suppliers of building products, a project's specifications can have many uses, some of which determine whether or not a particular product will be used on a specific project. With many building materials, manufacturers' representatives help the specifier prepare the specifications. Their knowledge of their product and their skills in preparing product specifications are helpful to the specifier. When assisting the spec writer however, product representatives go beyond the task of spec writing. They demonstrate to the specifier and the design team members the suitability of their product for the proposed application. In this connection, the manufacturer must be careful to point out options, custom features, and accessories that should be incorporated into the work. Then the product selected and the representative prepare the draft specification incorporating the precise features selected by the design team.

Product representatives, especially those who have taken the *Construction Document Technologist and Certified Construction Specifiers Representative* examinations offered by the Construction Specifiers Institute, are aware of their need to study the division requirements for the project. They will make proposed substitution requests and prepare their submittal in accordance with the requirements specified. They know that the division in which their product is specified is but part of the whole. This increases the likelihood that their substitution proposal or their submittal will be accepted.

Attorneys Need Specifications

Attorneys become involved with specifications primarily when preparing the contracts and resolving disputes. Many attorneys are not proficient at reading drawings. They rely heavily on the specifications and discussions with their clients to understand the project. The architects' preparation of contracts should be limited to filling in the blanks in the American Institute Of Architects forms or similar prewritten contract forms. Modification to suit the project or the preparation of contracts in custom form should be done only by attorneys. If disputes arise, the terminology, the punctuation, and the sentence structure all become subjects of intense scrutiny in determining what the various portions of the contract document said, what they meant, and how they were interpreted. At these times, the accuracy, completeness, and consistency of the specifications become critical.

Identifying Problems

A good inspection program helps the owner establish the need for maintenance, identify the causes of problems, and prescribe the appropriate maintenance and repair procedures. Regularly scheduled inspections of structures can detect small problems before they cause major damage and require costly repairs. The owner should make a visual inspection of the exterior of his building and talk to the tenants of the building to see if they are experiencing any problems. A visual inspection should be taken of the exterior skin to see if there is any cracking or shifting of the masonry, any aluminum flashing out of line on the coping, or around the windows. The A-E and the specialty contractor should perform the same basic inspection that the owner can do from the ground.

If the owner would like either of the two groups to hang a stage, they could give him a written report on the true condition of his building, with pictures and marked drawings at cost to the owner.

Making Recommendations for Repairs

Once the problems are located and the owner recognizes there are problems, the owner can either hire the A-E or go to the specialty contractor for repair recommendations. A list of problems to be repaired should be made and a priority made with a rough estimate on the costs for the repairs. The A-E or specialty contractor should base recommendations on the industries' latest standard for repair methods for each type of repair that must be made. Then a meeting should be held with the owner to discuss the recommendations.

Preparation of Quality Specifications

The preparation of the specifications occurs after all problems have been identified and is a compilation of the recommendations for repairs with selected materials. This is presented in combination with the execution portion of the specifications, which describes how all of the work items are to be performed and how the materials are to be used.

The Architect-Engineer can readily prepare specifications for competitive bidding processes or solely for the negotiated uses, usually on a fee basis. Certain consultants or specialty contractors also are capable of providing these specifications, sometimes on a fee basis or as part of a negotiated or design-build project. It is important to realize that there are many "canned" or standard specifications that are readily used in the industry. This standard verbiage may not be applicable to each building and it becomes very important for the owner and the A-E to determine what is pertinent to provide a clear and focused specification that will benefit all parties on the project.

The needs and uses of the specifications by various parties described are sometimes divergent, putting several obligations on persons who prepare specs. The Manual of Practice of CSI describes the process in detail. Here, we will touch on a few important items [3].

1. The specifications should be accurate and complete and should amplify, explain, and qualify that information.
2. The specification terminology should be consistent. Use of terms describing parties to the contract should be limited to the parties to the contract. Specifications should be checked carefully during preparation to ensure that cross-references are correct.
3. Division I requirements, such as substitutions, submittals, and record drawings, should not be repeated, or worse still, specified differently in subsequent divisions.
4. The specifications should be clear. As an example, if we mean a polyurethane sealant as an expansion joint sealant is to be used in a concrete joint, it must be spelled out specifically. A casual reference to the ASTM-C 920 specification dealing with joint sealants is not sufficient. The precise details, method of manufacture, the issue of the latest specifications, and the materials to be sealed or caulked should all be spelled out so there is no doubt as to what is intended.
5. Clauses that are not clear, concise, and consistent should be avoided. The contractor does not bid on *intent*. He bids on what he sees on drawings and what he reads in the specifications.

The joint publication of CSI and AIA, *Uniform Location of Subject Matter,* AIA document A521, contains detailed descriptions of what goes in specifications and what goes in drawings. By understanding the needs of each party and by placing information where it belongs in a clear, concise, and consistent manner, the specifications will enhance the smooth progress of the project as a whole and the interests of each party involved.

Competitive Bidding

Contractors can compete with one another for construction contracts by submitting proposals to the owner designating pricing for which they will do work. The owner then selects the bid of the contractor that is the most advantageous. The contractors prepare bids from the set of specifications so there is a fair chance for all who are competing. It is very important that a clear set of specifications be prepared so the owner receives good comparable bids for evaluation.

The contractor then performs only the work that was in the specifications, unless a change order has been executed by all parties, thus authorizing a change to the project documents. The competitive bidding process is meant to encourage efficiency and innovation by contractors in order to produce work of the specified quality at the lowest possible cost.

Negotiated Method

Instead of open competitive bidding, it can be greatly advantageous for the owner to negotiate a contract with a preselected contractor or small group of contractors. There are many types of negotiated contracts, because such agreements can include many different provisions that are mutually acceptable to both parties and that may be best suited for the type of work involved. After the owner performs a study, which includes the reputation, experience, financial stability, dependability, and ability to perform the work of one or more candidates, a contract then is negotiated. Normally this type of contract is limited to privately financed work, because competitive bidding is usually required for public projects, except under unusual circumstances.

Sealant Validation

The Sealant, Waterproofing & Restoration Institute developed and implemented its Sealant Validation Program to help clarify the information presented on manufacturers' standard material data sheets. Confusion in the data presented on standard information sheets can result from different manufacturers using modified versions of standard ASTM tests with notations listed within the data. It is generally recognized that ASTM C-920 is the accepted basis for testing. Sometimes a non-C-920 test is substituted, which makes design selections and product comparisons difficult for the design professional and contractor. Many times the product used for testing is not the standard distribution product quality, but is taken from a closely controlled laboratory sample specimen.

The SWRI Sealant Validation Program has established three priority performance characteristics of sealants and their respective ASTM tests:

- Adhesion and cohesion: ASTM C-719
- Adhesion in peel: ASTM C-794
- Hardness: ASTM C-661

These three represent the working performance profile of a sealant. However, other ASTM C-920 criteria should not be ignored by the design professional. Validation of a product is initiated by any manufacturer wishing to do so by calling an independent, certified laboratory and noting the intent to validate. An independent lab, at the manufacturer's direction, extracts the product sample from the current distribution system and performs the profile tests and others as desired. The manufacturer receives all results, which then may be sent with the application to SWRI. When the test results verify compliance with ASTM C-920 and are sent with application fees, the manufacturer is issued a certificate of validation for that product. In addition, the manufacturer will receive SWRI's exclusive copyright seal with a validation number and the copyrighted Priority Performance Profile Format for approved use on the manufacturer's standard data sheets. Actual uniform test results appear on the profile format. Thus, all products can be compared on a clear and equal basis by any end user. All validation information is available upon written request to SWRI, via fax or mail. The SWR Seal of validation is shown in Figure 18.1 [5].

CONSTRUCTION DOCUMENTS TECHNOLOGY MANUAL

Contractors, project architects, contract administrators, material suppliers, and manufacturers' representatives are realizing the advantages of being Construction Document Technologists. By understanding and interpreting written construction documents, CDTs can perform their jobs more effectively and demonstrate their commitment to improving communication among all construction industry professionals. This two-hour exam is based on CSI's Construction Documents Fundamentals and Format module of the Manual of Practice and its appendices and general conditions in common use. (AIA A 201, 1987 edition or EJCDC 1910-8 1990 edition).

FIGURE 18.1 SWR Institute seal of validation.

TYPES OF STANDARDS [5]

United States Federal Specifications

The National Bureau of Standards established a federal specification for the types of sealant and caulks available. While these specifications have recently been replaced with ASTM specs as the standard, they are indicated here for historical reference. The Federal specifications were last updated in the 1970s. Federal specifications are included because they are still used in construction to government specifications and were the core around which ASTM and Canadian specifications were written—the exception being U.S.-ASTM C-920 or Canadian 2-19-13 M82. In the Canadian specifications, there is a class for higher movement sealants including a ±40 percent movement class (total 80 percent movement), and a test requirement for demonstrating cold temperature flexibility.

SS-S-1401C, 8/15/84

Sealing compound, hot applied for concrete asphalt pavements.

Summary: No specific material; mixture of materials which form a resilient and adhesive compound. It will not flow from the joint or be picked up by tires at temperatures up to 125° F (51.7° C). Tests include safe handling temperature, penetration flow, resilience, bond, and compatibility.

SS-S-1614A, 8/15/84

Sealant, joint, jet-fuel resistant, hot applied, for portland cement and tar concrete pavements.

Summary: No specific material; mixture of materials which form a resilient and adhesive compound resistant to jet fuel. Pouring temperatures not over 450° F (232° C). Shall not be picked up by tires at 125° F (51.7° C). Tests include penetration, solubility, flow, and jet fuel immersed bond strength. The compound may be furnished as a solid or a liquid, both requiring field heating.

SS-S-200E, 8/23/88

Sealant, joint, two-component, jet-blast resistant, cold-applied, concrete paving.

TT-S-00230c (COM-NBS)

Elastomeric type, cold-applied single-component for caulking, sealing, and glazing in buildings, building areas, (plazas, decks, pavements), and other structures.

Type I—Flow, self-leveling for use in horizontal joints, i.e., floor joints.

Type II—Non-sag or gun-grade- for use in vertical joints, walls, windows, etc.

Class A*—Compounds resistant to 50 percent maximum total joint movement, often expressed as ±25 percent joint movement

Class B*—Compounds resistant to 25 percent maximum total joint movement, often expressed as ±12½ percent

TT-S-00227e (COM-NBS)

Elastomeric type, cold-applied multicompound for caulking, sealing, and glazing in buildings, building areas, (plazas, decks, pavements, etc.), and other structures.

Type I— Flow, self-leveling- for use in horizontal joints, i.e., floor joints

Type II—Non-sag or gun-grade, for use in vertical joints, walls, windows, etc.

Class A*—Compounds resistant to 50 percent maximum total joint movement, often expressed as ±25 percent joint movement

Class B*—Compounds resistant to 25 percent maximum total joint movement, often expressed as ±12½ percent

TT-S-001543a (COM NBS)

Single component, cold-applied silicone rubber base for caulking, sealing and glazing in buildings, building areas, and other structures.

Type II—Non-sag or gun-grade, for use in vertical joint walls, windows, etc.

Class A—Compounds resistant to 50 percent maximum total joint movement, often expressed as ±25 percent joint movement.

Class B—Compounds resistant to 25 percent maximum total joint movement, often expressed as ±12½ percent joint movement.

Note: Class A sealants are capable of 50 percent movement but only 25 percent in each direction if the temperature of installation is near the mid-range of 70° F (21.11° C). If installation takes place at excessively high or excessively low temperatures, total performance range may be out of phase with the performance range necessary. The same applies to Class B sealants. It is necessary to design the joint and select the sealant class according to the sealant's required performance after installation at the installation temperature.

TT-S-001657- 10/8/70

Single-component butyl rubber-based, solvent release type for sealing, caulking, and glazing operations in buildings and other types of construction.

Type I—This type shall be suitable for use in caulking guns at temperatures above 40° F (4.44° C).

Type II—This type shall be suitable for applications with a putty knife above 40° F (4.44° C).

The sealing compound covered by this specification shall be formulated to withstand joint movement of 10 percent (which could be expressed as ±5 percent) of the nominal width.

TT-C-00598C—Federal Specifications, Caulking Compound

One-component oil and resin base type for masonry and other types of structures.

Type I—For gun application—suitable for use in caulking guns above 40° F (4.44° C).

Type II—For knife application—suitable for application with knife above 40° F (4.44° C).

HH-F-341F, 6/677

Fillers, expansion joint, bituminous (asphalt and tar) and nonbituminous (preformed for concrete).

HH-P-119a, 2/16/67

Packing material, sewer joint, asphalt saturated cellulose fiber.

The above sealant designations define minimal acceptable performance for sealants as classified. Many sealants exceed the performance designated in the specifications. If the job conditions require properties in excess of those indicated of the classification, consult the sealant manufacturer and demand test data to certify the claims of superior performance and a letter stating the manufacturer's support of the claims.

The materials listed hereafter show both United States Federal Specification and Canadian General Standards Board Numbers. They are the most commonly used. There are, however, other products that may be used in certain applications.

ASTM STANDARDS IMPORTANT TO HIGH PERFORMANCE SEALANTS

ASTM C 510-90

Standard test method for staining and color change of single or multi-component joint sealants.

Summary: This method uses three samples of substrate. A ¼ inch-thick bead of sealant is placed on two of the substrates. One sealant sample is exposed to 100 hours of accelerated weathering, the other sample is exposed to room temperature air for 14 days, during which the sample is immersed in room temperature distilled water 10 times for one minute. Staining of the substrate is compared to the control sample.

ASTM C 603-90

Standard test method for extrusion rate and application life of elastomeric sealants.

Summary: This method reports the period of time necessary to extrude a 6-oz tube filled with 6 fluid ounces of sealant through a 0.540 in. diameter opening using a pressure of 50 psi and an air-powered caulking gun. Two-part sealants are hand-mixed for five minutes, placed in the tube, extruded at various intervals up to 3 hours, and the times are reported.

ASTM C 639-90

Standard test method for rheological (flow) properties of elastomeric sealants.

Summary: This method describes the flow properties of four types of sealants: one-part non-sag, one-part self-leveling, two-part non-sag, and two-part self-leveling, using a ¾ inch wide × ½ in. deep × 6-inch long channel filled with sealant and tested at both 40° F (4.44° C) and 122° F (50° C). Non-sag materials are reported as inches of sag after four hours in a vertical position. Self-leveling materials are reported as being or not being self-leveling when in a horizontal position.

ASTM C 661-93

Standard test method for indentation hardness of elastomeric-type sealants by means of a durometer.

Summary: This method measures indentation hardness with a Type A-2 Durometer, with units measuring 0-100. On a ¼ inch thickness of sealant for 14 days, (two-component sealants) or 21 days (one-component sealants), durometer readings can vary ±6 units within a laboratory and ±8 units between laboratories.

ASTM C 679-87

Standard test method for tack-free time of elastomeric sealants.

Summary: This method is a measure of a sealant's surface cure by lightly touching, at regular intervals, the surface of a curing sealant to determine when a polyethylene film can be cleanly pulled away from the sealant surface.

ASTM C 717- 94

Standard terminology of building seals and sealants.
Summary: Definitions and descriptions of terms.

ASTM C 719-93

Standard test method for adhesion and cohesion of elastomeric joint sealants under cyclic movement (Hockman Cycle).

Summary: This method subjects a ½ inch × ½ inch × 2 inch sealant joint installed between parallel plates of glass, aluminum, or mortar which are cured for 21 days to: 1) water immersion for 7 days; 2) a 60° F (15.56° C) flex test; 3) specified compression for 7 days at 158° F (70° C); 4) 10 cycles of specified movement at room temperature at a rate of ⅛ inch per hour; and 5) 10 cycles of specified movement at alternating temperatures of 158° F (70° C) and 15°F (-9.44° C). Test results are recorded as square inches of bond failure, cohesive or adhesive. One ½ square inch of failure is the maximum allowed.

ASTM C 792-93

Standard test method for effects of heat aging on weight loss cracking and chalking of elastomeric sealants.

Summary: A method of measuring weight loss and recording related visual observations for sealants cured for seven days at room temperature (¼ in. thickness) then exposed to a temperature of 158° F (70° C) for 21 days. A sealant that develops no cracks, chalking, or low weight loss in this test method does not necessarily assure good durability.

ASTM C 793-01-91

Standard test method for effects of accelerated weathering on elastomeric joint sealants.

Summary: A sample of sealant of ⅛ inch thickness is applied to an aluminum plate and allowed to cure 72 for hours. It is then placed in a carbon arc weatherometer (ASTM G26,102:18 light: water cycle) for 250 hours. The sample is chilled to $-15°$ F $(-26°$ C) and bent over a ½ inch diameter steel mandrel. Visual observations are recorded after weathering and bending.

ASTM C 794-93

Standard test method for adhesion-in-peel of elastomeric joint sealants.

This method records the peel strength of a cured-in-place elastomeric sealant by testing three 1 inch-wide peel strips after a 21-day period followed by seven days' water immersion. Peel strips are pulled back upon themselves at a rate of 2 in. per minute at a 180° angle. Glass, aluminum, and mortar are specified as substrates; however, manufacturers use this method on a variety of substrates to determine surface preparation and primer requirements. Accuracy of results between laboratories can be ±100 percent. It is a good method for determining surface preparation requirement for a particular sealant. It is a *poor* method for comparing the physical properties of one sealant with another.

ASTM C 920-87

Summary: A specification that establishes limits for a sealant for ASTM test methods C510, C603, C639, C661, CC679, C719, C792, C794, and C1183. It classifies sealants as Type S (single-component), Type M (multi-component), Grade P (self-leveling), Grade NS (non-sag), Class 25 (passes 25 percent movement) per C719, Use T (traffic area), Use NT (nontraffic area), Use M (passes C794 and C719 on mortar, Use G (passes C794 and 719 on glass), Use O (passes C794 and C719 on other substrates). The specification is applicable to silicone, polysulfide, and polyurethane sealant.

ASTM C 1087-87

Standard test method for determining compatibility of liquid-applied sealants with accessories used in structural glazing systems.
 Summary: This method involves placing a sealant adjacent to a gasket or accessory material and evaluating color change and adhesion to glass after a 21-day exposure to incandescent UV tanning bulbs at a temperature of 122° F (50° C). Current revisions are being made to change the light source to UVA-340 fluorescent bulbs. Structural sealants material fail if the material causes severe color change and adhesion loss.

ASTM C 1135-90

Standard test method for determining tensile adhesion properties of structural sealants.
 Summary: This method creates samples by placing a sealant bead ½ inch × ½ inch × 2 inch between two plates followed by curing for 21 days. Samples are then extended at two inches per minute. Data is recorded for ultimate tensile strength, ultimate elongation and stress at 10 percent, 25 percent, 50 percent and 100 percent extension. This is an effective method to determine changes in sealant physical properties through the indicated aging cycles and environmental conditions by comparing aged data to unaged data.

ASTM C 1183-91

Standard test method for extrusion rate of elastomeric sealants.
 Summary: This method measures the weight of sealant extrudes from a 6 ounce tube through a nozzle with a ⅛ inch diameter orifice using an air-powered caulking gun pressured at 40psi. Using the density of the sealant, the volume per minute based upon the measured weight per minute is calculated.

ASTM C 1184-91

Standard specification for structural silicone sealants.

Summary: This standard establishes requirements for silicone sealants for ASTM tests methods C639, C603, C661, C792, C679, and C1135. Test method C1135 uses a ½ inch per minute pull rate requiring a 50 psi minimum ultimate strength at standard conditions: 109° F (42.78° C), -20° F (-28.89° C), 7 days water immersion, and 5000 hours of accelerated aging in a QUV weatherometer (ASTM G-53).

ASTM C 1193-91

Standard guide for use of joint sealants.

Summary: This guide describes proper sealant joint design, sealant backing, surface preparation, and sealant application. The standard describes thermal expansion coefficients of typical building materials and includes a sample calculation to size joints based upon sealant movement capacity, specified movement, and material tolerances.

ASTM C 1247-93

Standard test method for durability of sealants exposed to continuous immersion in liquids.

Summary: Sealant beads ½ inch × ½ inch × 2 inch are placed between substrate plates and allowed to cure. The specimens are immersed in liquid at a temperature of 122° F (50° C) for six weeks and then subjected to three compression/extension cycles per ASTM C719 and any failures recorded. The specimens are re-immersed for an additional four weeks, then subjected to compression/extension cycles. The four weeks' immersion followed by three movement cycles is continued for as long as necessary. Standard substrates are portland cement, mortar, and anodized aluminum.

ASTM C 1248-93

Standard test methods for staining porous substrates by joint sealants.

Summary: This method requires 12 specimens of sealants ½ inch × 2 inch to be cured between 1 inch × 1 inch × 3 inch masonry plates for 21 days at room temperature. The samples are then compressed to the rated movement capability and: 1) four are exposed to room temperature, 2) four are exposed to 70° F (21.11° C), and 3) four are exposed to a QUV weatherometer cycling of four hours at 140° F (60° C) under UVA -340 bulbs and four hours condensation at 122° F (50° C). After two weeks, two of the four samples are removed from their environment, the stains measured on the exterior and interior of the plates, and data recorded. The remaining specimens are evaluated after four weeks' exposure.

ASTM C 1253-93

Standard test method for determining the outgassing potential of sealant backing

Summary: Visual observations are recorded after placing a sealant in a ½ inch × 2 inch joint backing rod, which was punctured six times followed by immediately placing the joint in a 122° F (50° C) temperature oven for one hour. The joint is compressed 12.5 percent and placed back in the oven for two hours, then removed from the oven and allowed to cure for two weeks at room temperature. The sealant is then slit with a razor along its entire length and observations of outgassing into the sealant are recorded.

ASTM C 1265-94

Standard test method for determining the tensile properties of an insulating glass edge seal for structural glazing applications.

ASTM D 412

Standard test method for tear strength of conventional vulcanized rubber and thermoplastic elastomers-tension.

ASTM D 624

Standard test method for tear strength of conventional vulcanized rubber and thermoplastic elastomers.

ASTM D 1190-80 (1980)

Concrete joint sealer, hot-poured elastic type.

ASTM D 1854-74 (1985)

Jet-fuel resistant, hot-poured elastic type.

ASTM D 2202-93

Standard test method for slump (rheology) of sealants.

ASTM D 2203-93

Standard test method for staining from sealants.

OTHER PERTINENT ASTM STANDARDS (UNITED STATES)

ASTM C 681-94

Test method for volatility of oil- and resin-based, knife-grade, channel-glazing compounds.

ASTM C 711-73

Test method for low-temperature flexibility and tenacity of one-part, elastomeric, solvent-release type sealants.

ASTM C 712-93

Test method for bubbling of one-part, elastomeric, solvent-release type sealants.

ASTM C 713-94

Test method for slump of an oil-base, knife-grade, channel-glazing compound.

ASTM C 731-93a

Test method for extrudability, after package aging of latex sealants.

ASTM C 732-82 (1987)

Test method for aging effects of artificial weathering on latex sealants.

ASTM C 733-93

Test method for volume shrinkage of latex sealants.

ASTM C 790-90

Guide for use of latex sealants.
   (Discontinued 1993—Replaced by Guide C 1193)

ASTM C 797-94

Practices and terminology for use of oil- and resin-based putty and glazing compounds.
   (Discontinued 1993—Replaced by C 1193)

ASTM C 834-91

Specification for latex sealing compounds.

ASTM C 836-89a

Specification for high solids content, cold-liquid applied elastomeric waterproofing membrane for use with separate wearing course.

ASTM C 898-89

Guide for use of high solids-content, cold-liquid applied elastomeric waterproofing membrane with separate wearing course.

ASTM C 910-93a

Test method for bond and cohesion for one-part elastomeric solvent-release type sealants.

ASTM C 919-84 (1992)

Practice for use of sealants in acoustical applications.

ASTM C 957-93

Specification for high solids-content, cold-applied elastomeric waterproofing membrane with integral wearing surface.

ASTM C 962- 86

Guide for use of elastomeric joint sealants.
    (Discontinued 1993—Replaced by C1193)

ASTM C 981-89

Guide for design of built-up bituminous membrane waterproofing systems for building decks.

ASTM C 1016-94

Test method for determination of water absorption by sealant backup (joint-filler) material.

ASTM C 1021-92

Practice for laboratories engaged in the testing of building materials.

ASTM C 1085-91

Specification for butyl rubber-based solvent release sealants.

ASTM C 1127-89

Guide for high solids-content, cold-applied elastomeric waterproofing membrane with an integral wearing surface.

ASTM C 1216-92

Test method for adhesion and cohesion of one-part elastomeric solvent release sealants.

ASTM C 1241-93

Test method for volume shrinkage of latex sealants during cure.

ASTM C 1250-93

Test method for nonvolatile content of cold liquid-applied elastomeric waterproofing membrane.

ASTM C 1257-94

Test method for accelerated weathering of solvent-release type systems.

ASTM D 638-89

Tensile properties of epoxy.

ASTM D 903-93

Peel or stripping strength of adhesive bonds.

ASTM D-1002-72

Strength properties of adhesives in shear by tension loading (metal to metal).

ASTM D 1876-93

Peel resistance of adhesives (T-Peel Test).

ASTM D 2240-91

Durometer hardness.

ASTM D 2377-94

Test method for tack-free time of caulking compounds and sealants.

ASTM D 2450-87 (1993)

Test method for bond of oil- and resin-base caulking compounds.

ASTM D 2452-94

Test method for extrudability of oil- and resin-base caulking compounds.

ASTM D 2453-94

Test method for shrinkage and tenacity of oil- and resin-base caulking compounds.

ASTM D 3236-88

Test method for apparent viscosity of hot-melt adhesives.

ASTM D 3405-78

Joint sealants, hot-poured for concrete and asphalt pavement.

ASTM D- 3406-85

Specification for joint sealant hot-applied elastomeric type for portland cement concrete pavements.

ASTM D 3569-85

Specification for joint sealant hot-poured elastomeric type, jet-fuel resistant for PCC airfield pavements.

ASTM D 3581-80 (1985)

Specification for joint sealant hot-poured, jet-fuel resistant type for PCC and tar concrete pavements.

ASTM E 84-94

Test method for surface burning characteristics of building materials.

ASTM E 119-88

Test methods for fire tests of building construction and materials.

ASTM E 136

Test method for behavior of materials in a vertical tube furnace at 1382° F (750° C).

ASTM E 176-93a

Terminology for fire standards.

ASTM E 330-90

Test method for structural performance of exterior windows, curtain-walls and doors by uniform static air pressure difference.

ASTM E 547-93

Test method for water penetration of exterior windows, Curtain walls, and doors by cyclic static air pressure differential.

ASTM E 621-84 (1991)

Practice for the use of metric (SI) units in building design and construction.

ASTM E 631-93a

Terminology for building construction.

ASTM E 773-88

Test methods for seal durability of sealed insulating glass units.

ASTM E 773-92

Specification for sealed insulating glass units.

ASTM E 814-94b

Test method for fire tests of through-penetration firestops.

ASTM E 833-92a

Terminology for building economics.

ASTM E 917-93

Practice for measuring life-cycle costs of building and building systems.

ASTM E 1399-91

Test method for cyclic movement and measuring the minimum and maximum joint widths of architectural joint systems.

ASTM E 1605-94

Terminology relating to the abatement of hazards from lead-based paint in buildings and related structures.

ASTM E 1612-94

Specification for preformed architectural compression seals for building and parking structures.

CANADIAN STANDARDS FOR HIGH PERFORMANCE SEALANTS [6]

CAN CGSB-19.13-M87

Sealing compound, one-component elastomeric, chemical curing.
　　Summary: This standard classifies one-component joint sealing compounds by:

Type I—Self-leveling
Type II—Non-sag

Use M (for metals), use C (for concrete), and use G (for glass). Movement classes are classified as ±25 percent movement capability and ±40 percent movement capability.

Glazing sealants are classified as:

Class A—Suitable for UV exposure on glass.

Class B—Not suitable for UV exposure on glass.

Note: *Sealant properties that are measured include tack-free time, extrudability, loss of mass, cracking and chalking after heat aging, hardness after heat aging, staining of concrete, low temperature flexibility, tensile adhesion after cycling, adhesive peel strength, adhesive peel strength after UV exposure through glass, and blister formation.*

CAN/CGSB-19.18-M87

Sealing compound, one-component, silicone-based, solvent curing.

Summary: This standard assumes that all silicones are single-component, not self-leveling or suitable for horizontal applications.

Note: *Testing to this standard includes tack-free time, flow extrudability, low temperature flexibility, hardness and low mass after heat aging, resistance to cracking and chalking, staining, color stability, tensile adhesion after cycling (±25 percent or ±40 percent), adhesive peel strength, and adhesive peel strength after UV exposure.*

CAN/ CGSB-19.24-M80

Sealing compound, multicomponent, chemical curing.

Summary: This standard applies to self-leveling and non-sag materials used in glazing and non-glazing applications that are chemically cured, multi-component sealants with movement capabilities of ±25%.

Materials are classified as:

Type I—Self-leveling

Type II—Non-sag

   Class A—Glazing

   Class B—Non-glazing

Note: *Properties of the sealants are tack-free time, extrudability, loss of mass, cracking and chalking after heat aging, flow, hardness, hardness after heat aging, staining of masonry, color stability, tensile adhesion after cycling, adhesive peel strength, adhesive peel strength after UV exposure through glass (class A only), and blister formation.*

OTHER CANADIAN STANDARDS: CAULKING AND SEALING COMPOUNDS

CAN/CGSB-19.0-M77

Methods for testing putty, caulking and sealing compounds.

CAN/CGSB-19.1-M87

Putty, linseed-oil type.

CAN/CGSB-19.2-M87

Glazing compound, nonhardening, modified oil type 19-GP-5M sealing compound, one-component, acrylic base, solvent curing.

CAN/CGSB-19.6-M87

Caulking compound, oil base.

19-GP-14M

Sealing compound, one component, butyl-polyisobutylene polymer base, solvent curing.

CAN/CGSB-19.17-M90

One component acrylic emulsion base sealing compound.

CAN/CGSB- 19.18

Sealing compound, one component, silicone base, solvent curing.

CAN/CGSB-19.20-M87

Cold applied sealing compound, aviation-fuel resistant.

CAN/CGSB-19.21-M87

Sealing and bedding compound acoustical.

CAN/CGSB-19.22-M89

Mildew-resistance sealing compound for tub and tiles.

CAN/CGSB-19.24-M80

Sealing compound, multicomponent, chemical curing.

CAN/CGSB-19.28-91

Glossary of terms related to sealants.

CAN 2 -51.32-M77

Sheathing, membrane, breather type.

CAN 4-S115M

Standard method of fire test of firestop materials.

CAN/ULC S101M

Standard test of fire endurance.

M-264-M-83 (Ontario Hydro)

Polyvinylchloride waterstop.

L-1219-88 (Ontario Hydro)

Standard specification for waterstop (styrene butadiene rubber).

AMERICAN CONCRETE INSTITUTE STANDARDS

ACI 504R-90

Guide to sealing joints in concrete structures:

    Chapter 1—General p.504R-2
    Chapter 2—How joint sealants function p.504R-4

Chapter 3—Sealant materials, p.504R-12
Chapter 4—Joint movement and design, p.504R-25
Chapter 5—Joint details, p.504R-31
Chapter 6—Installation of joint sealants, p.504R-31
Chapter 7—Performance, repair and maintenance of sealants, p.504R -36
Chapter 8—Sealants in the future and concluding remarks, p.504R-37
Chapter 9—References, p.504R-37
Appendix A—Layman's glossary of joint sealant terms, p. 504R-38
Appendix B—Key symbols used in figures, p. 504R-39
Appendix C—Sources of specifications, p.504R-40

PREFORMED SEALANTS AND TAPES

ASTM D 2628-81

Specification for preformed polychloroprene elastomeric joint seals for concrete pavement.

ASTM D 3542-85

Specification for preformed polychloroprene elastomeric joint seals for concrete bridges.

ASTM D 994-71

Specification for preformed expansion joint filler for concrete (bituminous) concrete).

ASTM D 1751-83

Specification for preformed expansion joint filler for concrete paving and structural construction (nonextruding and resilient bituminous types).

ASTM D 1752-84

Specification for preformed sponge rubber and cork expansion joint fillers for concrete paving and structural construction (nonextruding and bituminous types [similar to ASTM C 582-82].

ASTM C 509-84

Specification for cellular elastomeric preformed gasket and sealing material.

ASTM C 542-82 (1984)

Specification for lock-strip gaskets.

AMERICAN ASSOCIATION OF STATE HIGHWAY AND TRANSPORTATION OFFICIALS SPECIFICATION

M-33-81

Preformed expansion joint filler for concrete (bituminous type) same as ASTM D 994-71 (1982).

M-46-70 (1982)

Asphalt plank.

M-153-84

Preformed sponge rubber and cork fillers for concrete paving and structural construction.

M-173-60

Hot poured elastic type.

M-220-1985

Preformed elastomeric compression joint seals. (same as ASTM D 1752-67 [1978]).

M-213-81

Preformed expansion joint fillers for concrete paving and structural construction (nonextruding and resilient bituminous types).

U.S. ARMY CORPS OF ENGINEERS

CRD-C-513-74

Specifications for rubber waterstops.

CRD-C-572-74

Specifications for polyvinylchloride waterstops.

CRD-C-527-88

Specification for joint sealants, cold applied, Non-et-fuel resistant, for rigid and flexible pavements.

CRD-C-548-88

Specification for jet fuel and heat resistant preformed polyvinyl chloride blastomeric joint seals for rigid pavements.

AMERICAN ARCHITECTURAL MANUFACTURERS ASSOCIATION

AAMA 800-86, Voluntary Specification and Test Methods for Sealant

1. Back Bedding Glazing Compounds
   - 802.3-85 (type I) Ductile back bedding compound used to bed glass.
   - 802.3-85 (type II) Ductile back bedding compound used to bed glass.
   - 805.2-8 Bonding type bedding compound.
2. Back bedding mastic type glazing tapes.
   - 804.1-85 Back bedding glazing tape used to seal glass and/or panels to the surrounding aluminum.
   - 806.1-85 Back bedding glazing tape intended to prevent air infiltration and water leakage.
   - 807.1-85 Back bedding glazing tape intended to remain functional indefinitely and to air prevent infiltration and water leakage.
   - 803.3-85 Narrow joint seam sealer intended to provide a minimum level of quality for narrow joint seams sealers.

- 808.3-85 Exterior perimeter sealing compound intended to provide a minimum level of quality for exterior perimeter sealing compounds.
- 809.2-85 Nondrying sealant intended to provide a minimum level of quality for nondrying sealing compounds.
- 810.1-85 Expanded cellular glazing tape intended to provide a minimum level of quality and performance for non-interconnecting cellular (closed cell) products.

BOCA

Building Officials and Code Administration International Article 9, Section 921.

SBCCI

Southern Building Code Congress International; Section 705, Chapter 10.

ICBO

International Code of Building Officials; Uniform Building Code; Chapter 42, Section 4304-5.

ICBO

#4577 ICBO Evaluation Report.

NBCC

National Building Code of Canada.

NEC

National Electrical Code; Section 300-21, 800-52(b), 820-52(b).

NFPA 101

National Fire Protection Association Life Safety Code.

NFPA 70

National Fire Protection Association National Electrical Code.

NFPA 255

Test method of surface burning characteristics of building materials.

UL 723

Standard test method for surface burning characteristics of building materials.

UL 263

Fire tests for building construction and materials.

STATE OF CALIFORNIA

#4485-1372: 100-104

City of Los Angeles, CA

RR # 24962

CONSTRUCTION SPECIFICATIONS INSTITUTE

SPECTEXT SpecGUIDE Section 07900

Joint sealants
    Sealants and caulking, Section 07920
    Note: *Many architectural and engineering design firms use this CSI format for writing and developing their specifications. It is complete, concise, and universal.*

QUALITY ASSURANCE

Scope and Purpose

There have been many recent advances in the technology of construction sealants and caulks. Silicone or polyurethane with their engineered properties have resulted in a large number of possible combinations of materials for expansion and contraction joints in the construction of a crack-free building. The modified specifications have attempted to cover most of the movement occurrences that can be encountered in the structure, giving minimum criteria for quality control. This holds true for a glazing compound or sealant.

Some adverse effects of change in the manufacture and formulation of various sealants may be detectable only after long-term usage. These cannot be considered in a quality assurance program. As with any product fabricated from a number of components, workmanship, and attention to directions furnished by the material suppliers are paramount. When the properties of a component do not comply with the standards set out in the specifications, the supplier should be notified and requested to remedy and/or explain the condition. Noncomplying materials should not be used until the problem (or misunderstanding) has been resolved.

SUMMARY

The emphasis on product quality has placed the burden of specifications and product testing on all facets of the sealants and sealants industry. The process of developing specifications and product testing described by the various specifications listed will help the industry meet those requirements for consistent quality. Depending on the size of the company, much of this work can be done in-house. If a company does not have in-house facilities, outside laboratories can perform the evaluations, interpret the results, and provide reports to the client. The laboratories also can assist in developing the specifications for the various stages of production. Whatever vehicle you decide to use, specifications and testing are the standards to follow for consistent product quality and good, finished-product performance.

REFERENCES

1. Amstock, J.S. 1997. *Handbook of Glass in Construction*, McGraw-Hill, Inc., New York, NY.
2. Chapter III, Sealants, The Professional Guide. 1995. Sealants, Waterproofing and Restoration Institute. Kansas City, MO.
3. Chusid, M.T., *Current Approach to Specification Practice, Building Design & Construction*.
4. Amstock, J.S. *Handbook of Glass in Construction*, McGraw-Hill, Inc., New York, NY.
5. Sealant, Waterproofing & Restoration Institute, Kansas City, MO. 1995.
6. Ortech Corp., Building Performance Centre, Mississauga, Ontario, Canada. 1994.

CHAPTER 19
JOINT DESIGN, JOINT DETAILS, AND INSTALLATION

INTRODUCTION

Before 1950, building sealants and caulks still were being developed in the laboratory. Building joints were sealed with oil base caulks and putty, which provide adequate protection from the effects of weathering [1].

The development of flexible polysulfide rubber sealants in the early 1950s opened the door to sealants for curtain walls and other types of glass facades. It also opened a new chapter in joint design, movement, and sealant technology. Non-masonry substrates such as aluminum are significantly affected by temperature change. Materials with high coefficients of expansion made oil-based caulks obsolete. Those caulks would harden, crack, and fall out of joints, leaving a clear passage for water, dust, dirt, and air.

To meet these stringent requirements of the curtain-wall structure, new products were developed for the construction markets. A more sophisticated product, a new chemical compound cured to a synthetic rubber (elastomer) sealant, was born. Unlike their oil-base predecessors, sealants were formulated to expand and contract with constant joint movement. The first sealants or caulks were based on polysulfide liquid polymer, many also known as Thiokol LP. They were followed by acrylic, polyurethane, and silicone sealants. Each had unique characteristics, causing each to perform differently under varying weather conditions. The differing strengths and weaknesses demanded varying surface preparation techniques and uses. A thorough understanding of every material by the contractor was necessary to ensure that the most effective sealant would be specified for, and applied to, a particular project, a task that was not without its difficulties. The need to design joints whose dimensions were within the performance limitations of the sealant was critical. Failure to do so shortened the life of the sealant and often resulted in costly leakage.

Properly preparing joints for a sealant required a multitude of unfamiliar considerations.

Some of these considerations included:

To prime (use an adhesion promoter) or not to prime?
Which backup or bond breaker should be used?
Should impregnated fiber filler or cork joint filler board be used?
Should joints be cleaned manually or by sand blasting or other mechanical methods?
How to repair or replace existing joints?

Many manufacturers and raw material suppliers sought to educate the sealant industry in the use of its products. There was much to be learned as the number of available sealants steadily increased. Many products were formulated differently, despite apparent similarity. During this learning process, some applicators recognized the importance of communicating as a unified group with the building trades, particularly those who specified building materials.

Most joints, and some cracks in concrete structures, require sealing against the adverse effects of environmental and normal service conditions. This chapter will guide you to a better understanding of the properties of joint sealants and caulks and where and how they are used in present practice [2]. This book will attempt to guide the architect or design engineer to several options rather than one standard practice, because in most instances there is more than one choice available. Without knowledge of the structure, its design, service use, environment, and economic constraints, it is impossible to prescribe a "best joint design" or a "best joint sealant." We have attempted to provide information based on current practices and experience that are judged sound by the industry and used by several of the organizations consulted in the writing of this book. It should, therefore, be useful in making an enlightened choice of a suitable joint sealing system and to insure that it is properly detailed, specified, installed, and maintained.

No attempt has been made to reference the voluminous literature except for those specific examples necessary for an understanding of the subject. An extensive glossary of terms that might not be generally familiar is provided in Chapter 24.

WHY JOINTS ARE REQUIRED

Concrete normally is subject to changes in dimensions, plane, and volume as a result of exposure to the environment or by the imposition or maintenance of loads. The effect might be permanent contractions due to, for example: initial drying, shrinkage, and irreversible creep. Other effects are cyclical and depend on service conditions and differences in humidity and temperature, or the application of loads and might result in either expansions or contractions. In addition, abnormal volume changes, usually permanent expansions, might occur in the concrete because of sulfate attack, alkali-aggregate reactions, and other causes. The results of these changes are movements, either permanent or transient, of the extremities of concrete structural units. If, for any reason, contraction movements are excessively restrained, the movement might result in distortion and cracking within the unit, or crushing of its end and the transmission of unanticipated forces of the abutting units. In most concrete structures these effects are objectionable from the perspective of structural considerations. One of the means of minimizing them is to provide joints at which movement can be accommodated without a loss of integrity of the structure.

There might be other reasons for providing joints in concrete structures. In many buildings, the concrete serves to support or frame curtainwalls, cladding, doors, windows, partitions, mechanical, and other services. To prevent development of stress in these sections, it is often necessary for them to move independently of overall expansions, contractions and deflections occurring in the concrete. Joints also might be required to facilitate construction without serving any structural purpose.

WHY SEALING IS NEEDED

The introduction of joints creates openings that usually must be sealed to prevent passage of gases, liquids, or other unwanted substances into or through the openings. In buildings, to protect the occupants and the contents, it is important to prevent intrusion of wind and rain. In tanks, most canals, pipes, and dams, joints must be sealed to prevent the contents from being lost.

Moreover, in most structures exposed to the weather the concrete itself must be protected against the possibility of freezing and thawing, wetting and drying, leaching or erosion caused by any concentrated or excessive influx of water at joints. Foreign solid matter, including ice, must be prevented from collecting in open joints; otherwise, the joints cannot open and close freely later. Should this happen, high stresses might be generated and damage to the concrete might occur.

In industrial floors the concrete at the edges of the joints often needs protection of a filler or sealant between armored faces capable of preventing damage from impact of concentrated loads such as steel-wheeled traffic.

In recent years, concern about the spread of flames, smoke, and toxic fumes has made the fire resistance of joint sealing systems a consideration, especially in high-rise buildings. (See Chapter 15 for additional data on firestops, firestopping, etc.)

The primary function of sealants is to prevent the intrusion of liquids (sometimes under pressure), solids or gases, and to protect the concrete against damage. In certain applications, secondary functions are to improve thermal and acoustical installations, damp vibrations or prevent unwanted matter from collecting in crevices. Sealants often must perform their prime function while subjected to repeated contractions and expansions as the joint opens and closes and while exposed to heat, cold, moisture, sunlight, and sometimes aggressive chemicals. As discussed in other parts of this book, these conditions impose special requirements on the properties of the materials and the method of installation.

In most concrete structures, all concrete-to-concrete joints (contraction, expansion, and construction), and the periphery of openings left for other purposes, require sealing. Exceptions are contraction joints (and cracks) that have very narrow openings, for example, monolithic joints not subject to fluid pressure or joints between precast units used either internally or externally with intentional open draining joints.

JOINT DESIGN AS PART OF OVERALL STRUCTURAL DESIGN

In recent years it has become increasingly recognized that there is more to providing an effective seal at the joint than merely filling the "as constructed" gap with an impervious material. The functioning of the sealant, described in Chapter 20, depends as much on the movement to be accomplished at the joint and on the shape of the joint, as on the physical properties of the sealant. Joint design, which broadly covers the interrelationship of the factors, is discussed in some detail in Chapter 21, since it should be an important, sometimes governing, consideration in the joint design of most concrete structures. It is beyond the scope of this book to venture into the field of volume change in concrete (See Chapter 7) and the structural considerations that determine the location and movement of joints. However, many years of experience in trying to keep joints sealed indicate that joint movements might vary widely from those postulated by theory alone.

There are probably as many "typical joint details" in existence as there are structures incorporating them. To illustrate how they can be sealed, it seems best to present them in schematic form (as illustrated in Chapter 21), to bring out the principles involved for each of the three major groups:

1. Structures not under fluid pressure (most buildings, bridges, storage bins, retaining walls, and so on).
2. Containers subject to fluid pressure (dams, reservoirs, tanks, canal linings, pipelines, etc.)
3. Pavements (including highways and airfields).

From both structural and sealant considerations, irrespective of design detail and end-use, all the joints might be classified according to their principal function and configuration.

TYPES OF JOINTS AND THEIR FUNCTION

Contraction (Control) Joints

These are purposely made planes of weakness designed to regulate cracking that might otherwise occur because of the unavoidable, often unpredictable, contraction of concrete structural units. They are appropriate only where the net result of the contraction and any subsequent expansion during service is such that the units abutting are always shorter than at the time the concrete was placed. They are frequently used to divide large, relatively thin structural units, such as pavements, floors, canal linings, retaining, and other walls, into smaller panels. Contraction joints in structures are often called control joints because they are intended to control crack location.

Contraction joints might form a complete break, dividing the original concrete unit into two or more units. Where the joint is not wide, some continuity might be maintained by aggregate interlock. Where greater continuity is required without restricting freedom to open and close, dowels, and in certain cases steps or keyways, might be used. Where restriction of the joint opening is required for structural stability, appropriate tie bars or continuation of the reinforcing steel across the joint might be provided.

The necessary plane of weakness might be formed either by partly or fully reducing the concrete cross-section. This may be done by installing thin metallic, plastic, or wooden strips when the concrete is placed, or by sawing the concrete soon after it hardens.

Expansion (Isolation) Joints

These are designed to prevent crushing and distortion (including displacement, buckling, and warping) of the abutting concrete structural units that might otherwise occur because of the compressive forces developed by expansion, applied loads, or differential movement arising from the configuration of the structure or its settlement. They are frequently used to isolate walls from floors or roofs, columns from floors or cladding, pavement slabs and decks from bridge abutments or piers, and in other locations where restraint or transmission of secondary forces is not desired. Many designers consider it good practice to place such joints where walls or slabs change directions such as in L-, T-, Y-, and U-shaped structures and where different cross-sections develop. Expansion joints in structures are often called isolation joints because they are intended to isolate structural units that behave in different ways.

Expansion joints are made by providing a space for the full cross-section between the abutting structural units when the concrete is placed through the use of filler strips of the required thickness, bulk-heading or by leaving a gap when precast units are positioned. Provision for continuity or for restricting undesired lateral displacement may be made by incorporating dowels, steps, or keyways.

Construction Joints

These are joints made at the surfaces created before and after interruptions in the placement of concrete or through the positioning of precast units. Locations are usually predetermined by agreement between the design engineer and the contractor, to limit the work that can be done at one time to a convenient size with the least impairment to the finished structure. They also might be necessitated by unforeseen interruptions in concreting operations. Depending on the structural design, they might be required to function later as expansion or contraction joints having the features already described, or they might be required to be monolithic, with second placement soundly bonded to the first to maintain complete structural integrity. The design of the structure might require construction joints to run horizontally or vertically.

Combined and Special-Purpose Joints

Construction joints at which the concrete in the second placement is intentionally separated from that in the preceding placement by a bond-breaking membrane, but without space to accommodate expansion of the abutting units, also function as contraction joints. (See above.) Similarly, construction joints in which a gap is formed by bulkheading or by the positioning of precast units, function as expansion joints. Conversely, expansion joints often are convenient for forming non-monolithic construction joints. Expansion joints automatically function as contraction joints, although the converse is true to an amount limited by any gap created by initial shrinkage.

Hinge Joints

Hinge joints are joints that permit hinge action (rotation) but at which the separation of the abutting units is limited by tie bars or the continuation of reinforcing steel across joints. This term has wide usage in, but is not restricted to, pavements where longitudinal joints function in this manner to overcome warping effects while resisting deflections due to wheel loads or settlement of the subgrade. In structures, hinge joints are often referred to as *articulated joints*.

Sliding Joints

Sliding joints might be required where one unit of a structure must move in a plane at right angles to the plane of another unit, for example, in certain reservoirs where the walls are permitted to move independently of the floor or roof slab. These joints are usually made with a bond-breaking membrane, such as a bituminous compound, paper, or a felt that also facilitates sliding.

Cracks

Although joints are placed in concrete so that cracks do not occur elsewhere, it is extremely difficult to prevent occasional cracks between joints. As far as sealing is concerned, cracks might be regarded as contraction joints of irregular line and form. Treatment of cracks will be considered in Chapter 22.

JOINT CONFIGURATIONS

In the schematic, joint details for various types of concrete structures as shown in Figures 19.1 and 19.2, are two basic conditions that occur in the sealant. Figures 19.1a and 19.1b show how sealants perform, and Figure 19.2 illustrates the shape factors and strains that sealants undergo.

These are known as butt joints and lap joints. In butt joints, the structural units being joined abut each other and movement is largely at right angles to the plane of the joint. In lap joints, units being joined override each other and relative movement is in sliding. Butt joints, including most stepped joints, are by far the most common. Lap joints might occur in certain sliding joints (see combined and special-purpose joints) between precast units or panels in curtain walls, and at the junctions of these and of cladding and glazing with their concrete and other framing. As covered in Chapter 20, at least in part, the difference in the mode of the relative movement between structural units at butt joints and lap joints controls the functioning of the sealant. In many of the applications of concern to this book, pure lap joints do not occur, and the functioning of the lap joint is a combination of butt and lap joint action.

In regard to the sealant, two sealing systems should be recognized. First, there are open surface joints, as in pavements and buildings, in which the joint sealant is exposed to outside conditions on at least one face. Second, there are joints, as in containers, dams, and pipelines, in which the primary line of defense against the passage of water is a sealant such as a waterstop or gasket buried deeper in the joint. The functioning and type of sealant material that is suitable and the method of installation are affected by these considerations.

JOINT DETAILS

Figures 19.3 through 19.8 illustrate the application of joint sealants to a wide variety of design configurations that occur in concrete construction (see Appendix G for a key to symbols).

The details shown are representative of current practice and cover most standard variations, although other variations in use might not be shown. These details are presented in outline form, omitting for the sake of clarity structural

The behavior of field-molded sealants in service depends upon a combination of their elastic and plastic properties. Elastomeric sealants should behave largely elastically to regain after deformation their original width and shape, that is full strain recovery (no permanent set) is desirable. However due to plastic behavior some set, flow, and stress relocation occurs. The extent of its effect depends on the properties of the materials used and conditions such as temperature, repetition and rapidity of cycles of stress reversal and duration of deformation at constant strain. Largely plastic behavior, that is, returns to original shape by flow, is only acceptable for sealants used in joints with small and relatively slow movements.

① IN BUTT JOINTS

	(A) AS INSTALLED	(B) JOINT OPEN	(C) JOINT CLOSED
(i) Sealant is:	. . . .	Sometimes in tension and	sometimes in compression
and Sealant should:	. . . .	Change its shape without changing its volume	

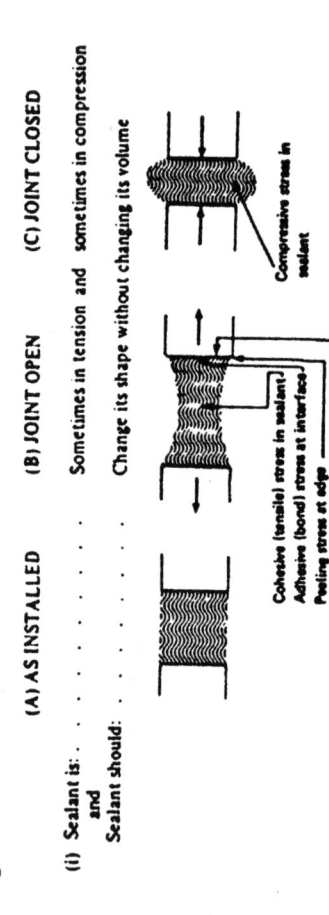

(ii) Material requirements for good performance:

(A)
(a) Ease of installation
(b) Good bond to faces
(c) Homogeneity
(d) Low shrinkage

(B)
(e) High ultimate strength in rubberlike materials
(f) Low elastic modulus in rubberlike materials
(g) Resistance to flow and stress relaxation

(C)
(g) Resistance to flow and stress relaxation
(h) Low compression set

Also required (1) Impermeability (3) Recovery (6) Resist flow (7) Not harden
(8) Not deteriorate

FIGURE 19.1a How sealants work.

(iii) Deficiencies in   (b) (c) (f) predispose towards adhesion failure
                          (c) (d) (e) predispose towards cohesive failure
                          (h) (3) (6) predispose towards permanent deformation
                          (g) (3) (6) predispose towards flow and stress relaxation
                          (a) (7) (8) accelerate failures due to above causes

② **IN LAP JOINTS**

(A) AS INSTALLED   (B) JOINT OPEN   (C) JOINT CLOSED

(I) Sealant is: . . . .   Always in shear (Note 1)   Always in shear (Note 1)

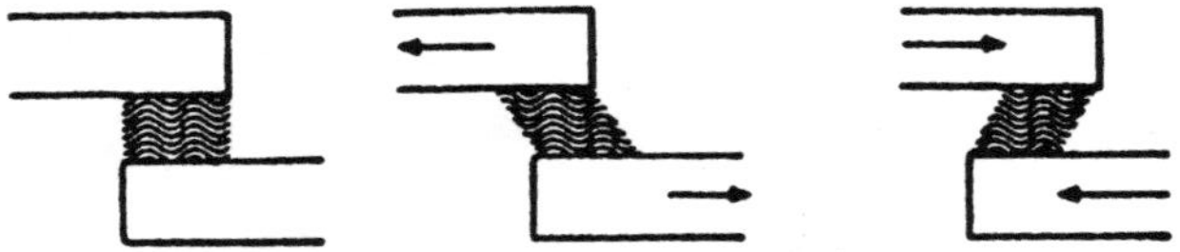

(II) **Material Requirements:** These are generally similar to those above for butt joints. Same materials used with thickness of sealant (distance between the overlapping faces) equal to ½ to 1 times the deformation of sealant in shear (which is the joint movement) depending on installation temperature

Note 1: If, as lap joint opens or closes, units move closer together or farther apart in plane at right angles to main movement then compression or tension of the sealant will also occur. This combination of movements is common in many applications to buildings. However tension-compression is not a governing criterion provided minimum thickness requirements of sealant to handle shear are met.

FIGURE 19.1b How sealants work.

Cases showing the effect of shape on the maximum strains 'S' that occur on the parabolic exposed surface of elastomeric sealants. Sealant assumed to be installed at mean joint width so that ½ change of width of sealant will be extension and ½ compression.

BUTT JOINTS

① **JOINT DEPTH TO WIDTH RATIO 2:1**

② **JOINT DEPTH TO WIDTH RATIO 1:1**

③ **JOINT DEPTH TO WIDTH RATIO 1:2**

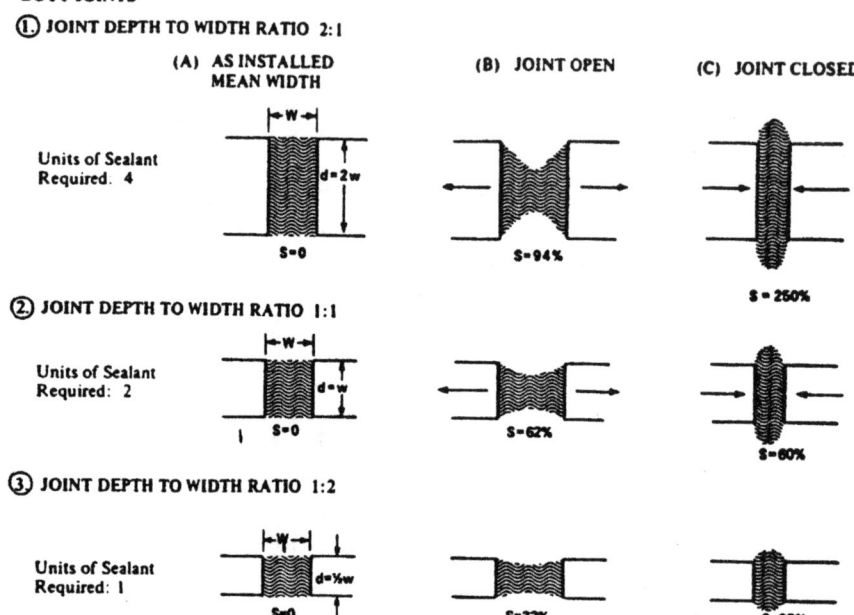

'CONCLUSION: Increasing the width and reducing the depth generally reduces strains and hence improves performance of field molded sealants. At the same time less sealant is required. Shape Factor is less important in mastic sealants since plastic not elastic behavior dominates.

④ **PURPOSE OF BOND BREAKER AND BACK UP:** In joints open on one face only the back face of the sealant must not adhere to the bottom of the sealant reservoir so that the sealant is free to assume the desired shape. See (A) below. Control of depth of sealant is achieved as shown in (B) where the joint is formed or sawn initially deeper than the required depth to width ratio. (Bi) and (Bii) present cases as to desirable shape of backup.

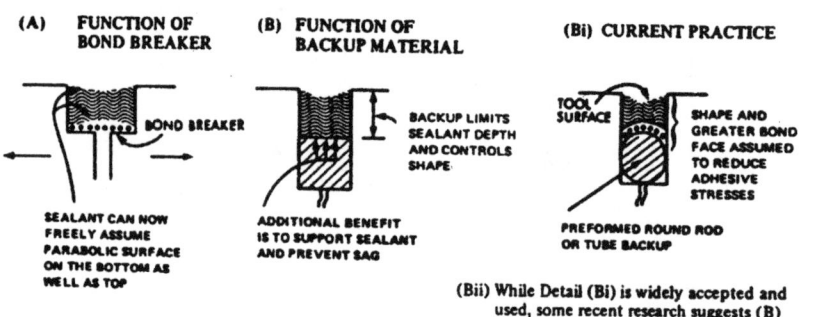

(Bii) While Detail (Bi) is widely accepted and used, some recent research suggests (B) may be better since, if backup material presents flat face to sealant, peeling stresses at corners are reduced.

FIGURE 19.2 Shape factor and strains in sealants.

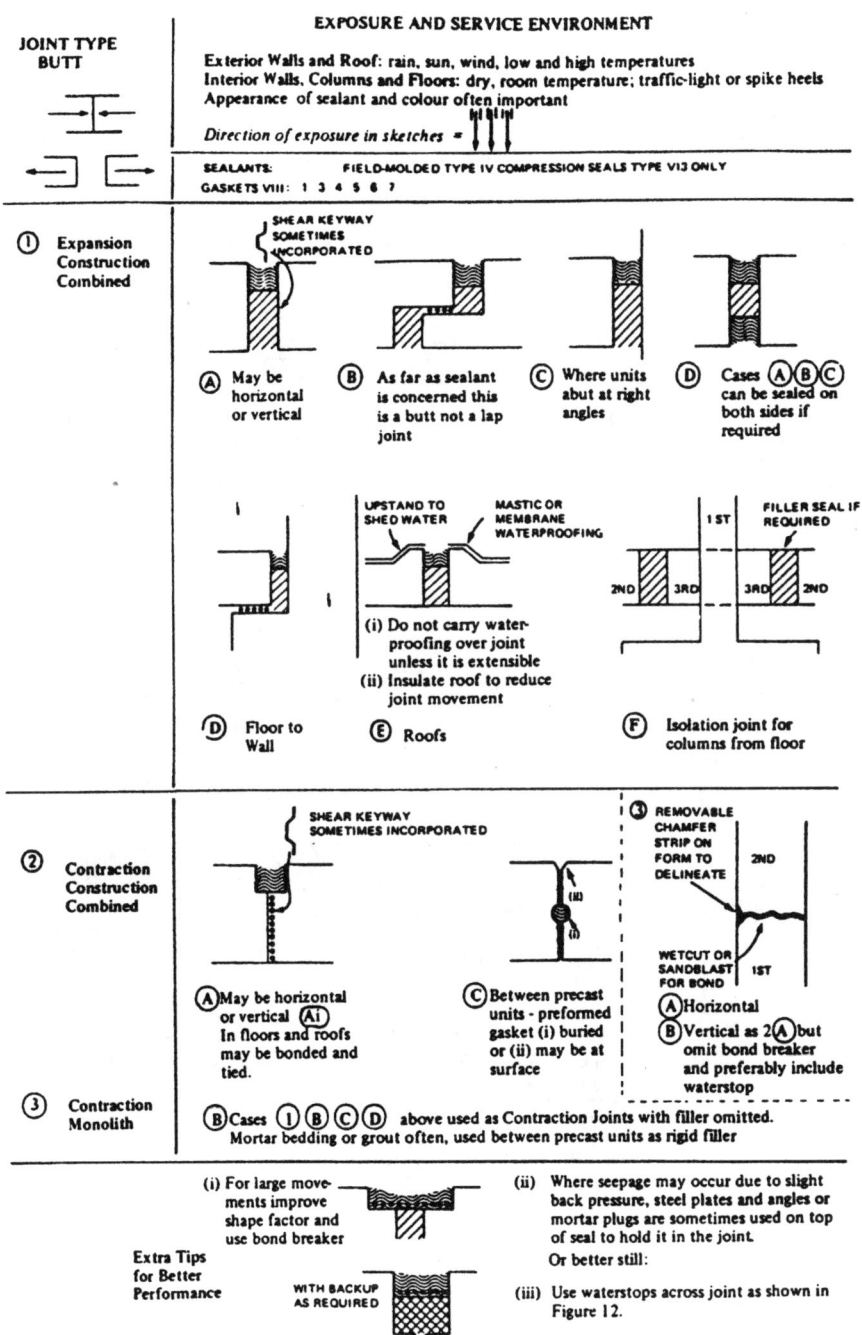

FIGURE 19.3 Joints for structures: concrete to concrete.

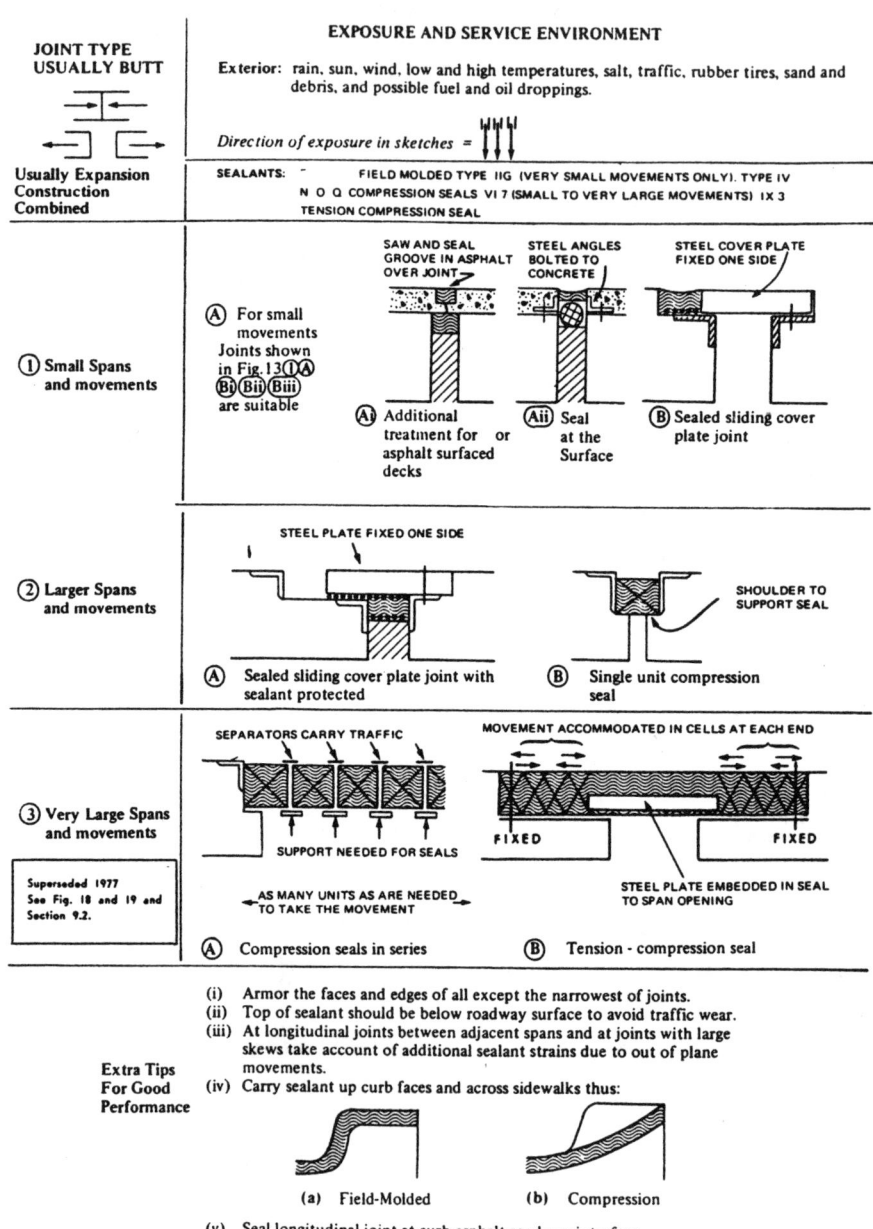

FIGURE 19.4 Joints for bridges.

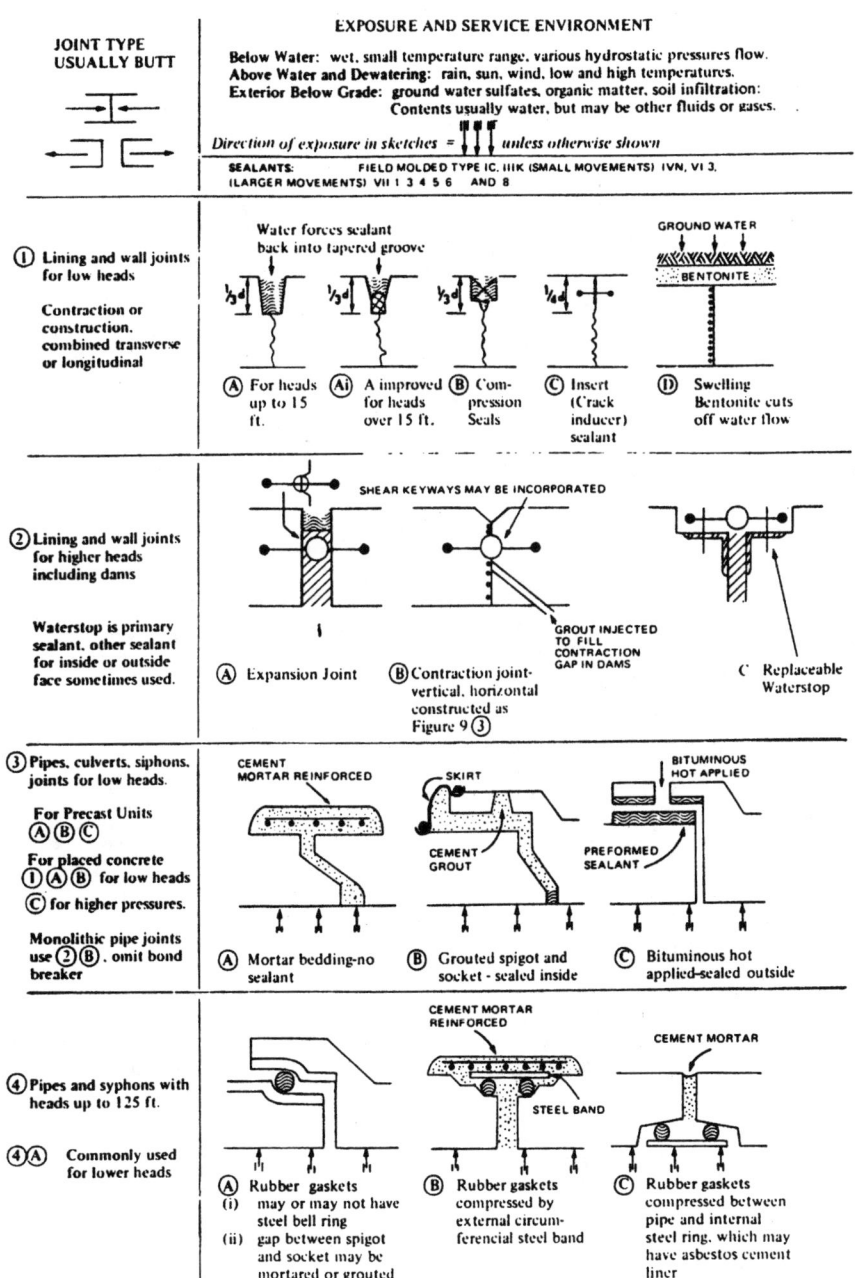

FIGYRE 19.5 Joints for containers: canal linings, walls, dams, pipes, culverts, syphons.

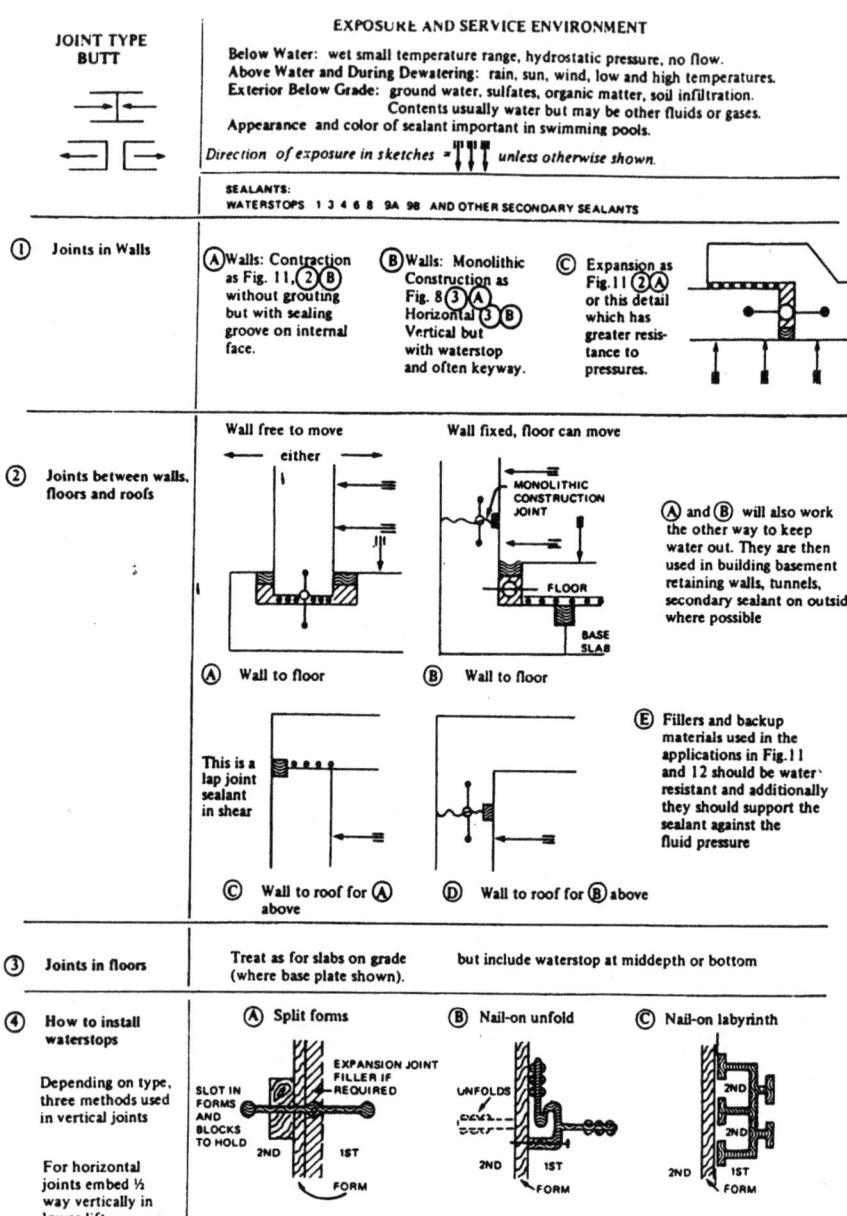

FIGURE 19.6 Joints for containers: tanks, reservoirs, swimming pools, waterstops.

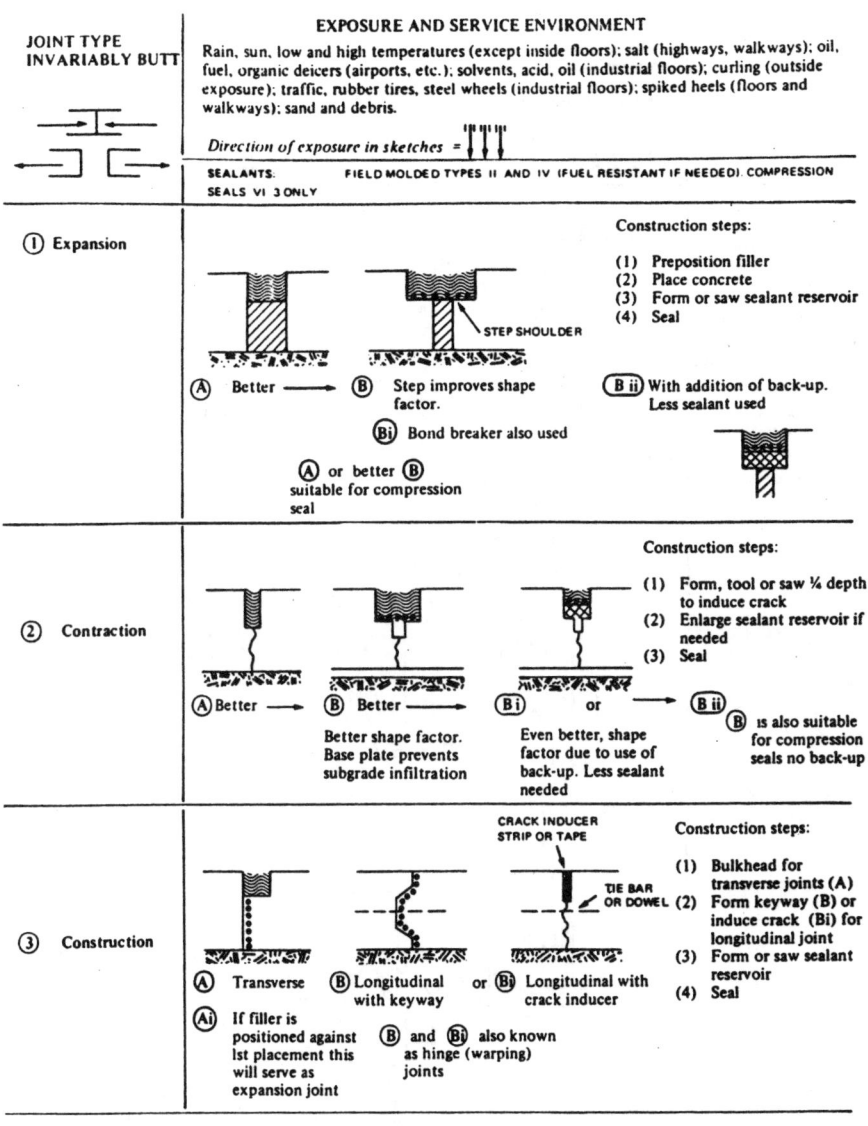

FIGURE 19.7 Joints for slabs on grade: highways, airports, walkways, floors.

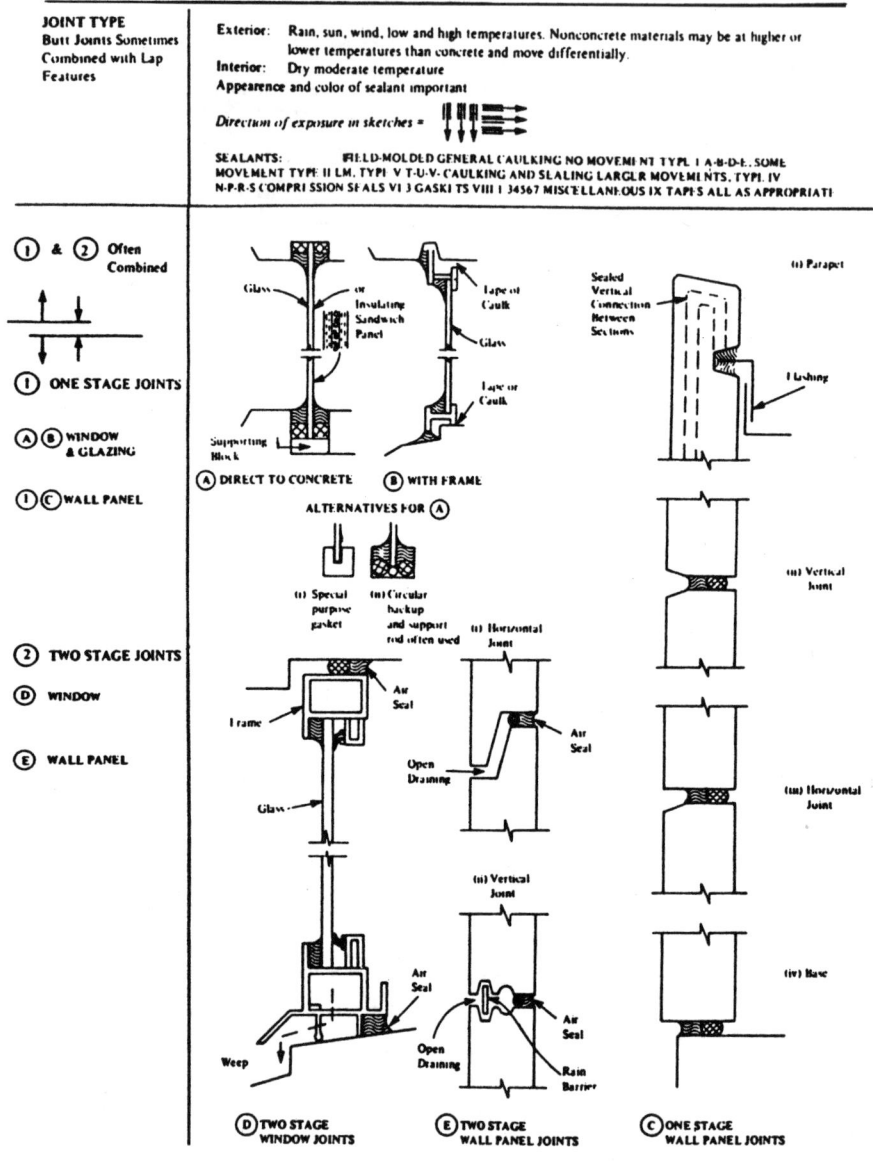

FIGURE 19.8 Joints for buildings: special purposes.

details, such as reinforcing steel, dowels, etc., not directly relevant to the sealing of the joint. The location of the joint is indicated only where this is significant to delineate the types of joint and sealant that might be suitable. Similarly, the sealant reservoirs (grooves) and expansion and contraction gaps are not dimensioned as to width or depth because differing sealants have a wide range of performance capabilities. The required configuration should be determined as outlined in this chapter, or might be obtained from the manufacturer's data sheets.

Exposure and environmental service conditions are shown for each group of applications, because this is an important consideration in selecting the correct sealant.

No endorsement is intended for selecting one detail, or choosing one sealant rather than another. This book does endorse and promote those standard design features documented throughout (i.e., improving the shape factor of field-applied sealants) that will ensure the best possible performance of any given seal or sealant.

STRUCTURES

Figures 19.3 and Figure 19.8 cover applications to structures in general and buildings in particular, where sealing against significant fluid pressure is not a consideration. Where groundwater must be excluded—for example, in basements or earth-retaining walls—reference should also be made to Figure 19.6, because additional sealing using waterstops might be indicated. Since the appearance of sealed joints is important in many buildings, additional architectural treatment not shown in the figures—for example, V-ing of joint edges—might be required.

Bridge deck joints are treated separately in Figure 19.4, because large movements and special sealing problems often are involved. Joint details suitable for bridge substructures generally follow those in Figure 19.3, and, where water pressure is involved, those shown in Figure 19.6. Containers of all types are covered in Figures 19.5 and Figure 19.6. Except where the head is small, the use of waterstops in "in-place construction" or gaskets under compression for precast pipes is almost essential if the contents are to be retained. In certain uses, for example, dams, a secondary waterstop might be used some distance behind the first as an additional line of insurance against premature failure. Many of the details shown in other figures also serve equally well for keeping water that is outside the structure from passing through the joint to the outside face. The exclusion of water from basements, subways and tunnels are examples of this application. Tunnel applications are discussed in greater detail in ACI 504 .1R.

SLABS ON GRADE, HIGHWAY AND AIRPORTS

Slabs on grade are shown in Figure 19.7. These might be outside as on highways, parking garages and airports, or they might be within a building or container with modifications indicated in the figures to these applications. Many highway authorities are specifying short contraction joint spacings in both plain and reinforced concrete pavements. Some are using a random spacing averaging between 15 and 20 feet (4.57 and 6.10m) in plain pavements for which the repeating series, 13, 18, 19, 12 feet (3.6m) is popular. Some also are skewing joints at 2 feet and in 12 foot (0.61m in 3.66m). While the objectives are reducing intermediate slab cracking and improving ride and load transfer, such designs place less demand on the sealant because of the smaller movement that occurs at each joint. More detailed information on the design and construction of joints in concrete pavements can be found in the reports of ACI Committee 325.

It also should be noted that experience points to better field-molded sealant performance if a shape factor of 2:1 is used. For pavements with short slabs a typical joint sealant reservoir configuration found to be satisfactory is: joint width ⅜ inch to ¹⁄₁₆ inch (9.5mm to 12.7mm), joint sealant depth 1 inch (25.4mm) using cord or tape as a backer material. In most applications, including airports, the sealant is installed ¼ inch- ¹⁄₁₆ inch (6.4mm-1.5mm) below the level of the pavement.

CONSTRUCTION AND INSTALLATION CONSIDERATIONS

The practical aspects of constructing the joint and sealing must be kept in mind when its details are being designed. The general construction steps for any expansion, contraction, and construction joints are shown in Figure 19.7(1), (2), and (3). The method of making monolithic construction joints is outlined in Figure 19.3(3). The positioning of waterstops is shown in Figure 19.6(4) and further discussion of their installations and that of sealants in general follows in this chapter. It must be remembered that a joint detail that makes it unnecessarily difficult to install the sealant is a poor one as it is likely to lead to premature failure.

INSTALLATION OF SEALANTS

The most appropriate technique for installing (applying) a joint sealant depends on the materials, the width, shape, inclination, and accessibility of the joint, and whether it is a small or large project. Each step in the construction and preparation of the joint to receive the sealant and for its installation requires careful workmanship and thorough inspection to avoid initial defects that might be costly and time-consuming to correct.

The specification for the work should state how the sealant is to be installed and any special features required in the construction or preparation of the joint to receive it. Before the containers of sealants are opened, their labels should be checked to make certain that the right sealant has been supplied and that there is no conflict between the specification and the manufacturer's instructions for installation. Any discrepancy should be referred to the architect or design engineer before work commences. The most favorable time for installing field-molded sealants, if the construction schedule permits, is on dry days when the temperature is close to the annual mean. Compression seals, especially large ones, are easiest to install on cold days. However, a satisfactory job usually can, and must, be done in less than ideal conditions provided that the effects of this are compensated for in the design of the joint.

Sealant storage and installation requirements are summarized later in this book. These operations are discussed in greater depth in the following paragraphs.

JOINT CONFIGURATION WITH SEALING IN MIND

Some of the defects resulting from improper concrete joint construction are shown in Figure 19.9. These and others can be avoided by the following steps:

1. Saw or form the joint to the required (and uniform) depth, width, and location shown in the plans. Manufacture precast units to close tolerances and position them carefully.
2. Align all joints with any connecting joints to avoid blockage of free movement.
3. Judge the time of sawing to avoid edge spalling or plucked aggregate (too early) or random cracks (too late).
4. Correctly position dowels and other joint hardware, fillers, waterstops, and bulkheads, and rigidly support them to avoid displacement during concreting.
5. Remove any temporary material or filler used to form the sealant reservoir by raking out or rotary cutting to the specified depth.
6. Keep the curing compound or other materials from contaminating joint faces. Apply supplemental curing where the original curing is broken by construction operations before the joint edges and faces have fully cured.

PREPARATION OF JOINT SURFACES

Joint faces must be clean and free of defects that would impair bond with field-molded sealants or prevent uniform contact of preformed sealants. Removal of contaminants might require washing out debris left by sawing, wire brushing, routing, and/or sandblasting. Though sandblasting is more expensive, it is more likely to succeed and therefore, is warranted where relatively expensive thermosetting, chemical curing field-molded sealants are used.

Solvents intended to remove oil, etc., usually have the opposite effect and carry the contaminants further into the pores of the concrete. Solvents are, however, distinctly useful in cleaning nonporous substrates such as glass or metal frames. Defects in the joint such as loose aggregate, embedded foreign material, and spalls in the case of compression seals or blockages to free movement require repair before cleaning. (See Chapter 23). Final cleanup to remove dust usually is required.

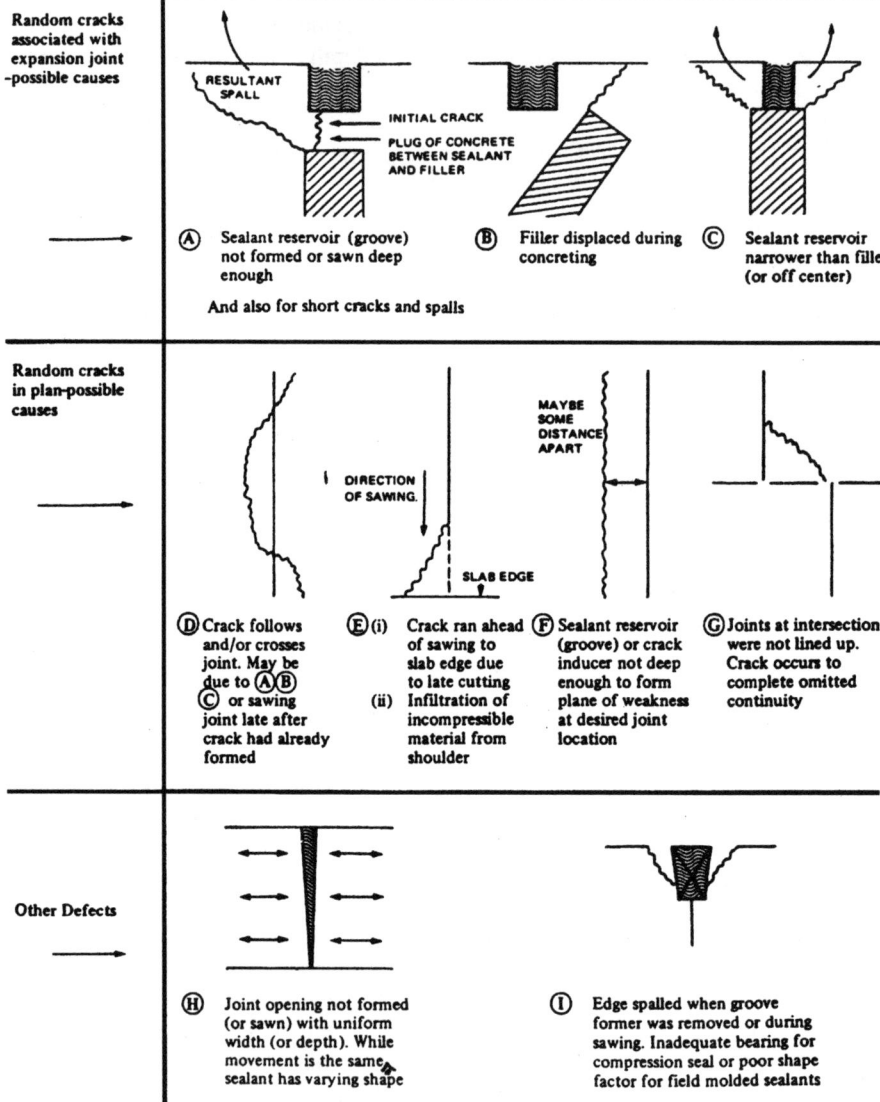

FIGURE 19.9 Defects in joint construction.

This is essential where a good bond must be developed with chemically curing thermosetting sealants. Final cleanup can be done by a brush, but the user of oil-free compressed air or vacuum cleaner is more likely to be successful.

As a rule, joint faces also must be dry because the sealant has to bond with the concrete. Exceptions are claimed by sealant suppliers. They include neoprene compression seals and emulsion, and certain elastomeric sealants formulated to displace water from contact faces. Notwithstanding, better results will be achieved if sealant installation is done in the dry.

INSPECTION OF READINESS TO SEAL

Inspection of each joint to ensure it is sufficiently clean and dry is essential prior to placing backup material, priming, or sealant installation. It is also wise to check the joint width and temperature (preferably that of the concrete rather than the ambient) against assumptions made in the joint design. Restrictions on joint width and temperature at the time of sealant installation should be shown on the plans. In the absence of these, installations at above 90° F (32° C) or below 40° F (4° C) generally should be avoided. Installation at temperatures above or below these values might lead to various failures. Extra strains might be induced on field-applied sealants (Selection of butt-joint widths for field-molded sealants). Problems might arise because of the shortened working life at higher temperatures or moisture condensation and frost at lower ones. In the case of compression seals, it is harder to install them properly in tight joints as the lubricant might be too fluid or viscous.

PRIMING, INSTALLATION OF BACKUP MATERIALS AND BOND-BREAKERS

When priming is required with the selected field-molded sealant necessary primer is usually supplied with the sealant and can be applied by brush or spray. (As a rule, priming is required for all porous surfaces such as concrete, wood, and possibly plastics if thermosetting, if chemical curing field-molded sealants are to adhere satisfactorily). Brushing can be tedious, and unless excess material is properly brushed out to insure a uniform film over the whole joint face, adhesion failures might result. For horizontal joints on larger projects, spray applicators might be more appropriate. Most primers require time to dry before the sealant is installed. Failure to permit this action might lead to adhesion failure or exudation of the primer.

Backup materials and/or bond breakers require positioning, usually by hand, before the sealant is installed. They might be set at the correct depth avoiding twisting or contamination of the cleaned joint faces.

INSTALLATION OF FIELD-MOLDED SEALANTS, HOT-APPLIED

As noted on Table 22.3, certain joint sealants are melted and hot-applied in the field. These hot-applied compounds usually are comprised of bituminous materials (either asphalt or tar) and might or might not contain rubber or other elastomeric substances. Each manufacturer's recommended pour point, as well as a safe heating temperature, should not be exceeded. The safe heating temperature usually is 20 ° F (11° C) above the recommended pour temperature. Subjecting sealants to temperatures above the safe heating temperature limits results in breakdown or setting up of the compound, which precludes good field performance and affects longevity.

These materials usually are heated in double-boiler-type melting kettles equipped with a suitable agitation system in the sealant melting chamber, a positive pressure delivery and re-circulation system, and a recording thermometer. The inner tank should be oil-jacketed and the temperature of the high flash-point heat transfer oil should be thermostatically controlled. Earlier models of melters were gravity-fed and some required the use of separate pouring devices. These were preferably oil-jacketed as well, but all suitable units were insulated. The newer melting kettles equipped with pressure discharge through hoses, wands, and nozzles should have the application lines insulated.

Hot-poured materials usually are suitable for installation only in horizontal joints. They can be placed in vertical joints but adequate dams are necessary to prevent the sealant from flowing to the bottom before it cools and sets. Horizontal joints should be filled to slightly below the pavement surface.

INSTALLATION OF FIELD-APPLIED SEALANTS, COLD-APPLIED

Except for extremely short joints or in touch-up work, cold-applied sealants usually are extruded under pressure from a hand-held caulking gun having a nozzle orifice sized and shaped to mold the required bead of sealant to suit the opening. The simplest piece of equipment for this purpose is the familiar hand-operated caulking gun. The sealant is supplied either prepacked in cartridges to suit the caulking gun or the chamber or cartridges are loaded on the job from bulk containers as required, or in the case of two component sealants, they are filled with the compound after mixing. If the size of the project warrants, more sophisticated pressure application equipment can be used, including models where two-component materials are brought by individual lines to the nozzle where they are mixed in a small chamber immediately prior to extrusion. Pumps, compressed air, or gas might be used to supply the necessary pressure for extrusion.

With two-component sealing, full and intimate mixing is essential if the material is to cure with uniform properties. Little can be done with patches of sealant that do not harden except to remove and replace them with properly mixed sealant. Small quantities of two component sealants can be mixed with a broad-bladed putty knife. However, for any significant quantity of material, mechanical means is required. For small batches, hand-held electric drill mixers with paddle blades can be used. Large batches need purposely made mixing machines. Figure 19.10 provides a reasonably accurate estimate of how much sealant is required for a given project based on joint dimensions.

Frequently, it is convenient to premix sealants at a location remote from the job site. To prevent curing until it is time to use them, sealants might be frozen at –40° F (-40° C) or below and held in storage. In some urban areas, frozen premixed cartridges are available from the sealant suppliers. On the jobsite, cartridges are thawed for about 30 minutes at a temperature of 70° F (21° C) (Additional heating to hasten thawing might be detrimental and should not be used.)

Application of a sealant to fill a joint reservoir requires a skilled caulker. The gun nozzle must be controlled at an angle (about 45°) and moved steadily along the joint so a uniform bead is supplied without dragging, tearing, or leaving unfilled spaces. A skilled applicator will be seen to push the bead, rather than draw it with the gun leading. In large joints, several runs might be needed, building up the sealants in roughly triangular wedges at each run.

For non-sagging materials, when the joint has been filled with the required amount of sealant, it is tooled to ensure intimate contact, or wetting, of the joint faces, to remove any trapped air or voids, to consolidate the material, and to provide a neat, uniform appearance. At the joint faces, the exposed face of the sealant usually should match the level of the edge of the concrete. An exception is in areas subject to traffic where self-leveling sealants are used. In this case the surface should be left slightly low. It must be remembered that two-component materials in particular have a limited working life (pot life), especially on hot and humid days. Once the catalyst is mixed, the curing reaction starts; therefore the batch size should be limited to what can be used within the pot life of the sealant. In most cases the sealants are supplied in pre-measured kits—$1^1/_2$-gallon kits packaged in a two-gallon container—so there is adequate room for mixing.

INSTALLATION OF COMPRESSION SEALS

Compression seals require a uniform joint width along the whole length with straight, smooth, spall free, properly cleaned joint faces to permit proper installation and to provide uniform contact. It is advantageous to remove sharp ridges at the joint edge or to form or saw the joint with a slight rounded or V-edge.

A neoprene-based or other lubricant (which might have adhesive properties for most applications) is applied in a bead to the upper edge of each joint face to facilitate installation of the seal. The lubricant is fluid at normal temperatures and usually is applied by a hand-pressure applicator. Where machine installation of

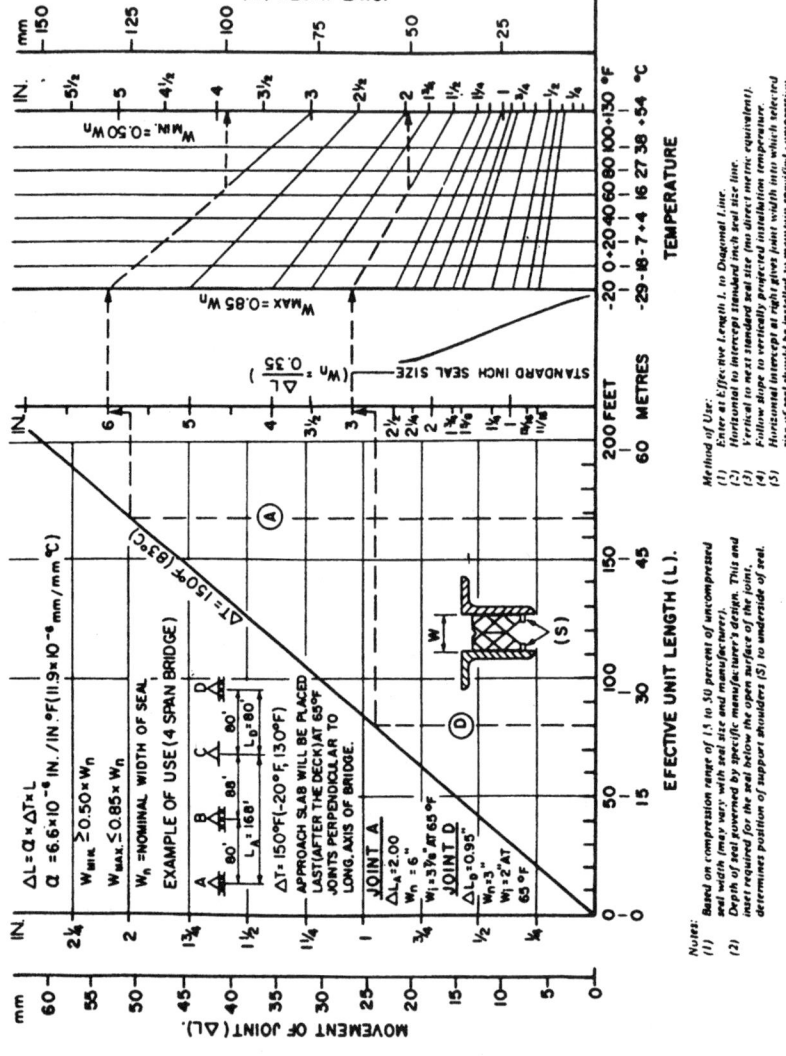

FIGURE 19.10 Chart for determining movement of joint and width or for selecting seal size; compression seals.

the sealant is used for pavement joints, the unit might also be designed to apply the lubricant, which then generally should be a thixotropic formulation. The lubricant must be applied immediately ahead of inserting the seal so that it does not prematurely dry out.

For installation by hand roller or with the machine, the seal is positioned vertically over the joint opening and then, by pressing down and forward, forced into the opening. The seal must not be twisted, folded over on itself or stretched during this operation. A small permissible amount of stretching, up to 5 percent, might occur as the seal is forced in. The seal must not be willfully stretched (thus reducing the cross-section) to make installation easier and the seal length go further. Near zero stretch can be achieved with both hand, and automatic-machine installation, which might be desirable because the seal will not be under tension along its length nor reduced in effective width.

It is important to install the seal at the specified depth. In highway pavements, this usually is slightly below the surface to keep it out of contact with traffic. The seal should be installed in as long a continuous piece as possible. If field splices cannot be avoided, they should be made in the least critical locations as far as maintaining a sealed joint is concerned. Usually, the seal is spliced by butting it against the next length with some lubricant adhesive. However, more sophisticated means are available and might be warranted where it is important that a splice should not split. Where a compression seal is to be installed between precast units, it may be attached to the face of one and compressed as the adjacent unit is positioned.

The polybutylene impregnated-foam type of compression seal is precompressed and inserted in the joint opening. To achieve a good bond, the joint faces might first require priming with an epoxy adhesive. Other cellular foams, such as ethylene vinyl acetate, are installed in a similar fashion.

INSTALLATION OF PREASSEMBLED DEVICES

Placing large seals 4 in. (100mm) and larger in bridge joints and other large movement joints presents special problems. First, these seals are not particularly easy to handle and they cannot be bent or formed to suit an abrupt change in direction. Second, they require considerable force to compress them as they are pushed and leveled into the opening, especially if it is a warm day and the joint is partly closed beyond its mid-range. For this reason (and because the seal must be sized to the joint opening), there is merit in joint devices that are installed as a complete unit prior to concreting with the seal precompressed or preset to the required width. The joint is activated after the concrete has set by releasing the ties connecting the joint faces. Strip (gland) seals and modular systems designed to accommodate large movements are similarly supplied precompressed or preset, ready for installation and subsequent activation.

Tension compression devices as used in bridge decks require setting flush with the pavement surface. Except where subsequent bituminous surfacing is to be laid, this requires a recess in the concrete surface on each side of the joint. Provision for the hold-down bolts required for the mechanical anchoring of the device can be made either by presetting inserts using a template when the concrete is placed, or by drilling and installing the anchors after the concrete is set. In cases when the anchorage units are pre-attached to the edge elements (strip seals and armored-type systems) the expansion joint is set to line and grade and then concrete is placed.

INSTALLATION OF WATERSTOPS

Three methods of positioning waterstops in vertical joints are shown in Figure 19.6(4). Of these, placing between spilt forms is still the most common, although nail-on types might be more convenient and economical. In horizontal joints, waterstops usually are embedded halfway into the first lift. In all instances, the waterstop should be held in position securely so it will not be displaced during concreting, and some care is required in placing and consolidating the concrete so that no voids or honeycombing occur adjacent to the waterstops to prejudice its sealing ability. Contamination of the waterstop surfaces, for example, by form coatings, should be avoided. While rubber and polyvinyl chloride waterstops are not susceptible to damage during normal handling in concreting operations, thin metal waterstops are bent or torn easily and require special care.

Waterstops might need splicing at intersections, need abrupt changes of direction or need to form long, continuous lengths. It is often convenient to order prefabricated junction pieces from the manufacturer so these can be joined to the main run by simple butt splices in the field. Polyvinyl chloride waterstops can be spliced by trimming their ends to the required matching shape and then butt-welding them together by softening under heat and pressing them together until cooled. Because excess heat or an open flame would char the material and destroy its resilience, thermostatically controlled electric heating tools should be used. Rubber waterstops can be joined by mitering the ends to mate and, after cleaning and roughening, cementing them together. They must be held in a mold under heat and pressure until cured (vulcanized). An alternative is to use premolded splicing sleeves into which opposite sides of the cemented ends of the waterstop are inserted.

INSTALLATION OF GASKETS

Gaskets are positioned either in the joint opening or prepositioned because they are attached to one of the units to be joined, for example, on the spigot end of the pipe. Positioning the unit in place closes the joint on the gasket, which, under pressure, then forms the seal.

INSTALLATION OF FILLERS

Most compressible fillers in expansion joints are installed ahead of the concreting operation in the required location and position and held either by the bulkhead, if it is a construction joint, or by some rigid device at other expansion joint locations until the concrete has been placed and set. Problems of the type illustrated in Figure 19.9 have occurred often because due care and attention has not been given to making sure that fillers are positioned accurately and/or are not displaced.

NEATNESS AND CLEANUP

Nothing looks worse on a new structure than a sloppy job of joint sealing, in which the sealant is uneven or is adhering to everything except the joint faces. Careful workmanship, such as uniform depth of installation, proper tooling, and lack of spilled or excess material on surfaces adjacent to the joint, is a sign of a good, conscientious job.

A very neat joint can be obtained with field-molded sealants if strips of masking tape are first placed on each side of the joint opening. These can be removed later, carrying any excess sealant with them. Proper cleaning of equipment and tools immediately after use for even a short period, will avoid contamination of the work or delays due to hardened sealant on surfaces. The instructions on the containers of cold-applied sealants usually list suitable solvents for this purpose. Unused hot-applied sealants must not be allowed to set up in heating vessels and applicator equipment.

SAFETY PRECAUTIONS

There are certain hazards in using joint sealant. These can be minimized, however, by taking simple precautions. Specific warnings are stated on the containers together with the actions or antidotes to use in case of accidents. In addition, material safety data sheets are required by law and should be available for all products. Users of joint-sealing materials should expect to receive MSDS with the best information regarding the use of the particular material they will be handling.

1. Hot-applied materials can cause serious burns, or fire might ignite if flammable materials are spilled. Excessive breathing of fumes or skin contact with hot coal-tar compounds might cause irritation.
2. Cold-applied materials (other than emulsions) and primers might contain flammable solvents. Containers should be kept closed and away from flames. Working areas must be well ventilated.

3. Toxic chemicals might be present in many elastomeric sealants. Skin, eye, or internal contact must be avoided. Protective gloves and masks and goggles are required. Lunch pails should not be opened until the operators have cleaned up.
4. Sealants containing poisonous chemicals, for example, lead dioxide, might not be appropriate in joints open to potable water or food processing areas.
5. Most liquid sealants are highly sensitive to liquid oxygen, creating a serious safety problem. The chemicals in sealants are in a state such that they easily react with oxygen to promote explosion and/or toxic gases. Special materials have been developed for exposure to LOX; however, they usually are not durable and a service life of 12 months is as long as the user/owner should anticipate before replacement. Loss of bond might result in a safety hazard by allowing infiltration of LOX where otherwise inert materials will become highly reactive due to contamination.
6. Since many joint sealants are combustible organic materials, attention should be given to their effect on the fire resistance of the structure.
7. Solvents used to clean up, or released during curing, might be restricted by some jurisdictions because they are deemed to be atmospheric pollutants even though nonhazardous.

SUMMARY

Two terms should be mentioned since they are in wide, though imprecise, use. Regardless of their type or configuration, joints are often spoken of as "working joints," where significant movement occurs, and as "nonworking joints," where movement does not occur or is negligible. Other joint sealing materials, including backer rods, fillers and preformed gaskets have been discussed thoroughly.

REFERENCES

1. *Sealants: The Professional Guide*, 1995. Sealant, Waterproofing & Restoration Institute, 1995
2. ACI 504R-90, "*Guide for Sealing Joints in Concrete Structures*," 1990. American Concrete Institute.

CHAPTER 20
HOW SEALANTS FUNCTION

INTRODUCTION

To function properly, a sealant must deform or expand in response to opening and closing joint movements without any change that would adversely affect its ability to maintain the seal.

The sealant material behaves in both elastic and plastic manners. Which type of action predominates at any time depends on the type and shape of the joint, the movement and rate of movement occurring, installation and service temperatures, and the physical properties of the sealant material concerned, which, in service, is either a solid or an extremely viscous liquid.

CLASSIFICATION OF SEALANTS

Sealants may be classified into two main groups:

1. Field-applied sealants that are applied in liquid or semi-liquid form, and thus are formed into the required shape within the mold provided at the joint opening.
2. Preformed sealants that are functionally preshaped, usually at the manufacturer's plant, resulting in a minimum of site fabrication necessary for their installation.

BEHAVIOR OF SEALANTS IN BUTT JOINTS

As a sealed butt joint opens and closes, one of three functional conditions of stress can exist. These are:

1. The sealant is always in tension. Some waterstops, Figure 20.1, function to a large degree in this way though compressive forces that may be present at their sealing faces and anchorage areas.
2. The sealant is always in compression. The principle, as illustrated in Figures 20.1a, 20.1b and 20.1c, is the one on which compression seals and gaskets are based.
3. The sealant is cyclically in tension or compression. Most field-molded and certain preformed sealants work in this way. The behavior of a field-molded sealant is illustrated in Figure 19.1, and an example of a preformed tension-compression seal is shown in Figure 20.2.

The behavior of a field-applied sealant is illustrated in Figure 19.1. The sealant was installed when the joint was in its fully closed position so that thereafter, as the joint opened, the sealant was extended. This is only possible with preformed sealants, such as waterstops that are buried in the freshly mixed concrete and have mechanical end anchors. Field-applied sealants cannot be used this way and the magnitude of the tension effects shown in Figure 19.1 probably would lead to failure as the joint opened in service. Most sealing systems used in open surface joints are therefore designed to function under either sealant-in-compression or a condition of cyclically-in-compression-and-tension to take best advantage of the properties of the available sealant materials and permit ease of installation.

MALFUNCTION OF SEALANTS

Malfunctioning of a sealant under conditions of stress consists of a tensile failure within the sealant or its connection to the joint face. These are known as cohesive and adhesive failures, respectively, as shown in Figure 20.3.

In the case of preformed sealants that are intended to be always in compression, malfunctioning usually results in failure to generate sufficient contact pressure with the joint faces. This leads to the defects shown in Figure 20.4.

Where secondary movements occur in either or both directions at right angles to the main movement, including impact at joints under traffic, shear forces occur across the sealants. The depth (and length) of the sealant required to accommodate the primary movement can more than provide any shear resistance required.

HOW SEALANTS FUNCTION   20.3

The behavior of these preformed sealants depends on a combination of their elastic and plastic properties under sustained compression (or for waterstops tension), full strain recovery, no permanent deformation (set) very desirable.

① COMPRESSION SEALS AND GASKETS

	(A) AS INSTALLED	(B) JOINT OPEN	(C) JOINT CLOSED
(i) Sealant is:	....	Always in compression	Always in compression
and Sealant must:	....	Change its shape as its width changes (Note 1)	

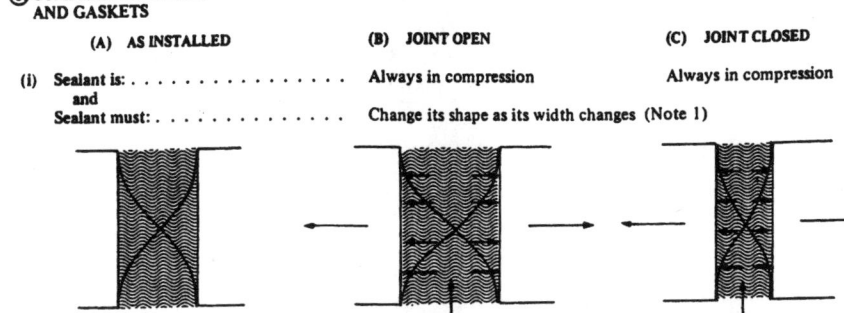

Outward pressure on faces maintains the sealing action

(ii) Material requirements for good performance:

(A)
(a) Good contact (bond not needed)
(b) Correct size
(c) Suitable configuration

(B)
(d) Rubber-like properties

(C)
(e) Low compression set
(f) Webs should not weld
(g) Should not extrude from the joint

Also required (see Section 3.1) (1) Impermeability (3) Recovery (7) Nonembrittlement (8) Not deteriorate

(iii) Deficiencies in (b) (d) (e) (f) predisposes to loss of contact pressure.

Note 1
Compression seals in working joints require to be compartmentalized or foldable to meet this criterion, gaskets in nonworking joints may not.

② WATERSTOPS

(A) WORKING JOINT AS INSTALLED   JOINT OPEN TO WATER   (B) NONWORKING JOINT

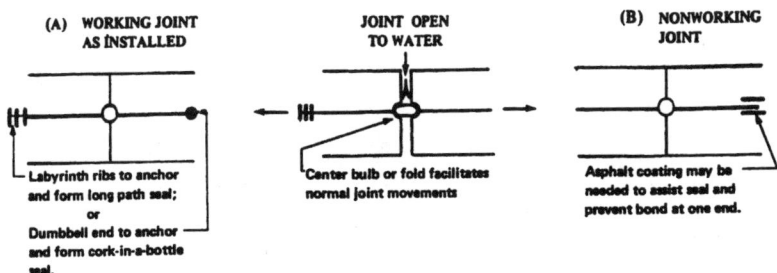

Labyrinth ribs to anchor and form long path seal; or Dumbbell end to anchor and form cork-in-a-bottle seal.

Center bulb or fold facilitates normal joint movements

Asphalt coating may be needed to assist seal and prevent bond at one end.

(ii) Material requirements

(A) (i) Flexible materials with properties similar to ① above

(ii) Rigid flat plates also used where movement is comparitively small (otherwise sliding end or fold needed to permit movement). Must resist deformation due to fluid pressure. High durability since replacement not practical

(B) (i) Rigid noncorrosive materials suitable, some ductility and flexibility may be desirable

(ii) Flexible materials may be convenient but not essential

(iii) Deficiencies lead to failures

FIGURE 20.1 How preformed compression seals, gaskets, and waterstops function.

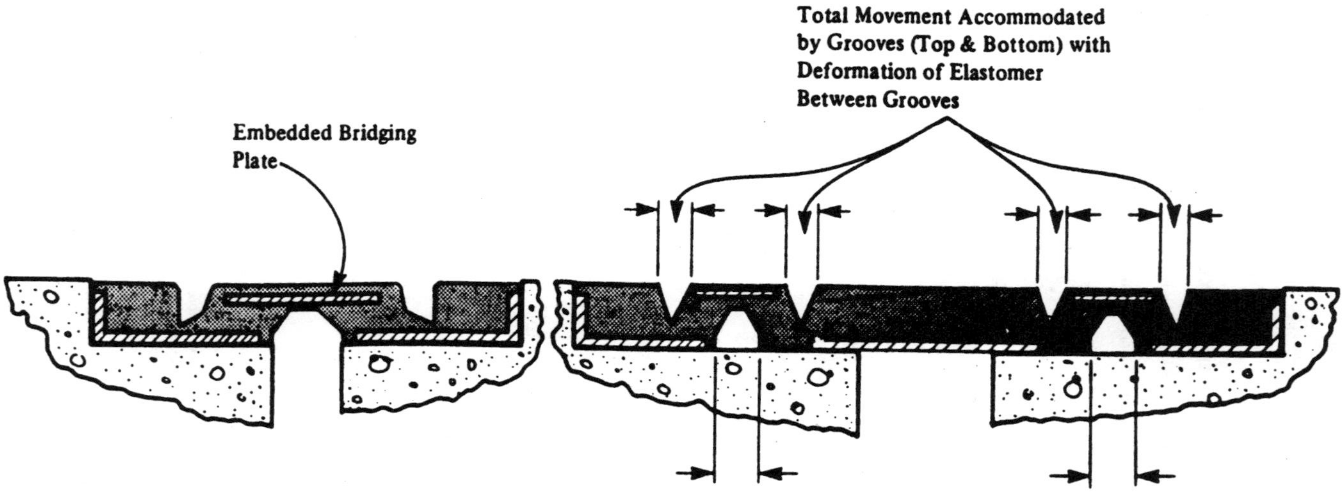

FIGURE 20.2 Representative cross-section of tension-compression seals.

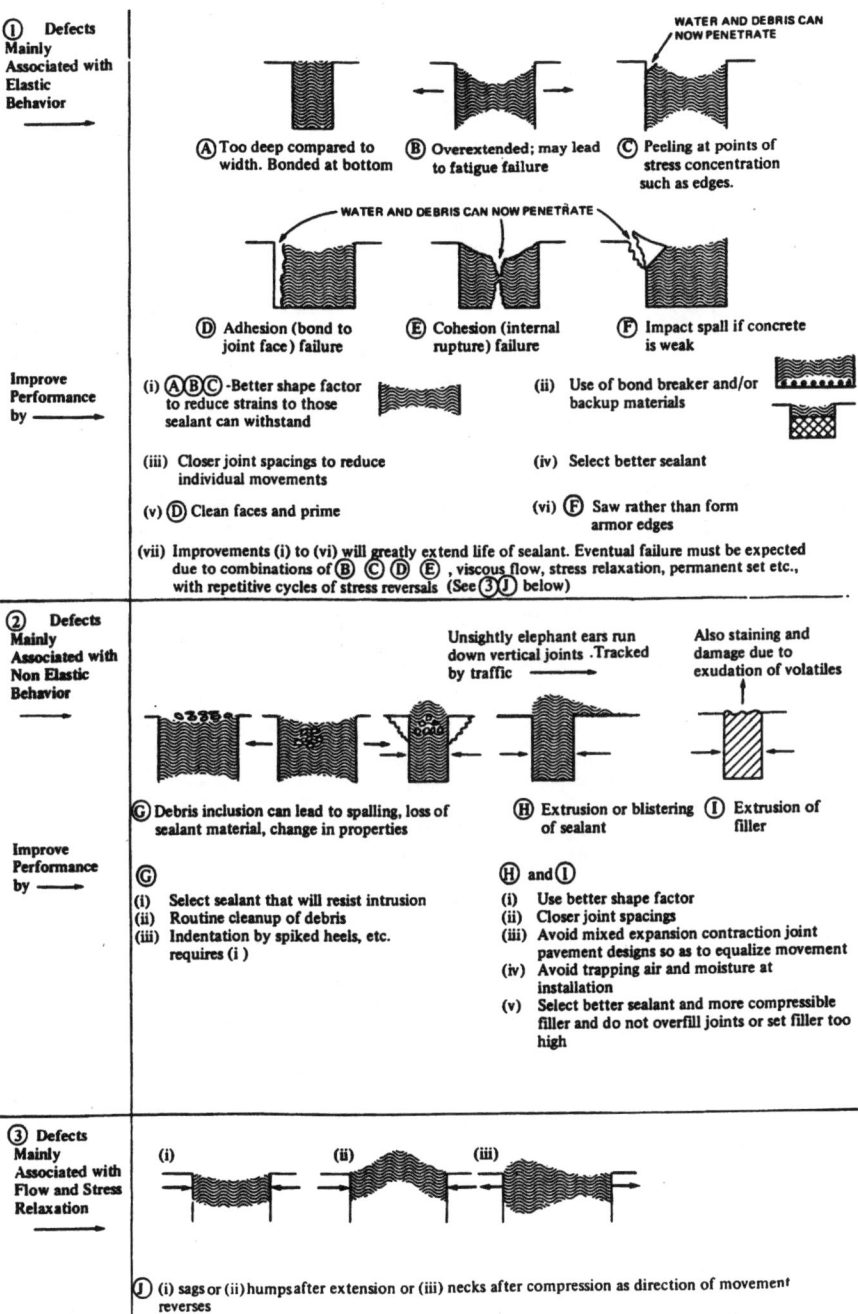

FIGURE 20.3 Defects in field-applied sealants.

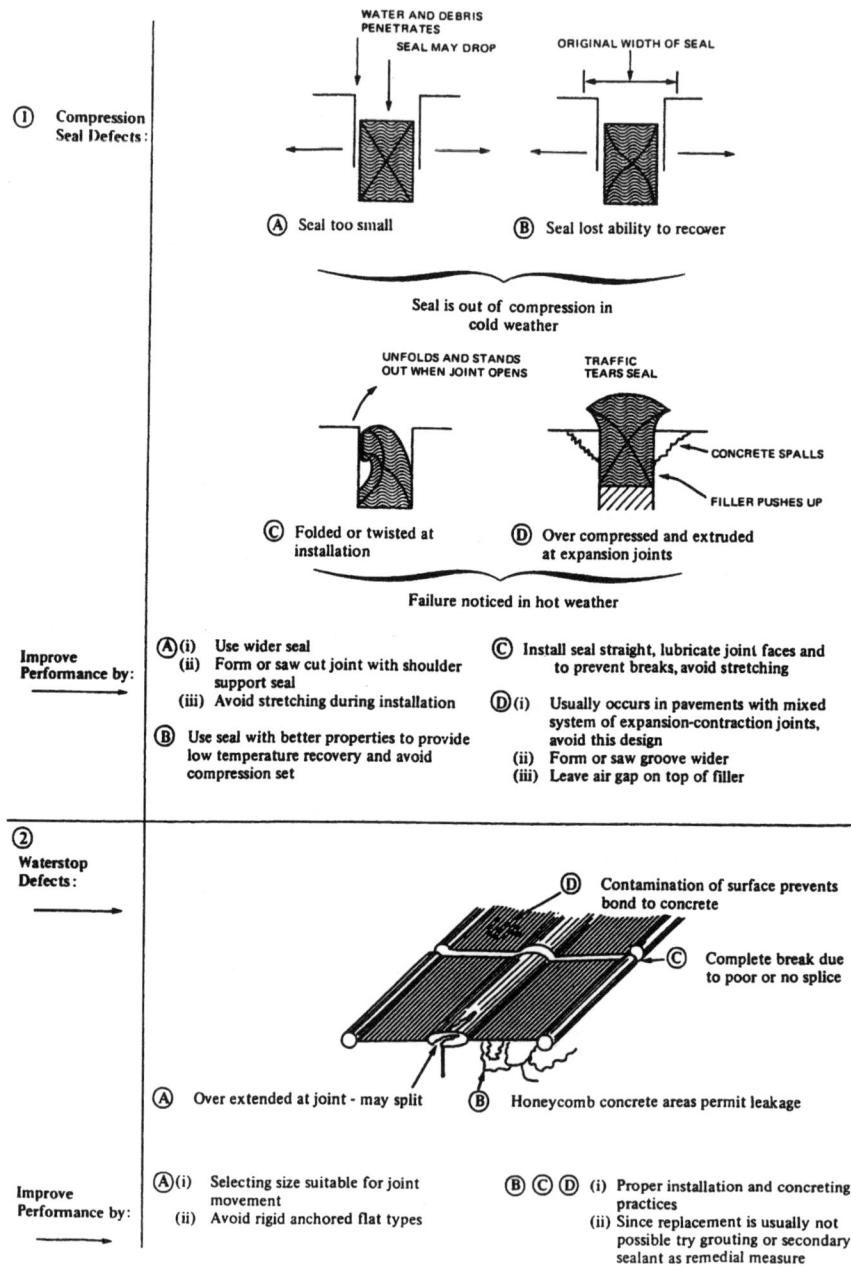

FIGURE 20.4 Defects in preformed sealants.

BEHAVIOR OF SEALANTS IN LAP JOINTS

The sealants as illustrated in Figure 19.1 are always in shear as the joint opens and closes. Tension and compression effects may, however, be added in the modified type of lap joint used in many building applications.

EFFECT OF TEMPERATURE

Changes in temperature between that at installation and the maximum and minimum experienced in service affect behavior. This is explained by the reference in Figure 20.5.

The service range of temperatures that affect the sealant is not the same as the ambient air temperature range. It is the actual temperature of the units being joined by the sealant that govern the magnitude of joint movements that must be accommodated by the sealant. By absorption and transfer of heat from the sun, and loss due to radiation, etc., depending on the location, exposure, and materials being joined, the difference between the service range of temperature and the range of ambient air temperature can be considerable.

For the purpose of the book, the service range of temperatures has been assumed to vary from −20° F to 130° F (−29° C to 54° C) for a total range of 150° F (83° C). In very hot or cold climates or where the joint is between concrete and another material that absorbs or loses heat more readily than concrete, the maximum and minimum values might be greater. This is particularly true in building walls, roofs, and pavements. On the other hand, inside a temperature-controlled building or in structures below ground, the range of service temperatures can be quite small. This applies also to containers below the water line. However, where part of a container is permanently out of water, or is exposed by frequent water removal, the effects of a wider range of temperatures must be taken into account.

The rate of movement due to temperature change for short periods (i,e., an hour, a day) is quite as important as the total movement over a year. Sealants usually perform better, that is, respond to and follow joint opening and closing, when this movement occurs at a slow and uniform rate. Unfortunately, joints in structures rarely behave this way. Where restraint is present, sufficient force to cause movement must be generated before any movement occurs. When movement is inhibited due to frictional forces, it is likely to occur with a sudden jerk that might rupture a brittle sealant. Flexibility in the sealant over a wide range of temperatures where undue hardening or loss of elasticity occurs otherwise would not be suitable. Generally all materials perform better at higher temperatures, although with certain thermoplastics, softening may lead to problems of sag, flow and indentation.

Furthermore, in structures having a considerable number of similar joints in a series, such as retaining walls, canal linings, and pavements, it might be expected that an equal share of the total movement might take place at each joint. How-

Hypothetical cases showing the effect of installation temperatures in relation to the range of service temperatures, assuming the joint width at mean temperature equals the total joint movement between fully open and fully closed positions. (for simplicity of analysis only temperature effects shown)

(1) SEALANT INSTALLED AT NORMAL TEMPERATURE

(A) INSTALLATION AT MEAN TEMPERATURES 55 F (13 C) (B) JOINT OPEN AT -20 F (-29 C) (C) JOINT CLOSED AT 130 F (54 C)

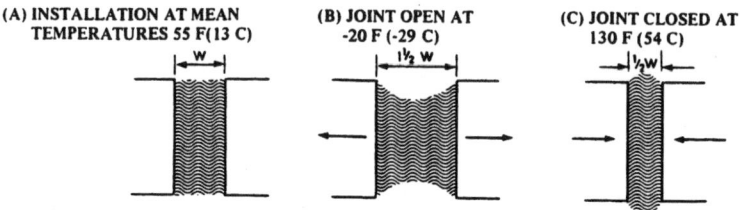

Sealant must extend or compress by 50 percent in service.

(2) SEALANT INSTALLED AT LOW TEMPERATURE

(A) INSTALLATION AT MINIMUM TEMPERATURES -20 F (-29 C) (B) JOINT HALF CLOSED AT 55 F (13 C) (C) JOINT CLOSED AT 130 F (54 C)

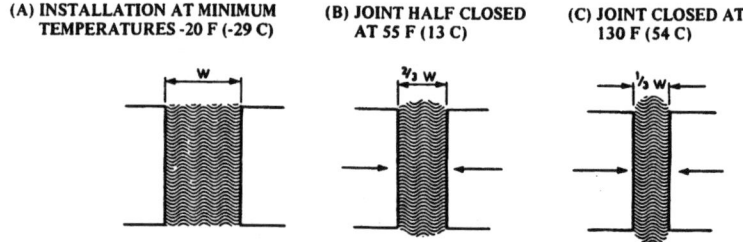

Sealant must compress by 66.66 percent in service.
Probability of Permanent Deformation or Extrusion. 50 percent more sealant needed.

(3) SEALANT INSTALLED AT HIGH TEMPERATURE

(A) INSTALLATION AT MAXIMUM TEMPERATURE 130 F (54 C) (B) JOINT HALF OPEN AT 55 F (13 C) (C) JOINT OPEN AT -20 F (-29 C)

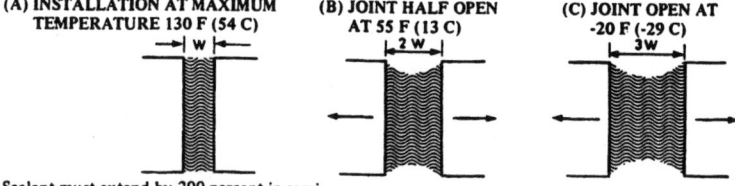

Sealant must extend by 200 percent in service.
Adhesion, cohesion, or peeling failure certain.

CONCLUSION: The closer the installation temperature is to the mean annual temperature the less will be the strain in the sealant in service and the better it will perform in butt joints. Taking into account practical considerations (see Chapter 4 and 6) an installation temperature range of from 40 to 90 F (4 to 32 C) is acceptable for most applications.

Note: (i) Though not illustrated, similar considerations govern the selection of the size of compression seals. Failure in case (3) above would however be by loss of contact with joint faces when seal passes out of compression.

(ii) Maximum deformation of a sealant in lap joints is also governed by installation temperature. Sealant thickness not less than joint movement acceptable for all temperatures may be reduced to ½ provided installation temperature is between 40 and 90 F (4 and 32 C) (movement approximately ½ inch wav.)

FIGURE 20.5 Effect of temperature on field-applied sealants.

ever, one joint in the series may initially take more movement than others and therefore the sealant should be able handle the worst combination. These considerations are discussed in detail in Chapter 21. They are, however, relevant at this stage in determining the properties and performance required of sealants.

SHAPE FACTOR IN FIELD-APPLIED SEALANTS

Field-applied sealants should be 100 percent solids (or semi-solids) at service temperatures. They alter their shape, but not their volume, as the joint opens and closes. These strains in the sealant, and hence the adhesive and cohesive stresses developed, are a critical function of the shape of the sealant. For a given sealant then, its elastic extensibility is a function of the shape of the mold in which it was installed as well as the physical properties of the material. A mathematical analysis of sealant deformation was made by Tons [1], whose laboratory measurements showed that the exposed surfaces of an elastically deformed sealant assume a parabolic shape until close to rupture. He concluded that total extensibility is increased directly with width and inversely with the depth of the sealant in the joint. From this data and that of Schultz [2], Figure 20.6 has been prepared to illustrate the critical importance (and economy) of using good shape factor especially with thermosetting, chemically curing field-applied sealants. Shape factor pertains to the ratio between the width of a sealant and its thickness (depth) determined by experience and lab tests.

It must be remembered that while selections of the shape factor are essentially based on accommodating cohesive stresses in the sealant, at the time of placement an adequate area must be provided at the joint face to accommodate adhesive (bond) stresses. For this reason, *experience* has indicated a preference in certain applications, such as in concrete pavements, for a minimum 3:2 (depth to width) shape factor rather than a *theoretically* more desirable ratio.

FUNCTION OF BOND-BREAKERS AND BACKUP MATERIALS

Bond-breakers and backup materials are used, as illustrated in Figure 20.7, to achieve the desired shape factor for field-molded sealants. The principal material requirement for a bond-breaker is that it should not adhere to the sealant. Important secondary benefits of a back-up are that it supports the sealant and helps resist indentation, and sag, and allows the sealant to take advantage of maximum extension. These often can be important considerations when selecting the appropriate type and shape of preformed backup material. The backup material also must be compressible without extruding the sealant, and must recover to maintain contact with the joint faces when the joint is open.

Cases showing the effect of shape on the maximum strains 'S' that occur on the parabolic exposed surface of elastomeric sealants. Sealant assumed to be installed at mean joint width so that ½ change of width of sealant will be extension and ½ compression.

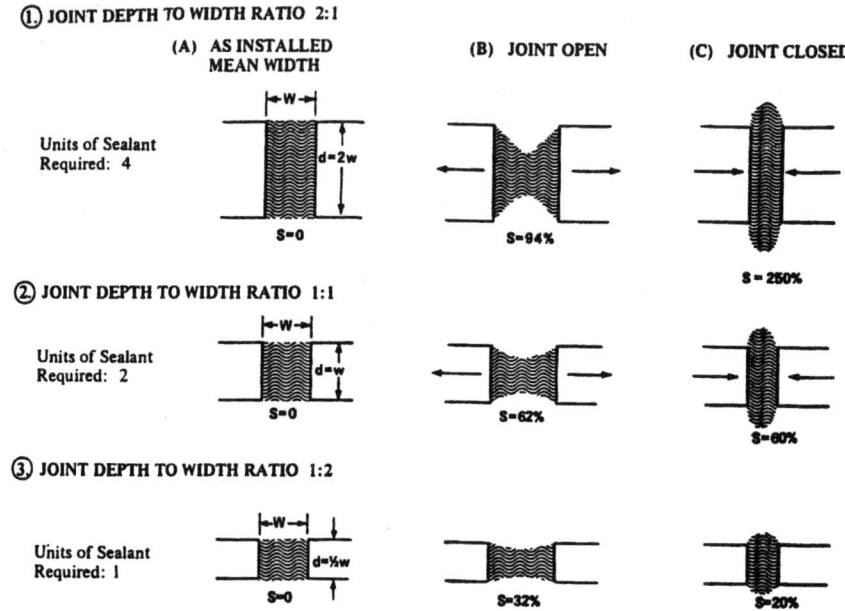

FIGURE 20.6 Shape factor and strains in field-applied sealants.

FUNCTION OF FILLERS IN EXPANSION JOINTS

Fillers are used in expansion joints to assist in making the joint and to provide room for the inward movement of the abutting concrete units as they expand. Additionally, they can be required to provide support for the sealant, or limit its depth in the same manner that backup materials do. These requirements usually are met by preformed materials that can be compressed without significant extrusion and, preferably, recover their original width when compression ceases. Stiffness to maintain alignment during concrete placement—and resistance to deterioration due to moisture and other service conditions also usually are required.

PURPOSE OF BOND BREAKER AND BACK UP: In joints open on one face only, the back face of the sealant must not adhere to the bottom of the sealant reservoir so that the sealant is free to assume the desired shape. See (A) below. Control of depth of sealant is achieved as shown in (B) where the joint is formed or sawn initially deeper than the required depth to width ratio. (Bi) and (Bii) present cases as to desirable shape of backup.

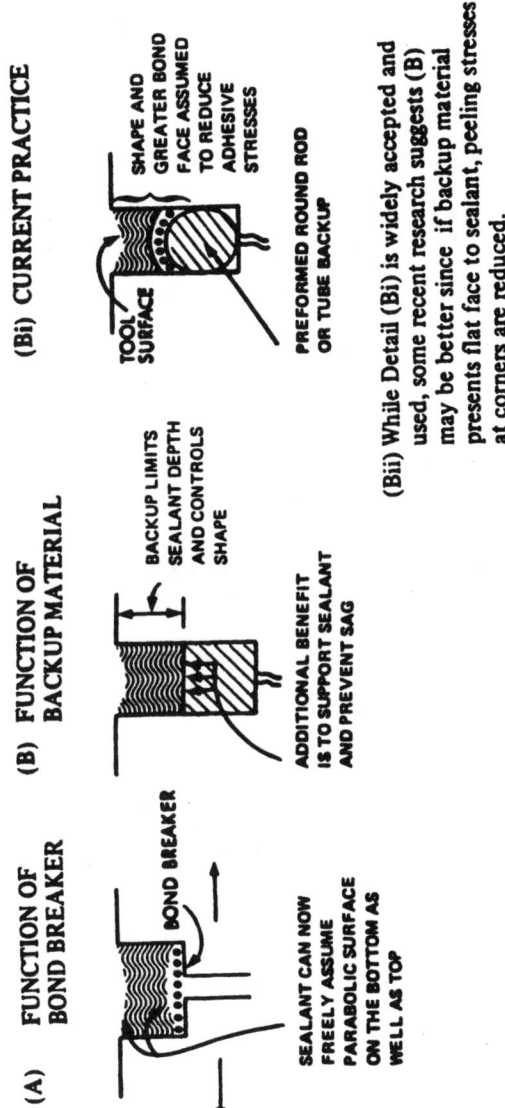

(Bii) While Detail (Bi) is widely accepted and used, some recent research suggests (B) may be better since if backup material presents flat face to sealant, peeling stresses at corners are reduced.

FIGURE 20.7 Bond breaker tape and backer rod.

FUNCTIONS OF PRIMER

Laboratory and field experience indicates that priming joint faces is essential for certain field-molded sealants and can generally improve their bond strength and hence extensibility. Depending on the sealant and conditions of the sealant-to-joint interface, the improvement in adhesion can result from one or more of the following: sealing and penetration of the concrete pores, precoating of the concrete pores, precoating of the dust particles, reduction in bubble formation, and reduction in the absorption of oils by the concrete.

SUMMARY

It is essential for all persons associated with the sealing process to be knowledgeable about joint design. Use specifics, such as the required mathematics for joint movement, bond breakers and backer materials when working on this project.

REFERENCES

1. Tons, Egons, 1959. "*A Theoretical Approach to Design of a Road Joint Seal*" Bulletin No. 229, Highway Research Board.
2. Schultz, Raymond J., October 1962. "*Shape Factor in Joint Design*" Civil Engineering, ASCE., V 32, No.10.

CHAPTER 21
JOINT MOVEMENT AND DESIGN SCHEMATICS

INTRODUCTION

The location and width of joints that require sealing can be specified only with the following considerations: Is a sealant available that will take the anticipated movement? What shape factor (or in the case of preformed sealants—*size*) is required? If the first answer is no, then the joint system for the structure must be designed to reduce the movement at the joints. Sealing systems currently available can accommodate (at increasing costs) movements to about 48 inches (1220 mm). With due forethought, it should be possible to design and specify a suitable sealed joint for almost any type of structure.

DETERMINATION OF JOINT MOVEMENTS AND LOCATIONS

The anticipated length changes within the structure must be determined and translated into joint locations and movements that not only fit the structural design and maintain the integrity between the individual structural units, but also take into account that each type of sealant imposes specific limits on both the shape of the joint that can be sealed and the movement that can be accommodated. It should be remembered that the sources and nature of the movement, both long and short term, can be very complex in other than simple structures (see Chapter 20) and that experience and judgment play a big part in designing joints that function satisfactorily [1]. A more complete discussion of this is beyond the scope of this chapter except to draw attention to the following simple facts which, if overlooked, result in poor joint sealant performance [2].

1. The movement of the end of a unit depends on its effective length, that is, on the length of the part of the units that is free to move in the direction of the joint.
2. Except where a positive anchor is a feature of the design, experience shows that the preferred safe assumption is that a joint between two units might be called upon to take the total movement of both units.
3. The temperatures of the materials being joined might vary from the ambient condition, affecting joint movements.
4. Where units to be joined are dissimilar materials they might not be at the same surface temperature and the appropriate coefficient for each material must be used in calculating its contribution to the joint movement.
5. Where knowledge exists of actual movements that have occurred in similar situations, these should be considered in the design to supplement those indicated by theory.
6. Allowance must be made for the practical tolerances that can be achieved in constructing joint openings or in casting and positioning precast units.
7. Butt-joint movement to which a sealant can properly respond is that at right angles to the plane of the joint faces. Shearing movements in the plane of the joint faces must be taken into account where they are large by comparison, for example, where very large skews (more than 30 degrees) or deflections occur.
8. The width of the joint sealant reservoir must always be greater than the movement that can occur at the joint.
9. When viewing a structure, the joints, either sealed or unsealed, tend to stand out. It is therefore desirable to locate and construct them as a purposeful feature of the architectural design or to hide them by structural or architectural details.

SELECTION OF BUTT-JOINT WIDTHS FOR FIELD-APPLIED SEALANTS

The selection of the width (and depth) of the joint for field-applied sealants—to accommodate the computed movement in a joint—is based on the maximum strain allowable in the sealant. This occurs in the outer fibers, usually when the sealant is extended, although in some cases maximum strain might occur while the sealant is compressed. The part of the total movement that extends the sealant is the difference between the width of the joint at the time the sealant is installed and the width of the joint at its maximum opening. The temperature difference between that at installation and that at maximum opening is the main contribution to the extension of the sealant. But any residual drying shrinkage or creep of the concrete that has yet to occur, and shrinkage in the sealant as it sets, will also impose additional extension of the sealant.

When suitability of a new joint sealant is first being considered and a precise determination of the dimensions of the sealant reservoir are required, the approach using Figure 21.1 from Schutz [3] may be followed. This figure relates to the maximum allowable strain in a sealant to an assumed joint width and various joint shape factors. First, the maximum allowable strain for the sealant under consideration must be determined by testing at a specified temperature. Usually this temperature is 0° F (−18° C) and the test is performed in accordance with the requirements of Federal Specification SS-R-406C. Second, a likely approximation to the joint width is assumed and the computed linear extension that the sealant would undergo between the as-installed width and the width at maximum opening of the joint, is calculated.

The various curves then permit the computed extension and shape factor to be interrelated so that the maximum allowable strain will not be exceeded. More than one solution usually is possible, and where the upper limits of the curves are approached, a wider assumed joint width should be tried. In practice, to allow for unforeseen circumstances, a safety factor of four should be applied in using this chart.

This detailed procedure is simplified for practical use by the aid of the percentage extension-compression shown in Table 21.1 for each type of field-applied sealant.

This table has been derived by considering the maximum allowable strains for materials of each type and applying the suggested safety factor. The percentage extension-compression is the percentage increase or decrease in the as-installed width of the sealant that can be safely accommodated as the joint subsequently opens and closes. The width of the joint to be formed, which becomes the sealant mold and thus determines the as-installed sealant width, then can be obtained by simple calculation, so that in service the permissible extension-compression range is not exceeded. This calculation should, of course, take into account:

1. The anticipated temperature at the time of forming the joint
2. The temperature of sealant installation
3. Any additional joint opening which will be caused by initial drying shrinkage of the abutting concrete units
4. The extremes of service temperature.

When the joint width is designed, a precise installation temperature usually cannot be known or specified; otherwise an intolerable restriction would be placed on the installation operation. All that can be done is to specify installation within the general temperature range. This can be done easily by ensuring that for the worst installation temperatures, the seal still will function as anticipated (for extension the top of the range is used, and for compression the bottom of the range). A practical range of installation temperatures taking into account this and other

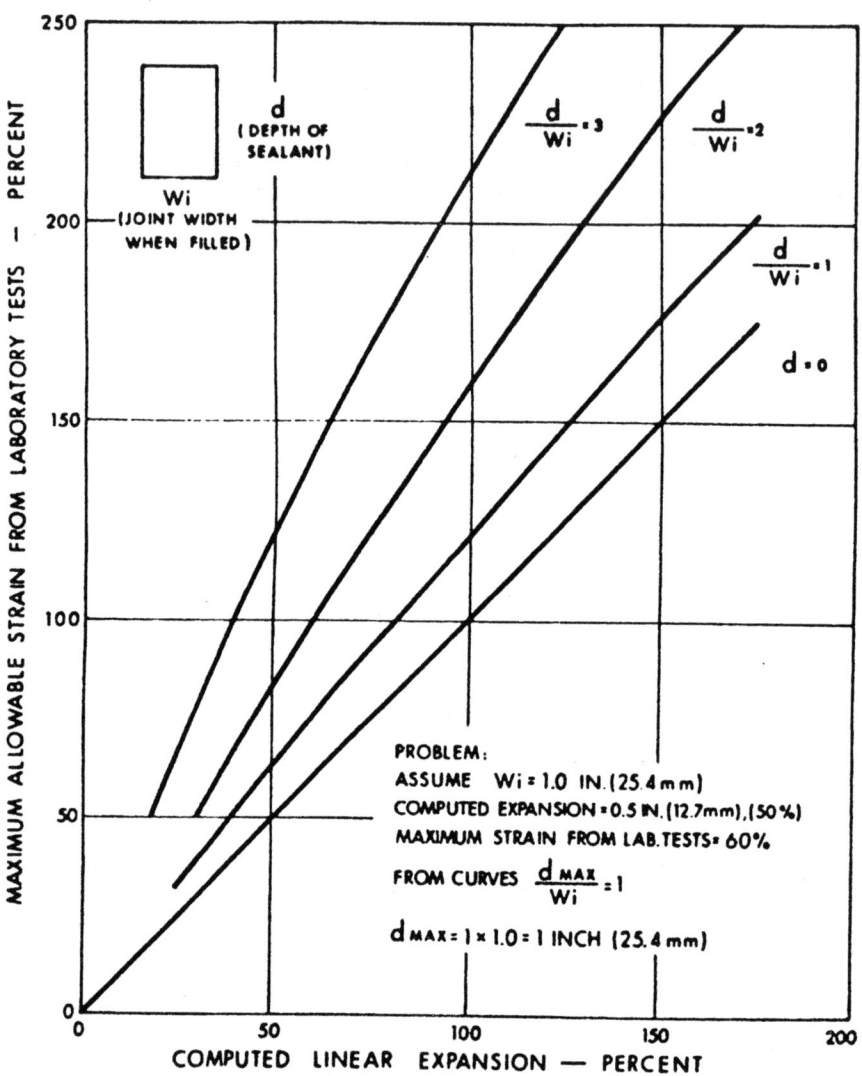

FIGURE 21.1 Selection of dimensions for sealants.

TABLE 21.1 Material Used for Sealants in Joints Open on at Least One Surface

Group						Preformed
Type		Field-molded				Compression
	I. Mastic	Thermoplastics		Thermosetting		VI. seal
		II. Hot-applied	III. Cold-applied	IV. Chemically curing	V. Solvent release	
Composition	(A) Drying oils (B) Non-drying oils (C) Low-melt. point asphalt (D) Polybutenes (E) Polyisobutylenes or combination of D and E All used with fillers such as asbestos fiber or siliceous materials, all contain 100% solids, except D and E, which may contain solvent	(F) Asphalts (G) Rubber asphalts (H) Pitches (I) Coal tars (J) Rubber coal tars; all contain 100% solids (W) Hot-applied PVC coal tar	(K) Rubber asphalts (L) Vinyls (M) Acrylics (K) contains 70-80% solids (L) (M) contain 75-90% solids All contain solvent, (K) may be an emulsion (60-70% solids)	(N) Polysulfide (O) Polysulfide coal tar (P) Polyurethane (Q) Polyurethane coal tar (R) Silicones (S) Epoxy (N) (R) contain 95-100% solids (O) (Q) (S) contain 90-100% solids (P) contains 75-100% solids (N) (P) (R) may be either 1- or 2-component system (O) (Q) (S) 2-component system	(T) Neoprene (U) Butadiene styrene (V) Chlorosulfonated polyethylene (T),(V) contain 80-90% solids (U) contains 85-90% solids (R) Silicones	(3) Neoprene rubber
Colors	(A),(B) Varied (C) Black only (D),(E) Limited	Black only	(K) Black only (L) (M) Varied	(N) (R) Varied (O) (P) Limited (Q) Black only	(T) Limited (V) Varied	Black, exposed surfaces may be treated to give varied colors
Setting or curing	Noncuring, remains viscous, A and B form skin on exposed surface	Noncuring, sets upon cooling. Softens on warming, hardens on cooling. (W) Resilient	Noncuring, sets on release of solvent or evaporation of water. (M) remains soft except for surface skin	2-component system catalyst 1-component moisture pickup from the air	Release of solvent	
Aging and weathering resistance	Low	Moderate (W) High resistance to weather	Moderate	High	High	High
Increase in hardness in relation to (1) age	High	High to moderate (W) No hardness	High	(S) High (N) (O) (P) (Q) (R) Moderate	High	Low

(Continued)

TABLE 21.1 Material Used for Sealants in Joints Open on at Least One Surface (Continued)

Group						Preformed
Type	I Mastic	Thermoplastics		Thermosetting		Compression
		II Hot applied	III Cold applied	IV Chemically curing	V Solvent release	VI seal
or (2) Low temp.	High	High to moderate (W) No hardness	High	(S) (N) (O) (P) (Q) (R) Low	High	Low
Recovery	Low	Moderate (W) High	Low	(N) (O) Moderate (P) (Q) (R) High (S) Low	Low	High
Resistance to water	Low	Moderate	Moderate	(P) (Q) (R) (S) High (N) (O) Moderate	Moderate	High
Resistance to indentation and intrusion of solids	Low	Low at high temperatures (W) High	Low at high temperatures	High	Low	High
Shrinkage after installation	High	Varies (W) None	High	Low	High	None
Resistance to chemicals	High except to solvents and fuels	(F) (G) High except to solvents and fuels (H) (I) (J) High and fuel resistant (W) High	(K) High except to solvents and fuels (L) (M) High except to alkalis and oxidizing acids	(N) (P) Low to solvents, fuels, oxidizing acids (O) (Q) Low to solvents, but moderate fuel resistance (R) Low to alkalis (S) High	Low to solvents, fuels and oxidizing acids	High
Modulus at 100% elongation	Not applicable	Low	Low	(R) (O) (P) (Q) Low (R) High and low (S) Not applicable	Moderate	....
Allowable extension and compression	± 3%	± 5% (W) ± 25% extension	± 7%	± 25% except (s) less	± 7%	Must be compressed at all times to 45–85% of its original width

(Continued)

TABLE 21.1 Material Used for Sealants in Joints Open on at Least One Surface *(Continued)*

Group		Field-molded				Preformed
		Thermoplastics		Thermosetting		Compression
Type	I Mastic	II Hot applied	III Cold applied	IV Chemically curing	V Solvent release	VI seal
Other properties	(A) (B) D) (E) non-staining (D) (E) pick up dirt; use in concealed location only	Due to softening in hot weather, usable only in horizontal joints (W) No flow at elevated temperatures	(K) Usable in inclined joints	(N) (P) (R) (S) Non-staining	(U) (V) Non-staining (V) Good vapor and dust sealer	
Unit first cost	(A) (B) (C) Very low (D) (E) Low	(F) (G) (H) (I) (J) Very low (W) Medium	(K) Very low (L) Low (M) High	(O) (Q) High (N) (P) (R) (S) Very high	(T) (U) (V) Low	(3) High

factors, such as moisture condensation at low temperatures and reduced working life at high temperatures, has been determined to be 40°-90° F (4°-32° C). This is generally because the tension case of the joint opens with fall of temperature and is the more critical to sealant behavior. (See Figures 19.2 and 19.1.) Joint sealants installed at the low end of this range might be expected to perform best. A warning note should be included on the plans or drawings that, if for any reason, sealing must take place at temperatures above or below the specified range, then a wider-than-specified joint might have to be formed, or changes might have to be made in the type of sealant or shape factor to secure greater extensibility.

Detailed calculations for selection of the joint width for the sealants with an expansion-compression range of ±25 percent (the most common range for the widely used class of thermosetting-chemical curing sealants) can be dispensed with by use of Figure 21.2. This has been prepared using the previous procedures to cover the range of service temperatures of −20° F to 130° F (-29° C to 54° C) and other conditions in the handbook. Similar charts can be prepared for other sealants and conditions. In addition, most sealant manufacturers publish data sheets, charts, tables, and training aids for the proper selection of joint widths to suit their products.

Where a reasonable joint width, *limitations on butt-joint widths and movements for various sealant types*, cannot be determined by previous considerations, nor those that follow as in, *selection of butt-joint shape for field-applied sealants*, as to sealant depth, the proposed joint layout for the structure must be redesigned to accommodate movements tolerable to the sealant.

SELECTION OF BUTT-JOINT SHAPE FOR FIELD-APPLIED SEALANTS

When a suitable joint width has been established (see page 21.2: *selection of butt-joint widths for field-applied sealants*), the appropriate depth of the sealant reservoir must be determined so that the sealant has a good shape. Figure 21.1 can be used for this purpose. Curves for depth-to-width ratios of 1:1, 1:2, and 1:3 are shown on this chart. Any depth-to-width ratio may be used provided that at the computed extension or compression expected in the sealant, the maximum allowable strain is not exceeded. The benefits of both better performance and economy of material by using the smallest possible depth-to-width ratio already have been pointed out. See Figure 21.2. The chosen depth generally should not be less than ½ in. (12.7mm); otherwise, with aging sealant, performance might be affected adversely. The depth of sealant is controlled by using a suitable backup material. To obtain full benefit of a well-designed shape-factor, a bond-breaker tape must be used behind the sealant.

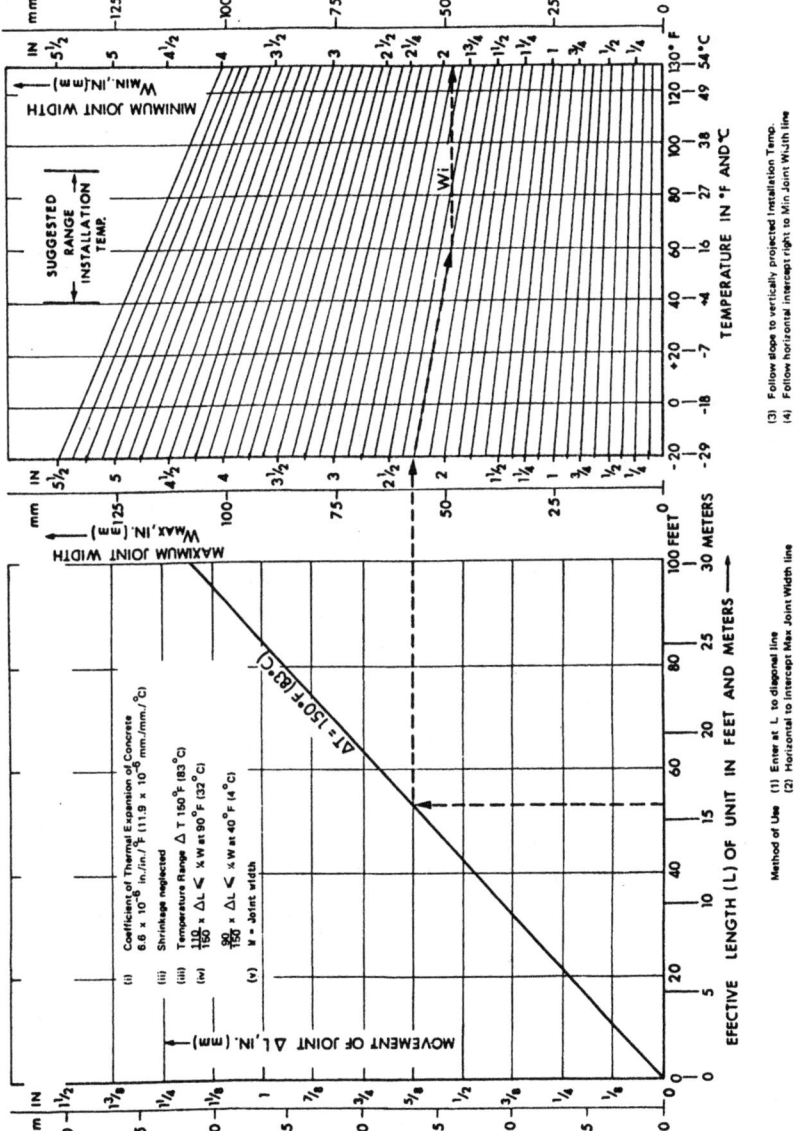

FIGURE 21.2 Chart for determining joint movement and width: filed-applied sealants.

SELECTION OF SIZE OF COMPRESSION SEALS FOR BUTT JOINTS

A positive contact pressure must be exerted against the joint faces at all times for compression seals to function properly. The development of suitable seal configurations to achieve this, while following the principles explained by Dreher [4], largely has been based on the results of trial and error, and laboratory and field experiments in the United States and Europe [2, 6, 7]. Compartmentalized compression seals must remain compressed approximately 15 percent [2] at 85 percent nominal width at maximum joint opening to maintain sufficient contact pressure for sealing and to resist displacement, and generally not be compressed more than 50 percent (50 percent nominal width) at maximum closing to prevent over-compression. This limit of compressibility has been established by the manufacturers of sealants and end users to be at a point when the pressure on the seal reaches 35 psi. Higher pressures tend to accelerate pressure decay. Pressure decay is the failure of the elastomeric seal to regain its original shape, thus losing its sealing pressure when the joint opens.

The allowable movement of compartmentalized compression seals is approximately 35 percent to 40 percent of the noncompressed seal width. The allowable movement for impregnated foams is less, on the order of 10 percent.

The critical condition for maintaining a positive contact pressure is when the joint is fully open at low temperature, since compression set or lack of low temperature recovery might affect the sealant performance adversely. The principle size selection of the field-molded sealants in that original uncompressed width of seal is that required to maintain the seal within the specified compression range, taking into account the installation temperature, the normal width opening and the expected movement. A detailed method for doing this has been described by Kozlov [5]. A simplified chart applicable to the conditions specified in this chapter is shown in Figure 19.10. For specified products, charts of seal sizes for various applications are available from the suppliers of sealants and caulks.

LIMITATIONS ON THE BUTT-JOINT WIDTHS AND MOVEMENTS FOR THE VARIOUS TYPES OF SEALANTS

The applicability of various sealants to joints of different movements in different types of structures is summarized in general terms in Table 21.2.

Field-applied sealants generally require a minimum joint width of ½ inch (6 mm) to provide an adequate reserve against loss of material due to extrusion (Figure 20.4) or to accommodate unexpected service conditions.

The upper limit of joint width and permissible movement varies with the type of material used. Mastics, thermoplastic and solvent-release thermosetting sealants might be used in joints up to 1½ in. (40mm) wide with a permissible movement of ¼ inch (6mm). Chemically curing thermosetting sealants have been used in joints up to 4 in (100mm) wide with movements in the order of 2 inch (50mm), although it is more usual to confine them to joints of half that size to insure good performance and economy in materials. In wide joints, increasing care with sealant installation is necessary and, where subject to traffic, protection of the upper surface against damage is required with a steel plate or other means.

Turning to preformed sealants, single unit compression seals are available in widths up to 6 in. (150mm) wide, permitting joints with movements of about 2½ inches (63mm) to be sealed. The smallest compression seal available can be installed in a ⅛ inch (3.2mm) wide joint, where the movement will be negligible. By placing compression seals or strip (gland) seals in modular series (see Figure 19.10), movements of up to 48 inches (1220mm), like in the longest suspension bridges, have been accommodated. Tension-compression seal systems have been used to accommodate movements of 13 inches (330mm) (Figures 21.3a and 21.3b).

LAP JOINT SEALANT THICKNESS

As mentioned earlier, shear governs sealant behavior in lap joints and its magnitude is related to both the movement that occurs and the thickness of the sealant between the two faces. Usually, for installations at normal temperatures of 40° to 90° F (4° to 32° C), the thickness of the sealant should be at least one-half of the anticipated movement and where higher or lower temperatures prevail at installation, the thickness of the sealant should be equal to the anticipated movement. Where there will be no movement, the sealant thickness can be as little as ⅛ in (3.2mm). However, in assembling concrete units, a minimum thickness of ¼ inch (6.4mm) is desirable to compensate for casting tolerance or any irregularities in the faces.

SHAPE AND SIZE OF RIGID WATERSTOPS

Metal waterstops might be either flat stock or folded in Z or M cross-sectional shapes. The choice depends on the movement at the joint. End anchored flat shapes permit little or no movement without inducing excessive stresses in the embedded portion of the waterstop. Coating one end with asphalt to permit sliding yet maintain sealing compatibility might not be entirely satisfactory because leaks might occur. The Z cross section can accommodate slight movements and the M cross section greater movements.

TABLE 21.2 Uses for Field-Applied and Preformed Sealants

| | | Field-molded | | | | | Preformed | | |
| | | Thermoplastics | | Thermosetting | | | | | |
Type of application	I Mastics	II Hot applied	III Cold applied	IV Chemical cure	V Solvent release	VI Compression seal	VII Waterstops	VII Gaskets	IX Misc.
Structures not under fluid pressure; e.g. buildings, bridges, storage bins, retaining walls									
Caulking & glazing	A B D E		L M	N P R S	T U V	3		1 3 4 5	1 2
Precast panels	A B D E		M	N P R	T V	3		1 3 4 5 7	12 10
Walls (vertical joints)	A B D E			N P R	T V	3 7		7	10
Roof deck (horiz. joints)	A B D E	F G W		N O P Q	T V	3 7		7	10
General floors				N O P Q		3			14 9d
Industrial floors		G H W	K	N P Q		3			13 9c
Floors w/ oil&solvents		H I J W		N O Q S		3			1
Services								3 8	11
Note 3									
Bridges		G W		N O Q		3 7	1 4		10 3 Note 2
Containers subject to fluid pressure; e.g., water containing or excluding structures									
Canal linings	C	G W	K			3		1 3 4 5 6 8	10
Precast pipes	C	G W				3	1 3 4 6 8 9a 9b		10
Tanks & monolithic pipe	C	G W	K			3	1 3 4 6		
Swimming pools		G W				3	4 9a 9b		
Dams									
Walls & floors w/ water outside		G W	K			3	1 3 4 6 9b		
Note 3									
Pavements									
Walkways	F G W		K	N O P Q		3			3 Note 2
Highway	G W			N O Q		3			
Airport	G W			O Q		3			
Areas w/ fuel spillage	H J W			O Q		3			
Grouting nonworking cracks			K	S					23
Suitable in above applications where joint movement is:									
None or very small*	A B C D E F G H I J		K L M	N O P Q R S	T U V	3	1 3 4 6 8 9a 9b	1 3 4 5 6 7 8	12 9c 9d
Small*		F G H I J	K L M	N O P Q R	T U V	3	1 3 4 6 8 9a 9b	7	
Large**				N O P Q R		3	3 4		
Very large**						3			3 Note 2
Note 4									

*Contraction joints   **Expansion joints

(Continued)

TABLE 21.2 Uses for field-applied and preformed sealants *(Continued)*

Type of application		Field-molded						Preformed		
		Thermoplastics		Thermosetting			VI Compression seal	VII Waterstops	VII Gaskets	IX Misc.
	I Mastics	II Hot applied	III Cold applied	IV Chemical cure	V Solvent release					
Storage life: Limited (l) Over 1 year (o) Emulsions are damaged by freezing	A B C D E(o)	F G H I J(o)	K L M(o)	N O P Q R S(l)	T(o) U V(l)		3 7(o)	1-9(o)	1-8(o)	1-11(o)
Installation: Knife or trowel (k) Insert (i), heat & pour (h) Mix if 2-component (m) Note 5 Hand gun (g), pressure gun (p) Preposition (pp)	A B(k)(g)(p) C(k)(g) D E (g)(p)	F G H I J(h) (W) (h)	K L (g)(k)(p) M(g) preheat to 100°F (40°C)	N O P Q R S (m)(k)(g)(p)	T U V (g)(k)(p)		3(i)	1-9(pp)	1-8(pp)	1234 9d 10(pp) 9c(i)(h) 11(pp)

Notes to table 21.2

Note 1: Table 21.2 is only a general guide. Before deciding on a particular material for a specific application all circumstances, in particular the joint movement to be expected and a suitable joint design and joint detail must be considered.

Note 2: 3 refers to tension-compression seals described in Section 3.6.

Note 3: Certain sealants may contain substances toxic to potable water or foodstuffs. Check local or national restrictions that may govern use in areas exposed to these.

Note 4: Certain materials are equally suitable for both vertical and horizontal joints. Others are not and while they may stay in place in horizontal joints, they would sag or flow out of vertical joints in hot weather. Asphalt and rubber-asphalt materials are examples of these. Some materials are available in two grades. One known as nonsag or gun grade is thixotropic and is suitable for vertical joints. The other known as self-leveling or pour grade is intended for use in horizontal joints.

Note 5: Pot life (time material still usable after mixing) is limited and correct proportioning and mixing is critical with 2-component materials.

Note 6: Field-molded sealants furnished as follows:

Liquid in drums, cans or cartridges	A B D E
Liquid in drums	C K W
Liquid in drums or cans	O Q
Liquid in cans	P S
Liquid in cans or cartridges	L N R T U V
Liquid in cartridges	M
Solid in cakes for melting	F G H I J
for preformed materials see table 3	

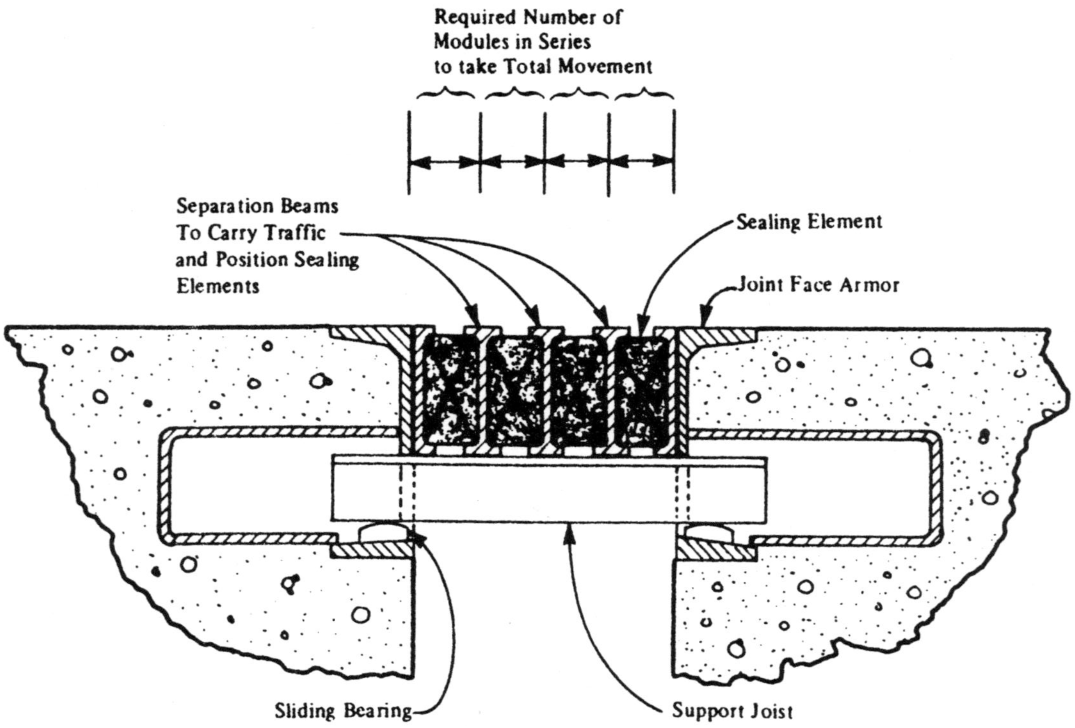

FIGURE 21.3a Modular compression of seal systems.

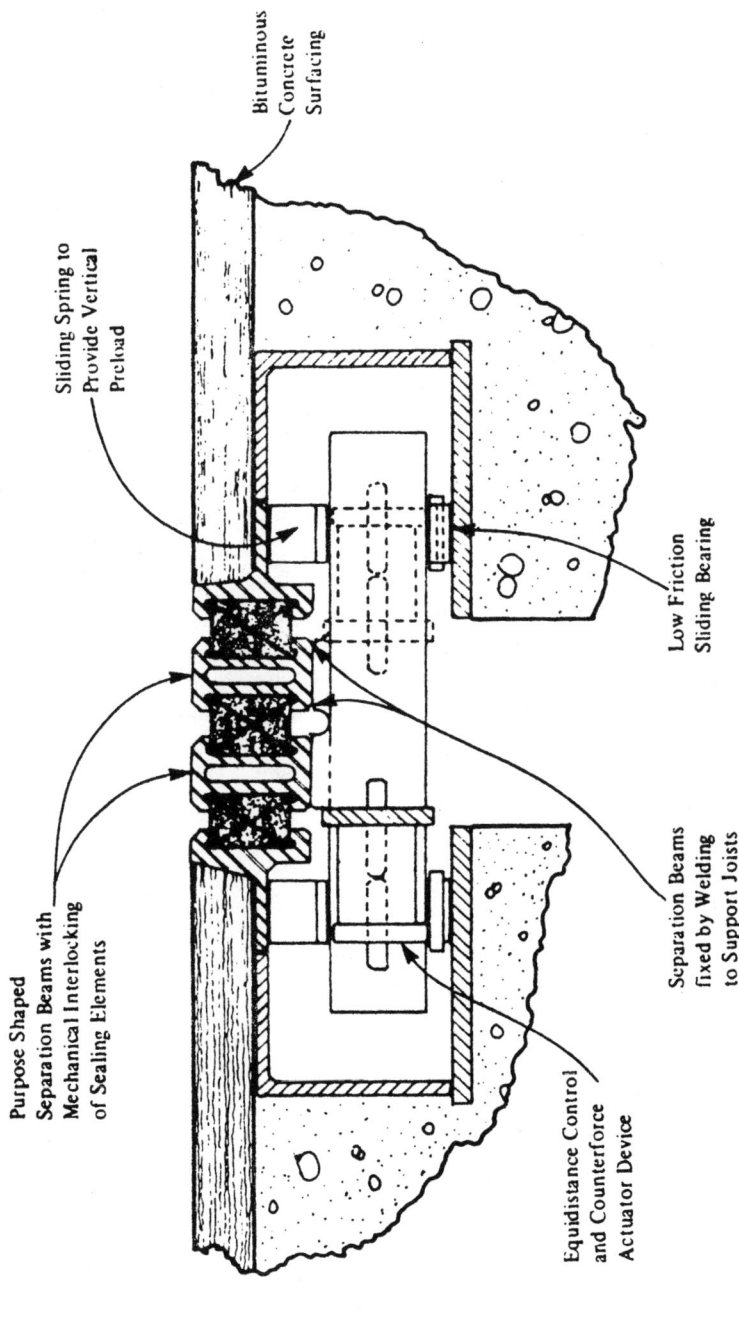

FIGURE 21.3b Modular compression of seal systems.

SHAPE AND SIZE OF GASKETS AND MISCELLANEOUS SEALS

Seals used for concrete pipes or building components usually are sized and shaped to suit the joint configuration, including the irregularity of the surfaces being joined. Since the movement is small, the width of the sealant might not be the primary consideration. Square, rectangular, trapezoidal, O-ring, and H, U, and W shapes, some with ribs, flanges, and serrations, are used, depending on the application and how they can be installed. Pressure-sensitive tapes of suitable widths are used as auxiliary materials to make window and door frames or panels weather-tight.

MEASUREMENT OF JOINT MOVEMENTS

A better understanding of in-service joint movements in all types of structures is needed to confirm the theories and laboratory experiments upon which the design prediction of joint widths and sealant performance are based. The factors that influence the movement of joints are the functional performance of sealants discussed in Chapters 19 and 20, as well as earlier in this chapter.

In view of the many variables involved, it is impossible to specify a standard procedure for the observation and assembly of data on joint movements, the causative factors, and sealant behavior. However, it is important that both the short-term rates of movement over a matter of hours or days and the long-term extremes of movements over the annual environment cycle, together with any permanent changes in interfacial joint distance, are established.

MEANS OF MEASURING JOINT MOVEMENTS

Hand gauges, either a simple vernier caliper or reference bar with a dial gauge, can be used to measure the distance between reference plugs set on each side of the joint. While this system is simple, it only provides a discontinuous record and requires an operator to make each reading. To overcome these disadvantages, a scratch gauge may be employed. These gauges have a scratch probe fixed to one side of the joint opening and a plate or a hand or power rotated disc attached to the other side. The trace of the movement cycle is then measured. The next step of sophistication is to use an electronic gauge. Usually this is a transducer for greater precision at a greater cost, and it measures the movement, which then is recorded on a strip chart or recorded digitally for later analysis.

In most structures, the movement of concern to sealant performance is horizontal (across the plane of the joint). In skewed joints, lateral movement (along the plane of the joint) also might need to be measured or calculated, because a skew introduces shear in the joint. Vertical movements (at right angles to the plane of the joint) can be measured by fast-response transducers where, as in pavements, moving loads cross the joint. Measurement of other dynamic effects, such as vehicle braking, impact, and noise generation, require specialized instruments. Absolute measurements of the relative positions of structural members can be made using standard survey practice techniques against a reference datum clear of the structure.

CORRESPONDING MEASUREMENT OF TEMPERATURE AND MOISTURE CONTROL

Corresponding data on the thermal- and moisture-dependent behavior of abutting structural units are needed to fully interpret joint movement measurements. Response to ambient temperature change and solar radiation is much greater and faster than that due to seasonal changes in moisture content in the concrete. Since moisture content is difficult to measure and unlikely to significantly affect the overall findings, it often is ignored.

When a continuous record is required, ambient, surface, or internal temperatures are easy to measure using thermocouples, recorded on a strip chart, or digitally, for analysis. It must be remembered that while surface temperature changes induce warping and curling fairly rapidly in thin sections, the internal temperatures of a structural unit control its overall dimension and hence the end movements at joints. Especially in massive concrete sections there is a considerable time lag between change in external and internal temperatures. This must be taken into account in determining any relationship between temperature and movements. In massive sections, or where differential heating because of sun and shade is significant, it might be prudent also to measure heat flow and solar radiation. Notwithstanding all these cautions, as a minimum observation, a thermometer reading of ambient shade temperature should accompany any single measurement of joint width.

SURVEY OF JOINT SEALANT PERFORMANCE

In addition to the measurement of the movements and the factors that cause these movements, it is important to note the conditions of the installed sealant, joint hardware, and abutting concrete as part of any overall appraisal of joint performance.

REFERENCES

1. deCourcy,J.W., No.6, June 1969, No.7, July 1969,and No.8, August 1969. *"Movement in Concrete Structures,"* Concrete (London).
2. Watson, S.C., V.5 No. 7 August 1968. *"Compression Seals for Bridges,"* ACI journal, Proceedings.
3. Schutz, R.J., V 32 No. 10, October 1962. *"Shape Factors in Joint Design."* Civil Engineering ASCE.
4. Dreher, D. 1965. "A Structural Approach to Sealing Joints in Concrete," Highway Research Record No.200, Highway Research Board.
5. Kozlov, G.S. 1967. *"Preformed Elastomeric Bridge Joint Sealers,"* Highway Research Record No 200, Highway Research Board.
6. Graham, M.D.; Burnett, W.C.; Hiss F., Jr. and Lambert, J.R., 1965. *"New York State Experience with Concrete Pavement Joint Sealers,"* Highway Research Record, NO.80, Highway Research Board.
7. Watson , S.C. 1965. *"Performance of a Compression Joint Seal,"* Highway Research Record No.80, Highway Research Board.

CHAPTER 22
PERFORMANCE, DEFECTS, REPAIR AND MAINTENANCE OF SEALANTS

QUESTIONABLE PERFORMANCE

Much experience of poor sealant performance and resulting damage to a wide variety of structures exists. Concern with problems arising from the use of low-grade asphalts and asphaltic sealants spurred the development and introduction of higher grades of sealants, both field-applied and preformed. Failures have continued to occur, however, often within days or weeks of installation, rather than months or years, for five main reasons:

1. Design of the joint geometry was insufficient to accommodate the movement.
2. Unanticipated service conditions resulted in greater joint movements than those considered when the joint design and type of sealant were determined.
3. The wrong type of sealant for the particular conditions was selected, often on the false grounds of economy.
4. New sealants sometimes have been over-promoted and used before their limits were documented.
5. Poor workmanship occurred during joint construction and preparation to receive the sealant or sealant installation.

Some of the more common joints are shown in Figures 19.9, 20.3 and 20.4, with advice as to how these defects might be avoided in future work.

REPAIR OF CONCRETE DEFECTS AND REPLACEMENT OF SEALANTS [1]

At joints—Minor touchup of small gaps and soft or hard spots in field-molded sealants usually can be made with the same sealant. However, where the failure is extensive, it usually is necessary to remove the sealant and replace it.

Where the sealant generally has failed, but has not come out of the sealing groove it can be removed using hand tools, or on larger projects such as pavements, by routing or plowing with suitable tools. Alternatively, especially where widening is required to improve the shape factor, the sealant reservoir can be enlarged by sawing.

SAW-CUTTING JOINTS IN CONCRETE

Before 1925, most concrete slabs were built without joints of any kind except for construction joints. Sawed joints were first used as a substitute for formed joints in 1949 on concrete pavements in Kansas. Today sawed joints are used widely and specified for both paving and industrial floor construction [3]. That is why it is important for contractors to know the why, where, when, and how of saw-cutting joints.

WHY SAW-CUT JOINTS

Saw-cut joints minimize random cracking due to drying, shrinkage and temperature changes. The weakened sections created by the joint cause the cracks to form at these locations. Cracking occurs beneath the sawed slot when the shrinkage stress exceeds the tensile strength of the concrete. Sawed joints are widely used in both paving and industrial construction for efficient, cost-effective, crack control. Benefits from saw-cutting are:

- Consistent joint dimensions (same width and depth at each cut)
- Straight clean cuts
- Good joint sealant performance
- No interference with concrete finishing operations
- Cost effective for concrete placements of any size
- A larger window of sawability than the window of jointing by hand

WHERE TO SAW-CUT

Saw-cut joints in the same location as hand-tooled joints, as shown in Figure 22.1. Contraction and control joints should be saw-cut on or at the center of column lines, with intermediate joints between column lines as necessary to keep the maximum distance between joints at 24 to 36 times the slab thickness. The resulting panels should be as square as practical, dividing a large floor area into relatively small panels. Never make the long side of a panel more than 1½ times the short side.

In general, joint spacing to control drying and thermal shrinkage should range from 12 to 25 feet in unreinforced and lightly reinforced concrete floors. Variation in joint spacing results from differences in local conditions such as concrete materials and mixes, climate, construction practices, and sub-grade or sub-base restraint. However, avoid elongated or L-shaped panels, re-entrant corners, and sharp corners. Also, decrease joint spacing when using concrete suspected of having excessive shrinkage. Make saw-cuts continuous, not staggered or offset. If an engineer has designed the slab, locate the joints according to the plans. In concrete floors that contain continuous steel reinforcement, don't continue the bars unless the bars are close enough to the surface to be cut by the saw. Unless the slab is only lightly reinforced, the saw-cut will be relatively ineffective in controlling crack location if continuous bars are not cut. Joint spacing on industrial or commercial floor slabs subject to forklift traffic is usually wide, about 36 times the slab thickness. Though a few random cracks might appear because of the wide joint spacing, this is preferable to the cost of maintaining a large number of joints. Typically the more joints a floor contains, the more it costs to maintain the joints. The joints are more likely to spall under heavy traffic.

WHEN TO SAW-CUT

The timing of sawing joints in concrete is crucial, as shown in Figure 22.2. Correct timing depends on many variables, such as weather conditions, concrete-mix design, aggregate size and hardness, blade type and size, curing, and subgrade conditions. Sawing too early causes raveling, or the dislodging of aggregate, which results in joint spalls. Sawing too late results in uncontrolled cracking. Industrial floor slabs usually are saw-cut 4 to 12 hours after finishing. In hot weather, the saw-cut time is about four hours after finishing. In cold weather, joints are sometimes not sawed until 48 hours after final finishing. The operator typically makes trial cuts a few hours after finishing to determine the looseness of particles. Proper timing is a factor in preserving the life of the blades.

Use an experienced saw operator to assure that sawing is done at the proper time. A good operator relies on experience to determine when the concrete is ready to cut. The operator should be prepared to saw at any time of the day or night in any type of weather. Equipment should be available for any emergency; including a standby saw in case of equipment failure. To ensure a quality job and avoid large labor costs, many concrete contractors subcontract the saw-cutting to specialist firms.

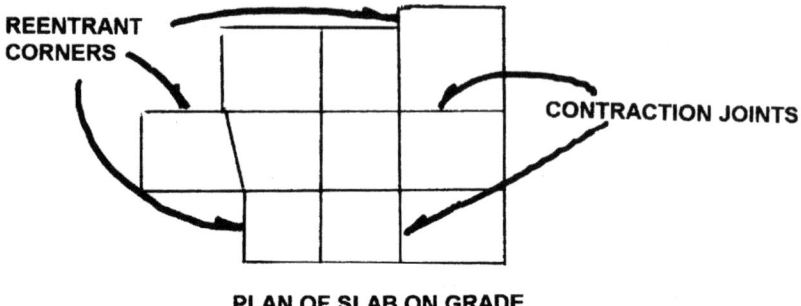

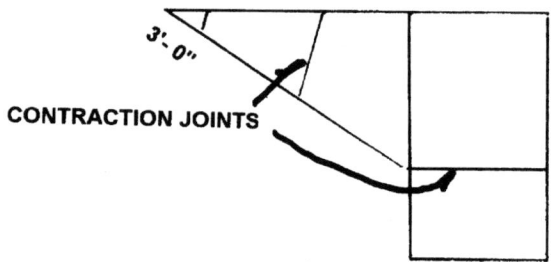

Be sure to locate a contraction joint at all reentrant corners to prevent radial cracking of the slab.

FIGURE 22.1 Saw-cut joints in the same location that hand-tooled joints require.

UNACCEPTABLE RAVELING

Saw-cutting too early in concrete, before the cement completely hydrates, causes unacceptable raveling. Most saw operators determine the earliest time to saw-cut by judging the degree of raveling in trial cuts made in the slab. Recent studies, however, provide the approximate minimum compressive strength of concrete required before joints can be cut with minimal raveling (0.12 square inch per 24 lineal feet of saw-cut joint). The maturity method and the pulse velocity technique were also found to provide reasonable field estimates of concrete compressive strength to determine the earliest time to saw joints. The earliest saw-cutting occurs in the con-

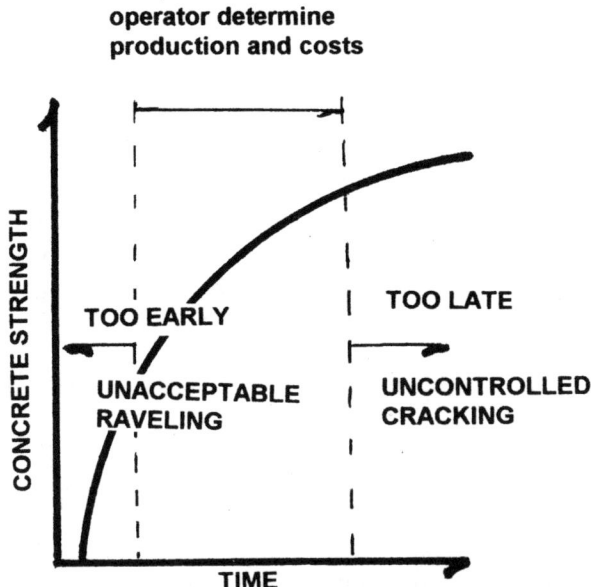

FIGURE 22.2 Saw-cut window.

crete with high cement content and rounded, soft, coarse aggregate. Saw-cuts through these aggregates are less likely to break the surrounding strong bond created by the high cement content. As the saw blade cuts through crushed, hard aggregate, the concrete needs a higher compressive strength to maintain paste-aggregate bond and to minimize undesirable raveling (see Table 22.1).

Uncontrolled Cracking [2]

Joints must be saw-cut before random cracking occurs. Experience and field studies show random cracking occurs as the top slab surface cools during early evening and night immediately following the concrete placement. Random slab cracking also can occur if the surface is cooled by rain after the concrete placement. Test results show cracking occurs when the concrete immediately below the surface cools more than 15° F (9.44° C). Because of cooling rates and safety

TABLE 22.1 Compressive Strength, Pulse Velocity, and Maturity Values for Acceptable Saw-cut Joints

Aggregate geometry	Aggregate hardness	Cement content, pcy	Compressive strength, psi	Pulse velocity, fps	Arrhenius maturity, hrs	Nurse-Saul maturity, °F-H
Crushed	Soft	500	730	11,101	18.4	530
		650	530	10,376	14.5	440
Crushed	Hard	500	1270	12,353	33.8	817
		650	1010	11,835	25.1	667
Rounded	Soft	500	470	10,105	13.5	414
		650	310	9163	10.8	343
Rounded	Hard	500	920	11,624	22.7	621
		650	690	10,973	17.6	512

Maturity relationships based on equivalent age at 68° F and a datum temperature of 32° F.

factors [4], reference suggests saw-cutting all joints before the concrete surface cools 7° F (−13.89° C) in one hour. Instead of waiting for a 7° F (−13.89° C) surface cooling, consider waiting until the latest time possible to saw-cut joints after the surface starts to cool. Measuring surface temperature is easy and inexpensive with a thermometer.

Equipment Affects Timing

The saw and blade types are also important in the timing of joint sawing. Diamond blades last longer when the concrete is stronger. Early sawing dislodges diamond particles. Some contractors delay sawing as long as possible, just before random cracking occurs, to minimize excessive blade abrasion. Silicon carbide blades, on the other hand, wear less when sawing is performed sooner. However, be sure to wait until the concrete can be sawed without unacceptable raveling. Occasionally, contractors are faced with a situation where concrete is too green to saw-cut without raveling, but random cracking is occurring. When this occurs, make sure the blade is matched to the concrete. Most experienced operators carry different saw blades to each job so the blade can be matched to the concrete and timing requirements. Sometimes even an asphalt blade works best.

Even if the concrete strength is adequate, joint raveling can be caused by using the wrong equipment or an inexperienced operator. If the concrete seems strong enough, make certain the saw operator isn't causing joint raveling by:

- Using an improperly tensioned or balanced blade that creates vibrations
- Pushing the blade too hard through the cut
- Running the saw blade at high revolutions per minute (slow speeds are preferred for green concrete)
- Using a saw blade with a loose or bent spindle

HOW TO SAW-CUT

Mark all proposed joints to be cut with a chalk line. After checking the equipment and examining the blade, position the saw at the line. Start the saw, but turn the water cooling system on before the blade contacts the concrete. Maintain adequate water flow at all times; usually 2 to 5 gallons per minute are required to cool the blade. Let the blade contact the concrete, and guide the saw along the chalk line. Without forcing, allow the blade to sharpen itself on the material being cut. Maintain steady even cutting pressure. Avoid twisting the blade in the cut and never force beyond its cutting capacity. If forced, the blade will become polished, stop cutting, and might become distorted. Also, don't let the blade spin in the cut. This practice, called *babying* or *sandbagging* the blade, increases wear on the bond, causing diamond chips to pop out before they've done much work. Stand at the rear of the saw, not in the front or at the side, while the machine is operating. Wear safety glasses, hearing protection, hard hat, and safety shoes. At times, it might be necessary to wear respirators.

DEPTH OF CUT

Current accepted practice is to cut the joint one-quarter to one-third the slab thickness. This forms a plane of weakness in which the crack forms. Vertical loads are transmitted across the joint by aggregate interlock between the opposite faces of the crack, providing the crack is not too wide. Always check cutting depth. If the joint is too shallow, random cracking can occur; if the joint is too deep, aggregate interlock might be insufficient to transfer vertical loads. A worn blade or riding up of the saw over coarse aggregate can also cause cuts to be too shallow. Though little research has been done to verify the effectiveness of these guidelines, experience indicates satisfactory joint performance when these recommendations are followed. The depth of the joint is important, but other factors such as time of sawing, slab design (thickness, base type, slab length and width), curing conditions, and saw-cutting techniques also influence the initiation and propagation of uncontrolled cracking. One statistical study [4] verifies the adequacy of currently acceptable saw joint depths. At a saw-cut depth of about 30 percent of the slab thickness, cracks occur below saw-cuts with a probability of 88 percent for shrinkage stresses and about 98 percent for warping or curling stresses.

SAW-CUT JOINT SEQUENCE

Transverse joints perform best when joints are sawed consecutively, starting where the concrete was first placed and finished. This permits all joints to begin opening about the same time and makes joint movements more uniform, improving sealant performance and load transfer. In hot weather, it might be necessary to saw every third or fourth joint, spacing them no more than 60 feet (18.25 meters) apart to create a relief joint to prevent early random cracking. The intermediate joints can be sawed later, but usually no later than three days after the relief joints are cut. This is only an emergency procedure to reduce uncontrolled cracking and has the disadvantage of causing larger joint openings at the relief joints.

JOINT CURING

Saw-cut joints are particularly susceptible to damage if concrete strength is low because of insufficient curing. Spray the joint with a curing compound if the joint won't be sealed. If the joint is to be sealed, curing compounds can't be used because they prevent bonding between the joint sealant and the concrete. One common method of curing exposed concrete in the joint is to apply the joint sealant immediately after the joints are sawed and cleaned. However, if the joint is sealed immediately after sawing, be prepared to reseal the joint six to 12 hours later, because the joint widens once the majority of concrete shrinkage has occurred. Other methods of curing joints to ensure maximum strength and reduce potential curling include a wet burlap covering, wet sand, plastic coverings, and wet rope inserted into the joint. If a wetting method is used, be sure to rewet the materials as required. After proper preparation to ensure clean joint faces and additional measures designed to improve sealant performance—such as the improvement of shape factor, provision of backup material, and possible selection of a better type sealant—the joint might be sealed as described earlier.

Minor edge spalls to concrete joint faces might be repaired with suitable repair materials, an essential operation if a compression seal is being used. Otherwise most repairs to correct defects in the original construction of the joint involve major, exacting, and often expensive work. The reason for the failure must be identified and, depending on the cause, continuity must be restored in the joint system by either removal of whatever is blocking the free working of the joint or by cutting out the whole joint and rebuilding it.

At cracks—Where cracks have occurred because of a non-working or absent joint, or because of unanticipated deformation of the structure, they can be routed out and sealed with a suitable field-applied sealant to prevent damage to the structure. ACI Committee 224 has done considerable work in this area and their information regarding repairing of cracks would be of significant help. (See ACI 224.1R). An additional problem occurs where water is flowing through the crack and the upstream face cannot be reached for sealing. Before sealing can be successfully undertaken, the water flow must be stopped. If the source of water cannot be cut off by removal, then, depending on the circumstances, one of many alternatives—cutting the crack deeper and plugging with a quick setting or dry-pack mortar or cement, chemical, or epoxy resin grouting might be tried. External plates sometimes are bolted to the concrete structure, or keyed grooves are filled with mortar to hold the sealant in case water pressure redevelops as the joint moves. Successful execution of any of these procedures usually requires specialized knowledge, experience and workmanship.

PREPARING CRACKS FOR SEALING

No matter how well a pavement is designed or constructed, it eventually will crack. If left unattended, cracks allow water air and dirt to enter the pavement structure. Water will weaken the base material, air will oxidize and age the asphalt, and debris will cut off the thermal movement, all causing more cracks. Fatigue-type alligator cracking and potholes then can develop from the advanced failure.

Crack-sealing prior to advanced failure can be the most cost-effective preventive maintenance to concrete pavement. If done properly and in a timely way, crack-sealing can add many years to the functional life of the concrete pavement. There are many types of crack-sealing materials available, but none will function effectively if the cracks are not prepared properly. Before preparing cracks for sealing, it is necessary to recognize what type of cracks you will be sealing. Some cracks (fatigue) are too numerous to seal cost-effectively and need base rehabilitation or strengthening. Other types (thermal) shrink and expand much more when they are spaced far apart. The distance between the cracks can directly affect the type and amount of reservoir and sealer needed. (See Figure 22.3.)

You also must consider the long-term goals of the maintenance efforts. As an example:

- Numerous large meandering cracks in an older pavement should be approached differently than small, scattered thermal cracks in a newer pavement.
- It might not be cost-effective to fill cracks less then ½ inch wide.
- It might be impractical to fill a crack wider than 2 inches.

  Evaluate each situation on an individual basis.

RESERVOIR OR DAM

The general intention of any crack preparation is to create a clean, dry, warm reservoir area for the crack sealant to bond to. If the crack sealant is applied to a dirty surface, the sealant will not adhere to the walls of the crack.

Research has found that sealing of a dry crack can last two to four times longer than sealing of damp surfaces. Crack sealants should not be applied less than 24 hours after a rain. Surface moisture will evaporate, but the crack will hold moisture much longer. Mechanical means might be used to dry damp concrete,

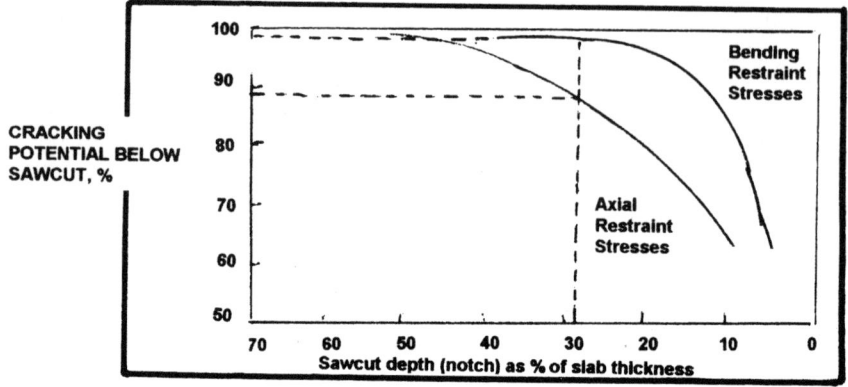

Probability that a crack will occur below sawcut when the depth of cut is about 30% of slab thickness: 88% for shrinkage stresses and about 98% for warping or cutting stresses.

FIGURE 22.3 Crack probability.

but the moisture will resurface quickly. It has also been demonstrated that a crack with a surface temperature of a least 40° F (4.44° C) will have greatly increased average life expectancy. 40° F (4.44° C) is a commonly accepted minimum, but some research sets the minimum at 60° F (15.56° C). When the surface temperature is below the minimum temperature, the sealant might cool too quickly to bond to the walls adequately.

Cracks in the northern climates, expansion joints, and cracks that are spaced far apart generally experience more thermal expansion and contraction movement. The more a crack expands and contracts, the more sealant material is required and/or the more flexible the material must be. A crack with insufficient sealant, even if properly installed, can experience cohesive failure. This is not a failure of the sealant to adhere to the walls, which is adhesion failure, but a tearing or separation of the sealant material itself. Cohesive failure occurs when there is insufficient flexible material available to absorb the stresses related to contraction of the placement.

CREATING A DAM

Smaller cracks ⅛ in. to ½ in. (3.525 mm to 12.7 mm) wide and expanding and contracting cracks might need to be routed first to create an ample dam. Generally, the shape factor of the routed area is 1 inch (25.4 mm) deep. However, because each crack must contain enough material to absorb the stresses of expansion and contraction, some cracks require a reservoir that is wider than it is deep. Routing can produce a smooth, clean material dam. But the pounding, chewing action of the router can loosen the aggregate. The loosened aggregate might eventually pull loose and cause even more problems. In some instances it is difficult for the equipment and tiring for the operator to follow a meandering crack. This can result in two cracks instead of one. Routers are available designed specifically to follow and widen random cracks.

Routing the cracks also leaves a coating of fine dust on the walls of the crack. Clean this dust off before applying the sealant. Consider economics, rate of thermal movement, and site conditions when determining if routing is necessary. These factors also affect the shape of the routed dam. A saw also might require water. As noted, the surface must be dried thoroughly before applying the sealant.

CLEANING THE DAM

Not all cracks will require routing. Larger cracks might need only to be cleaned. In some cases, such as in older pavements with a short life expectancy and numerous larger cracks, it might be more cost effective to just "blow and go," Routed or not, there are many ways to clean:

- By stiff wire bristle brush
- By compressed air only
- By sand-blasting
- By high pressure water blast
- By heat lance (heat and air combined).

Whatever method of cleaning you choose, make certain you select the approach that will cause the least damage to the surrounding pavement as well as providing optimum results to the cracks. In some cases you will be required to apply a weed killer or soil sterilizer to cracks. The weed killer is applied and days later, after it has had time to take effect, the vegetation can be cleaned out. A water-based herbicide, rather than an oil-based herbicide, is recommended. An oil-based material will leave a residue of oil. Therefore the sealant will not adhere. On the other hand, water-based materials will evaporate. When the crack is clean, apply the sealant as soon as possible to minimize the chance of contamination.

When using the heat lance, advantages are derived because the lance burns and the air also dries out the surface and drives the moisture inward. However, the moisture will reappear as the pavement cools, causing capillary pressures, which reduce sealant bond. The use of the heat lance might be necessary on cool or damp pavements, but it is not adequate if there is standing water. Proper operation of a heat lance is critical because it can do more harm than good. With extreme heat, it can burn up the concrete, ravel the edges of the crack, and loosen the aggregate, leaving nothing solid for the sealant to bond to. When a heat lance is operated properly, there should be no exposed flame. Exposed flame or too much heat can leave a coating of oxidized concrete, which is like charcoal dust. If this situation occurs, you will be required to clean the surface prior to sealing. Proper use of a heat lance requires a diligent and knowledgeable operator. Because of the heat and noise involved, protective clothing and other safety precautions are a must for the operator. Ideally the sealant should be applied immediately after lance cleaning. This could pose a problem if there is residue to be cleaned.

NORMAL MAINTENANCE

Few exposed sealants have a life as long as that of the structure whose joints they are intended to seal. Fortunately, buried sealants such as waterstops and gaskets have a long life because they are not exposed to weathering and other deteriorating influences.

Most field-applied preformed sealants, however, will require renewal sooner or later if an effective seal is to be maintained and deterioration of the structure is to be avoided. The time at which this becomes necessary is determined by service conditions, by the type of material used, and whether any defects of the kind already enumerated were built in at the time of the original sealing.

The opportunity should be taken when inspections are being made for other purposes, or in the case of buildings when the facade is cleaned, to establish the condition of sealed joints and whether resealing is required immediately or is likely to be required in the near future. Far too often in the past, resealing has been postponed either because of lack of knowledge that it was needed or failure to budget ahead, with inevitable costly consequences.

Sealant renewal follows closely the methods listed under repair of defects (repairs of concrete defects and replacement of sealants). When renewal is required prematurely, consideration must be given to improving the sealing system from that originally used, otherwise money will be wasted since failure soon might re-occur. Ways of accomplishing this have already been discussed.

SEALING IN THE FUTURE—CONCLUDING REMARKS

What is Possible Now

This conclusion covers Chapters 19, 20, 21, and 22. The cost of producing well-sealed joints using the best possible sealants carefully specified and installed in joints of the correct type, size, and location, is usually only a small fraction of the total cost of a concrete structure. The available sealants and knowledge of the criteria for joint sealing are now adequate to ensure success for joint sealing in at least 9 out of 10 situations. There is no justification for poor sealing practices continuing on a wide scale when the very integrity and service life of a structure might be at stake.

Advancements Still Are Required

Research and development work is still required to improve:

1. Knowledge of the movements that occur in every type of concrete structure.
2. The design details of joints and their location so that unnecessary demands are not made on the sealant.
3. The matching of sealant materials to the design of the joint and vice-versa.
4. The materials available for use as joint sealants. The challenge is to achieve good performance in wide joints that are wet and dirty when the sealant is installed.
5. The methods by which sealants might be installed so that human error is avoided as far as possible.
6. Techniques for resealing leaking joints and cracks.

Educating the Public

Public authorities, sealant manufacturers, and suppliers have been the source of copious technical data and advice that has greatly benefited the art of joint design and sealing. Many of the current sealant problems, however, will continue unless improvements are made in disseminating and applying available knowledge and upgrading specifications and application skills. Improvements are required, for example, in:

1. Making designers more aware of the importance of joint design and the selection of suitable sealants.
2. Providing clear instructions on the plans and specifications and on the sealant containers so the user on the job can understand and execute what is required.
3. Education and training at all levels so that joint sealing is no longer regarded as a necessary evil to be left to do at the last minute by the lowest worker on the scaffold.

Future Codes, Standards, Recommended Practices, and Specifications

Appropriate criteria should be included for joints in concrete and other types of structures (location, type, movement, determination, width, shape, and sealant selection and installation).

NEW DEVELOPMENTS

Field-Applied Sealants

The importance of shape factor in successful joint sealing where joint sealants are used is discussed in Chapter 20 (Shape Factor in Field-Applied Sealants). While selection of shape factor essentially is based on accommodating cohesive stresses in the sealant, at the same time, an adequate area must be provided at the joint face to accommodate adhesive (bond) stresses. For this reason, experience is indicating a preference in certain applications, such as joints in concrete pavements, for a 2:1 shape factor over the theoretically more desirable 1:1 or 1:2 to achieve better service performance overall. Practice in this respect for pavement contraction joints is further discussed in this section.

Polyvinyl Chloride Coal Tar

To the category of thermoplastic, hot-applied materials as described in Chapter 9, (Thermoplastics, Hot-Applied), a new class of material "polyvinyl chloride-coal tar" should be added.

In relation to materials described as Type II in Table 22.2, polyvinyl chloride-coal tar sealants have the following characteristics and properties:

1. Do not flow at elevated service temperatures
2. Are resilient
3. Have good resistance to weathering and aging
4. Are resistant to jet fuels or other similarly aggressive chemicals
5. Have an allowable extension and compression of ± 25 percent
6. Unit cost is medium.

Polyvinyl chloride-coal tar sealants are finding extensive use as sealants in pavement and canal liner joints, respectively. Installation procedures are similar to those for other field-molded, hot-applied sealants as described in Chapter 19.

Specifications and Warranties

Current generic specifications for most types of sealants are to be found in Chapter 18. Some brand manufacturers, in addition, offer specific warranties relative to the in-service performance of their products for periods of 10 to 15 years. Specifying authorities, in considering the benefit of such warranties, should examine the written terms and conditions in light of the sealant application in mind and any documented performance in similar circumstances.

TABLE 22.2 Materials Used in Sealants

Group		Field-molded				Preformed
		Thermoplastics		Thermosetting		Compression
Type	I. Mastic	II. Hot-applied	III. Cold-applied	IV. Chemically curing	V. Solvent release	VI. Seal
Composition	(A) Drying oils (B) Non-drying oils (C) Low-melt. point asphalt (D) Polybutenes (E) Polyisobutylenes or combination of D&E All used with fillers such as asbestos fiber or siliceous materials, all contain 100% solids, except D and E, which may contain solvent	(F) Asphalts (G) Rubber asphalts (H) Pitches (I) Coal tars (J) Rubber coal tars; all contain 100% solids (W) Hot-applied PVC coal tar	(K) Rubber asphalts (L) Vinyls (M) Acrylics (K) Contains 70-80% solids (L) (M) contain 75-90% solids All contain solvent, (K) may be an emulsion (60-70% solids)	(N) Polysulfide (O) Polysulfide coal tar (P) Polyurethane (Q) Polyurethane coal tar (R) Silicones (S) Epoxy (N) (R) contain 95-100% solids (O),(Q), (S) contain 90-100% solids (P) contains 75-100% solids (N) (P) (R) may be either 1- or 2-component system (O) (Q) (S) 2-component system	(T) Neoprene (U) Butadiene styrene (V) Chlorosulfonated polyethylene (T), (V) Contain 80-90% solids (U) Contains 85-90% solids (R) Silicones	(3) Neoprene rubber
Colors	(A),(B) Varied (C) Black only (D) (E) Limited	Black only	(K) Black only (L) (M) Varied	(N) (R) Varied (O) (P) Limited (Q) Black only	(T) Limited (V) Varied	Black, exposed surfaces may be treated to give varied colors
Setting or curing	Noncuring, remains viscous, A & B form skin on exposed surface	Noncuring, sets upon cooling. Softens on warming, hardens on cooling. (W) Resilient	Noncuring, sets on release of solvent or evaporation of water. (M) remains soft except for surface skin	2-component system catalyst 1-component moisture pickup from the air	Release of solvent	....
Aging and weathering resistance	Low	Moderate (W) High resistance to weather	Moderate	High	High	High
Increase in hardness in relation to (1) age	High	High to moderate (W) No hardness	High	(S) High (N) (O) (P) (Q) ,(R) Moderate	High	Low

(Continued)

TABLE 22.2 Materials Used in Sealants *(Continued)*

Group		Field-molded					Preformed
		Thermoplastics		Thermosetting			Compression
Type	I Mastic	II Hot applied	III Cold applied	IV Chemically curing		V Solvent release	VI seal
or (2)Low temp.	High	High to moderate (W) No hardness	High	(S) (N) (O) (P) (Q) (R) Low		High	Low
Recovery	Low	Moderate (W) High	Low	(N) (O) Moderate (P) (Q) (R) High (S) Low		Low	High
Resistance to water	Low	Moderate	Moderate	(P) (Q) (R) (S) High (N) (O) Moderate		Moderate	High
Resistance to indentation and intrusion of solids	Low	Low at high temperatures (W) High	Low at high temperatures	High		Low	High
Shrinkage after installation	High	Varies (W) None	High	Low		High	None
Resistance to chemicals	High except to solvents and fuels	(F) (G) High except to solvents and fuels (H) (I) (J) High and fuel resistant (W) High	(K) High except to solvents and fuels (L) (M) High except to alkalis and oxidizing acids	(N) (P) Low to solvents, fuels, oxidizing acids (O) (Q) Low to solvents, but moderate fuel resistance (R) Low to alkalis (S)High		Low to solvents, fuels and oxidizing acids	High
Modulus at 100% elongation	Not applicable	Low	Low	(R) (O) (P) (Q) Low (R) High and low (S) Not applicable		Moderate	
Allowable extension and compression	± 3%	± 5% (W) ± 25% extension	± 7%	± 25% except (s) less		± 7%	Must be compressed at all times to 45-85% of its original width

(Continued)

TABLE 22.2 Materials Used in Sealants *(Continued)*

Group		Field-molded					Preformed
		Thermoplastics			Thermosetting		Compression
Type	I Mastic	II Hot applied	III Cold applied	IV Chemically curing		V Solvent release	VI seal
Other properties	(A) (B) D) (E) non-staining (D) (E) pick up dirt; use in concealed location only	Due to softening in hot weather, usable only in horizontal joints (W) No flow at elevated temperatures	(K) Usable in inclined joints	(N) (P) (R) (S) Non-staining		(U) (V) Non-staining (V) Good vapor and dust sealer	
Unit first cost	(A) (B) (C) Very low (D) (E) Low	(F) (G) (H) (I) (J) Very low (W) Medium	(K) Very low (L) Low (M) High	(O) (Q) High (N) (P) (R) (S) Very high		(T) (U) (V) Low	(3) High

Installation of Hot-Applied Sealants

Installation of hot-applied sealants is covered in Chapter 19. Experience continues to indicate the need for better care and vigilance in installing sealants if they are to live up to performance expectations.

Preformed Sealants

There have been significant developments in two classes of preformed sealants, compression seals, and tension-compression seals. In addition, a new type of sealing system known as strip (gland) seals has come into use.

Compression Seals

Compartmentalized compression seals have found continued and increasing use in recent years. While practice has not changed greatly in pavement and other narrow joint applications, there have been important design, material, and fabrication changes where these seals are used either singularly or as components of modular systems, in bridges, and other applications in which larger movements occur.

The configuration of the compartmentalized compression seals has been modified in some of the larger movement designs to provide for mechanical interlocking of the seals to their supporting hardware. Web arrangements have been altered in some cases to provide better performance at lower stress levels during cyclical movement of the joint. One present design incorporating these features is shown installed in Figures 22.4a and 22.4b. Otherwise, the main changes have been in the engineering design and fabrication of the assemblies that support and connect the compression seals to the structure.

Joint Face Armoring and Mechanical Locking of Seals

To reduce mechanical damage to the concrete and provide a tighter seal, armoring of the joint faces and mechanical locking or support of the seals are now widely used in joints subject to vehicular traffic. Figure 22.4 illustrates joint face armoring and mechanical interlocking for compression seals. Features having a similar purpose are also used for strip (gland) seal applications discussed in the paragraph *strip* (or *gland*) *seals position* and illustrated in Figure 22.6.

Steel angle and channel sections, sometimes with added locking strips or support brackets to hold the compartmentalized seal in position, have been used to armor, but purpose-shaped steel extrusions with lipped recesses to lock in place low-stress sealing elements having corresponding lugs are now in common use. (See Figure 22.4a.) The mechanical locking of compression seals improves performance because direct compression alone is not relied on to keep the sealing

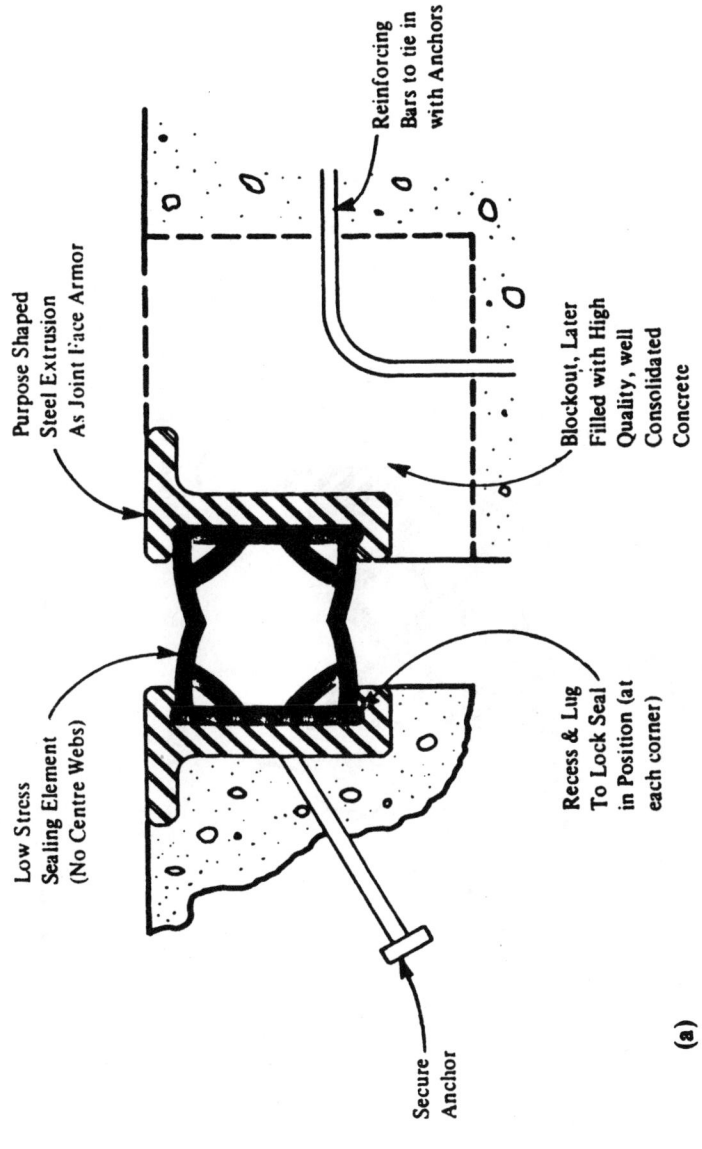

FIGURE 22.4a Compression seal installation.

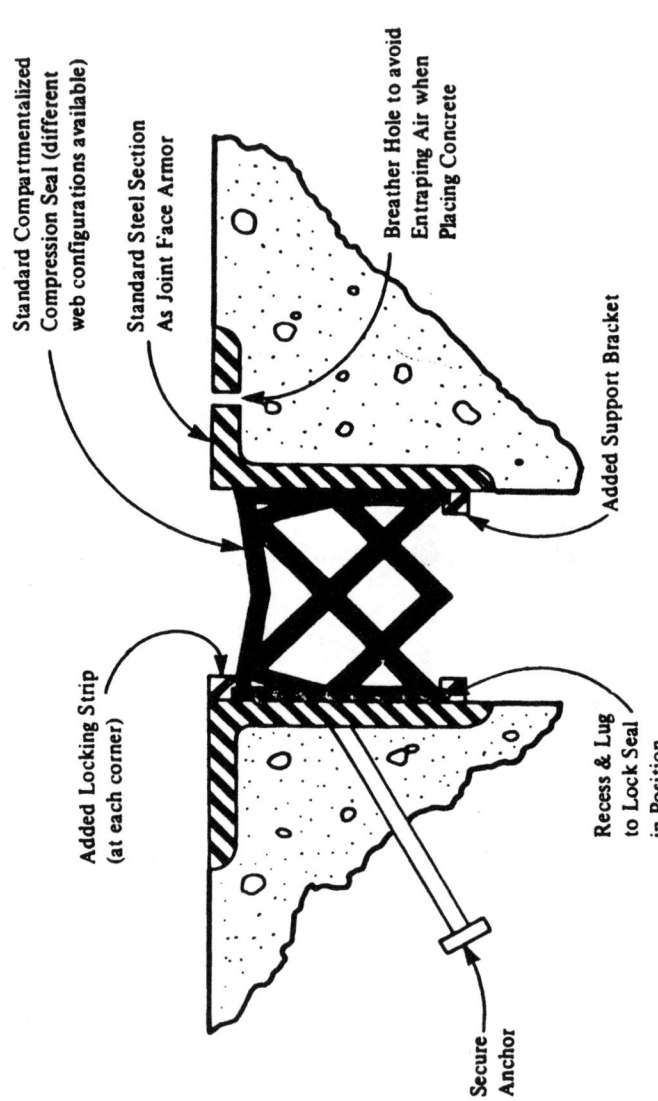

FIGURE 22.4b Compression seal installation. (b) — (with Mechanical Interlocking shown on LH side)

element in place and to maintain the sealing action against the joint face. It is important that the armoring system be designed to properly connect with the concrete by adequate ties, bonded anchors, or bolts, and that the concrete be placed and compacted under re-entrant angles with great care; otherwise the whole assembly might well break loose from the structure under the impact of traffic.

Modular Compression Seal Systems

Modular systems have found wide use in bridges and other structures where large joint movements must be accommodated. The main features of modular systems are shown in Figures 22.5a and 22.5b. The sealing elements are positioned in a series by separation (supporting) beams resting on support joists (bars). Each module takes part of the total movement at the joint and movements of from 2.3 inches (64 mm) to 80 inches (2030 mm) have been designed into modular installations.

Improvements in sealing performance and ride-ability based on experience have been achieved by refinements in design, materials, and fabrication, illustrated in Figure 22.8b for one design. Purpose-shaped high strength, corrosion resistant, steel extrusions might be used for the separation of beams to provide a greater strength-to-weight ratio and higher torsional rigidity. These extruded separation beam sections and the end sections armoring the joint faces, also provide for mechanical locking of the sealing elements described in the paragraph on *Joint Face Armoring and Mechanical Locking of Seals*. Steel used in these extrusions should at least meet the requirements of ASTM A 242. To prevent lifting, wear, and tilting of the separation beams under traffic or cyclical movement of the joint, they each can be welded to one (or more) of the support joints. Traffic can actuate undesirable noise and vibration in joint assemblies and isolation techniques have been developed to significantly reduce these by pre-loading (pre-stressment) of the support joist system through sliding elastomeric block springs acting vertically over sliding bearings. To further improve the performance of individual sealing elements and to equalize the movements and forces in each one in a series, innovations include: use of low friction materials on sliding surfaces, low stress sealing elements as described in the paragraph on compression seals, and a device located between, and connected to, adjacent support joists, known as equidistance control (to center the modules) and counter-force actuator (to provide a built-in return force).

Adhesion Lubricants

The use of liquid materials to facilitate the installation and bonding of compression seals (see paragraphs *Compression Seals and their Uses* and *Installation of Compression Seals*), has proved to be important to good performance. The material most widely used at present is a single-component, high-solids, high lubricity polyurethane-based adhesive.

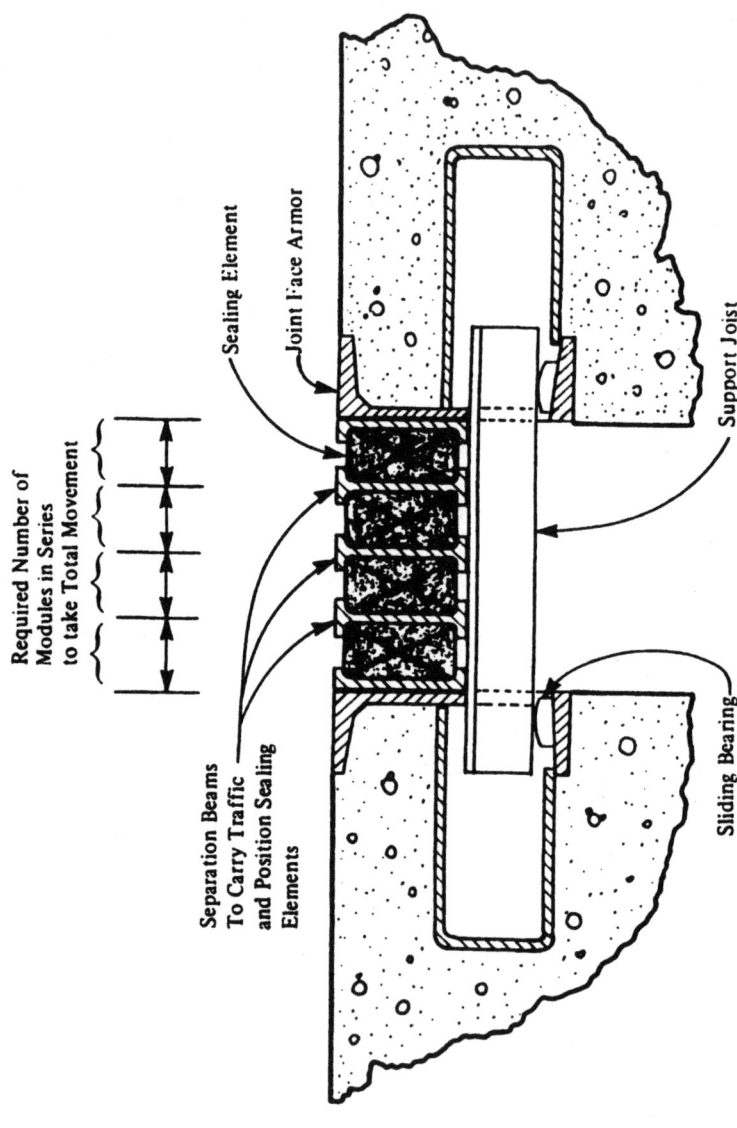

FIGURE 22.5a Modular compression seal systems.

(a) — **Basic Features**

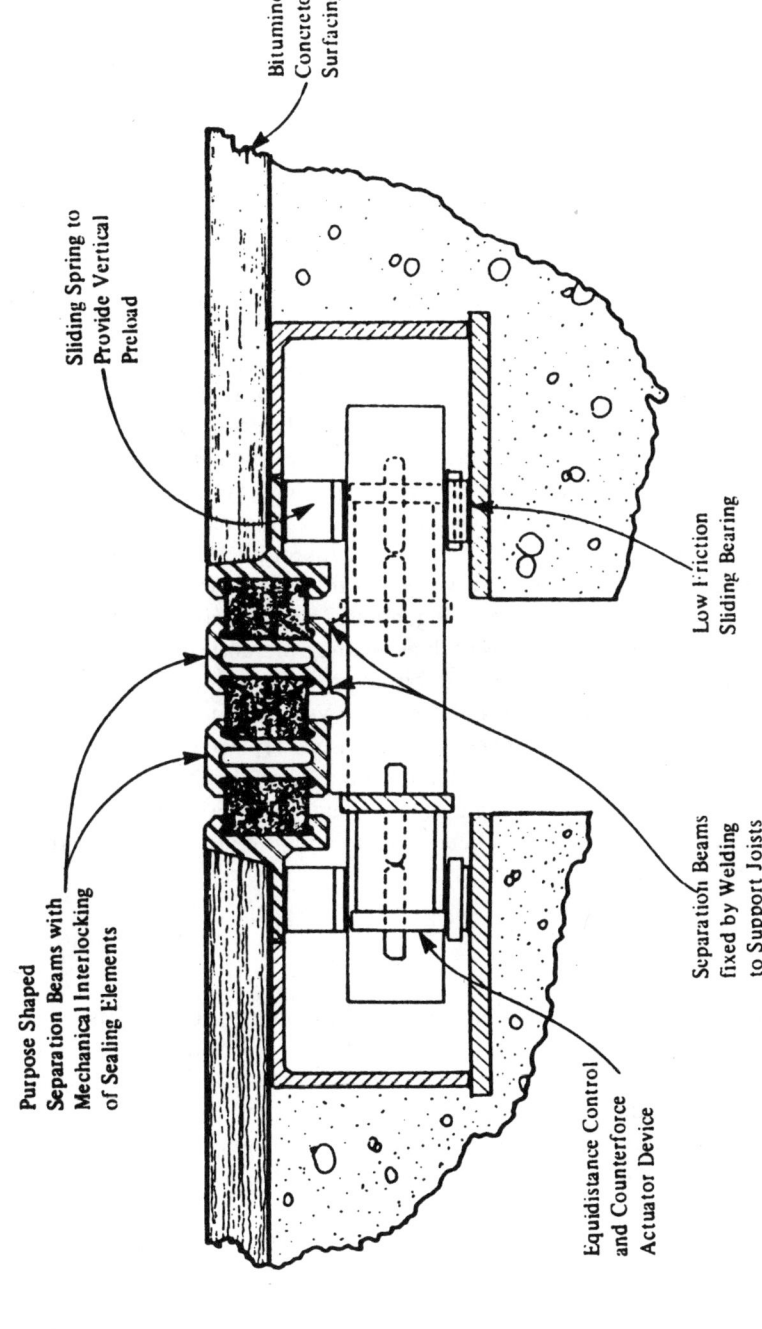

FIGURE 22.5b Modular compression seal systems.

Tension-Compression Seal Improvements

Tension-compression seals were a relatively new system when data about tension-compression seals and their uses (see Figure 22.6 section 3B) originally were published. Since then, many design, fabrication, and installation innovations have taken place and significant reductions in undesirable cyclical stresses have been achieved in the improved versions of tension-compression seals now available. As a result, these systems are finding wide use, particularly in bridges, for movements up to 13 inches (330 mm).

Two changes from the system shown in Figure 22.6, section 3b are that the bridging plate (to span the joint opening) is not necessarily embedded within the lower part of the elastomeric material, and the grooves are used to permit change of width as the joint opens and closes.

These new features are illustrated in Figure 20.3 for representative cross-sections of tension-compression seals on the market. In cooperation with the positioning of the bridging and other plates, it is possible to change the configuration of the elastomeric material to better accommodate movements by using open grooves in the upper and lower surfaces of the sealing element rather than the enclosed voids or cells illustrated in Figure 22.7, section 3b. In some designs the plates have been located so as to serve also as the riding surface. This then helps resist wear due to traffic, snow-plowing, abrasives, and studded tires.

Improved fabrication methods such as welding prefabrication and use of special moldings have been developed to better provide for the sealing of joints at curbs and gutters and for changes of direction. Vulcanized connections between each elastomeric molded unit are coming into use to replace troublesome tongue-and-groove connections. Field installation procedures also have been improved with respect to bedding the sealing unit to the underlying concrete to establish an impervious connection and by jacking of the moldings end-to-end to improve the seal at each juncture.

Strip (or Gland) Seals

During the last few years, strip (gland) seals have come into wide use in joints on concrete bridge decks and other large horizontal slab structures. They consist of a preformed elastomeric sealing element mechanically locked or bonded at each side of the joint face. Movement is accommodated by fold or bulb within the element. Though placed at the exposed surface of the concrete, their mode of functioning is similar to that of embedded waterstops as shown in Figures 22.7a and 22.7b, section 2. Waterstops might come under tension as a joint opens in service because they are embedded at the time of construction in a joint which is usually fully closed. Tension can be avoided with strip (or gland) seals because they can be set to predetermined initial joint widths. Now, tension in the elastomer does not necessarily occur when the joint subsequently opens beyond its original width.

PERFORMANCE, DEFECTS, REPAIR AND MAINTENANCE OF SEALANTS

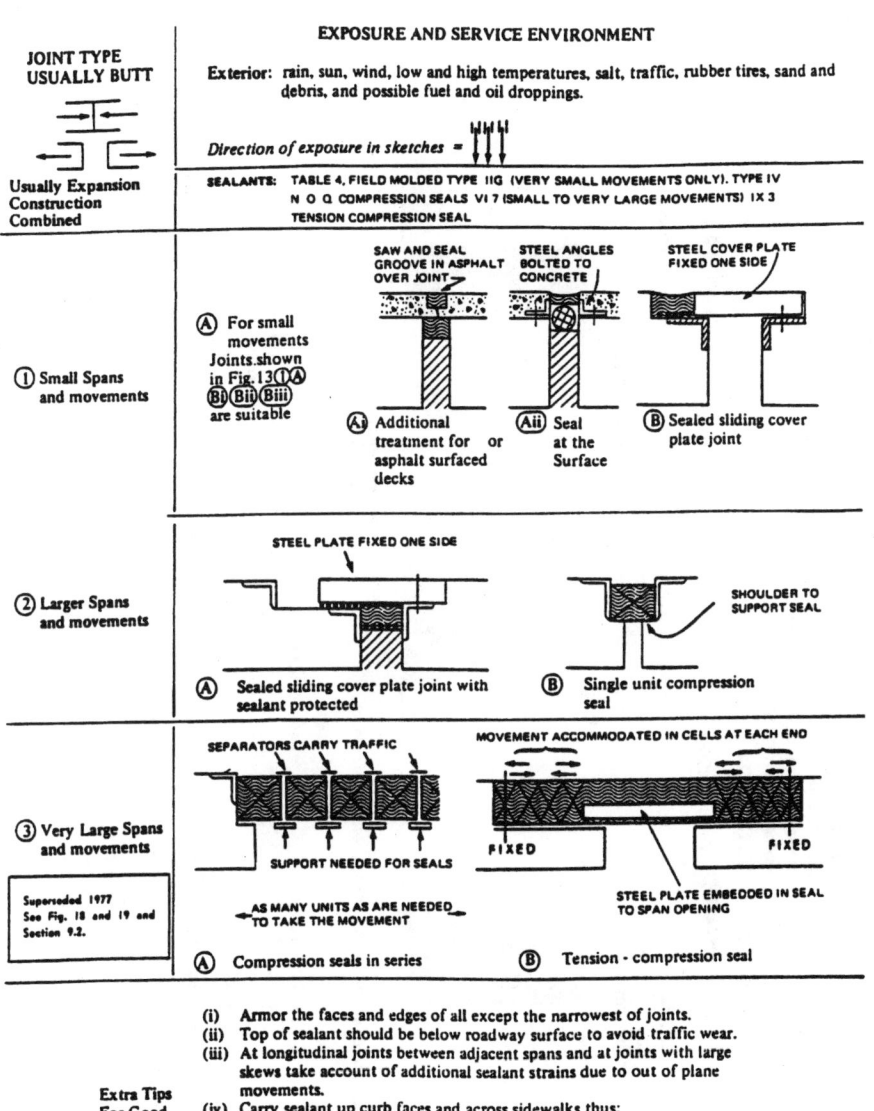

FIGURE 22.6 Joints for bridges.

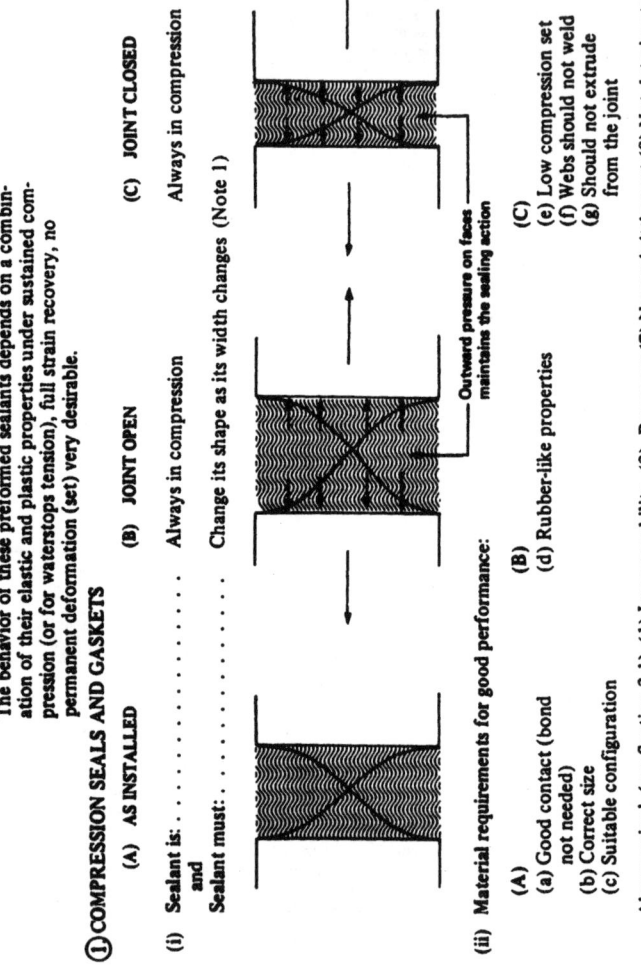

The behavior of these preformed sealants depends on a combination of their elastic and plastic properties under sustained compression (or for waterstops tension), full strain recovery, no permanent deformation (set) very desirable.

① COMPRESSION SEALS AND GASKETS

	(A) AS INSTALLED	(B) JOINT OPEN	(C) JOINT CLOSED
(i) Sealant is:	and	Always in compression	Always in compression
Sealant must:		Change its shape as its width changes (Note 1)	

Outward pressure on faces maintains the sealing action

(ii) Material requirements for good performance:

(A)
(a) Good contact (bond not needed)
(b) Correct size
(c) Suitable configuration

(B)
(d) Rubber-like properties

(C)
(e) Low compression set
(f) Webs should not weld
(g) Should not extrude from the joint

Also required (see Section 3.1) (1) Impermeability (3) Recovery (7) Nonembrittlement (8) Not deteriorate

(iii) Deficiencies in (b) (d) (e) (f) predisposes to loss of contact pressure. See Fig. 16 ① for consequences

Note 1
Compression seals in working joints require to be compartmentalized or foldable to meet this criterion, gaskets in nonworking joints may not.

FIGURE 22.7a Representative cross-section of tension-compression seals.

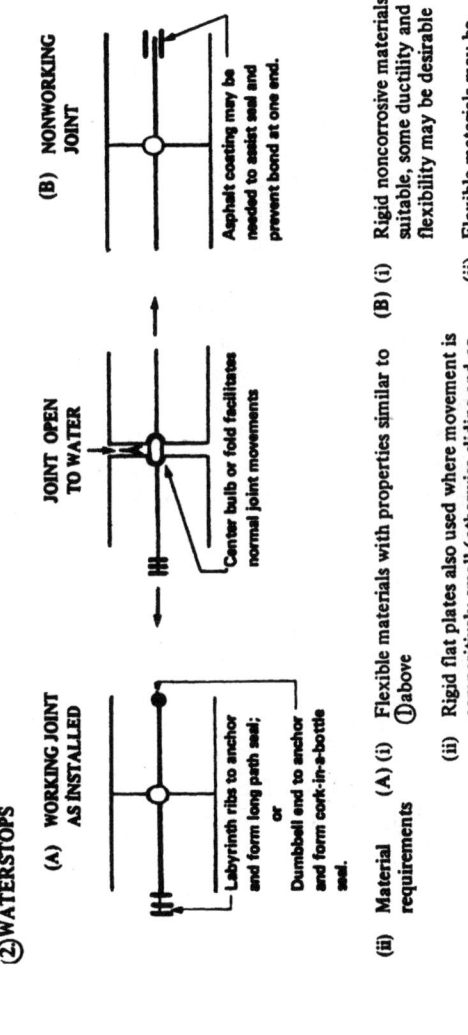

FIGURE 22.7b Representative cross-section of tension-compression seals.

Many different design details exist and cross sections of a representative range of strip (or gland) seals are shown in Figure 22.8. Movements between 0 to 4 inches (0 to 101mm) can be accommodated by these systems which are furnished in prefabricated form to suit a particular application. In many designs the sealing element might be installed, removed and replaced on site as necessary between the steel joint armor faces that are anchored to the concrete.

Trends in Contraction Joint Practice for Slabs on Grade, Highway, and Airports

While the practices for joints in highway and airfield pavements remain essentially the same, two changes are occurring that affect joint sealants.

First, many highway authorities are specifying shorter contraction joint spacings in both plain and reinforced concrete pavements; some are using random spacing averaging between 15 and 20 feet (4.57 m and 6.10 m) in plain pavements for which the repeating series 13, 18, 29, 12 ft (3.66m) is popular. Some also are skewing joints at 2 feet 12 feet (0.61m in 3.66m). While the objectives are to reduce intermediate slab cracking and improve ride and load transfer, such designs place less demand on the sealant because of the smaller movement that occurs at each joint. Fuller information on the design and construction of joints in concrete pavements will be found in the reports of ACI Committees 316 and 325.

Second, experience points to better field-applied sealant performance if a shape factor of 2:1 is used. See paragraph *Additional thought on shape factor*. For pavements with short slabs, a typical joint sealant reservoir configuration found satisfactory is: joint width ⅜ inch to ½ inch (9.5 mm to 12.7 mm), joint sealant depth 1 inch (25.4 mm) using cord or tape as a backup material. In most applications, including airports, the sealant is installed ¼ inch ±1/16 inch (6.4 mm ±1.5 mm) below the level; of the pavement, In addition to the Type II and Type IV field-applied sealant materials referred to in Table 22.2 and Table 22.3 and described in paragraphs *Thermoplastics, Chemically Curing* and *Thermosetting, Chemically Curing*, respectively, polyvinyl-chloride coal-tar sealants described in paragraph *Polyvinyl Chloride-Coal Tar* are finding considerable use in this application.

TWO-STAGE JOINTS FOR BUILDINGS

The idea behind two-stage joints is to give a building more then one weather-proofing line of defense. Near the exterior wall face, a rain barrier is provided to shed most of the rain from the joint while airborne water that penetrates the comparatively narrow gap at the rain barrier enters a wider, vented, and drained expansion chamber where the remaining water that drops out of the air as kinetic energy is dissipated by eddying. The sealing system is completed by an air seal between the expansion chamber and the interior wall face. This air seal demarks

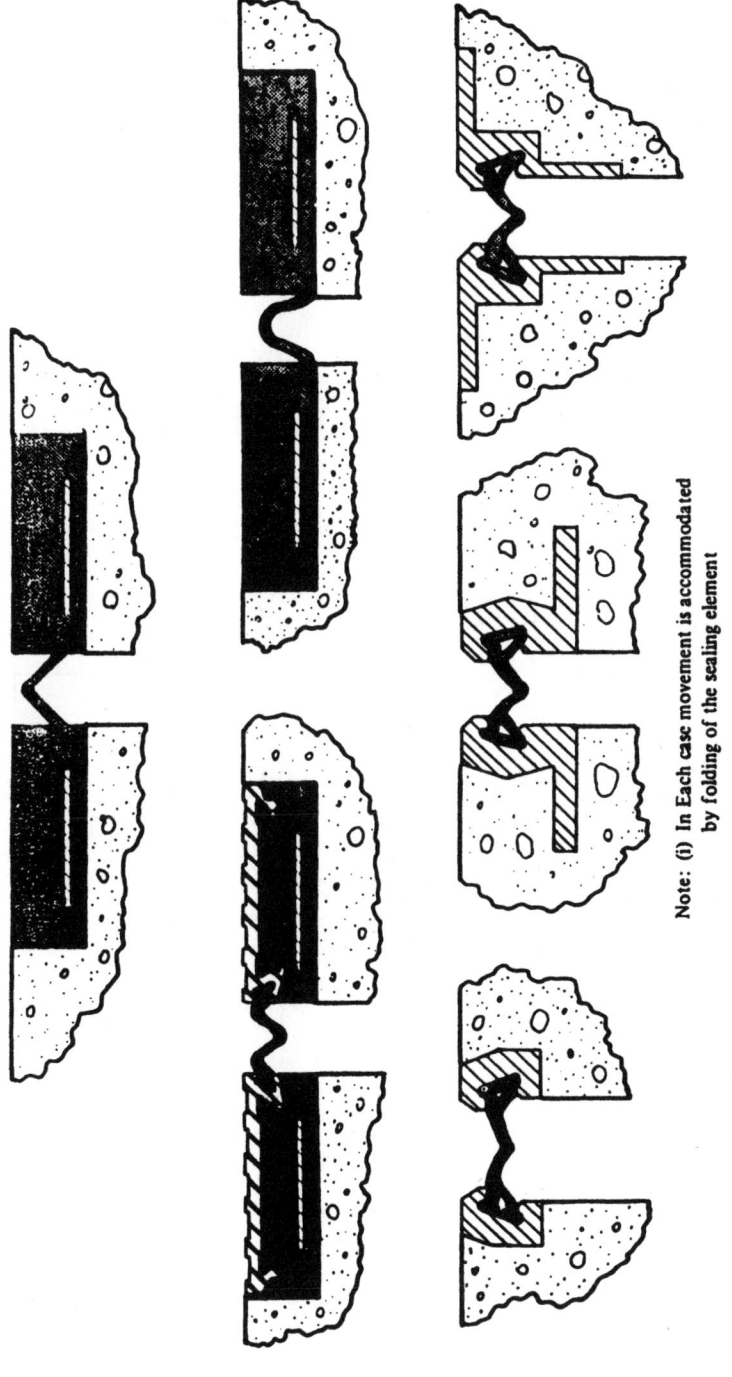

FIGURE 22.8 Typical cross-sections of strip or gland seals.

TABLE 22.3 Uses for Field-Applied and Preformed Sealants

Type of application			Field-molded						Preformed		
			Thermoplastics		Thermosetting			VI Compression seal	VII Waterstops	VII Gaskets	IX Misc.
		I Mastics	II Hot applied	III Cold applied	IV Chemical cure	V Solvent release					
Structures not under fluid pressure; e.g. buildings, bridges, storage bins, retaining walls	Caulking & glazing	A B D E		L M	N P R S	T U V		3		1 3 4 5	1 2
	Precast panels	A B D E		M	N P R	T V		3		1 3 4 5 7	1 2 10
	Walls (vertical joints)	A B D E			N P R	T V		3 7		7	10
	Roof deck (horiz. joints)		F G W		N O P Q	T V		3 7		7	10
	General floors		G H W		N O P Q			3			1 4 9d
	Industrial floors			K	N P Q			3			1 3 9c
	Floors w/oil and solvents		H I J W		N O Q S			3			1
	Services									3 6	11
Note 3	Bridges		G W		N O Q			3 7			10 3 Note 2
Containers subject to fluid pressure; e.g., water containing or excluding structures	Canal linings	C	G W	K	O			3	1 4		
	Precast pipes	C	G W								
	Tanks & monolithic pipe	C	G W	K	N O P			3	1 3 4 6 8 9a 9b	1 3 4 5 6 8	10
	Swimming pools		G W		N P			3	1 3 4 6		10
	Dams				O P			3	4 9a 9b		
	Walls & floors w/ water outside		G W	K	N O P			3	1 3 4 6 9b		3 Note 2
Note 3											
Pavements	Walkways		F G W	K	N O P Q			3			
	Highway		G W		N O Q			3			
	Airport		G W		O Q			3			
	Areas w/ fuel spillage		H J W		O Q			3			
Grouting nonworking cracks				K	S						23
Suitable in above applications where joint movement is: Note 4	None or very small*	F G H I J		K L M	N O P Q R S	T U V		3	1 3 4 6 8 9a 9b	1 3 4 5 6 7 8	1 2 9c 9d
	Small*	F G H I J		K L M	N O P Q R	T U V		3	1 3 4 6 8 9a 9b	7	
	Large**				N O P Q R			3	3 4		
	Very large**							3			3 Note 2

*Contraction joints   **Expansion joints

(Continued)

TABLE 22.3 Uses for Field-Applied and Preformed Sealants *(Continued)*

Type of application		Field-molded						Preformed		
		Thermoplastics		Thermosetting						
	I Mastics	II Hot applied	III Cold applied	IV Chemical cure	V Solvent release	VI Compression seal	VII Waterstops	VII Gaskets	IX Misc.
Storage life: Limited (l) Over 1 year (o) Emulsions are damaged by freezing	A B C D E(o)	F G H I J(o)	K L M(o)	N O P Q R S(l)	T(o) U V(l)	3 7(o)	1-9(o)	1-8(o)	1-11(o)
Installation: Knife or trowel (k) Insert (i), heat & pour (h) Mix if 2-component (m) Note 5 Hand gun (g), pressure gun (p) Preposition (pp)	A B(k)(g)(p) C(k)(g) D E (g)(p)	F G H I J(h) (W) (h)	K L (g)(k)(p) M(g) preheat to 100°F (40°C)	N O P Q R S (m)(k)(g)(p)	T U V (g)(k)(p)	3(i)	1-9(pp)	1-8(pp)	1234 9d 10(pp) 9c(i)(h) 11(p)

Notes to table 22.3

Note 1: Table 22.3 is only a general guide. Before deciding on a particular material for a specific application all circumstances— in particular the joint movement to be expected and a suitable joint design and joint detail must be considered.

Note 2: 3 refers to tension-compression seals described in Section 3.6.

Note 3: Certain sealants may contain substances toxic to potable water or foodstuffs. Check local or national restrictions that may govern use in areas exposed to these.

Note 4: Certain materials are equally suitable for both vertical and horizontal joints. Others are not and while they may stay in place in horizontal joints, they would sag or flow out of vertical joints in hot weather. Asphalt and rubber-asphalt materials are examples of these. Some materials are available in two grades. One known as nonsag or gun grade is thixotropic and is suitable for vertical joints. The other, known as self-leveling or pour grade, is intended for use in horizontal joints.

Note 5: Pot life (time material still usable after mixing) is limited and correct proportioning and mixing is critical with 2-component materials.

Note 6: Field-molded sealants furnished as follows:

Liquid in drums, cans or cartridges	A B D E
Liquid in drums	C K W
Liquid in drums or cans	O Q
Liquid in cans	P S
Liquid in cans or cartridges	L N R T U V
Liquid in cartridges	M
Solid in cakes for melting	F G H I J
for preformed materials see table 3	

only between the inside and outside air pressure and is not drained away from the rain barrier or the expansion chamber to the exterior face of the building.

The principle of two-stage joints and their application is illustrated in Figure 22.9 details D and E. The main use of two-stage joints is for both horizontal and vertical joints where architectural panels are used in buildings.

APPLICATION AND PERFORMANCE

Experience indicates that two-stage joint systems will give very satisfactory performance provided proper attention is give-to panel and joint design details, panel fabrication and installation, and, of course, to the choice and installation of the sealant. Water penetrating the rain barrier must be drained away by a flashing installed at appropriate horizontal joints. Such flashing installed at appropriate horizontal joints. Such flashings have a second purpose—to act to avoid vertical air movement in the expansion chambers of vertical joints. For this purpose, they might be required at each floor level or, where the exposure to wind is less severe, at every two or three floors. Two-stage joint systems can be designed to accommodate relatively large external openings between panels, where this is desired, and are accommodating to deviations in as-constructed joint widths from those designed, provided a reasonable uniformity in joint width is maintained. A minimum of 4 in. (100 mm) panel thickness is required for the system where sealants are used for the air seal together with accessibility from the interior wall face to install the sealant. Where compression seals or gaskets are used, the panel thickness might require up to 5 in. (125 mm) and provision should be made in the panel connecting details at vertical joints to close the joint opening slightly after panel erection so as to compress the air seal which is usually pre-installed against the joint face of one of the abutting panels.

Unsatisfactory performance in two-stage joints is likely to arise from gaps in the air seal, making the rain barrier airtight, or from not venting or draining the expansion chamber properly. A special problem might occur in buildings with high interior humidity, especially if they are tall and the internal air pressure is occasionally above atmospheric, in that moisture might pass outward through the air seal and condense. This can be overcome by using cavity wall construction in which the exterior panel overlaps serve as the rain barrier, the cavity serves as the expansion chamber, and the outside face of the internal wall is insulated so that this wall is maintained at room temperature. Though initially more costly, cavity wall construction of this type has an excellent general- and sealant-performance record and low maintenance costs.

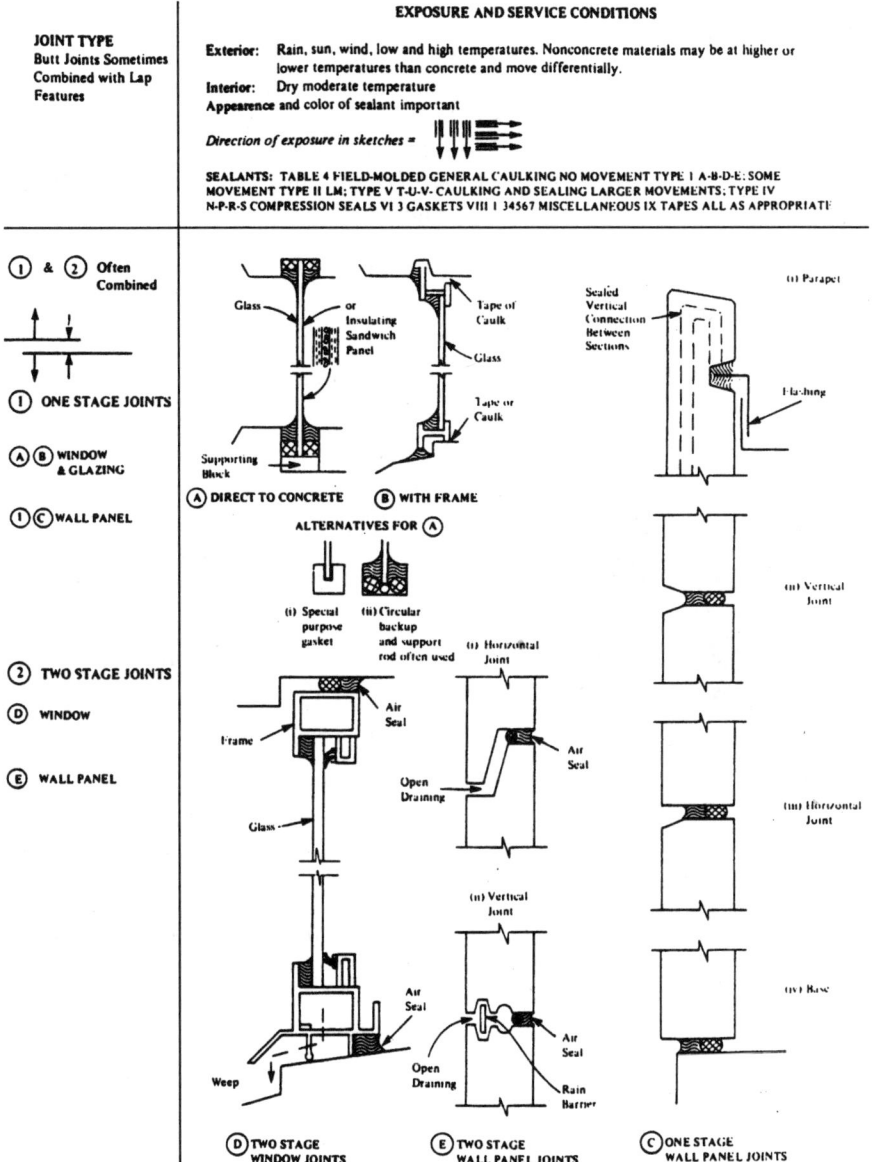

FIGURE 22.9 Joints for Special Purpose buildings.

SUMMARY

Chapters 18, 19, 20, 21, and 22 give concise and current data on specifications, joint design, installation of sealants, performance, and maintenance. It is hoped that these chapters will give the architect or designer a better understanding of the properties of joint sealants and of where they should be used in practice today. Other factors included are new sealant developments and design concepts for both field-applied and preformed materials.

REFERENCES

1. American Concrete Institute—Guide to Joint Sealants for Concrete Structures, ACI-504-90, 1990
2. Pelletier, P.G. and Schmidt, P.H. March 1993. *"Preparing Cracks for Sealing,"* Progressive Maintenance.
3. Suprenant, B.A., January 1995. *"Saw-cutting Joints in Concrete."* Concrete Construction.
4. "Guidelines for Timing Contraction Joint Sawing and Earliest Loading for Concrete Pavements" Vol. I and ll: Final Report, FHWA-RD-91-079, February 1994. Federal Highway Administration, McLean, VA.

Note: The author wishes to thank the American Concrete Institute for permission to use much of its material.

CHAPTER 23
TRADE ORGANIZATIONS, PUBLICATIONS, GOVERNMENT, AND SUPPLIERS

The trade organizations, trade publications, governmental groups and adhesive and sealants raw material suppliers associated with the building trades industry, curtain-wall manufacturers, general contractors and owners etc., have contributed to making this book a potentially successful tool for readers in this industry. The author wishes to thank them for their unselfish contribution of time. Each trade organization and trade-associated magazine offers a list of suppliers (too voluminous to be printed in this publication); these groups can be contacted directly.

Abatron, Inc.[1]
5501-95th St.
Kenosha, WI 53144

A & C Catalysts, Inc.[1]
7-33 Amsterdam St.
Newark, NJ 07105

Acoustical Society of America[3]
500 Sunnyside Blvd.
Woodbury, NY 11797

Adhesives Age Magazine[4]
6151 Powers Ferry Rd., NW
Atlanta, GA 30339

Adhesive and Sealant Council, Inc.[3] (ACS)
1627 K St. Suite 1000
Washington, DC 20006

Adhesive and Sealant Manufacturers Association of Canada[3]
208 Brimorton Dr.
Scarborough, Ontario M1H 2C6
Canada

Adhesive Manufacturers Association[3] (AMA)
401 N. Michigan Ave.
Chicago, Il 60611-4267

Adhesives and Sealants Industry[3]
755 W. Big Beaver Rd.
Troy, MI 48084

Adhesive Systems, Inc.[1]
9411Corsair Rd. PO Box 518
Frankfort, IL 60423-0518

Air Products and Chemicals, Inc.[1]
7201 Hamilton Blvd.
Allentown, PA 18195-1501

Akzo Nobel Chemicals, Inc.[1]
300 S. Riverside Plaza
Chicago, IL 60606

Alumi-News[4]
109-4920 de Maisonneuve Blvd. W
Westmont, Quebec H3Z 1N1
Canada

Aluminum Association[3] (AA)
900 19th St. NW #300
Washington, DC 20006

American Arbitration Association[3]
140 West 51st St. 9th floor
New York, NY 10020

American Bar Association[3]
750 N. Lake Shore Dr.
Chicago, IL 60611

American Chemical Society[3] (ACS)
1155 16th St., NW
Washington, DC 20036

American Consulting Engineers Council[3] (ACEC)
1015 15th St.
Washington, DC 20005

American Institute of Architects[3] (AIA)
1725 New York Ave., NW
Washington, DC 20006

American Institute of Architects Library[3]
1735 New York Ave., NW
Washington, DC 20006

American Institute of Chemical Engineers[3]
345 E. 47th St.
New York, NY 10017

American Institute of Chemists[3]
501 Wythe St.
Alexandria, VA 22314-1917

American Institute of Certified Planners[3] (AICP)
11776 Massachusetts Ave., NW
Washington, DC 20036

American Architectural Manufacturers Association[3] (AAMA)
2400 E. Dundee Road
Palatine, IL 60067

American Concrete Institute[3] (ACI)
PO Box 19150, Redford Station
Detroit, MI 48219

American Concrete Pumping Association[3]
Box 4307
Vallejo CA 94590

Americans with Disabilities Act Information[2] (ADA)
U.S. Department Of Justice
Civil Rights Division
P.O. Box 66738
Washington, DC 20075

American Federation of Labor and
Congress of Industrial Organization[3] (AFL-CIO)
815 16th St.
Washington, DC 20006

American Institute of Real Estate Appraisers[3] (AIREA)
155 East Superior St.
Chicago, IL 60611

American Institute of Steel Construction[3] (AISC)
1 E. Wacker Dr. Suite 3100.
Chicago, IL 60611

American Institute of Timber Construction[3] (AITC)
333 West Hampden Ave.
Englewood, CO 80110

American Iron and Steel Institute[3] (AISI)
1101 17th St. NW
Washington, DC 20036

American National Standards Institute[5] (ANSI)
11 W. 42nd St.
New York, NY 10036

American Plywood Association[3] (APA)
Box 11700
Tacoma, WA 98411

American Society of Civil Engineers[3] (ASCE)
345 E. 47th St.
New York, NY 10017

American Society of Concrete Construction[3]
111 East Wacker Dr. Suite 600
Chicago, IL 60611

American Society of Highway Engineer[3]
151 Old Ford Drive
Camp Hill, PA 17011

American Society of Testing Materials[5] (ASTM)
100 Barr Harbour Dr.
Conshohocken, PA 19428-2959

American Society of Heating, Refrigeration,
and Air Conditioning Engineers, Inc.[3] (ASHRAE)
791 Tullie Ave.
Atlanta, GA 3032

American Society of Landscape Architects[3] (ASLA)
1733 Connecticut Ave., NW
Washington, DC 20009

American Society of Mechanical Engineers[3] (ASME)
345 East 47th St.
New York, NY 10017

American Solar Energy Society[3] (ASES)
2400 Central Ave., Suite B-1
Boulder, CO 80361

American Subcontractors Association[3] (ASA)
8401 Corporate Dr.
Landover, MD 20785

American Wood-Preservers' Association[3] (AWPA)
7735 Old Georgetown Rd.
Bethesda, MD 20014

American Wood Preservers Bureau[3] (AWPB)
2772 S. Randolph St.
Arlington, VA 22206

American Wood Council (AWC)
National Forest Products Association[3]
1250 Connecticut Ave., NW Suite 230
Washington, DC 20036

American Hardboard Association[3] (AHA)
887B Wilmette Rd.
Palatine IL 60067

American Plywood Association[3] (APA)
7011 S. 19th St.
Tacoma, WA 98466

Ameripol Synpol Corporation[1]
Mallard Creek Polymers
146 S. High St.
Akron, OH 44308-1493

Amoco Chemical Company[1]
200 E. Randolph Dr. MC-4106
Chicago, IL 60601-7125

BP Amoco Chemicals, Inc.[1]
150 W. Warrenville Rd.
Naperville, IL 60563

Asphalt Institute[3]
University of Maryland
College Park, MD 20740

Air-Conditioning and Refrigeration Institute[3] (ARI)
Architectural Testing, Inc.
130 Derry Ct.
York, PA 17042-9405

Applicator, The[4]
Sealant and Waterproofers Institute
3101 Broadway, Suite 300
Kansas City, MO 64111

Architects' Guide to Glass, Metal & Glazing[4]
2701 Union Ave. Ext. Suite 410
Memphis TN 38112-4479

Architectural Precast Association[3]
825 East 64th St.
Indianapolis, IN 46220

Architectural Record[4]
Two Penn Plaza
New York, NY 10121-2298

Architectural Specifier[4]
6213 Howard St.
Niles, MI 60714

Architectural Woodwork Institute[3]
2310 S. Walter Reed Dr.
Centreville, VA 22020-8550

ARCO Chemical, Inc.[1] (now Lyondell Chemical)
3801 W. Chester Pike
Newtown Square, PA 19074

Arizona Chemical[1]
PO Box 550850
Jacksonville, FL 32255

Ashland Specialty Chemical Co.[1]
Drew Industrial Div.
One Drew Plaza
Boonton, NJ 07005

Assembly Magazine[4]
Cahners Business Information
191 South Gary St.
Carol Stream, IL 60188-2086

Associated Builders and Contractors[3] (ABC)
729 15th St.
Washington, DC 20005

Associated Construction Distributors International[3]
2110 Powers Ferry Rd. NW Suite 300
Atlanta, GA 30339

Associated General Contractors of America[3] (AGC)
1957 E. St. NW.
Washington, DC 20006

Association of Physical Plant Administrators
of Universities & Colleges[3]
1446 Duke St.
Alexandria, VA 23412

BASF Corp.[1]
11501 Steel Creek Rd.
Charlotte, NC 28273

Bayer Corporation[1]
Fibers, Additives, and Rubber Division
2603 Market St.
Akron, OH 44313

Bayer Corp., Polyurethanes Div.[1]
100 Bayer Rd. Bldg. 4
Pittsburgh, PA 15205-9741

BF Goodrich Coatings, Inc.[1]
9911 Brecksville Rd.
Cleveland, OH 44141

Bostik, Inc.
211 Boston St.
Middleton, MA 01949-2128

Bowser-Morner, Inc.[5]
4518 Taylorsville Rd.
Dayton, OH 45401

Brick Institute of America[3] (BIA)
11490 Commerce Park Dr.
Reston, VA 22091

British Adhesive and Sealants Association[3]
33 Fellows Way
Stevenage
Hertfordshire, SG2 8BW
UK

British Standards Institute[3]
389 Chewick High Rd.
London, W4 4AL
UK

Building Stone Institute[3]
Box 5047
White Plains, NY 10602-5047

Building Research Establishment[3]
Garston Watford
Herts, WD2 7JR
UK

Building Research Advisory Board[2] (BRAB)
2101 Constitution Ave., NW
Washington, DC 20418

Building Officials and Code Administrators International[3] (BOCA)
4051 West Flossmoor Rd.
Country Club Hills, IL 60477

Building and Construction Information Exchange[3] (BACIX)
c/o Thomas Martineau AIA
Productivity House, Inc.
3476 Valley Creek Dr.
Tallahassee, FL 32312-3633

Building Systems Magazine
2820 W. 21st St.
Erie, PA 16506-2970

Bureau of Construction Quality Control[5]
Materials and Testing Division
Pennsylvania Department of Transportation
Harrisburg, PA

Brick Institute of America[3] (BIA)
1750 Old Meadows Rd.
McLean, VA 22101

Cabot Corp.[1]
125 High St.
Boston, MA 02110

California Glass Association[3]
1154 N. Knollwood Circle
Anaheim, CA 90601

Canadian General Standards Board[2]
Standards and Specification Branch
Ottawa, Ontario K1A 0S9
Canada

Canadian Mortgage and Housing Corp.[2]
700 Montreal Rd.
Ottawa, Ontario K1A 0PA
Canada

Ciba Specialty Chemicals Corp. Performance Polymers[1]
281 Fields La.
Brewster, NY 10509

Cities Service Company[1]
Columbian Div.
3200 W. Market St.
Akron, OH 44313

CM News[4]
National Concrete Masonry Association
2302 Horse Pen Rd.
Herndon, VA 20171

Compliance Engineering[4]
11444 W. Olympic Blvd., Suite 900
Los Angeles, CA 90064

Concrete Sourcebook[4]
Concrete Construction Publications, Inc.
426 South Westgate
Addison, Il 60101

Concrete Repair[4]
Aberdeen Group
426 South Westgate
Addison, Il 60101-4546

Concrete Repair Builder[4]
International Concrete Repair Institute
1323 Shepard Dr. Suite D
Sterling, VA 20164

Concrete Sawing and Drilling Association[3]
729 15th St.
NW Washington, DC 20005

Construction Canada[4]
46 Midland Ave.
Scarborough, Ontario N0M 1S0
Canada

Construction Consulting Laboratory, Inc.[5]
4751 W. State St., Suite E
Ontario, CA 91762

Construction Dimensions[4] (AWCI)
803 West Broad St., Suite 600
Falls Church, VA 22046

Constructioneer Magazine[4]
30 Technology Parkway, South Suite 100
Norcross, GA 30092

Construction Specifiers Institute [Ref. 23.3,4] (CSI)
601 Madison Ave.
Alexandria, VA 22314

Construction Index Architex[4]
410 S. Michigan Ave., Suite 1008
Chicago, Il 60605

Council Of American Building Officials[3] (CABO)
2233 Wisconsin Ave., NW
Washington, DC 20007

Concrete Reinforcing Steel Institute[3] (CRSI)
933 North Plum Grove Rd.
Schaumburg, IL 60195

Construction Industry Manufacturer's Association[3] (CIMA)
525 School Street NW
Washington, DC 20024

Consumer Products Safety Commission[2]
5401 Westbard Ave.
Bethesda, MD 20207

Ceramic Tile Institute[3] (CTI)
700 North Virgil Ave.,
Los Angeles, CA 90029

Deguusa[1]
2 Penn Plaza
New York, NY 10001

Design/Build Business[4]
c/o McKellar Publishers
333 E. Glenoaks Blvd. #204
Glendale, CA 91207-2074

Design Research Library[4]
5777 West Century Blvd.
Los Angeles, CA 90045

DIN[2]
Burggrafen Strasse 6
D-10782 Berlin (Tiergarten)
Germany

Doors and Hardware Institute[3]
14150 Newbrook Dr. Suite 200
Chantilly VA 20151-2223

Dow Chemical Corp[1]
Emulsion Polymers/Epoxy Products
2040 Dow Center
Midland, MI 48674

Dow Corning Corporation[1]
2040 Dow Center
Midland MI 48686-0994

DuPont Dow Elastomers—Adhesives[1]
300 Bellevue Pkwy.
Wilmington, DE 19809

Eastman Chemical Products, Inc.[1]
Adhesives and Sealant Raw Materials
PO Box 431
Kingsport, TN 37662

Electrical Testing Laboratories, Inc.[5] (ETL)
3933 US Route 11
Cortland, NY 14624

Elf Atochem North America, Inc.[1]
2000 Market St.
Philadelphia, PA 19103-3222

Engineering News Record[4]
The McGraw-Hill Co.
1221 Avenue of the Americas
New York, NY 10020

Engineering Society Library[3]
345 E. 47th St.
New York, NY 10017

Environmental Protection Agency[2] (EPA)
401 M Street
Washington, DC 20460

Epoxy Technology, Inc.[1]
14 Fortune Dr.
Billerica, MA 01821-3972

Epoxies, Inc.[1]
717 East Jericho Turnpike, Suite 126
Huntington Station, NY 11746

E-poxy Industries, Inc.[1]
14 West Shore Street
Ravena, NY 12143-1698

Factory Mutual Engineering Corporation[5]
1151 Boston-Providence Turnpike
Norwood, MA 02062

FMC Corp.[1]
100 Niagara St.
Middleport, NY 14105

FEICA Association of European Adhesive Manufacturers[3]
Ivo-Beucker Str. 43
D-40237 Dusseldorf
Germany

Flat Glass Manufacturers Association[3] (FGMA)
3310 SW Harrison St.
Topeka, KS 66611

HP Fuller Co.
3530 Lexington Ave. N
St. Paul, MN 55126-8076

General Electric Co.[1]
Silicone Products Department
Waterford, NY 12188

Georgia Pacific Chemical
55 Park Ave., 19th Floor
Atlanta, GA 30333

Glass Association of North America[3]
3310 SW Harrison St.
Topeka, KS 66611

Glass Digest[4]
18 E 41st St.
New York, NY 10017

Glass International[4]
Queensway House
2 Queensway, Redhill
Surrey, RH1 1QS
UK

Glass Production Technology International[4]
c/o MMG
81 North Forest Ave.
Rockville, Centre, NY 11570

Glass Magazine[4]
National Glass Association
8200 Greensboro Rd.
Suite 302
McLean, VA 22102

Glass Tempering Association[3]
3310 SW Harrison St.
Topeka, KS 66611

BF Goodrich Coatings Div.[1]
9911 Brecksville Rd.
Cleveland, OH 44144

Goodyear Tire and Rubber Co[1]
Chemical Div.
1144 E. Market St.
Akron, OH 44316

Gougeon Bros., Inc.[1]
PO Box 908
Bay City, MI 487-7-0908

Great Lakes Chemical Corp[1]
1 Great lakes Blvd.
West Lafayette, IN 47906

Groliers Multimedia Encyclopedia[4]
90 Sherman Turnpike
Danbury CT 06816

Gypsum Association[3]
1603 Orrington Ave.
Evanston, IL 60201

Hercules, Inc.[1]
Hercules Plaza, 10145 Market St.
Wilmington, DE 19894-0001

Illuminating Engineering Society of North America[3]
345 West 47th St.
New York, NY 10017

Insulating Glass Manufacturers Association of Canada[3] (IGMAC)
27 Goulburn St.
Ottawa, Ontario H1N 8C4
Canada

Institute for Research in Construction/
National Research Council[2] (IRC/NRC)
Building M 24
Ottawa, Ontario K1A 0R6
Canada

International Conference of Building Officials[3] (ICBO)
5360 S. Workman Mill Rd.
Whittier CA 90601

International Masonry Institute[3] (IMI)
823 15th St. NW, Suite 1001
Washington, DC 20005

Japan Adhesive Industry Association[3]
1-15-10 Uchikanda
Chiyodaku, Tokyo 101
Japan

Journal of Acoustic Emissions[4]
16350 Ventura Blvd. #106
Encino, CA 91436

Journal of Polymer Science[4]
John Wiley & Sons
605 3rd Ave.
New York, NY 10158

Kraton Polymers[1]
One Shell Plaza
PO Box 2463
Houston, TX 77252

Lawrence Berkeley Laboratory[5]
Mail Stop 90-31-11
1 Cyclotron Rd.
Berkeley, CA 94720

Magazine of Masonry[4]
Aberdeen Group
426 South Westgate
Addison, IL 60101-4546

Manufactured Housing Institute[3]
1745 Jefferson Davis Hwy.
Arlington, VA 22202

Marble Institute of America[3] (MIA)
33505 State St.
Farmington, MI 48335

Masonry Institute of America[3] (MasIA)
2550 Beverly Blvd.
Los Angeles, CA 90057

Masonry Society, The[3] (TMS)
2619 Spruce St.
Boulder, CO 80302

Merck Chemical Division[1]
126 E. Lincoln Hwy.
Rahway, NJ 07065

Mobay Chemical[1]
Baychem Corp.
Penn Lincoln Pkwy, West
Pittsburgh, PA 15205

Miami Testing Laboratory, Inc.[5]
1640 W 32nd Pl.
Hialeah, FL 33012

Metal Construction News[4]
109 Portage St.
Woodville, OH 43469

W.R. Meadows, Inc.[1]
PO Box 543
Elgin, IL 60121

Morton International, Inc.[3]
Rohm & Haas Co., Industrial Adhesives
100 N. Riverside Plaza
Chicago, IL 60606-3328

National Building Museum[2]
Pension Building at Judiciary Square, NW
Washington, DC 20001

National Association of Home Builders of the United States[3] (NAHB)
15th and M Streets, NW
Washington, DC 20005

National Association of Minority Contractors[3] (NAMC)
1333 F Street NW Suite 500
Washington, DC 20004

National Association of Women in Construction[3] (NAWIC)
327 S. Adams St.
Fort Worth, TX 76104-1002

National Bureau of Standards[2] (NBS)
Center for Building Technology and Center for Fire Research
Gaithersburg, MD 20899

National Concrete Masonry Association[3] (NCMA)
2302 Horse-Pen Rd. Box781
Herndon, VA 22070

National Constructors Association[3] (NCA)
1101 15th St. NW
Washington, DC 20005

National Electrical Manufacturers Association[3] (NEMA)
2101 L St. NW
Washington, DC 20037

National Fire Protection Association[3] (NFPA)
1 Batterymarch Park
Quincy, MA 02269

National Forest Products Association[3] (NFPA)
1619 Massachusetts Ave., NW
Washington, DC 20036

National Fenestration and Rating Council[3] (NFRC)
952 Wayne Ave., Suite 750
Silver Spring, MD 20910

National Glass Association[3] (NGA)
8200 Greensboro Rd. Suite 302
McLean, VA 22102

National Institute of Building Sciences[3] (NIBS)
1015 15th St. NW
Washington, DC 20005

National Industrial Institute of Japan[2]
Nagoya, Japan

National Lime Association[3] (NLA)
3601 North Fairfax Dr.
Arlington, VA 22200

National Ready Mixed Concrete Association[3]
900 Spring St.
Silver Spring, MD 20910

National Research Council of Canada[2] (NRC)
Division of Building Research
Montreal Rd.
Ottawa, KIA OR6 Ontario
Canada

National Roofing Contractors Association[3] (NRCA)
300 W. Washington St.
Chicago, IL 60120

National Safety Council[3]
425 N. Michigan Ave.
Chicago, IL 60611

National Sand and Gravel Association[3]
900 Spring St
Silver Spring, MD 20910

National Society of Professional Engineers[3]
1420 King St.
Alexandria, VA 22314

National Starch & Chemical Co.[1]
10 Finderne Ave.
Bridgewater, NJ 08807

National Stone Association[3]
1415 Elliot Place NW
Washington, DC 20007

National Wood Window and Door Association[3] (NWWDA)
1400 E Toughy Ave. Suite G 54
Des Plaines, Il 60018

Neville Chemical Company[1]
2800 Neville Rd.
Pittsburgh, PA 15225-1496

North American Association of Mirror Manufacturers[3]
9005 Congressional Ct.
Potomac, VA 20854

Norwegian Building Research Institute[2] (NBRI)
Hogskoleringen 7
Trondheim, Norway

NuSil Technology[1]
1050 Cindy La.
Carpenteria, CA 93013

Occupational Safety and Health Administration[2] (OSHA)
United States Department of Labor
200 Constitution Ave.
Washington, DC 20210

OSI Specialties, Inc.[1]
PO Box 7429
Endicott, NY 13761-7429

Polymeric Systems, Inc.[1]
723 Wheatland St.
Pheonixville, PA 10460

Portland Cement Association[3] (PCA)
5420 Old Orchard Rd.
Skokie, IL 60077

Polygem, Inc.[1]
1105 Carolina Drive
West Chicago, Il 60185

Polyurethane Manufacturers Association[3]
Bldg. C, Suite 20
800 Roosevelt Rd.
Glen Ellyn, IL 60137-1083

Post-Tensioning Institute[3]
301 West Osburn, Suite 301
Phoenix, AZ 85103

Primary Glass Manufacturers Association[3]
3310 SW Harrison St.
Topeka, KS 66611-2279

Professional Women in Construction[3]
342 Madison Ave.
New York, NY 10173

Prestressed Concrete Institute[3]
201 N. Wells St.
Chicago, Il 60606

Reichhold, Adhesive Raw Materials[1]
PO Box 13582
Research Triangle Park, NC 27709-3582

Rohm & Haas Co[1]
100 Independence Mall, W.
Philadelphia, PA 19106

HM Royal[1]
689 Pennington Rd.
Trenton, NJ 08601

Sartomer Company, Inc.[1]
468 Thomas Jones Way
Exton, PA 19311

Schnee-Morehaed, Inc.[1]
111 N. Nursery Rd.
Irving, TX 75060

Sealant, Waterproofing and Restoration Institute[3] (SWRI)
3101 Broadway, Suite 585
Kansas City, MO 64111

Sealed Insulating Glass Manufacturers Association[3] (SIGMA)
401 N. Michigan Ave.
Chicago, Il 60611-4264

Sherman Williams , Inc.[1]
601 Canal Rd.
Cleveland, OH 44113

Simpson-Strong-Tie Co., Inc.[3]
4637 Chabot Dr., Suite 200
Pleasanton, CA 94588

Sika Corporation[1]
PO Box 300680
4800 Blue Parkway
Kansas City, MO 64130-2880

Society of Plastics Industry[3] (SPI)
355 Lexington Ave.
New York, NY 10017

Southern Building Code Congress International[3] (SBCCI)
900 Montclair Rd.
Birmingham, AL 35213

Southwestern Labs/ Huntington[5]
2200 Gravel Dr.
Ft. Worth TX 76118-7755

Steel Joist Institute[3]
1703 Parham Rd.
Richmond, VA 23229

Superior Epoxies,& Coatings, Inc.[1]
2511 Lantrac Court
Decatur, GA 30035

Sun Oil Co.[1]
1608 Walnut Sy
Philadelphia, PA 19103

Texaco Chemical Co.[1]
Austin TX 78761

Tile Council of America[3]
Box 326, Route 1
Princeton, NJ 08542

Union Carbide Plastic Co.[1]
39 Old Ridgebury Rd., Section N1
Danbury, CT 06817-0001

United States Army Construction[2]
Engineering Research Laboratory
PO Box 4005
Champaign, IL 61820

USDA Forest Service[2]
Southeastern Forest Experimental Station
Asheville, NC

Underwriters Laboratory[5] (UL)
333 Pfingsten Rd.
Northbrook, Il 60062

United States Department of Agriculture[2] (USDA)
Forest Products Laboratory
PO box 5130
Madison, WI 53705

United States Department of Commerce[2]
Washington, DC 20230

US Glass, Metal Glazing and Glass and Windows[Ref. 23.3]
385 Garrisonville Rd Suite 116
Stafford VA 22554

Vanchem, Inc.[1]
One North Transit Rd.
Lockport, NY 14094

RT Vanderbilt Company, Inc.[1]
30 Winfield St.
Norwalk, CT 06855

Warnock Hersey International, Inc.[5]
530 Garcia Ave.
Pittsburg, CA 94565

Witco[1]
OrganoSilicones Group
One American lane
Greenwich, CT 06831-2559

Wood & Fiber Magazine[4]
Forest Products Center
Virginia Polytechnic Institute
Blacksburg, VA 24061-0503

Wyandote Chemicals Corporation[1]
Market Development, Research Division
Wyandote MI

NOTES

1 — Supplier
2 — Government agencies, U.S. and foreign
3 — Trade organizations
4 — Publications
5 — Testing laboratories

CHAPTER 24
GLOSSARY OF TERMS[1234567]

Å Abbreviation for Angstrom Unit.
a The symbol for a repeating unit in a polymer chain.
ABA copolymers Block copolymers with three sequences, but only two domains.
abherent (adhesive) A coating or film applied to one surface to prevent or reduce its adhesion to another surface whith which it has been brought into intimate contact. Abherents applied to plastic films are often called anti-blocking agents.
abhesion A material that resists adhesion. A film or coating applied to surfaces to prevent sticking, heat sealing, and so on, such as a parting agent or mold release agent.
ablate Removal of a material by melting or burning away, or both.
ablation The degradation and decomposition of a material caused by high heat friction.
ablative A self regulating heat and mass transfer process in which incident thermal energy is expended by sacrificial loss of material. Materials that also provide fire resistance by gradually eroding to the flame front at a known or predictable rate.
ablative plastic A term applied to any polymer or resin which decomposes layer by layer when its surface is heated, leaving a heat resisting charred material which eventually breaks away to expose the virgin material.
abrasion The wearing away of a material surface by friction. A combined cutting, shearing, and tearing action detaches particles. Important factor in tire treads, soles, and conveyer belts.
abrasion cycle The number of repetitive abrading motions to which a specimen is subjected in an abrasion resistance test.

abrasion index   A value expressing abrasive resistance.
abrasion resistance   The ability of a material to withstand mechanical action such as rubbing, scraping, or erosion, that tends progressively to remove material from its surface.
abrasive tester   A machine for determining abrasion loss quantitatively.
ABS   Acronym for *acrylonitrile-butadiene-styrene*. A combustible thermoplastic resin used in the manufacture of certain non-metallic pipes used primarily for DWV applications. They are rigid, hard, and tough but not brittle, and possess good impact strength, heat resistance, low temperature properties, chemical resistance and electrical properties.
absolute humidity   The weight of water vapor present in a unit of air, such as grams per cubic foot, or grams per cubic meter.
absolute specific gravity   The ratio of weight of a given volume of a substance to that of an equal volume of water at the same temperature, as determined by an apparatus that provides correction for the effects of air buoyancy. See *specific gravity*.
absolute units   A system of units based on the smallest possible number of independent units. Specifically, units of force, work, energy and power not derived from, or dependent on, gravity.
absolute zero   The temperature at which a gas would show no pressure if the general law for gases would hold for all temperatures. It is equal to $-459.72°$ F ($-273.18°$ C).
absorption   1. The dissipation of sound energy by heat and other forms of energy loss. 2. Penetration of a substance into the body of another. 3. Transformation into other forms suffered by radiant energy passing through a material substance.
absorption factor   The ratio of the intensity loss by absorption to the total original intensity of radiation. If I_o represents the original intensity, I_r, the intensity of reflected radiation, I_t, the intensity of the transmitted radiation, the absorption factor is given by the expression $[I_o - (I_r 1 I_t)]/I_o$
Also called *coefficient of absorption*.
absorption, index of k' is given by the relation $k = (4\pi (k' n)/ \lambda$ (where n is the index of refraction and λ (the wavelength in vacuo. The **mass absorption** is given by k/d when d is the density. I/I_o gives the transmission factor.
absorption, Lambert's law   If I_o is the original intensity after through a thickness x of a material whose absorption coefficient is k, then $I_o = I_{oe}{}^{-kx}$
absorption spectrum   The spectrum obtained by the examination of light from a source, itself giving a continuous spectrum, after this light has passed through an absorbing medium in the gaseous state. The absorption spectrum will consist of dark lines or bands, being the reverse of the emission spectrum of the absorbing substance. When the absorbing medium is in the solid or liquid state, the spectrum of the transmitted state shows broad regions that are not resolvable.
accelerate   To hasten or quicken the natural progress of a reaction or event. For example, the drying rate of an adhesive or sealer is hastened or accelerated by increasing the temperature. An accelerated test is usually a severe test that

determines the comparative durability in a shorter length of time than required under service conditions

accelerated aging Any set of test conditions designed to introduce in a short time the result obtained under normal conditions of aging. In accelerated aging tests, the usual factors considered are heat, light, and oxygen, either separately or combined. Sometimes called accelerated life, it is most often accomplished by heating samples in an atmosphere of oxygen at 300 lbs per square inch pressure and 70° F (22° C), (Biermer-Davis), or by heating in an oven provided with circulating air, maintained at 158° F to 212° F (70° C to 100° C), (Gear).

accelerated weathering Machine-made means of duplicating or reproducing weather conditions. Such tests are particularly useful in comparing a series of products at the same time. No real correlation between test data and actual service is known for resins and rubbers used in many products.

acceleration The change of velocity. Cgs unit—centimeter per second, per second (cm/sec/sec). Dimensions—[lt'^2].

acceleration loss Energy used in an air-moving system to accelerate the air to the required velocity.

accelerator An agent used to hasten a reaction, to reduce the curing or hardening time of a thermosetting resin by entering into a reaction.

accelerator (concrete) An admixture that speeds the rate of hydration of cement, shortens the normal time of setting, or increases the rate of hardening, strength development, or both, of concrete, mortar or grout.

accelerator (of vulcanization) Any substance that hastens the vulcanization of rubber, causing it to take place in a shorter time or at lower temperature. In the past, basic oxides such as lime, litharge, and magnesia, were recognized as having this function. The important accelerators now are organic substances containing either nitrogen or sulfur or both. According to potency, or speed of action, accelerators are sometimes classified as slow, medium, rapid, semi-ultra, and ultra-accelerators. Most accelerators enhance tensile properties, and many improve age resistance.

accessory materials Filler boards, bond breakers, and backup materials and primers.

acetate (1) A derivative of acetic acid. (2) A generic name for cellulose acetate plastics, particularly for fibers thereof, where at least 92 percent of the hydroxyl groups are acetylated. The term triacetate can be used as the generic name of the fiber.

acetone Dimethyl ketone, CH_3-CO-CH_3. Used to extract most of the non-rubber constituents in natural rubber and the free sulfur in vulcanizates.

acetyl plastics Plastics based on polymers having a predominance of acetyl linkages in the main chain.

achromatic A term applied to lenses signifying their more or less complete correction for chromatic aberration.

acid Substance whose molecules ionize in water solutions to produce the hydrogen ion from their constituent elements. The strength of an acid is proportional to the concentration of hydrogen ions present.

acid- and alkali-resistant grout   Grout that resists the effects of prolonged contact with acids and alkalis.
acid cleaning   Washing concrete with a 5 to 10 percent solution of muriatic acid. This can be done after the cement has cured for at least two weeks.
acidity   The property of being acid or containing ionizable hydrogen. In more common language, the property of reddening litmus, or of combining with bases to form salts. The terms are also used with reference to the quantity of acid substances present in any materials, for example, naturally occurring organic acids in crude rubber.
acid polishing   The polishing of a glass surface by acid treatment.
acid rain   Rainfall that becomes acidic because of air pollution.
acoustic   Pertaining to hearing or sound.
acoustical double glazing   Two monolithic glass panels, set in a frame, with an air space between them usually larger than one inch and generally not hermetically sealed.
acrylate resins   Polymerization products of certain esters of acrylic and methacrylic acid as methyl or ethyl acrylate. They possess great optical clarity and a high degree of light transmission, the nearest approach to an organic glass.
acrylic   A group of thermoplastic resins or polymers formed by polymerizing the esters of acrylic acid.
acrylic latex   Latex-modified acrylic caulk.
acrylic resin synthetic   Resin used as a bonding agent or sealer, principally in concrete construction.
acrylic rubber (AR)   A synthetic rubber at least partially made from acrylonitriles. Or made from ethyl acrylate copolymerized with many of the monomers or block polymers of the synthetic rubber family.
acrylonitrile   A monomer within the structure (CH_2-CH-CN).
acrylonitrile-butadiene-styrene (ABS)   This three-way copolymer includes a family of tough, hard, chemically resistant resins with many grades and varieties, depending on variations in constituents. See *ABS*.
activation   (1) The process of inducing radioactivity in a specimen by bombardment with neutrons or other types of radiation. (2) The process of rendering a thermoplastic surface more receptive to adhesives by means of chemical treatments, corona discharge, or flame treatment.
activator   A substance that, by chemical interaction, promotes a specific chemical action of a second substance. Most accelerators require activators to bring out their full effect in vulcanization (e.g., zinc oxide or other metallic oxides). Some accelerators require a fatty acid, especially with zinc oxide. In the case of two-part systems, activators will speed-up or initiate the curing mechanism.
active mass   Of a substance is the number of gram molecular weights per liter in solution, or in gaseous form.
active solar heat gain   Solar heat that passes through a material and is captured by mechanical means.

addition   A substance that is inter-ground or blended in limited amounts into a hydraulic cement during manufacture—not on the job-site—either as a processing addition to aid in manufacturing and handling the cement, or as a functional addition to modify the use properties of the cement.

addition polymerization   A chemical reaction wherein molecules combine through interaction of unsaturated groups without splitting off by-products. The monomer can, by self-adaptation, produce the polymer. See *polymerization*.

additives   Components added to a formulation to achieve a specific polymer (or polymer mixture) and to modify the properties for the polyurethane. Also called auxiliary components.

adduct   A chemical addition product.

adhere   That property of a sealant compound that measures its ability to bond to the surface to which it is applied.

adherence numbers   The adherence number is a measure of the adhesion properties of solid particles. Experimentally, it is obtained by allowing a heterodisperse mixture of particles to settle in a tube, on a quartz plate. On inversion of the system, some of the particles will fall from the plate and the residual number adhering to the plate is assessed by microscopic observation.

adherend   A body that is held to another by an adhesive.

adhesion   The clinging or sticking of two material surfaces to each other. In rubber parlance, the strength of the bond or union between two rubber surfaces or plies, cured or uncured. The bond between a cured rubber surface and non-rubber surface, e.g., glass, metal, wood, or fabric.

adhesion [1]   A property of matter by which close contact is established between two or more surfaces when they are brought together intimately; they take up an equilibrium position in which the mutual potential energy is minimum. The forces acting at the interfaces are unsaturated and electrical in origin, and tend to bind the surfaces together so that force is required to separate them. For this reason we speak of the energy of adhesion, which, for two liquids in contact, depends on the surface and interfacial tensions; for liquids on solids, the surface free energies, as well as the equilibrium contact angle, define the work involved.

adhesion failure   Failure of a compound by pulling away from the surface with which it is in contact. See *cohesive failure*.

adhesion-in-peel   Force required to peel a sealant, caulk, adhesive, or coating from the surface.

adhesion, mechanical   Adhesion due to the physical interlocking of the adhesive with the base irregularities.

adhesion of rubber to metal   The strength of a bond formed between a metal surface and natural or synthetic rubber must be known regardless of the method used in obtaining the bond. Adhesion is expressed quantitatively as the tension per unit area required to cause a rupture of the rubber-to-metal bond (ASTM D 429).

adhesion peel-back test The separation of a bond, whereby the material is pulled away from the matting surface at a 90° angle or a 180° angle to the plane to which it is adhered. Values are generally expressed in pounds-per-inch width and as to whether the failure was adhesive or cohesive.

adhesion specific Adhesion due to valence forces at the adhesive-base surface interface. Such valence or attraction forces are of the same type that gives rise to cohesion.

adhesive Substance capable of holding materials together by surface attachment. This is a general term that includes cement, glue, mucilage, paste, etc. Various descriptive adjectives are used with the term *adhesive,* to indicate different types, such as:

Physical form: Liquid adhesive, tape adhesive, etc.
Composition: Resin adhesive, Rubber adhesive, etc.
End use: Paper adhesive, label adhesive, etc.
Application: Sprayable adhesive, hot melt adhesive, etc.

Also see *dry seal adhesive*; *pressure sensitive adhesive.*

adhesive, assembly An adhesive that can be used to bond parts together, such as in the manufacture of boats, concrete to concrete, and so forth.

adhesive, cold setting An adhesive that sets at temperatures below 68° F (20° C).

adhesive, dispersion A two-phase system in which one phase is suspended in liquid.

adhesive, hot setting An adhesive that requires a temperature at or above 212° F (100° C) to set it.

adhesive, intermediate temperature setting An adhesive that sets in the temperature range of 87°-211° F (31°-99° C).

adhesive, organic (mastic) A prepared organic material, for interior use only, that is ready to use with no further addition of liquid or powder, and which cures or sets by evaporation.

adhesive, pressure sensitive An adhesive made so as to adhere to a surface at room temperature by briefly applied pressure alone. Adhesives include:

Contact An adhesive that forms a strong, instantaneous bond between two plies of a substrate when the surfaces are brought together.
Laminating An adhesive that forms a slowly developing bond between two plies of gypsum board, or between masonry or concrete and plasterboard.
Stud An adhesive suitable for attaching wood or plaster building boards to studs.

adhesive setting Classifies the conditions to convert the adhesive from its packaged state to a more useful form. Chemically setting: Requires addition of accelerator or catalyst to cure.

Cold setting: Sets at temperatures below 68° F (120° C).
Hot setting: Sets at temperatures above 212° F (100° C).

Intermediate temperature setting: Sets in the temperature range of 87°-212° F (31°-99° C).

Room-temperature setting: Sets in the temperature range of 68°-86° F (20°-30° C).

adhesive spreader A notched trowel or special tool that aids in the application of adhesives.

adiabatic A body is said to undergo an adiabatic change when its condition is altered without gain or loss of heat. The line on the pressure volume diagram representing the change is called an *adiabatic* line.

adiabatic extrusion A method of extrusion in which the sole source of heat is the conversion of the drive energy, through viscous resistance of the plastics mass in the extruder.

admixture A material, other than water, aggregate, or cement, used as an ingredient in concrete, mortar, or plaster, and added to the mix either immediately before, or during, the mixing.

adobe Unfired clay bricks that are dried in the sun. A common building material in the Southwest and closely simulated by a concrete masonry unit called a slump block.

adsorbent A highly porous solid that has the ability to concentrate and hold gases and vapors in contact with a solid. This includes moisture as well as many other organic and inorganic molecules.

adsorption [1] The action of a body in condensing and holding gases, dyes, or other substances. The action usually is considered to take place only at or near the surface. Power-of-adsorption is one of the characteristic properties of matter in the colloidal state and is associated with the surface energy phenomena of colloidally dispersed particles.

aerosol A suspension of liquid or solid particles in a gas under pressure. Used in the packaging industry where the propellant is supplied by a liquefied gas, such as Freon.

aesthetic Pleasing to the eye.

affinity With respect to an adhesive, affinity is an attraction or polar similarity between the adhesive and an adherend.

affix To attach physically.

age resistance Resistance to aging by oxygen and ozone in the air, by heat and light. Antioxidants help, although there is no non-toxic substance against ozone for natural rubber, GR-S neoprene, and nitrile rubbers.

aggregate finish The various sizes of stone, rock, pebbles, etc., used in the make-up of a panel that would be placed on the facade of the building. Not a smooth finish, the aggregate may be bonded with concrete or mortar to give the desired or designated matrix.

aging A progressive change in the chemical and physical properties of rubber, especially vulcanized rubber, usually marked by deterioration. With natural rubber it is due primarily to oxidation or reversion or both; with GR-S, it is primarily to continued co-polymerization. May be retarded by the use of anti-oxidants. (See *shelf aging*; *accelerated aging*).

aging tests   Accelerated tests of rubber specimens to find out their endurance by heating them in air, under pressure, or similarly in oxygen.

agitating truck   Another term for a concrete truck. It is usually a truck with a rotating drum that continuously agitates fresh concrete during transit from the ready-mix plant to the work-site.

air content   The volume of air voids in cement paste, concrete, or mortar. Entrained air adds to the durability of hardened concrete and the workability of fresh concrete and mortar mixes.

air curing   Vulcanization at ordinary room temperatures, or without the aid of heat.

air drying   A material is said to be air drying when it can be dried at ordinary room temperature, without the use of artificial heat.

air-entraining agent   An admixture for concrete or mortar that causes the formation of air bubbles in the mix in order to improve workability and frost resistance.

air entrainment   The introduction of air in the form of minute bubbles (usually smaller than 1 millimeter in diameter) into the concrete or mortar during the mixing to improve the flow and workability of fresh mixes and the durability of the hardened concrete.

air infiltration   The amount of air that passes between a window sash and frame, or a door panel and frame. For windows it is measured in terms of cubic feet of air per minute/per square foot, and for doors it is measured in terms of cubic feet of air per minute/per foot of crack.

air-knife coating   A knife-coating technique especially suitable for thin coatings such as adhesives, wherein a high pressure jet of air is forced through orifices in the knife to meter or control the thickness of the adhesive coating.

air locks   Surface depressions on a molded part caused by trapped air between the mold surface and the plastic material.

air, saturated   Air that is fully saturated with water vapor (100% humidity) with the air and water vapor at the same temperature.

air side   In the float process, the upper side is called the air side.

air, standard   Air at 72° F (21° C) and standard atmospheric pressure of 29.92 in. (101.3 kPa) of mercury, and weighing about 0.075 lb/ft^3 (1.20 kg/m^3).

air-void   An entrapped or entrained air pocket in concrete or mortar. Entrapped voids usually are larger than 1 millimeter in diameter; entrained air voids are smaller. Entrapped air voids should be removed with vibration, power screeding or rodding.

alcohol   A molecule containing a hydroxyl group, which is an oxygen (O), and a hydrogen (H) reacting as a unit (-OH). It is in the *B* side of a (poly) urethane reaction.

aliphatic   Organic compounds (hydrocarbons) in which carbon atoms are arranged in an open or straight chain. More commonly called napthas, they are prepared by straight-run, overhead distillation of petroleum. Familiar examples include gasoline, kerosene, paraffin, and natural gas. Aliphatic solvents are generally confined to reclaimed, natural GR-S, and butyl rubber formulas. Of the common solvents they are about the lowest in price and the least toxic.

aliphatic glue   A yellow wood glue that is strong, but not waterproof.

aliphatic isocyanate   One of the two types of isocyanate. Does not contain a benzene ring. Used in light-stable weatherable polyurethanes.

alkali   Substance that neutralizes acids to form salt and water. Yields hydroxyl (OH-) ions in water solutions. Proton acceptor. Turns red litmus paper blue.

alkyd plastics   Plastics based on resins composed principally of polymeric esters, in which the recurring ester groups are an integral part of the main polymer chain and in which ester groups occur in most cross-links that may be present between chains.

alkyd resin   A polyester convertible into a cross-linked form, requiring a reactant of functionality higher than two, or having double bonds.

alkyl epoxy stearate   A plasticizer for PVC and some other resins, with good low temperature properties. Also acts as a stabilizer.

alligatoring   Term describing the appearance of a film that is cracked into large segments resembling the hide of an alligator. When alligatoring is fine and incomplete, it is usually called *cracking*. Alligatoring may be caused by one coat applied over another before the bottom coat is thoroughly dry and hard and/or having the material skinning so that the lower portion of the film is still soft and elastic, or by a less elastic material being applied over a more elastic undercoating. When these conditions are present, and the finished article is exposed to actinic rays or changes in temperature and moisture content, expansion and contraction of the film cracks the hard crust while the softer core gives without breaking. With excessively heavy coats of dilute materials, this cracking of the outer crust can take place without temperature change by the shrinkage action of the bottom portion—much like clay mud is cracked under the summer sun. Other causes of alligatoring include too-rapid evaporation of solvents or thinners and excessive air being forced into the film during spraying.

alloy   A term sometimes used in the plastics or adhesive industry to denote blends of polymers or copolymers with other polymers or elastomers.

allyl resin   A thermosetting resin, one that can be heated only once and then cooled into a temperature-resistant solid. Used in laminate adhesives.

ambient noise   The all-encompassing noise associated with a given environment, being usually a composite of sounds from many sources near and far.

ambient temperature   The environment temperature in general degrees surrounding the object under construction.

amide   A curing agent for epoxy resins. An organic compound containing the -$CONH_2$ group. It is closely related to the organic acids with COOH grouping, and a common example is urea; $CO(NH_2)_2$.

amine   An organic compound containing nitrogen, hydrogen, or complex groups. Often used in polyurethane formulations as chain extenders or catalysts.

amine catalysts   Chemicals that accelerate an isocyanate reaction.

amine extender   An addition to the *B* side of a polyurethane formulation that extends the polymerization and improves certain properties, such as hardening and resistance to heat and solvents.

amine group   A functional group, reacting as one unit, that contains one nitrogen (N) and two hydrogens (H_2).

amino plastics   Plastics based on amino resins.

amino resins   A resin made by poly-condensation of a compound containing amino-groups such as urea or melamine, with an aldehyde, such as formaldehyde, or an aldehyde yielding material.

amplitude   The maximum value of displacement in an oscillatory motion.

anaerobic   Living without air or free oxygen.

anchor bolt   A headed or threaded metal bolt or stud that is either cast-in-place, grouted in place, or cemented into a drilled hole, and used to attach steel or wooden structural members to concrete.

angle   The ratio between the arc and the radius of the arc. Units of angle—the radian, the angle subtended by an arc equal to the radius; the degree, 1/360 part of a circumference.

Angstrom unit (Å)   A unit of length. Originally and for practical purposes, an Angstrom is one hundred millionth (10^{-8}) of a centimeter. It is now defined more exactly in terms of the wave length of the red line of cadmium (6438.4696 Å).

angular dimensions [t '1]   If the angle described in time t is θ (, the angular velocity, ω (= θ (/t, where θ (in radians and t in seconds gives ω (in radians per second.

angular velocity   The time of angular motion about an axis. Cgs unit—one radian per second.

anhydride (of acid or base)   An oxide, which when combined with water gives an acid or base.

anhydrous   Having no water content.

anion   A negatively charged ion.

annealing   The process that prevents glass from shattering after it has been formed. The outer surfaces of the glass shrink faster than the glass between the surfaces, causing strain, which can lead to shattering. By reheating the glass and allowing it to cool slowly, this can be avoided.

annulus   The opening around the penetrant. For pipe penetrations, the ring around the pipe where the fill material is applied.

ANSI (American National Standard Institute)   A nonprofit group that tests building materials with the American Society of Testing and Materials (ASTM).

anticatalyst   An inhibitor to a chemical reaction; the opposite of a catalyst, which encourages a chemical reaction.

anticracking agents   Softening agents are usually anticracking agents. Some antioxidants work well also. Chiefly a good vulcanizate resists cracking, high tensile, high tear resistance, high resilience, and good resistance to heat aging. Antifoaming agent product that greatly increases the surface tension, thereby reducing the tendency to foam during mixing or application.

antifouling   A material or substance applied to objects used under water, to prevent marine growth.
antigelling agent   An additive that prevents a solution from forming a gel.
antimony trioxide $Sb_2\text{-}O_3$ A white powder widely used as a flame retardant in sealants, adhesives and plastics.
anti-walk blocks   Rubber blocks that prevent glass from moving sideways in the glazing rabbet from thermal effects or vibration.
antioxidant   Compounding ingredient used to retard deterioration caused by oxidation.
APA (American Plywood Association)   An association of plywood manufacturers that, among other things, sets the grading standards for plywood.
application   The act of going over the area to be coated or sealed one time. For example, when the operator spray-coats the entire surface without repetition, he has made one application. If he immediately goes over the same work again, he has made another application. For adhesives and coating, the principal methods of application are brushing, dipping, stenciling, flowing, stamp-padding, roll-coating, knife-coating, squeegeeing, spatula, and notched trowel. For sealers, the principal methods of application are spatula, caulking gun, flow gun, pressure extrusion units, and spray gun.
application life   The period of time during which a sealant or caulk, after being mixed with a catalyst or exposed to the atmosphere, remains suitable for application.
applicator   Any device used to mechanically apply molten sealant that is capable of delivering the required volume at the specified application temperature.
applied skin   A thin surface layer of elastomeric material applied to a cellular product.
apron   Trim or facing on the side or front of a counter top, table edge or windowsill.
aqueous   Having water as part of its makeup.
Archimedes principle   A body wholly or partly immersed in a fluid is buoyed up by force equal to the weight of the fluid displaced. A body of volume $V\ cm^3$ immersed in a fluid of density ρ (grams per cm^3 is buoyed up by a force in dynes. $F = \rho\ (gV$, where g is the acceleration due to gravity. A floating body displaces its own weight of liquid.
architect   A person trained in building design and strength analysis, thereby being licensed to practice architecture.
architectural concrete   Concrete that is permanently exposed to view and requiring special care in selection, placing, and finishing in order to get the desired architectural effect.
arc resistance   Total time in seconds that an intermittent arc may play across a plastic surface without rendering the surface conductive.
area, unit of   The square centimeter. The area of a square whose sides are one centimeter in length. Other units of an area are similarly derived. Dimensions— $[l^2]$.
argon   An inert gas with an atomic symbol Ar, atomic number 18, and the atomic weight of 39.948.

aromatic compounds   Organic compounds that include a benzene ring.
aromatic hydrocarbon   A hydrocarbon with a molecular structure involving one or more rings of six carbon atoms, having properties similar to benzene, which is the simplest of the aromatic hydrocarbons.
aromatic isocyanates   One of the two major types of isocyanates, and by far the most widely used. So named because they contain benzene rings. Aromatic isocyanates include MDI and TDI.
art glass   Art glass goes by many names. It is called opalescent glass, cathedral glass or stained glass and is usually produced in small batch operations.
asbestos   A natural mineral fiber formerly used for fireproofing.
ashlar   Usually a term referring to square stones but also used as the name of a pattern in masonry construction.
A-side   The isocyanate side of a two-component polyurethane formulation. The pall portion is designated as the B side.
asphalt   Naturally occurring solid or semi-solid mineral pitch or bitumen, more or less soluble in carbon disulfide, naphtha, and turpentine, and fusible at varying temperatures. Gilsonite, glance-pitch, manjak, Grahamite, and Trinidad pitch are examples. Also bituminous residues left from the distillation of petroleum or coal tar.
assembly time, closed   The interval between completing the assembly and the application of pressure or heat or both.
assembly time, open   The maximum interval between spreading the adhesive on the adherent and completion of the assembly for bonding.
ASTM (American Society of Testing and Materials)   A nonprofit organization made up of a variety of professionals who decide the level of quality that building materials must have to be used on particular jobs.
atmospheric pressure   The pressure of air exerted equally in all directions. The standard pressure is that at which mercury stands at 14.7 lbs per square inch (760 mm).
atmospheric or ozone cracking   A fissured surface condition that develops on stretched rubber exposed to the atmosphere or to ozonized air. When the fissures are minute, the condition is called *checking*.
atomic number   The number of electrons in an atom; equal to the number of protons in the nucleus.
atomic weight   The sum of the number of protons and neutrons in the nucleus.
atoms   The smallest particles of an element that have all the properties of the element. They contain neutrons (with no electrical charge), protons (with a positive electrical charge), and electrons (with a negative electrical charge).
atrium   A glass enclosed room or courtyard, usually in the center of a building that is open to the sky.
attenuation   The reduction of sound pressure level, usually expressed in decibels.
auto-bonding   The ability of fresh material to adhere to previously installed (cured or dried) material of the same type.

autoclave A vessel that employs heat and pressure. In the glass industry, used to produce a bond between glass and PVB or urethane sheet, thus creating a laminated sheet product.

autogenous volume change Change in volume produced by continued hydration of cement, exclusive of the effects of external forces or change of water content or temperature.

average molecular weight (viscosity method) The molecular weight of polymeric materials determined by viscosity of the polymer in solution as a specific temperature. This gives an average molecular weight of the molecular chains in the polymer independent of the specific chain length. The value falls between weight average, number average, and molecular weight.

avoirdupois weight A weight measurement based on 1 pound equaling 16 ounces and an ounce equaling 16 drams.

awning window A type of window in which the sash is hinged at the top and swings open at the bottom.

back bedding See *bedding*.

backer rod foam A flexible, compressible strip of plastic foam inserted into a joint to limit the depth to which the sealant can penetrate. See *sealant backing*.

back fill (glass) Placing material into the opening between glass and glazing. (Concrete) Earth, rubble etc., used to correct over-excavation or to replace earth in a trench or around a foundation wall.

back-mounted Mounted tile with perforated paper, fiber mesh or other suitable bonding material applied to the backs or edges of the tile so that a relatively large proportion of the area is exposed to the setting.

back plastering Applying a backup coat (or coats) of plaster to the backside of a solid plaster partition after wall mortar or plaster on the opposite side has hardened.

backup A material such as cotton, mop yarn, glass-fiber insulation, oakum, polyethylene, etc., that is pressed into an opening or joint so that the applied sealant will exert pressure on, and form good contact against, the sides of the joint or opening.

bag A measure of cement equal to 94 pounds.

bag molding A method of molding or laminating that involves the application of fluid pressure; usually by means of air, steam, water, or vacuum, to a flexible material that transmits the pressure to the material being molded or bonded.

bait A webbed metal frame used to draw molten glass.

bake test An accelerated test wherein a product, such as a sealant, is subjected to predetermined conditions of elevated temperature and time intervals. The test serves as an indication of what may be expected in respect to certain characteristics when the product is subjected to extended periods of exposure under normal conditions.

bandage joint Sealant joint composed of bond-breaker tape over the joint movement area. With an overlay of sealant lapping either side of the tape sufficiently to bond well to the surfaces bandage joints are often used where extreme movement occurs and conventional joint design is not possible.

barrier zone A subsurface zone said to be made unwettable by treatment with chemical reaction, water-repellent, products.

base Any substance, organic or inorganic, having an alkaline character of the ability to combine with, or neutralize an, acid.

base coat Plaster coat or coats applied before the final coat.

batch The mixed raw materials that are used to make glass, concrete, mortar, adhesives or sealants.

batching Weighing or volumetrically measuring and introducing into the mixer the ingredients for a batch of concrete, mortar, sealant or adhesive, etc.

Baumé (Bé) A system of specific gravity units devised by the French chemist Antoine Baumé for the graduation of hydrometers. The relations to Sp.Gr. (at 60/60° F) are:

$°Bé = 145-145/Sp. Gr.$, for liquids heavier than water

$°Bé = 140/Sp. Gr. - 130$, for liquids lighter than water

bead A sealant compound after application in a joint. Also a molding or stop used to hold a glass product in position.

bed, or bedding The bead of compound applied between sight bar glass or panel and the stationary stop or sight bar of the sash or frame, and usually the first bead of compound to be applied when setting glass or panels.

bedding stop The application of compound at the base of channel, just before the stop is placed in position, or buttered on the inside face of the stop.

bed joint (glass) A caulking joint of glazing material having a sloped surface to drain water away from the glass or panel. (Concrete) A horizontal joint in the masonry wall.

bend test A means of testing the flexibility of a compound at a specified temperature. The compound is applied to metal, cured or dried, and, after conditioning at the specific temperature, bent over a mandrel to determine the product's resistance to rupture.

bent glass Flat glass that has been shaped while hot into cylindrical or other curved shapes.

benzene ring The basic structure of benzene, the most important aromatic chemical. It is an unsaturated, resonant, six-carbon ring having three double bonds. One or more of the six hydrogen atoms of the benzene may be replaced by other atoms or groups.

beryllium A hard, lightweight, gray metallic element; atomic number 4; atomic weight 9.0122; atomic symbol Be.

beta (1) A prefix, usually abbreviated as the Greek letter ß, denoting the location of a substituting group of atoms in the main group of a compound, or (2) a type of radiation.

beveling The process of edge finishing flat glass to a beveled angle.

bicottura An Italian expression signifying *twice-fried*.

binder (1) In a reinforced plastic, the continuous phase that holds together the reinforcement. (2) The agent applied to mats or preforms to bond the fibers prior to laminating or molding.

biphenol A $(CH_3)_2 \, C(C_6H_4OH)_2$ is an intermediate used in the production of epoxy, polycarbonate and phenolic resins.

biscuit A mass of clay from which a tile unit is pressed.

bite The dimension by which the edge of a glass product is engaged into the glazing channel.

bitumen Originally, mineral pitch or asphalt. Now, any of a number of flammable mineral substances, consisting mainly of hydrocarbons, including the hard, brittle varieties of asphalt, the semi-solid maltha and mineral tars, the oily petroleum, or even the volatile naphthas. Asphalt and coal tar are the two bitumens used in construction.

black (1) A term referring to the several types of carbon black utilized in the rubber industry, and to the absence of color. (2) Zero brilliance and 100 percent saturation of light waves.

bleed To seep out, as in a stain or pigment seeping to the surface.

bleeding (glazing) Migration to the surface of plasticizers, waxes or similar materials to form a film or beads. Exudation with possible absorption by porous surfaces of a component of a sealant. (Concrete) The flow of mixing water within, or its emergence on the surface of newly placed concrete, caused by the settlement of the solid materials within the mass. Also see *bloom*.

blend (1) A physical mixture of silica gel and molecular sieve. (2) To mix together.

blending A step in reclaiming in which the de-vulcanized stock is mixed with reinforcing and processing agents prior to refining to ensure uniform distribution of all particles.

blister An imperfection, a rounded elevation of the surface of a plastic, with boundaries that may be more or less sharply defined, somewhat resembling in shape a blister on the human skin.

block (1) Rectangular, cured sections of neoprene or other approved materials, used to position the glass product in the glazing channel. (2) A concrete masonry unit, usually containing hollow cores.

blocking An undesired adhesion between touching layers of a material, such as occurs under pressure during storage.

block polymer A large polymer formed when chains of homopolymers link in a pattern.

bloom A discoloration or change in appearance of the surface of a rubber product caused by the migration of a liquid or solid to the surface. Examples are sulfur bloom and wax bloom. Not to be confused with dust on the surface from external sources. See *efflorescence*.

blowing Porosity or sponginess occurring during cure. In latex goods, a permanent deformation caused when the deposit leaves the form during curing or drying.

blowing agent A compounding ingredient used to produce gas by chemical or thermal reaction, or both, in manufacture of hollow or cellular articles.

blow molding A method of glass fabrication in which a parison (hollow tube) is forced into the shape of the mold cavity by internal-air pressure.

blueprint A term used to denote a construction drawing. It is so named because the original common mode for reproduction of such drawings resulted in a blue drawing with white lines.

blush See *bloom*. Also whitish surface appearance where moisture has condensed before solvent is all evaporated.

body The structural portion of a ceramic product. It also refers to the mixture or mixture from which it is made. The viscosity of a liquid.

body putty A paste-like mixture of resin, often polyester, and a filler such as talc, used in the repair of metal surfaces such as auto bodies.

boiling point The temperature at which the vapor pressure of a liquid is equal to the pressure of the atmosphere.

bond (n.) (1) The attachment at the interface between an adhesive and an adherent. See also *adhesion, joint*. (2) A coat of finishing material used to improve the adherence of succeeding coats.

bond (v.) To join materials together with adhesives. To adhere.

bond breaker A material to prevent adhesion at a designated interface.

bond coat A material that is used between the back of the tile and the prepared surface.

bond face That part of the joint face to which a sealant is bonded.

bonding agents Substances or mixtures of substances that are used in attaching rubber to metal. Generally the rubber compound is vulcanized by heat in the process. Cyclized rubber or rubber isomers, halogenated rubber, rubber hydrochloride, the reaction products of natural rubber and acrylonitrile, and polymers containing diisocyanates, are also used. In concrete, a substance applied as a coating to a suitable substrate to create a bond between the substrate and the succeeding layer of concrete.

bond strength The force per unit area or length necessary to rupture a bond. See also *tensile test, shear test*, and *peel-back test*.

boric acid A compound used in making borosilicate glass.

bow A continuous curve of the sheet, either vertical or horizontal.

Boyer-Beaman Rule A statement of the relationship between the glass transition temperature T_g and the melting temperature T_m of a polymer. The ratio of T_g to T_m (with T expressed in degrees Kelvin) usually lies between 0.5 and 0.7. For symmetrical polymers such as polyethylene the ratio is close to 0.5. For unsymmetrical polymers such as polystyrene and polyisoprene it is approximately 0.7.

branched polymers Polymers formed when three molecular chains come together in a Y-shaped intersection.

breather tube A small diameter tube placed into the space of an insulating glass unit through the perimeter wall for the purpose of equalizing the air pressure within the unit when shipping into areas where the units will change elevation by 5000 feet. These tubes are to be sealed on the jobsite prior to unit installation.

brick A rectangular building unit, not less than 75 percent solid, made from burned clay, shale, or a mixture of these materials.

brick paver A relatively thick, heavy-duty, load-bearing vitreous ceramic tile, rarely tinted and never glazed; packinghouse tile.

bridge abutments A portion of the bridge construction designed to withstand a thrusting force. Abutments are at the ends of a bridge.

bridge joints A sealant joint composed of a bond-breaker tape over the joint movement area with an overlay of sealant lapping either side of the tape sufficiently to bond well. It is used where extreme movement occurs and conventional joint design is not possible, i.e., metal joints, deep V-joints.

British Thermal Unit (BTU) The amount of heat required to raise one pound of water 1° F.

brittle point The highest temperature at which an elastomer fractures in a prescribed impact test procedure.

brittle temperature The lowest temperature at which a plastic material withstands given conditions without failure.

bronze glass A glare or heat-reducing glass intended for applications where glare control and reduction of solar heat is desired or where color can contribute to design.

Brookfield synchro-lectric viscosimeter A portable form of the rotating cylinder apparatus for measuring viscosity.

Brookfield test A means of determining the viscosity of liquids or semi-pastes by measuring the drag produced upon a cylinder or disc rotated at a definite constant speed and at a specified temperature, while immersed in the material tested.

brown coat The second coat of three coat plastering.

B-side The pall side of a two-component polyurethane formulation. The isocyanate side is designated as the *A* side.

B-stage An intermediate stage in the reaction of certain thermosetting resins in which the materials swells when in contact with certain liquids and soften when heated, but may not entirely dissolve or fuse.

bubble A globule of air or other gas trapped within a plastic.

buck Framing used around an opening in a wall where a door or window will be placed.

buckling Crimping of the fibers in a composite material, often occurring in glass-reinforced thermosets due to resin shrinkage during cure.

bug holes Small holes in concrete caused by air bubbles trapped in the surface of formed concrete during placement and compaction.

building construction The process of assembling materials to form a building.

building design The process of providing all information necessary for the construction of a building, to meet its owner's requirements and to satisfy public health, welfare, and safety requirements.

bulk density The density of a molding material in loose form (granular, modular, etc.) expressed as a ratio of weight to volume.

bulk factor The ratio of the volume of loose molding compound to the volume of the molded part made from it.

bulking The increase in the volume of a quantity of sand in a moist condition compared to the volume of the same quantity of sand in a dry state.

bulking value Solid volume of a unit weight usually expressed as gallons-per-pound.

bullet resisting glass A multiple lamination of glass with a tough clear sheet of plastic usually at least $1\frac{3}{16}$ inch thick overall, which is designed to stop bullets from ordinary firearms other than high-powered rifles.

bull float A tool with a large flat, rectangular piece of aluminum, wood, or magnesium measuring 8 inches wide (42 to 60 inches long) with a 4 to 16 foot handle used to smooth concrete slabs.

bullnose Trim tile with a convex radius on the one side. It is used to finish the edge of a tile surface of the original plane, or to turn an outside corner.

bull's eye The round, whorl shape in the center of old panes of glass.

burned Showing evidence of thermal decomposition through some discoloration or destruction of the surface of the plastic. See also *discoloration*.

bush hammer A hammer with a serrated face to roughen a concrete surface.

buttering Application of putty or sealant compound to the flat surface of some member before placing the member in position, such as the buttering of a removable stop before fastening the stop in place. (Concrete) The process of spreading mortar on a masonry surface with a trowel.

butt glazing The installation of glass products where the vertical glass edges are without supporting mullions.

butt joint A joint having opposing faces, which may move toward or away from one another; a joint in which the receiving surfaces stress the sealant in tension or compression.

butylene plastics Plastics based on resins made by the polymerization of butene or copolymerization of butene with one or more unsaturated compounds; the butenes being in greatest amount by weight.

butyl rubber A copolymer of about 98 percent isobutylene and 2 percent isoprene. It has the poorest resistance to petroleum oils and gasolines of any rubber. Excellent resistance to vegetable and mineral oils, to solvents such as acetone, alcohol, phenol and ethylene glycol, and to water and gas absorption. Heat resistance is above average. Sunlight resistance is excellent.

CADD See *computer aided design and drafting*.

calcination The process of breaking down chemical compounds by heat.

calcium A white metallic element that forms part of limestone, chalk and gypsum; atomic number 20; atomic weight 40.08; atomic symbol Ca.

calcium chloride A compound mixed with concrete or mortar to decrease the time it takes to set.

calcium oxide A compound used in glassmaking that comes from lime.

calender To prepare sheets of material by pressure between two or more revolving rollers. Used in connection with preparation of films and coating of materials.

caliper The thickness of tape, measured as the perpendicular distance between the opposite surfaces and expressed in English units (thousandths of an inch) or metric units (millimeters).

camber A slight upward curvature intentionally built into a structural element or form to improve appearance by offsetting the deflection of the element.

cant To place something at an angle.

capillary action Movement of liquid through the adhesion of small-diameter capillary pore structures of stone, concrete, or brick.

capillary tube units An insulating glass unit where a small metal tube of specific length and inside diameter is factory-placed into the unit's spacer to accommodate both the pressure differences to the point of installation, and the pressure differences encountered daily after installation.

capped pall Refers to a pall in which the terminal end groups are different from those making up the bulk of the polyol. For example, a PO-based EO-capped pall.

carbon An element that combines with other elements, and exists in pure form as graphite or diamond; atomic number 6; atomic weight 12.011; atomic symbol C.

carcinogen A substance known to cause cancer.

carnival glass An iridescent coloration obtained by firing metallic salts applied onto a glass body.

casement window A window that has side-hinged sashes that swing open.

cast To form a plastic into a definite shape by pouring it into a mold and letting it harden without applying external pressure. Can be accomplished with or without application of external heat either before or after pouring.

cast film A film made by depositing a layer of plastic, either molten in solution, or in a dispersion, onto a surface, solidifying and removing the film from the surface.

cast-in-place Concrete or mortar that is deposited in the place where it will harden as part of a structure (opposite of pre-cast).

Castor-Severs Test Performed with Castor-Severs Rheometer. Measures the rate of flow of a compound under controlled conditions of temperature, pressure, and orifice size. Of value for the measurement of the internal cohesive force of pumpable sealants.

catalyst Substance that markedly speeds up the cure of an adhesive when added in minor quantity as compared to the amounts of primary reactants.

caulk (1) See *sealant*. The term caulk has traditionally referred to non-elastomeric sealant compounds used where little or no movement capability is required (usually less than 10 percent); caulking compounds are a type of sealant. (2) To install or apply a sealant across, or into a point, crack or crevice.

caulking The process of sealing a joint or the materials used to. Most often refers to linseed oil and lead compounds and to cotton or oakum strands used in back of seams rather than to the more recently developed sealing materials.

caulking cord Flat, soft, belt-like band segmented into six beads of caulk of various widths. This material is used for setting sink drains (it is less messy than plumbers' compound) and for permanent and temporary caulking.
caulking gun A tool that holds a cartridge of caulk and has either an air or mechanical trigger plunger that forces the caulk from the cartridge.
caulking iron A tool for caulking lead or oakum into joints.
cell A small partially or completely enclosed cavity.
cell, open A cell not totally enclosed by its walls, and hence interconnected with other cells.
cellular plastics Plastics containing numerous small cavities (cells), interconnected or not, distributed throughout the mass.
cellulose acetate butyrate The butyrate copolymer is softer and more flexible than cellulose acetate and requires less plasticizer to achieve a given degree of softness. It is made in clear film or a molding powder.
cellulose nitrate One of the toughest of plastics. It is widely used in tool handles etc., requiring high impact strength.
Celsius The centigrade temperature scale, based on the freezing and boiling points of pure water. See *centigrade*.
Cement The dispersion or solution of unvulcanized rubber compound in a suitable solvent such as petroleum naphtha or aromatic hydrocarbons. Cements are also made from latex or water dispersions with or without the addition of organic solvents. See *adhesive*.
cement, masonry A cement for use in plaster or mortars containing one or more of the following materials: portland cement, blended cement, natural cement, slag cement, or hydraulic lime, and usually containing one or more materials such as hydrated lime, limestone, or chalk, as prepared for this purpose.
cementitious Made from, or composed of, portland cement.
centigrade A temperature scale with 100 divisions between the freezing point of water at 0 degrees and the boiling point of water at 100 degrees. Temperatures using the centigrade scale are indicated by the letter C, as in 30° C.
centimeter One one-hundredth (0.01) of a meter in the metric measurement system, and equal to 0.3937 inch.
centipoise $1/100$ of a poise, which is a value for viscosity. The viscosity of water at 64° F (20° C) is approximately 1 centipoise.

$$\text{Millipoise} = 1/1000 \text{ poise}$$

$$\text{Mullipoise} = \frac{\text{centist}}{10dt}$$

where dt = density of a substance at the same temperature t.

centistoke $1/100$ of a stoke, which is the unit of kinematic viscosity.
centrifugal Moving outward from the axis of rotation.

centrifuge A machine that employs centrifugal force and the difference in densities of materials to separate those materials.

ceramic tile A ceramic surfacing unit, usually relatively thin in relation to facial area, made from clay or a mixture of clay and other ceramic material. Has glazed or unglazed face, and is fired above red heat in the course of manufacture to a temperature sufficiently high to produce specific physical properties and characteristics.

certified IG unit An insulating glass unit constructed like a test model, which has successfully passed ASTM E 773 and ASTM E 774 tests of insulating seal durability performance at specific levels (pursuant to the administrative guidelines of a certification program).

chain alignment A factor determining the physical properties of a polymer and the end product of a polyurethane. Closely aligned chains produce high density, stiff polymers

chain extender A type of short chain diol or diamine that links a portion of the urethane molecule together and gives rigidity to the polymer.

chain shape A factor determining the physical properties of a polymer and the end product of a polyurethane. Small monomers that don't align with themselves make flexible polymers; bulky monomers make more rigid polymers.

chain stopper A material which, when added during the polymerization process, will terminate or stop the molecules from continued growth to still longer lengths.

chalk Soft limestone. Calcium carbonate.

chalking Formation of a powdery surface condition caused by disintegration of the surface binder or elastomer due to weathering or other destructive environmental factors.

channel A three-sided U-shaped opening in sash or frame to receive light or panel, with or without removable stop or stops. Contrasted to a rabbet, which is a two-sided, L-shaped section, as with face-glazed window sash.

channel black A form of carbon black made from a natural gas by the channel combustion process. The gas is burned with insufficient air in the jets, the flames from which are allowed to impinge on a cool metallic surface (channel) from which the deposited carbon is scraped. For many years it was the best reinforcing agent for natural rubber and made it highly abrasion-resistant.

channel depth The measurement from the bottom of the channel to the top of the stop, or the measurement from the sight-line to the base of channel.

channel glazing The sealing of the joints around lights or panels set in a U-shaped channel employing removable stops.

channel width The measurement between stationary stops (or stationary stop and removable stop) in a U-shaped channel.

charge The measurement or weight of material either solid or liquid, preformed or in powder form used to load a mold.

checking Development of shallow cracks at closely spaced but irregular intervals in concrete, plaster or mortar surfaces; also known as *crazing*. Checking may be described as *visible* (as seen by the naked eye) or as *microscopic* (as seen under magnification of 10 diameters).

checking sunlight The development of minute surface fissures as a result of exposing rubber articles to sunlight, generally accelerated by bending or stretching.

chemical attack An internal (alkali-aggregate reaction) or external sulfate attack or an aggressive service environment.

chemical bond The bond between materials resulting from *cohesion* and *adhesion* developed by chemical reaction.

chemical cure Curing (hardening) by chemical reactions usually involving the formation of cross-linked polymers.

chemical formula A way of designing a compound, expressed as a ratio of the number of atoms of each element present. For example, H_2O is the compound for water, containing two atoms of hydrogen (H) and one atom of oxygen (O).

chemically foamed polymeric material A cellular material in which the cells are formed by gases generated from thermal decomposition or other chemical reaction.

chemically strengthened glass Glass that has undergone ion-exchange to produce a comprehensive stress layer.

chemical reaction An interaction among atoms or molecules that results in the formation of new atoms or molecules by making and/or breaking chemical bonds.

chemical reaction water repellent products Products intended to function through chemical reaction or bonding as opposed to the deposition of a resin.

chemical resistance The resistance offered by elastomer products to physical or chemical reactions as a result of contact with or immersion in various solvents, acids, alkalis, salts, and so forth.

chemical symbol A letter or letters representing the name of the element, or one atom of it. Examples: hydrogen (H) and iron (Fe).

chloride ion Smallest component of soluble salt that can migrate through substrate in water solution.

chlorinated poly (vinyl chloride) plastics Plastics based on chlorinated poly (vinyl chloride) in which the chlorinated (vinyl chloride) is in the greatest amount by weight.

chlorinated rubber When chlorine is passed into a solution of crude rubber, a white fibrous product of the appropriate formula $(C_{10} H_{13} CL_7) X$ is obtained.

chlorofluorocarbon plastics Plastics based on polymers made with monomers composed of chlorine, fluorine, hydrogen, and carbon.

circuit In filament winding, the winding produced by a single revolution of mandrel or form.

circumference The perimeter of a circle.

circumscribe To draw a line around a figure or object.

classified UL terminology for products that in and of themselves have no listing or approval. Products that are UL classified have predictable ratings and performance only after installation in a construction condition similar to UL published designs.

clay A natural soil or mineral aggregate in hydrous aluminum silicate that dries hard, but can be worked or molded when moist.
clear glass As its name states, transparent or clear.
cleavage membrane Membrane that provides a separation and slip sheet between the mortar setting bed and the backing or base surface.
clinker A partially fused product of a kiln that is ground to make cement.
clips Wire spring devices to hold glass in a rabbet sash, without stops, and face glazed.
closed cell foamed plastic Cellular plastics in which almost all the cells are non-interconnecting.
closure A whole or partial masonry unit that is used to complete a course in a masonry wall.
coagulation A physical or chemical change inducing transition from fluid to a semi-solid or gelatinous state.
coat, double Generally applied to two successive coats applied to one surface. In spraying, it means to spray first a single coat with vertical strokes and then across with horizontal strokes, or visa versa. In the woodworking industry the term means one coat applied to each of the two mating surfaces. In adhesives and coatings, the terms mean a single coat applied to each surface.
coat, single One layer of applied material on a surface, whether applied by brush, spray gun, flow gun, etc. In spray jargon, it means two strokes of the gun, the first down and second up, or the first to the right, and the second to the left. The second stroke should cover the first completely. The third and fourth strokes should cover ½ of the first two, while the fifth and sixth should cover ½ of the third and fourth, etc. See *coat double*.
coating A material, usually liquid, used to form a covering film over a surface. Its function is to decorate and/or protect the surface from destructive agents or environments (abrasion, chemical action, solvents, corrosion, and weathering).
cobalt A metallic element; atomic number 27; atomic weight 58.933; atomic symbol Co.
cobblestone A rock fragment that is usually rounded or semi-rounded and averaging 3 to 12 inches in diameter.
coefficient of expansion (linear and volume) The coefficient of linear expansion is the ratio of the change in volume per degree to the length at 32° F (0° C). The coefficient of volume expansion for solids is three times the linear coefficient. The coefficient of volume expansion for liquids is the ratio of the change in volume per degree of the volume at 32° F (0° C).
coefficient of friction A measure of the resistance to sliding of one surface in contact with another surface.
coefficient of thermal expansion The change in volume per unit volume produced by a one-degree rise in temperature.
cohesion Ability of a material to bond together.
cohesive failure Splitting and opening of a compound from over-extension of the compound caused by excessive movement. See *adhesive failure*.

cold flexibility   Flexibility during exposure to the predetermined low temperature for a predetermined time.

cold flow   A permanent deformation under constant stress. Also defined as the continuing dimensional change under static load that follows initial instantaneous deformation. If subjected to pressure long enough, no organic material will return exactly to its original shape. Compression set is the amount by which a small cylinder fails to return. See also *creep*.

cold molding   A process of compression molding in which the molding is formed at room temperature and subsequently baked at elevated temperatures.

cold resistant   Withstands the effect of cold or low temperatures without loss of serviceability.

color   A property or quality of the visible phenomena distinct from form and from light and shade, depending on the effect of different wave lengths on the retina of the eye. The term color is sometimes used inaccurately to denote hue, tint, shade, pigment, dyestuff, and so forth.

colored cast glass   Includes many kinds of cast and rolled glass. There are more than 100 colors of *dalle* glass (*dalle* is French for *tile*).

color fast   Non-fading color.

combustible penetrants   Pipes, cables, or other penetrants that may burn or melt out during a fire.

compaction   The process of reducing the volume of freshly placed concrete or mortar by vibration, tamping, or rodding to ensure complete embedment of all reinforcement and to eliminate voids.

compatible   Two or more substances that can be mixed or blended or in close proximity without separating, reacting, or affecting the material adversely.

component   A part of a whole.

compound   An intimate admixture of a polymer with all the ingredients necessary for the finished article.

compression   Pressure exerted on a compound in a joint, as by placing a light or panel in place against bedding, or placing a stop in position against a bead of compound.

compression modulus   The ratio of the compressive stress to the resulting compressive strain (the latter expressed as a fraction of the original height or thickness in the direction of the force). Compression modulus may be either static or dynamic.

compression molding   The method of molding a material in a confined cavity by applying pressure and usually heat.

compression set   The residual decrease in thickness of a test specimen measured 30 minutes after removal from a suitable loading device, in which the specimen had been subjected for a definite time to compressive deformation under specified conditions of load application and temperature.

   Method A measures compression set of vulcanized rubber under constant load.

   Method B employs constant deflection (See ASTM Method D-395).

compressive strength   The ability of a material to resist a force that tends to crush.
concave bead   Bead of compound with a concave exposed surface.
concrete, green   Concrete that has been placed and has set but is not completely hardened.
condensation   Moisture that forms on surfaces when they are colder than the dew point.
condensation polymer   A polymer made by condensation polymerization.
condensation polymerization   Polymerization in which monomers are linked together with the splitting-off of water or other simple molecules.
conduction   The transfer of heat through matter, whether solid, liquid, or gas.
conductive mortar   The mortar to which specific electrical conductivity is imparted through the use of conductive additives.
conductive tile   Tile made from special body compositions, or by methods that result in specific properties of electrical conductivity. Other normal physical properties of ceramic tile are retained.
consistency   (1) The relative mobility or ability of freshly mixed concrete to flow, measured by slump; (2) That property of a liquid adhesive by virtue of which it tends to resist deformation.
construction adhesive   General term for adhesives used in various construction tasks. A wide variety of adhesives for building are available. They are available in tubes, cartridges, sausage packs, and one and two-gallon containers, suitable for use with hand or pneumatic caulking guns.
construction joint   The surface where two successive concrete placements meet.
construction specifications   See *specifications*.
contact cement   A type of adhesive that is applied to both objects to be joined and then allowed to set up. When the desired state of dryness is reached, the objects are touched together and the coated surfaces bond on contact.
container stability   Period of time a compound will remain in satisfactory condition when stored in unopened containers. Synonymous with *shelf life*.
continuous film   A coating film that is free of breaks or pinholes.
contraction   A decrease of volume occurring as the result of any or all processes affecting the volume.
contractors   The persons generally known as building contractors whose crews construct the buildings. They hire laborers, and craftsmen, who are engaged to carry out the task of construction.
control   A product of known characteristics that is included in a series of similar service or bench tests to provide a basis for evaluation of one or more unknown products.
control joint   (1) A joint between sections of a structure designed to permit differential movement of the structures. (2) A groove in a masonry wall designed to control cracking.
convection   A transfer of heat through a liquid or gas.

conventional rubber cure   See *vulcanization*.

conversion chart   A chart showing equivalent units of measurement from one standard to another, such as from feet to meters.

convex   A surface that curves outward at the center.

coolant   A liquid for cooling engineering work.

coping   A cap or finish on the top of a wall, pier, chimney, or pilaster to prevent penetration of water to masonry below.

co-polymer   A polymer formed when two or more different monomers alternate.

corrosion   The deterioration of a metal by chemical or electrochemical reaction resulting from exposure to weathering, moisture, chemicals, or other agents or media.

corrosive   The quality of a substance that can cause the chemical deterioration of a material.

cove   Trim tile with one edge a concave radius used to form a junction between the bottom wall course and the floor or form an inside corner.

CPVC   Acronym for *post-chlorinated polyvinyl chloride*. A combustible thermoplastic resin used in the manufacture of certain nonmetallic pipes used primarily for sprinkler applications

cracking   A fissured surface condition that develops on rubber articles exposed to the atmosphere, light, heat, or repeated bending or stretching. When the fissures are minute, the condition is called *crazing*.

crater   A small shallow surface imperfection.

crawling   Parting and contraction of the glaze on the surface of ceramic-ware during drying or firing. It results in unglazed areas bordered by coalesced glaze.

crazing   Fine, random cracking that occurs in fired glazes or other ceramic coatings due to critical tensile stresses and shrinkage. Fine line in concrete surfaces. See also *checking*.

creep   The dimensional change with time of a material under load, following the initial instantaneous elastic deformation. Creep at room temperature is sometimes called *cold flow*.

critical temperature   The temperature above which a gas cannot be liquefied by pressure. The term is loosely used, sometimes to denote an approximate temperature below which a reaction such as vulcanization does not take or proceeds only very slowly. Also used to denote a temperature below which an accelerator does not function properly.

cross-laminate   A laminate in which some of the layers of material are oriented approximately at right angles to the remaining layers with respect to the grain or strongest direction in tension.

cross-linking   The curing of an adhesive compound by physically bridging the molecular chains of the compound to increase its temperature, shear, chemical and solvent-resistant properties.

crown glass   Obsolete way of making glass. Also a very clear optical glass.

crown wool   A process used in making glass fiber.

cryogen   A substance that produces very low temperatures, such as liquid nitrogen, liquid oxygen, or liquid hydrogen. A liquid gas is very cold and takes up much less volume than when in gas form. Nitrogen becomes liquid at −320° F (-195.56° C) and oxygen at −298° F (−183.33° C).
crystalline   Resembling small crystals, or crystal-shaped.
C-stage   The final stage in the reaction of certain thermosetting resins in which the material is practically insoluble and infusible.
cullet   Broken glass, excess glass from the previous melt, or edges trimmed off when cutting glass to size. Cullet, in regular proportion, is an essential ingredient in the raw batch charge in glassmaking in that it facilitates melting. Recycled or waste glass.
cure   To change the properties of a material by chemical reaction, which may be condensation, polymerization, or vulcanization. Usually accomplished by the action of heat and catalysts, alone or in combination, with or without pressure. See also *overture*; *semi-cure*.
cure time   The time required to produce vulcanization at a given temperature. The cure time varies widely, being dependent on the type of compounding used, the thickness of the product, etc.
curing, (glass)   Similar to annealing. (Concrete) Keeping freshly placed concrete or mortar moist and at a favorable temperature for a suitable length of time to assure satisfactory hydration.
curing agent   Generally the second of a two- part system which, when added to the base materials, cures or solidifies the base material by chemical reaction.
curing range   In vulcanization, an approximate range of curing times at a given temperature, over which the physical properties of a vulcanizate do not change materially.
curing temperature   The temperature at which the rubber product is vulcanized.
curling   Distortion of an essentially straight or flat member into a curved, warped, or dished shape due to creep or to internal differences in temperature or moisture content.
curtain-wall   An exterior building wall that carries no roof or floor loads and consists entirely or principally of metal or a combination of metal, glass, and other surfacing materials supported by a metal framework. There are two basic types:
>*Custom:* Walls designed specifically for one project and using parts and details specially made for this purpose.
>*Standard:* Walls made up principally of parts and details standardized by their manufacturer and assembled in accordance with either the architect's design or the manufacturer's stock pattern.

In a masonry, a variation on a *veneer wall*, instead of veneer tied to a backup wall, the curtain wall is tied to a framework of metal framing members and encloses the space behind the framing.
cushion-edge tile   Tile on which the facial edges have a distinct curvature that results in a slightly recessed joint.

cut-back   A product, made of an asphalt or tar-pitch, that is made liquid by solvents, natural or added oils, or both.

cut joint   Masonry joints that have been cut flush with a trowel.

cycle   One complete operation of a molding press from closing time to closing time.

damping   The dissipation of sound energy in a medium over time or distance.

damp-proofing   Application of clear or mastic water repellent material. Also a treatment to resist water penetration without hydrostatic pressure.

dangerous chemicals   Any substance or mechanical mixture or chemical compound of substances that is volatile or unstable or that tends to oxidize or decompose spontaneously, thus creating fire or explosion, or that may generate flammable or explosive gas, and that is capable of creating hazard to life, or property, when designated as such by rule or regulation.

darby   A hand-manipulated straight edge, usually three to eight feet long, used in the early stage leveling of concrete or plaster surfaces.

D-cracking   The progressive formation on a concrete surface of a series of fine cracks at rather close intervals, often of random patterns, but in slabs on grade paralleling edges, joints, and cracks and usually with a radius at slab corners. May be accompanied by formation of calcium carbonate.

dead load   Load force due to glass weight.

decibel (dB)   A standard measure of sound pressure level.

decorative tile   Tile with a ceramic decoration on the surface.

deflection   The degree of inward or outward movement in lights of glass that are exposed to unequal pressures on their faces. This condition may be permanent or temporary. The wind or temperature, elevation, or barometric pressure changes can cause deflection. Air absorbing desiccants can contribute to deflection.

deformation   Any change of form or shape in a body; the linear change of dimension of a body given direction produced by the action of external force.

deformed bar   A reinforcing bar with ridges that serve to lock into surrounding concrete.

degassing   Opening and closing of a mold to allow the escape of gases and/or moisture vapor early in the molding cycle. Also referred to as *breathing*.

degradation   Deterioration, usually in the sense of a physical or chemical process rather than a mechanical one.

degree   (1) A unit of measurement used in temperature. (2) A 360th part of the circumference of a circle, the unit of measurement from which angles are determined.

degree of   Fineness of subdivision of dispersed particles.

dehydration   Removal of water as such from a substance, or after formation from a hydrogen and hydroxyl group in a compound, by heat or dehydrating substance.

deicer   A chemical such as sodium chloride used to melt snow and ice on pavements.

delaminate   To split a laminated material parallel to the plane of its layers. Sometimes used to describe cohesive failure of an adherent in bond strength testing.

delamination Separation or splitting, usually as a lack of adhesion in plied goods. A separation along a plane parallel to a surface. In the case of a concrete slab or wall panel, the splitting, cracking, or separation in a plane roughly parallel to the surface. Similar to spalling, scaling, or peeling except that delamination affects large areas and can often only be detected by tapping.
density The ratio of the mass of a body to its volume. For ordinary and practical purposes, when using grams per cubic centimeter units, density and specific gravity may be regarded as equivalent.
depth In the case of a beam, the dimension parallel to the direction in which the load is applied.
desiccant Porous crystalline substance used to absorb moisture and solvent vapors from the air space of insulating glass units. (More properly called *absorbents*.)
desiccate To dehydrate or remove moisture.
devil's float A wooden float with two nails protruding from the toe used to roughen the surface of the brown coat in plastering and stuccoing.
dew point The temperature at which air is saturated with respect to a condensable component, such as water vapor or solvent.
dew point (IG Units) The temperature above 32 °F (0° C) at which visible vapor or other liquid vapor begins to deposit on the air-space glass surface of a sealed insulating glass unit in contact with the measuring surface of the dew-point apparatus.
diameter The distance across the center of a circle.
diamond Pure carbon formed under great pressure into a very hard crystal.
diamond mesh Sheets of metal that are slit and pulled out to form diamond shaped openings; used as metal reinforcement for plaster. Also known as expanded metal lath.
diaphragm A thin body that separates two areas; in sound, the skin of a partition or ceiling that separates the room from the structural space in the center of the partition or ceiling assembly.
diatomaceous earth A fine white or gray silicon material resembling chalk, which is composed of the fossilized remains of small marine life. It is used for fillers and as a paint extender.
difunctional An atom or molecule with two active sites.
diisocyanates Isocynates containing two -N=C=O groups.
dilute To weaken the strength of a fluid by mixing it with another.
diluent Liquid that lowers viscosity and increases the bulk but is not necessarily a solvent for the solid ingredients; a thinner.
dimensional stability Ability of a plastic part to retain the shape to which it was molded, cast or otherwise fabricated.
dimer Two molecules of the same type chemically bonded together (an impurity when pure MDI reacts with itself).
diol An alcohol with two hydroxyl (-OH) groups; also called *glycol*.
dip coat A thin coat on a surface obtained by the dipping of the material to be coated into the coating material.

discoloration Staining. Changing or darkening in color from the standard or original.

dispersion The act of causing colloidal particles of matter to separate and become uniformly scattered throughout a medium. Any system of matter in which finely divided colloidal particles of one or more phases (components) are uniformly scattered throughout another phase or medium; the components or phases may be solid, liquid, or gaseous (a wholly gaseous system is not considered a dispersion, but a simple mixture). Colloidal solutions are examples of liquid dispersions; the dissolved or dispersed particles are not subdivided to the molecular state as in true solutions and dispersion.

divider strips Nonferrous or plastic strips used in terrazzo work and embedded to depths of ⅜ inch to 1¼ inch. Also may be 1 (4 or 2 (4 redwood, cypress, or cedar left in concrete slabs permanently as a decorative or control joint.

doctor-bar or blade A scraper mechanism that regulates the amount of adhesive on the spreader rolls or on the surface being coated.

doctor-roll A roller mechanism that revolves at a different surface speed, or in the opposite direction, resulting in a wiping action for regulating the adhesive supplies to the spreader roll.

dome In reinforced plastics, an end of a filament-wound cylinder container.

double bonds A linkage between two atoms of carbon or other atoms, or a covalent linkage in which the atoms share two pairs of electrons. Double bonds in a formula are represented by the symbols "=" or ":" as in ethylene ($H_2C = CH_2$) or ($H_2C:CH_2$).

double bullnose Type of trim with the same convex radius on two opposite sides.

double glazing Windows in which two pieces of glass have been installed, usually ⅛ inch to ½ inch apart.

double-headed nail A nail with two heads used in formwork. The nail is driven in to the first head and removed by its protruding second head.

double-hung windows A window with an upper and a lower sash, both of which slide vertically, so that the window can be opened at the top or bottom.

double laminated insulating glazing Two laminated glass panels set in a frame that provides an air space between them. Such units may or may not be hermetically sealed. Air space thickness can vary with acoustic and thermal requirements.

double strength glass In Float glass approximately ⅛ inch thick.

dowel Steel pin extending into the adjoining portions of concrete to prevent shifting of the concrete. Also made of wood and used in wooden window frames.

draft Taper or slope of vertical surfaces of mold designed to facilitate removal of molded parts.

draw To stretch a sheet of plastic material to fit a mold; to cup.

drawing tower Used in the sheet glass process for drawing molten glass.

driers Additives that shorten the drying time of paints, coatings, and sealants.

dry To change the physical state of an adhesive on an adherend by the loss of solvent constituents by evaporation or absorption, or both.

dry-blend A free flowing dry compound prepared without fluxing or addition of solvent, also called powder blend.

dry glazing A method of securing glass in a frame by use of a dry preformed resilient gasket, without the use of a compound.

drying time The time required for solvent dissipation after a film is spread. The drying process includes several states. The first is known as *tacky* and starts almost immediately after application. The optimum point in the tacky state is the earliest period when the adhesive may be touched lightly without transferring material to the finger. The bond should be made at this point. The second state, *dust free*, is the time required for the film to reach the condition where, if dust settles on it, the dust will not become imbedded but may be wiped off after the material has been hardened. The *tack-free* stage is the time required for the film to reach the condition where it can be touched with the finger without feeling the slight retention of surface stickiness. *Print-free* is closely related to this, although one stage later. It refers to the time required for the film to reach the condition where it may be touched with the finger without retaining the imprint of the finger on the surface of the film. The final state, *hard-dry*, is the time required for the film to become thoroughly hard so that it may be handled and polished if necessary.

dry seal adhesive One that is non-blocking except to itself. Two adherents may be precoated, dried, then bonded at any time using only normal pressure.

dry-set system (mortar, grout) Sometimes called dry-mix. A mixture of portland cement and sand with organic additives, developed for quick setting ceramic tile over a variety of sub-surfaces, possessing, in certain circumstances, significant advantages in installation efficiency over the full mortar-bed system.

dry shake A dry mixture of cement and fine aggregate (usually colored) that is spread on a concrete surface after bleed water has disappeared following strikeoff and worked into the surface during floating.

dry-spot An imperfection in reinforced plastics; an area of incomplete surface film where the reinforcement has not been wetted with resin.

dry time Lack of adhesion; more specifically, poor contact of adhering surfaces. See *starved joint*.

dual sealed units Sealed insulating glass units fabricated with an inner seal and an outer secondary seal. Generally, each of the two seals has been selected for its special performance characteristics, i.e. adhesion and moisture vapor transmission properties.

durability The ability of portland cement concrete and mortar to resist weathering action, chemical attack, abrasion, and so forth.

durometer An instrument for determining the hardness of rubber by measuring its resistance to the penetration (without puncturing) of a blunt indentor point impressed on the rubber surface against the action of a spring. A special scale indicates resistance to penetration or *hardness*. The scale reads from zero to 100, zero being very soft and 100 being very hard.

dust-free   A coated surface that has dried sufficiently so that dust will not stick to it.

dusting   1. The application of dry portland cement to a wet floor or deck mortar surface; a pure coat is thus formed by suction of the dry cement. 2. The development of powdered material on the surface of concrete that has hardened.

DWV   Acronym for *drain*, *waste*, or *vent*. Pipes that are used in plumbing applications e.g. drain pipes, waste, or soil pipes and vent stacks.

dynamic elongation test   Elongation or stretching of a material under continuous movement.

edge clearance   Normal spacing between the edge surface of the glass product and the glazing channel base.

edge joint   A joint made by bonding the edge faces of two adherends.

edging   Grinding the edge of flat glass to a desired shape or finish.

effective length   The length of that section of a structural unit that is free to move toward, or away from, a joint.

efflorescence   A leeching outward of mineral salts from within a concrete or masonry wall to the surface. Efflorescence cannot be sealed or caulked over. The powder must be brushed off and the residue cleaned off with a dilute solution of muriatic acid.

E-glass (electrical glass)   A low alkali borsilicate glass. This type is the most widely used in fibers for reinforcing plastics. Its high resistivity makes E-glass suitable for electrical laminates.

elasticity   The property of matter that tends to return it to its original shape after deformation such as stretching, compression, or torsion. The opposite of plasticity. It is often loosely employed to signify the "stretchiness" of rubber. As applied to rubber, it usually refers to the distance to which vulcanized rubber can be stretched without losing its ability to return very nearly to its original shape; in this respect, rubber is the most elastic substance known. See also *modulus of elasticity*.

elastic limit   The extent to which a body may be deformed and yet return to its original shape after removal of the deforming force. Steel has a well-defined elastic limit (yield point) below which it is perfectly elastic; vulcanized rubber, on the other hand, shows no definite elastic limit, but takes more or less *set* depending on the stretch given to it.

elastomer   Elastomer is the name of a substance that can be stretched to at least twice its original length, and after having been stretched and the stress removed, returns to approximately its original length in a short time. (ASTM D 883-65T).

elastomeric   Having the property of returning to the original shape and position after deformation by tensile or compressive forces that are within the limits of its yield strength.

electrical strength (dielectric strength)   That property of an insulating material that enables it to withstand electric stress. The highest electric stress that an insulating material can withstand for a specified time without the occurrence of electrical breakdown by any path through its bulk.

elongation Increase in length expressed numerically as a fraction or percentage of initial length.

elongation at rupture test The amount a material has stretched at the time it breaks apart. Test to be run at specified conditions.

embossed Bearing a decoration in relief or excised on the surface.

emissivity The relative ability of a surface to radiate heat. Emissivity factors range from 0.0 (0 percent) to 1.0 (or 100 percent). Emittance is the ratio of the total radiant energy emitted by a given surface to that emitted by an ideal black body at the same temperature. Heat energy radiated by the surface of a body, usually measured per second, per unit area.

emulsion A suspension of microscopic particles in water. Oils and other substances normally not soluble in water also can be emulsified.

enamel A glossy paint.

enamel, glass A soft glass compound of flint or sand, soda potash and red lead.

endothermic A chemical reaction that absorbs heat energy is said to be endothermic. A compound, the formulation of which absorbs heat, is an endothermic compound. Such compounds are less stable than exothermic compounds, many being explosive.

ENR *Exterior noise rating.*

environmental stress cracking The susceptibility of a thermoplastic resin to crack or craze when in the presence of surface-active agents or other environments.

epoxy A thermosetting resin formed by combining epichlorohydrin and biphenols. Requires a curing agent for room temperature or elevated temperature hardening. Has outstanding adhesion and excellent chemical resistance. A family of resins containing:

epoxy systems (mortar, adhesives, and grout). Two-part mortar or adhesive setting systems, using epoxy resins plus catalytic hardening agents, combining chemical resistance with high-bond strength and impact strength.

equivalency The relationship between approved production insulating glass units and the prototype test units. In the United States under IGCC regulations, the production unit must have at least as much desiccant per perimeter inch as the test unit. In Canada, equivalency regulations are under discussion and have not been set to date.

equivalent of combined load Combination of the instant applied load of wind and factored long-term loading of glass weight and snow accumulation.

erosion Destruction of metal or other material by the abrasive action of liquid or gas. Usually accelerated by the presence of solid particles of matter in suspension and sometimes by corrosion.

ester groups   A linkage resulting from bonding a carboxylic acid (-COOH) and an alcohol (-OH), contained in a polyester pall.
etch   To eat away part of a surface by chemical means. Etching is used to create decorative patterns on glass or metal.
ether group   A carbon-oxygen-carbon (C-O-C) linkage that results from bonding an oxide to an alcohol (-OH). Contained in a polyether pall.
ethylene plastics   Plastics based on polymers or ethylene or copolymers of ethylene being in greatest amount by mass.
EWNR   *Exterior wall noise rating.*
exotherm   The temperature/time curve of a chemical reaction giving off heat. The term has not been standardized with respect to sample size, ambient temperature, degree of mixing, etc.
exothermic   A chemical reaction in which heat energy is liberated is termed exothermic, and a compound of which the formation involves the evolution of heat is called an exothermic compound. Exothermic compounds are more stable than endothermic compounds. Vulcanization of rubber with sulfur is an exothermic reaction.
expandable plastic   A plastic in a form to be made cellular by thermal, chemical, and mechanical means.
expansion   An increase in the volume occurring as the result of any or all processes affecting the bulk volume.
expansion and compression   The percentage increase and decrease in width from the installed width tolerable to a sealant in service.
expansion joint   A joint formed to separate concrete units to allow it to crack under controlled conditions because of temperature changes.
exposed joint   Any mortar joint on the face of masonry that is above ground level.
extender   A relatively inexpensive material, added to a compound for the purpose of reducing the cost and/or for the purpose of improving certain desirable characteristics.
extensibility   The ability of a material to extend or elongate upon application of sufficient force, expressed as a percent of the original length.
exterior glazed   Glass set from the exterior of the building.
exterior stop   The removable molding or bead that holds the light or panel in place when it is on the exterior side of the light or panel, as contrasted to an interior stop located on the interior side of the light.
extruded   A material forced through a die or continuous mold for shaping.
extruded tile   Tile or trim units that are formed when plastic clay mixtures are forced through a pug mill opening (die) of a suitable configuration and result in a continuous ribbon of formed clay. A wire cutter or similar cut-off device is then used to cut the ribbon into appropriate tile lengths and widths.
extrusion   A process whereby heated or unheated plastic forced through a shaping orifice becomes one continuously formed piece.

fabricate To work a material into a finished form by machining, forming, or other operation, or to make flexible film or sheeting into end-products by sewing, cutting, sealing or other operations.

façade The face of a building. Façade is normally used to describe building fronts.

face glazing On a rebated sash without stops, the triangular bead of compound applied with a glazing knife after bedding, setting, and clipping the light in place.

face-mounted tile Tile mounted with paper applied to the faces of sheets. The water-soluble adhesive can be removed easily prior to the grouting of the joints.

fading Any lightening of an initial color possessed by a plast. Measured by accelerating the process by subjecting the plastic to high intensity ultra-violet rays approximately the same wavelength as those found in sunlight.

Fahrenheit A temperature measurement scale in which pure water freezes at 32° F (0° C) and boils at 212° F (100° C) at sea level.

faience tile Tile with variations in the face, edges, and glaze that give a hand-crafted, non-mechanical, decorative effect.

failed insulating glass unit An insulating glass unit failure exhibits permanent material obstruction of vision through the unit due to accumulation of dust, moisture or film on the internal surface of the glass. Surface numbers 2 or 3 in dual-pane units; surface numbers 2, 3, 4 or 5 in triple-pane units.

falsework Temporary structures such as shoring, formwork, for beams and slabs etc., built to support work in progress.

fastener General term for hardware used to connect building materials.

fatigue The failure or decay of mechanical properties after repeated applications of stress.

faulting Differential vertical displacement of a slab or any other member adjacent to a crack or joint.

faying surface The surfaces of materials in contact with each other and joined or about to be joined.

feeder The part of the furnace that ensures that the glass has a consistent temperature throughout.

fiberglass Glass in fibrous form made by drawing molten glass.

field erected system Firestop materials assembled at the jobsite into a particular configuration utilizing materials that are intended to lend themselves to a variety of different conditions.

field molded sealant A liquid or semi-solid material molded into the desired shape in the joint into which it is installed.

filler Relatively non-adhesive substance added to an adhesive to improve its working properties, permanence, strength, or other qualities.

fillet A rounded bead or concave junction of sealing material over, or at the edges, of structural members.

fill material The firestop material used to seal the opening around the penetrant.

film Thin layer of material, not necessarily visible. *Free films* are not attached to any body. They are often called *unsupported films*. *Supported films* have flexible backing, usually cloth or paper, to give the film greater strength.

fin A narrow projection on a concrete surface caused by mortar flowing between cracks in the forms.

finish coat The last coat of plaster, the decorative surface, usually is colored and frequently textured.

finishing The leveling, smoothing, compacting and treatment of the surface of concrete, plaster, or sealants to attain desired texture or surface.

fire-proofing To cover with fire-resistant and heat-resistant covering.

firestop A solid, tight closure of a concealed space, placed to prevent the spread of fire and smoke.

firestop devices Firestop materials assembled into their final form at a point of manufacture. Example is a firestop collar.

firestop system An assembly consisting of: 1. The penetrant or penetrants, 2. The opening through a particular type of wall or floor, and 3. The firestop materials and design that are used to seal the opening or protect the penetrants.

firing A step during the manufacture of ceramic tile that uses a kiln or furnace to develop desired properties through controlled heat treatment.

fixed window A window with no moving parts, also called a stationary window.

flame resistant A material that does not burn readily, when the source of flame is removed.

flame retardants Substances mixed with rubber to retard its burning (i.e., highly chlorinated hydrocarbons). Neoprene is less flammable than natural rubber and GO-S.

flammability Measure of the extent to which a material will support combustion.

flammable A volatile liquid or gas that has a flash point of 30° F (−2° C) or lower. Flammable is synonymous with *inflammable*.

flange The projection around the exterior perimeter of some sashes.

flanking The passage of sound between two rooms through a medium other than the partition dividing them. The sound may pass via the sidewalls, floor, ceiling, or other structures. This term also refers to air leakage paths through partitions and to sound bypassing the main leaves of a partition via the studs or other connecting framework.

flashing Metal or other suitable material placed to shed water.

flash point The temperature to which a liquid must be heated before its vapors will flash or burn instantly when a small flame is applied. This ignition will not take place unless there is a spark or open flame. There are several standard methods for determining flash point, most of which may be classified as *open-cup* or *closed-cup*.

flat glass Pertains to all glass produced in a flat form.
flex life The time of heat-aging that an insulation material can withstand before failure when bent around a specific radius (used to evaluate thermal resistance).
flexural strength The resistance of a material to being broken by bending stresses.
float A rectangular hand tool, usually of wood, aluminum, steel, magnesium or plastic, used to impart a relatively even but still open texture to a concrete surface. Other types are cork, sponge, rubber, and so forth.
floated A method of using a straight edge to align mortar with float strips or screeds.
float glass Transparent glass with flat, parallel surfaces formed on the surface of a pool of molten glass tin.
flow Time-dependent irrecoverable deformation. (1) Movement of resin under application of pressure allowing it to fill all parts of a mold. (2) Flow or *creep*. The gradual continuous distortion of a material under continued load, usually at high temperatures.
flow marks Wavy surface appearance of an object molded from thermoplastic resins, caused by improper flow of the resin into the mold.
flow out The ability of a material to level after application (whether brushed, sprayed, roll-coated or applied through pressure units). *Orange-peel* is the surface appearance when a sprayed material does not flow or level. Excessive flowing on vertical surfaces is termed *sagging*, usually caused by too much material or solvent that is too slow drying.
fluorohydrocarbon plastics Plastics based on polymers made with monomers composed of fluorine, hydrogen and carbon only.
fluoroplastics Plastics based on polymers with monomers containing one or more atoms of fluorine or copolymers of such a monomer, with other monomers. The fluorine-containing monomer(s) contain the greatest amount by mass.
flush glazing A system of installing glass in which the member that holds the glass in place (the glazing bead) is recessed within and flush with the edge of the frame. These systems are also called *pocket-glazed* and *center-glazed* systems.
fluxes Oxides used to produce glass that is easy and inexpensive to shape.
foam core Various nonmetallic pipes where porosity has been induced into the pipe's cross-section to reduce weight and cost. Sometimes referred to as coax piping.
fog curing The application of a fine mist of water during the curing of concrete etc.
fogged unit A permanent deposit of contaminates on the interior surfaces of the insulating glass unit.
fore-hearth An extension of the furnace where glass is made at the same temperature throughout.
forming Shaping, or molding into shape.

forming material   The packing material used to dam an opening prior to the application of fill material. In some systems, packing materials may contribute to the fire rating of the system.

formulating   The systematic process of blending useful, active compounds for a specific end-use application.

F-rating   The amount of time that an assembly has successfully been tested to resist the penetration of fire.

free -N=C=O   The amount of isocyanate left unreacted in a prepolymer.

freeze thaw cycle   Accelerated weathering caused by cycles of freeze-expansion-fracture-thaw-water erosion, etc.

frequency   The number of times an action occurs in a given time period. In sound, the number of complete vibration cycles per second, represented by the Hertz (Hz).

frog   A depression in the bed surface of a brick or other masonry unit.

front putty   The putty forming a triangular fillet between the surface of the glass and the front edge of the rabbet.

frost line   In winter, the level above which the ground is frozen. The line varies greatly depending on the part of the country.

frost point   The temperature below 32° F (0° C) at which visible frost begins to deposit on the air-space surface of a sealed insulating glass unit in contact with the measuring surface of the frost-point apparatus.

FRPP   Acronym for *fire-retardant-polypropylene*. A combustible thermoplastic resin used in the manufacture of certain nonmetallic pipes used primarily for high-end DWV applications such as acid waste lines.

full mortar bed   A setting bed of Portland cement ¾ to 1¼ inches thick laid over a cementitious scratch coat and, if required, a vapor-proof membrane. Sometimes called the *mud method*. A thin cement bond coat is laid on top of the setting bed to hold the tile. Thin bed systems can also be installed over a full mortar bed.

fully tempered glass   Transparent or patterned glass with a surface compression of not less than 10,000 psi or edge compression of not less than 9700 psi, specified by ASTM C 1048 (Kind FT). Fully tempered glass, if broken, will disintegrate into many small pieces (dice) which are more or less cubical. Fully tempered glass is four to five times stronger than annealed glass of the same thickness.

furan plastics   Plastics based on furan resins.

furan resins   A resin in which the furan ring is an integral part of the polymer chain, the furan being in the greatest amount.

furan systems (mortar, grout)   Two-part installation systems using furan resins plus catalytic curing agents combining the highest level of resistance to chemical damage, with good bond strength and impact resistance. Furan mortars cure in minutes. This material is used less since the advent of high temperature, chemical-resistant, 100 percent epoxy mortars and grouts.

gallon   A unit of liquid capacity equal to 4 quarts or 231 cubic inches.

gas-filled unit   Insulating glass unit with a gas other than air in the air space to decrease the unit's thermal conductivity, U-value and sound insulating value.
gasket   Pre-formed shapes, such as strips, grommets, etc., of rubber and rubber-like composition used to fill and seal a joint or opening either alone or in conjunction with a supplemental application of a sealant.
gather   A small amount of glass taken from a melting pot.
gel   A semi-solid, jelly-like condition of matter. A form of colloidal dispersion in which the dispersed component and the dispersing medium are associated to form a jelly-like mass. A solution of gelatine or glue in warm water is liquid, and is termed a sol; on cooling, the liquid changes to a jelly, and is termed a gel.
gel point   The stage at which a liquid begins to exhibit pseudo-elastic properties. (ASTM D 883-65T).
gel time   The time required for a liquid material to form a gel under specific conditions of temperature as measured by a specific test.
geodesic   Pertaining to the shortest distance between two points on a surface.
girders   Heavily loaded beams or horizontal members that support other beams.
girts   The light horizontal members that span between columns to support walls.
glass   A transparent, brittle substance formed by fusing sand with soda or potash or both; it often has lime, alumina or lead oxide. A variety of glass types are available. The following is an abbreviated list:
   Single strength (SS): This is $\frac{3}{32}$ inch thick and comes in a variety of standard sizes or can be cut to order.
   Double strength (DS): This is standard single-strength glass but $\frac{3}{16}$ inch thick—twice as thick as single strength.
   Plate: This comes ¼ inch thick in various sizes and with beveled edges. Commonly used on coffee tables, countertops, and other surfaces. It can be cut the same way as other glass.
   Processed glass and rolled figured glass: These are general classifications of obscure glass. There can be many patterns and configurations.
   Obscure wire glass: The type of glass generally specified for its fire-retarding properties.
   Polished wire glass: More expensive than obscure wire glass, polished wired glass is used where clear vision is desired, such as in school or institutional windows and doors.
   Tempered glass: Looks like regular glass, however, it breaks into small, relatively dull-edged pieces.
   Bullet-resisting glass: Made of three or more layers of plate glass laminated under heat and pressure. Thickness varies from ¾ inch to 3 inches. The most common thickness is $1\frac{3}{16}$ inch to resist medium powered small arms.
   Laminated glass: This consists of two or more layers of glass laminated together by one or more coatings of a transparent film.

glass block Building block made of solid glass.

glass finish A material applied to the surface of glass fibers used to reinforce plastics and intended to improve the physical properties of such reinforced plasticsas compared to that obtained using glass reinforcement without finish.

glass transition The reversible change in an amorphous polymer or in amorphous regions of a partially crystalline polymer from (or to) a viscous rubbery condition to (or from) a hard and relatively brittle one.

glass transition temperature (Tg) The approximate midpoint of the temperature range over which the glass transition takes place.

glaze A hard-fired glass finish on pottery and tile, sometimes called *topping*.

glazing The securing of glass in prepared openings in windows, door panels, screens, partitions, etc.

glazing bead A strip surrounding the edge of the glass in a window or door; applied to the sash on the outside. It holds the glass in place.

glazing channel A three sided U-shaped sash detail into which a glass product is installed and retained by a removable stop.

glazing channel depth The measurement from the bottom of the glazing channel to the top of its stops.

glazing channel width The measurement between the stationary stop and the removable stop.

glazing compound Material used to seal the joint between window glass and the frame. This is commonly called *putty*, but glazing compounds are different, chiefly in that they will stay flexible, and therefore it is not likely that they will dry out as putty does.

glory hole An opening in a small furnace used to reheat glass articles.

gloss Brightness or luster of a plastic resulting from a smooth surface.

glue Originally, a hard gelatine obtained from hides, tendons, cartilage, bones and other connective tissues of animals. Also, an adhesive prepared from these substances by application of water and heat. Chemically known as a collagen.

glycerine A type of alcohol with three functional (-OH) groups.

glycol A type of alcohol with two functional hydroxyl (-OH) groups (a diol). Some are also referred to as chain extenders.

gobs Short lengths into which glass is cut for forming.

gradation The size distribution of aggregate particles, determined by separation with standard screen sieves.

graft copolymer A copolymer in which polymeric side chains have been attached to the main chain of a polymer of different structures.

greenhouse glass This is a translucent rolled glass with a special surface design to scatter light.

green strength The mechanical strength that, even though cure is not complete, allows removal of the mold and handling without tearing or permanent distortion.

groove joint A control joint made by forming a groove in the concrete surface before hardening to control crack location.

grout A mixture of cementitious material and water with or without aggregate, that is usually proportioned to produce a pourable consistency that will not segregate.

grouted masonry Masonry in which the interior joints are filled by placing grout into them as the work progresses.

GR-S One of the synthetic rubbers most nearly like crude rubber and is the product of styrene and butadiene copolymerization. It is one of the most widely used synthetic rubbers.

gum Any of a class of colloidal substances, exuded by or prepared from plants, sticky when moist, composed of complex carbohydrates and organic acids, which are soluble or swell in water.

gun consistency Compound formulated in a degree of softness or hardness suitable for application through the nozzle of a caulking gun.

gusset A piece used to give additional size or strength in a particular part of an object.

hardener The catalyst to cure epoxy. Also an additive added to concrete to set up hard.

hardness Property or extent of being hard. Measured by the extent of failure of an indenter point of a standard testing instrument to penetrate the product.

harsh mixture A mixture that lacks desired workability and consistency due to deficiencies in the cement paste, aggregate fines, or water.

hawk A tool to hold and carry plaster or mortar. Generally a flat piece of metal approximately 10 to 14 inches square, with a wooden handle fixed to the underside.

haze The cloudy or turbid aspect or appearance of an otherwise transparent specimen caused by light scattered from within the specimen or from its surfaces.

haze surface Indefinite cloudy appearance on the surface of plastic not describable by the terms chalking or lubricant bloom.

head The horizontal member that forms the top of a frame.

headers Are used to support structural members around openings in floors, roofs, and walls.

head of wall The gap between the top of a wall assembly and the lower surface of the floor assembly above it.

heat Energetic effect of accelerated molecular vibration.

heat-absorbing glass Glass (usually tinted) formulated to absorb an appreciable portion of solar energy. Values of absorption are defined in ASTM C 1036. It is obtainable in either float or patterned forms.

heat break This occurs at 90° to the surface of the glass and resembles smooth curve heat seal. Also to bond or weld a material to itself or to another material by the use of heat. This may be done with or without the use of adhesive, depending on the nature of the materials.

heat distortion point The temperature at which a standard test bar deflects under a stated load.

heat mark Extremely shallow depression or groove in the surface of a plastic visible because of a sharply defined rim or a roughened surface.
heat-resisting glass Glass able to withstand high thermal shock, generated because of a low coefficient of expansion.
heat seal To bond or weld a material to itself or to another material by heat alone.
heat strengthened glass Transparent or patterned glass with a surface compression of not less than 3500 psi or greater than 10,000 psi, or an edge compression of not less than 5000 psi.
heat treated Term sometimes used for both fully-tempered glass and heat strengthened glass.
heel bead Sealant applied at the base of channel, after setting light or panel and before the removable stop is installed, one of its purposes being to prevent leakage past the stop. Sealant must be able to bridge the gap between the glass and frame. See ASTM C 1048 (KindHS).
hermetically sealed A double section (two pieces) of glass sealed with a vacuum between. Thermopane® is an example. It is filled with a moisture absorbing material.
Hertz (Hz) The unit of frequency, representing cycles per second; named for Heinrich R. Hertz, the noted German scientist.
heterogeneous Differing or opposite in structural quality.
hide glue A non-waterproof glue made from animal hides, that is very strong, even with joints that are not well-fitted.
hiding power The power of a paint or pigment as used to obscure or render invisible a surface over which it is applied. It is one of the most important physical properties of a white pigment. It is determined by the difference in the index of refraction between the material and its surrounding medium.
high transmission glass Glass that transmits an exceptionally high percentage of light.
holiday In coated fabrics, a place not coated by coating compound.
homogeneity Uniformity of composition throughout the material.
homogeneous The opposite of heterogeneous. Consisting of the same element, ingredient, component, or phase throughout; or of uniform composition throughout. Crystalloid and crystalloid solutions (true solutions) are usually considered homogeneous systems of matter as opposed to colloids, which are heterogeneous or polyphasic.
hose stream test A test of the physical integrity of an assembly after a specified period of burning whereby it is removed from the furnace and exposed to a blast of water from the fireman's hose. ASTM E 119 specifies the nozzle size, pressure, duration and distance from the assembly.
hydrocarbon plastics Plastics based on resins made by the polymerization of monomers composed of carbon and nitrogen only.
hydrophilic Having a strong affinity for water.
hydrophobic Lacking affinity for water.

hydrostatic pressure A state of stress in which all the principal stresses are equal (and there is no shear stress), as in a liquid at rest. The product of the unit weight and the difference in elevation between the given point and the free liquid elevation.
hydrous Containing water.
hygroscopic Capable of absorbing and retaining atmospheric moisture.
hypalon A synthetic rubber base compound.
ignition point This is the temperature at which the liquid gives off sufficient vapors to burn continuously upon application of a flame, this temperature ordinarily being 40 degrees to 80 degrees higher than the flash point of the liquid. Most fire departments arbitrarily designate solvents as "extremely flammable," "flammable," and "combustible." These refer to solvents with flash points below 20° F (–2° C), and between 80° F and 150° F (25° C and 66° C) respectively. Liquids with no flash points are termed *non-flammable*.
immersion Placing an article into fluid, generally so it is completely covered.
impact The single instantaneous stroke or contact of a moving body with another either moving or at rest, such as a large lump of material dropping on a conveyor belt.
impact strength Measure of toughness of a material, as the energy required to break a specimen in one blow.
impervious tile It has a water absorption of 0.5 percent or less.
inert Not chemically reactive.
infrared (IR) Infrared radiation is used for the spectroscopic examination of high polymers. Infrared absorption spectra of high polymers are usually obtained for wavelength numbers 1800 to 600 cm and the percent of absorption recorded. Comparisons for structure are made of low molecular weight substances, which contain what are assumed to be similar groups.
inhibitor A substance that prevents a chemical reaction.
initial drying shrinkage The difference between the as-cast length of a specimen and its length when first cast.
injection molding The process of forming a material by forcing it, under pressure, from a heated cylinder through a sprue (runner, gate) into the cavity of a closed mold.
insert A part consisting of metal or other material that may be molded into position or pressed into the mold after completion of the molding operation.
insulating glass Multiple panels of glass separated by an air space and hermetically sealed to form a single glazed unit. It is also called *double glazing*.
intaglio A light engraving on the surface of glass.
intensity The rate of sound energy passing through a unit area. Normally expressed in dB, re 10-12 watts/sq. m.
interface The common boundary surface between two substances. Sometimes described as two surfaces with no air space between them (i.e., where the air contacts this paper is the air-paper interface).
interior glazed Glass set from the interior of the building.

interior muntins   Decorative grid installed on the glass lights that do not actually divide the glass.
interior stop   The removable molding or bead that holds the light in place when it is on the interior side of the light (as contrasted to an *exterior* stop which is located on the exterior side of a light or panel).
interlayer   The transparent damping material used in laminated glass.
intumescent   Materials that expand in volume when exposed to heat or flames exceeding a specified temperature.
iso   A prefix denoting an isomer of a compound.
I.S.O.   An acronym for the International Standards Organization.
isocyanate   A molecule composed of one nitrogen (N), one carbon (C), and one oxygen (O) in the functional configuration -N=C=O. It is generally the *A* side of a (poly)urethane reaction.
isolation joint   In building construction, a formed or assembled joint specifically intended to separate and prevent bonding of one element of a structure to another and having little or no transference of movement or vibration across the joint.
isomer   A molecule with the same number and kind of atoms and the same molecular formula as another molecule, yet is configured or arranged differently, often resulting in different chemical properties.
isostere   A plot of dew point versus temperature for the given values of water capacity of a desiccant.
isotastic   Pertaining to a type of polymeric molecular structure containing a sequence of regularly spaced asymmetric atoms arranged in like configuration in a chain polymer.
isotherm   A plot of the water capacity of a desiccant versus water partial pressure or dew point at a given temperature. Isotherms are often shown as a family of curves at various temperatures.
jalousie window   A type of window composed of a number of rectangular panes of glass placed in slots in a frame, one above the other. The panes are held in place by a pinned connection that rotates, allowing the panes to be opened simultaneously to provide a flow of air. When closed, the panes overlap slightly, preventing rain from entering the structure. They do not form a tight seal.
jambs   The vertical members of a frame adjacent to the structural members of a building.
join   To fasten two or more pieces together.
joint   The location at which two adherents are held together by an adhesive. See also *lap joint*; *starved joint*.
joint filler   See *backup*.
joint, lap   A joint made by placing one adherend partly to another.
joint, scarf   A joint made by cutting away similar angular segments of two adherends and bonding the adherends with the cut areas fitted together.
joint sealer   Caulking compounds, sealants, gaskets, and tapes used as part of a joint sealing system.

joint sealing system A combination of joint cleaners, backer rods, bond breakers, caulking compounds, sealants, gaskets, or tapes used to close joints between building components, sections, panels or dissimilar materials.

joint, starved A joint that has an insufficient amount of adhesive to produce a satisfactory bond.

joists Closely spaced members to carry light loads.

KD An abbreviation for *knock down*.

kicker Synonymous with the word *activator*, and sometimes actually added as a third material in a three part system.

knife consistency The sealant characteristics that describe the degree of hardness suitable for application with a glazing knife.

knuckle areas In reinforced plastics, the area of transition between sections of different geometry in a filament-wound part.

lack of fillout An area, occurring usually at the edge of a laminated plastic, where the reinforcement has not been wetted with resin.

laminate A product made by bonding together two or more layers of material or materials.

laminated glass Two or more lights of glass bonded together with a plastic interlayer.

laminated insulating glazing A laminated glass panel and a monolithic glass panel set in a frame that provides an air space between them.

lap A part that extends over itself or a like part.

lap joint A joint made by overlapping adjacent edge areas of two adherents to provide facing surfaces that can be joined with an adhesive.

lap seam A seam made by placing the edge of one piece of material extending flat over the edge of the second piece of material.

latex A colloidal dispersion of a rubber resin (synthetic or natural) in water which coagulates on exposure to air.

latex additive Rubber or resins in water, which coalesce to form a continuous film that imparts specific properties to Portland cement products. It improves adhesion, frost resistance, color retention, flexural and impact strength, and stain resistance.

latex/cement systems (mortars, grouts) Basically cementitious mixtures to which latex has been added to increase flexibility and reduce permeability.

lathe cutter Equipment used to slit rolls of tape into narrower widths.

lattice pattern In reinforced plastics, a pattern of filament winding with a fixed arrangement of open voids.

lay The length of twist produced by stranding—singly or in groups—filaments, such as fibers, wires, or roving; or the angle that such filaments make with the axis of the strand during a stranding operation.

lay up In reinforced plastics, an assembly of layers of resin-impregnated material ready for processing.

Ldn Day/night average sound level in dB.

legging The stringing out (filamentation) of a pressure-sensitive adhesive when a tape or label is drawn away from a surface, its release liner, or its matrix.
lehr Similar to an oven, used for reheating glass and allowing it to cool slowly.
Leq Equivalent sound level.
level The condition of perfect horizontal alignment.
lifting Softening and penetration of a film by the solvents of another film resulting in raising and wrinkling.
light (1) A term used for optical glass having a low index of refraction. (2) An architectural term for a panel or sheet of glass. See also *lite*.
light reducing glass Glass formulated to reduce the transmission of visible light.
light transmittance Clear glass depending on its thickness, it allows 75 to 92 percent of the visible light to pass through.
linear A molecule having one or two reactive sites, as with some polyols, isocyanates, or chain extenders, or polyurethanes made from them.
linseed oil Oil made from flax plants, used as the principal ingredient in paints or coatings.
lintel A horizontal structural member that spans an opening at the head to carry the weight of construction above the opening.
lite Another term for a pane of glass used in windows.
liter A measure of capacity in the metric system equal to 61.022 cubic inches, 0.908 US quarts dry and 1.0567 US quarts wet.
load, dynamic A loads that vary at times. They include impact and repeated loads.
load, impact The force that requires a structure or its components to absorb energy in a short interval of time.
load, live Load force due to the weight of non-permanent attachments: people, glazing rigs, washing rigs.
load, repeated The force that is applied a number of times, causing variation in the magnitude, and sometimes also in the internal forces.
load, seismic The force that produces maximum stresses or deformation in a building during an earthquake.
load, snow The maximum force that may be applied by snow accumulation in a mean recurrence interval.
load, static The force that is applied slowly and then remains nearly constant.
load, wind The maximum force that may be applied to a building by wind in a mean recurrence interval.
low-emissivity or low-E glass Glass with a transparent metallic or metallic oxide coating applied onto, or into, a glass surface, which reflects long-wave infrared energy and thus improves the U-value.
low pressure molding Distributing a relatively uniform low pressure over a resin-bearing fibrous assembly of cellulose, glass, asbestos, or other material, with or without application of heat from an external source, to form a structure possessing definite physical properties.
low temperature flexibility The ability of a rubber product to be flexed, bent, or bowed at low temperature.

L-rating A fairly recent addition to UL 1479 that provides a measure of the ability of a firestop sealing design to prevent leakage when subjected to a known over pressure and air flow. Test is conducted at ambient temperatures and at elevated temperatures of 400° F (204.44° C).

lubricant An addition to the plasmic mix, or an old coating, to prevent sticking of the molded pieces.

Lucite A trade name for a clear tough plastic used for items such as skylights and windows.

lugs Inserts attached to tiles to maintain even spacing for grout lines.

luminous transmittance The ratio of the luminous flux transmitted by a body to the flux incident upon it.

lumps Surface protrusions, usually of the basic material as distinguished from foreign material.

mandrel A forming or shaping tool used to make glass tubing.

manganese A metal powder used as a flux in glass making.

marver To roll molten glass on an iron slab.

masonry, ashlar Masonry composed of rectangular units usually larger than brick and properly bonded, having sawed, dressed, or squared beds.

mastic An adhesive of such a consistency that it must be applied by notched trowel, gob, or buttering. It is especially advantageous for lightweight installation of non-load bearing surfaces.

mat A fibrous material of randomly oriented, chopped, or swirled filaments loosely held together with a binder.

matte Dull finish.

matter Anything that has mass and occupies space.

mechanically formed plastics A cellular plastic in which the cells are formed by the physical incorporation of gases.

melt index The amount in grams of a thermoplastic resin that can be forced through a 0.0825 in. (2mm) orifice when subjected to 76.2 oz. (2160 gms) for 10 minutes.

melting point The temperature at which a resin or polymer loses its crystalline character, as evidenced by X-ray diffraction studies. For low molecular weight solids, it is the temperature at which a solid melts and becomes liquid.

membrane penetration In hollow wall or floor construction, an opening made through one side (membrane) of the assembly.

metal lath See diamond mesh.

metal spacers Roll-formed metal shapes used at the edges of an insulating glass unit to provide the designated air space thickness.

meta-substitution A substitution on a benzene ring at Site 1 and a location one site removed (1, 3 substitution).

methylene diphenyl diisocyanate (MDI) An aromatic isocyante produced in two forms: a polymeric and pure MDI.

migration Spreading or creeping of oil or vehicle from a compound onto adjacent non-porous surfaces, as contrasted to bleeding which refers to absorption into adjacent porous surfaces.

milliliter One-thousandth of a liter, equal to one cubic centimeter.

miscible Capable of being mixed to form a homogeneous mass.

mitred corners Usually a 45-degree mitred joint produced in some sash where vertical jamb members meet horizontal head and sill members. For mitred glass joints, see setting No.38.

mixer A machine, other than a mill, for mixing rubber compounds, dough, or a covered chamber or trough in which two blades or rotors revolve in opposition to each other. The axis of the blades may be horizontal as in the Banbury mixer, or vertical as in some types of cement mixers. In the latter case, the mixing chamber may rotate as well as the blades.

mixing The process of incorporating the ingredients of a rubber compound into rubber, usually done on a mixing mill or in an internal mixer. The mixing process consists of: (1) breaking down the rubber, (2) gradual incorporation of compound ingredients, (3) final working of the rubber (*cutting back*) after all the ingredients are in, and (4) removing the mixed compound from the mill in sheets. *Mixing*, or simply *mix*, also denotes the completed mixture.

modified epoxy mortar/grout A mortar/grout system employing water-cleanable emulsified epoxy resins and hardeners with portland cement and silica sand. Not designed for chemical resistance or exterior use.

modifier Any chemically inert ingredient added to an adhesive or sealant that changes its properties.

modulus In the physical testing of rubber, the ratio stress to strain, i.e., the load in pounds per square inch or kilos per square centimeter of the initial cross-sectional area necessary to produce a stated percentage-elongation. It is a measure of stiffness, is influenced by pigmentation, state of cure, quality of rubber and other factors.

moisture For all practical purposes, moisture may be considered as very finely divided particles of water. Moisture in the form of steam or a jet of water is sometimes used in a kiln to regulate the humidity.

moisture absorption The absorption of moisture or water vapor from air by a material. It relates only to vapor withdrawn from air by a material and must be distinguished from water absorption, which is the gain in weight due to the take-up of water by immersion.

moisture vapor transmission rate (MVTR) The steady water vapor flow in unit time through a unit area of a body, normal to specific parallel surfaces, under specific conditions of temperature and humidity at each surface.

mold The cavity or matrix into or on which the plastic compound is placed and from which it takes its form.

molding The shaping of a plastic composition within or on a mold; normally accomplished with the application of heat and pressure.

molding, contact pressure A method of molding or lamination in which the pressure is only slightly more than necessary to hold the materials together during the molding process. This pressure is generally less than 10 psi (69 kPa).

molding, high-pressure Molding or laminating in which the pressure used is greater than 200 psi (37 MPa).

molding pressure, compression The calculated fluid pressure applied to the material in the mold.
molding pressure, injection The pressure applied to the cross-section area of the material cylinder.
molding pressure, transfer The pressure applied to the cross-sectional area at the material pot or cylinder.
mold marks Defects or parting lines in the mold imparted to the molded material.
mold shrinkage The immediate shrinkage a molded part undergoes when removed from a mold and cooled to room temperature.
molecular sieve (13-X variety) An adsorbent. Any of a class of zeolite having small, precisely uniform pores in their crystal lattices that can absorb molecules small enough to pass through the pores. The cage structure is such that the pore openings are approximately 8.5 Angstroms.
molecular sieve (4-A variety) As above, but with a crystalline cage structure such that pore openings are approximately 4 Angstroms.
molecular weight The sum of the atomic weights of all atoms in a molecule.
molecule The result of combining two or more atoms.
mono A prefix denoting a single radical.
mono cottura An Italian expression signifying *once-fired*. Color-rich, textured, or patterned glazes can be fired into thin tile units at relatively vitrifying temperatures in one precisely controlled pass.
monolithic A glazing assembly construction consisting of only one light or pane of glass, polycarbonate, acrylic or plastic.
monolithic glass Glass having a single uniform thickness.
monomer A substance or simple chemical that can be polymerized, yielding a much larger molecule called a polymer.
mortar A mixture of cement paste and fine aggregate.
mortar bed A setting bed of mortar, especially cementitious mortar. See *full bed mortar*.
mosaic A small vitreous tile, dust-pressed from fine dense clay and fired at high vitrifying temperatures. Usually 4 square inches or less in size.
mottling A film defect associated with spraying. It appears as a uniform series of approximately circular imperfections.
mucilage An adhesive prepared from a gum and water. Also, in a more general sense, a liquid adhesive that has a low order of bonding strength.
mullion A horizontal or vertical member that holds together two adjacent lights of glass, or units of sash, or sections of curtain wall.
multiple glazed units Units of three lights (triple-glazed) or four lights (quadruple-glazed) with two and three air spaces, respectively.
muntin In sash having horizontal and vertical bars that divide the window into smaller lights of glass, the bars are termed muntin bars. Similar to mullion but lighter weight.
nanometer-(nm) A term sometimes used in place of millimicron (Mμ () in expressing the wavelength of light.

natural clay tile Ceramic, mosaic, or paver tile made by the dust-pressed or plastic methods. Made from clays that produce a dense body with a distinctive, slightly textured surface.

natural rubber The elastomer obtained from the hevea tree. The basic polymer is also present in other shrubs and trees. The first truly elastomeric type of product known.

necking The localized reduction in cross-section that may occur in a material under tensile stress.

needle glazing Application of a small bead of compound at the sight line by means of a gun nozzle about ¼ inch (⅛ inch in opening size).

neoprene A synthetic rubber having physical properties closely resembling those of natural rubber but not requiring sulfur for vulcanization. It is made by polymerizing chloroprenes (CH), and the latter is produced from acetylene and hydrogen chloride.

neutrons A particle in the nucleus of an atom with a mass slightly greater than that of a proton. Neutrons have no electrical charge.

nitrile rubber A class of rubber-like copolymers of acrylonitrile with butadiene. There are many types and a few of the trade names are: Buna-N, Butaprene, and Chemigum. It has high resistance to solvents and oils, greases, heat and abrasion.

noise reduction between rooms In decibels, the amount by which the mean square sound pressure level averaged throughout the source room exceeds the sound pressure level averaged throughout the receiving room.

non-ceramic tile Materials such as precast terrazzo, marble, or slate that are cut in the shapes and sizes characteristic of tile, and set by using the same methods and materials.

noncombustible Materials that will not burn or melt during a fire.

nondrying Descriptive of a compound that does not set up hard.

nonoxidizing Descriptive of a compound that withstands accelerated weathering, the equivalent of 20 years of normal weathering without oxidizing. Does not become hard after exterior exposure.

nonskinning Descriptive of a product that does not form a surface skin after application. Usually remains tacky and sticky.

nonstaining Characteristic of a compound that will not stain a surface by bleeding or migration of its oils or vehicle content.

nonvented Piping systems that do not allow the free passage of air, e.g. hot or cold water supply pipes, and electrical conduit. Also referred to as closed.

nonvitreous With water absorption lower than 7 percent, relatively inexpensive, generally non-load bearing, ceramic tiles that are dry-pressed from soft clays under lower temperatures than vitreous tiles.

nonvolatile Any substance that does not evaporate or volatilize under normal conditions of temperature and pressure.

notched trowel Trowel with a serrated or notched edge used for spreading mortar or adhesive in ridges of a specific thickness. These ridges collapse as a tile is beaten in and form a suction aiding in holding the tile in place until set.

These trowels are available in varying sizes to ensure that enough mortar is placed on the substrate to fully bed the tile.

novolak   A phenolic-aldehyde resin that, unless a source of methylene groups is added, remains permanently thermoplastic.

NR   Noise reduction.

nucleus   The central core of an atom containing protons and neutrons around which electrons move.

nylon plastics   Plastics based on resins composed principally of a long-chain synthetic polymeric amide, which has recurring amide groups as an integral part of this main polymer chain.

olefin plastics   Plastics based on polymers made by the polymerization of olefins or copolymerization of olefins with other monomers, the olefins being at least 50 percent.

oleoresinous   A mixture of natural or synthetic resins mixed with drying oils.

opalescence   The limited clarity or vision through a sheet of transparent plastic at any angle, because of diffusion within, or on the surface of, the plastic.

opaline glass   This glass is closely related to opaque. It is an opaque cast with ground and polished surfaces.

opaque glass   Glass that transmits no light.

open-cell foamed plastic   A cellular plastic in which there is a predominance of interconnected cells.

open time   Time interval between spreading the adhesive and completing the bond.

optimum cure   State of vulcanization at which the maximum desired property is attained.

orange peel   A surface defect caused by vortex currents set up during evaporation of solvents.

organic   Compounds that consist of carbon and, generally, hydrogen, with a restricted number of other elements, such as oxygen, nitrogen, sulfur, phosphorous, chlorine, etc., but not containing atoms or molecules, generally known as metals.

organic-metal chemistry   One or two types of catalysts used in a polyurethane formulation that activate polyols to react quickly with isocyanates.

organosol   Essentially a plastisol that contains solvent that must be evaporated prior to exposing the material to the elevated temperature necessary for fusion or curing.

orifice ring   Bowl-shaped, with a hole in the bottom.

ornamental glass   Rolled glass with the surface figured by shaping or embossing rolls.

outside glazing   A method in which glass is secured in an opening from the exterior of the building.

overcure   A state of excessive vulcanization resulting from overstepping the optimum cure i.e., vulcanizing longer than necessary to attain full development of physical strength. Manifested by softness or brittleness and impaired age-resisting quality of the vulcanizate.

oxides A component in polyol production, generally propylene oxide or ethylene oxide, that affects the properties of the polyol.

ozone resistant Withstands the deteriorating effects of ozone (generally cracking).

paint A pigmented liquid composition that is converted to an opaque solid film application in a thin layer. An oil base paint contains drying oil or oil varnish as the basic vehicle. A water base paint contains a water emulsion or dispersion as the vehicle. The term is loosely used, sometimes designating the whole coating field.

pane A single section of window. See *light*.

parallel laminate A laminate in which all the layers of material are oriented approximately parallel with respect to the grain or strongest direction in tension.

parison The rough shape of a glass item. It is sometimes known as a *blank*.

parlon The trade name for a chlorinated natural rubber. (Hercules Powder Co.)

parting line Mark on a molded piece where the sections of mold met in closing.

pass Term used in spraying to refer to the movement of a spray gun in one direction. A double pass, back and forth, is actually a single coat.

passive materials Materials that do not react with heat. Non-intumescent materials.

passive solar heat gain Solar heat that passes through a material and is captured naturally, not by mechanical means.

paste An adhesive composition having a characteristic plastic consistency, that is, a high order of yield value, such as that of a paste prepared by heating a mixture of starch and water and subsequently cooling the hydrolyzed product.

pattern cracking Fine opening on concrete surfaces in a pattern, resulting from a decrease in volume of the material near the surface or increase in volume of the material below the surface, or both.

patterned glass Rolled glass having a distinct pattern on one or both surfaces. Used extensively for light control and decorative glazing.

paver tile Dust-pressed vitreous (ceramic, mosaic) tile larger than 2 or 3 inches and somewhat thicker than an ordinary mosaic tile; can be extruded or glazed.

peel back A method of separating a bond of two flexible materials or a flexible and rigid material, whereby the flexible material is pulled from the mating surface at 90° or 180° angle to the plane to which it is adhered. The stress is concentrated only along the line of immediate separation. Strengths are expressed in pounds per inch width (piw).

peeling The loosening of a rubber coating or layer from a base material, such as cloth or metal, or from another layer of rubber.

penetration The entering of an adhesive or sealant into an adherend.

permanence The resistance of an adhesive bond to deteriorating influences.

permanent set The amount by which an elastomeric material fails to return to its original form after a deformation. In the case of elongation, the difference between the length after retraction and the original length, expressed as a percentage of the original length is called the permanent set. Permanent set is dependent on quality and type of rubber, degree and type of filler loading, state of vulcanization, and amount of deformation. Also see *adhesive setting*.

permeability   The time of water vapor or gas transmission through a unit area of the material of unit thickness, induced by unit vapor pressure differences between two specific surfaces under specified temperature and humidity conditions.

permeance   The time rate of water vapor or gas transmission through a unit area of a body, normal to specific parallel surfaces, under specific temperature and humidity conditions.

pervious   Cracks, crevices, leaks or holes larger than capillary pores, that permit a flow or leakage of water, are present. The material may or may not contain capillary holes.

phenol formaldehyde   This material provides the greatest variety of thermoplastic materials. They are hard and rigid and they change slightly, if at all, on aging indoors but, on outdoor exposure, lose their bright surface gloss.

phenolic plastics   Plastics based on resins made by the condensation of phenols such as phenol and cresol, with aldehydes.

phenolic resin compound, single-stage   A phenolic material in which the resin, because of its reactive groups is capable of further polymerization by application of heat.

phenolic resin compound, two-stage   A phenolic material in which the resin is essentially not reactive at normal storage temperatures, but contains a reactive additive, which causes further polymerization upon the application of heat.

phenyl group   Refers to a benzene ring which is substituted onto another functional group.

pigments   A finely divided material added to the formulation to add color or to increase chemical or physical resistance.

pimple   An imperfection, a small sharp, or conical elevation on the surface of a plastic product.

pinholing   A film defect characterized by the presence of tiny holes. The term is rather generally applied to holes caused by solvent bubbling, moisture, other products, dry spraying, or the presence of extraneous particles in the applied film.

pit   An imperfection, a small crater in the surface of the plastic, with its width approximately the same order of magnitude as its depth.

pitch   A black or dark heavy liquid or solid substance left as residue after distilling tar, oil, and similar materials; also found naturally as asphalt. Pitches are named according to the source from which they are obtained as *bone pitch* from bone oil, *petroleum pitch* from petroleum, etc.

plastic cracking   Cracking that occurs in the surface of fresh concrete soon after it is placed and while it is still plastic.

plastic deformation   Deformation beyond the elastic limit.

plastic flow   Gradual time-dependent deformation due to sustained load. See *creep*.

plasticity A property of plastics that allows the material to flow and be deformed continuously and permanently without rupture upon the application of a force that exceeds the yield value of the material. (The opposite of elasticity).

plasticizer Materials added to the formulation to reduce the hardness of the polymer.

plastics Natural and artificially prepared organic polymers of low extensibility as compared with rubber, which can be molded, extruded, cut and worked into a great variety of objects, rigid or non-rigid, and used as substitutes for wood, metal, glass, rubber, leather, fibers, and textile materials. Many are also referred to as synthetic resins. The first commercial plastic was celluloid, introduced by Hyatt in 1869. The first commercial thermo-setting resin was introduced by Baekland in 1909. There are two general methods of formation: (1) condensation polymerization, as in the case of phenol-aldehyde resins, and (2) vinyl polymerization, as in the case of polyvinyl chloride resins. Certain plastics are derived from casein. Some of the more recent products are organo-inorganic, such as the silicones.

plastisol A physical mixture of resin (usually vinyl) compatible plasticizers, stabilizers and pigments. A well-known example is Bakelite.

plumb The condition of perfect vertical alignment.

pointing The process of repairing mortar joints by removing old mortar and adding new mortar—also called tuckpointing.

points Thin, flat, triangular or diamond-shaped pieces of zinc used to hold glass in a wood sash by driving them into the wood.

Poisson's Law The ratio of transverse to axial strain, also is measured in tension tests. It may be taken as 0.30 in the elastic range and 0.50 in the plastic range for structural steels.

polybutene base Compound made from polybutene polymers.

polycarbonate A polyester polymer in which the repeating structural unit in the chain is of the carbonate type.

polycarbonate plastics Polyester plastics based on polymers in which the repeating structural units in the chain are essentially all of the carbonate type.

polyester There are many types of polyester resins and they are manufactured by reacting together a dicarboxylic acid and a dihydroxy alcohol. Polyesters are used in one- and two-part systems for coatings, molding compounds, and the manufacture of Dacron, a well-known polyester fiber.

polyether One of the two types of poylols (and by far the most widely used). Named because they contain an ester group.

polyethylene A straight chain plastic polymer of ethylene (gaseous hydrocarbon) used for containers, packing, etc.

polyisobutylene Polymer manufactured from gaseous hydrocarbons. The polymer is a major portion of butyl rubber, which also contains a small percent of isoprene.

polyisocyanates Isocyanates containing more than one -N=C=O group.

polymer A very long chain of units of monomers prepared by means of an addition and/or a condensation polymerization. The units may be the same or different. There are copolymers, tri- or ter-polymers, quadri-polymers and high-polymers.

polymerize To unite molecules of the same kind into another compound having the elements in the same proportion but possessing much higher molecular weight and different physical properties.

polymerized Treated by heating or cooking, chemically induced so that molecules of different substances unite into larger molecules of a different substance with individual characteristics.

polyolefin A polymer prepared by the polymerization of an olefin(s) as the sole monomer(s).

polysulfide elastomer A synthetic rubber-like elastomer practically insoluble in oils and solvents, prepared from ethylene chloride and sodium tetra sulfide, commonly called Thiokol. It was the first commercial synthetic elastomer (USA 1930). Other dichlorides used are di-(chloroethyl) ether and di-(2-chloroethyl). These are not vulcanized with sulfur but by heating with zinc oxide. Also see *Thiokol*.

polyterephthalate A thermoplastic polyester in which the terephthalate group is a repeated structural unit in the polymer chain.

polyurethane A family of polymers ranging from rubbery to brittle. Usually formed by the reaction of a di-isocyanate with a hydroxyl.

poly (vinyl acetate) Polymers prepared by polymerization of vinyl acetate as the sole monomer.

poly (vinyl alcohol) (PVA) Polymers prepared by the essentially complete hydrolysis of polyvinyl esters.

poly (vinyl chloride) (PVC) A polymer prepared by the polymerization of vinyl chloride as the sole monomer.

pontil A metal rod, to which a glass article is attached during its production.

porosity Presence of numerous visible or invisible voids.

Portland cement mortar A mixture of Portland cement and sand roughly in proportions of 1:6 on floors; and of Portland cement, sand, and lime in proportions of 1:5:0.5 to 1:7:1 for walls.

post cure In certain resins the complete cure and ultimate mechanical properties are attained only by exposure of the cured resin to higher temperatures. This second stage is the post cure and is necessitated by the fact that the higher temperatures would result in excessive reaction if used throughout the entire cure.

potash Potassium oxide (a flux).

pot furnace A pot-shaped furnace made of clay.

pot life The rating in hours of the time interval following the addition of accelerator before the material will become too viscous to pass predetermined viscosity (consistency) requirements. Closely related to *working life*.

power factor In an insulating material, the ratio of total power loss (watts) in the material to the product of voltage and current in the capacitor in which that material is a dielectric.

preform   A preshaped fibrous reinforcement formed by distribution of chopped fibers by air or water flotation.

pre-grouted tile (edge bonded)   Surfacing units consisting of an assembly of ceramic tile bonded together at the edges by a material, generally elastomeric, which seals the joints completely. Such materials (grout) may fill the joint completely or partially, and may cover all, a portion, or none of the back surfaces of the tile. The perimeter of the factory-grouted sheets may include all, part of, or none of the joint between sheets. The term *edge-bonded* is sometimes used to designate a particular type of pregrouted tile sheet having the front and back surfaces completely exposed.

premix   In reinforced thermosetting plastics, the admixture of resin reinforcements, fillers, which are not in web or filamentous form, to get ready for molding.

prepolymer   The result of prereacting anisocyanate with some of the polyols as an intermediate step in formulating polyurethanes.

pressure break   Occurs at angles to the surface, usually starting at a corner.

pressure sensitive adhesive   Type of adhesive that retains its tack even after the complete release of the solvent.

primary sealant   A sealant applied to the inner shoulders of a spacer with the principal purpose to minimize moisture, gas, and solvent migration into the unit air space.

primer   Special coating designated to provide adequate adhesion of a coating system to a new surface. In the case of new wood, it is used to allow for the exceptional absorption of the medium. Metal priming coatings for steel work contain special anti-corrosive pigments or inhibitors, such as red lead, white lead, zinc powder, zinc chromate, etc.

priming   Sealing of surfaces to produce adhesion of sealants, that the compound will not stain, lose elasticity, or shrink, excessively because of the loss of oil or vehicle into the surround. A sealant primer or surface conditioner may be used to promote adhesion of a curing-type sealant to certain surfaces.

processed glass   Glass in which the surface has been altered by etching, sand blasting, chipping, grinding, ceramic-enameling, etc., to change its light diffusion or to give it decorative effects. Either or both surfaces may be so treated. Also, glass which has been further treated such as, edge work, tempered, stained, and so forth, after forming.

product expectations   The expectations of performance that are raised by review of product literature, samples, and the like.

profile glass   A U-shaped rolled glass for architectural use.

project needs   The specific requirements that are found to exist for a given project.

propylene plastics   Plastics based on polymers of propylene or copolymers of propylene being in the greatest amount by mass.

proton   A particle that, with neutrons, forms the nucleus of an atom. Protons have a positive electrical charge.

purlins   Structural members, generally horizontal, on sloped glazing frames.

putty General term for a variety of soft, plastic materials used to seal or fill building products. The following are types of putty:
> *Electrician's putty*: Used to fill gaps where conduit passes through openings.
> *Painter's putty*: Used to fill small holes.
> *Plumber's putty*: A bead of this is laid around the opening where a sink is to go. When the sink is in place, the putty acts as a seal.
> *Wood putty*: Material used to fill gaps in woodwork or cover countersunk nails or screws.
> *Window putty*: Formerly used to seal glass in a frame, but now displaced by glazing compounds and sealants.

Pyrex Trade name for borosilicate glass.

pyrolytic deposition A process in which metallic oxide is added to glass while the glass is still hot. Because they are actually part of the glass sheet, pyrolytic or *hard coat* surfaces are very durable and require no special handling.

quality assurance (QA) A program of actions and procedures that are implemented to achieve and maintain a desired level of quality.

quality control (QC) A system designed to ensure that a desired level of quality is maintained in a product being produced under a defined standard. The process involves periodic inspection of the manufacturing procedures and materials to ensure that minimum standards are being met.

rabbet A two-side L-shaped recess in sash or frame to receive lights or panels. When no stop or molding is added, such rabbets are faced glazed. Addition of a removable stop produces a three-sided U-shaped channel.

racking Movement and distortion of sash or frames because of the lack of rigidity, or caused by adjustment of ventilator sections. Puts excessive strain on the sealant and may result in joint failure.

radiation Transfer of energy in wave form, from a hot body to a colder body independent of any matter between the two bodies.

rafters Structural members; vertical in sloped glazing frames.

raking The removal of slightly hardened mortar from a joint in a masonry wall to accent the line of the joint.

raw glass It is rolled glass with the surface rolled smooth or slightly figured.

reaction A mutual action of chemical agents upon each other resulting in a chemical change.

reaction injection molding (RIM) A process by which a glycol and isocyanate are rapidly mixed and injected into a mold where they co-react to form a shaped urethane product.

reactor A substance undergoing a reaction or chemical change. Also refers to the equipment use in the polymerization process.

rebar Reinforcing steel bars or rods.

reflective coated glass Glass with metallic or metallic oxide coatings applied onto or into the glass surface to provide reduction of solar radiant energy, conductive heat energy and visible light transmission.

reglet Any slot cut into masonry or formed into poured or precast stone. May also be an open mortar joint left between two courses of bricks or stones, or a slot cut or cast into other types of building materials.

reinforced plastic A plastic with high strength fillers embedded in the composition, resulting in some mechanical properties superior to those of the base resin.

reinforcement Material used primarily for strengthening a plastic product, i.e. fibrous glass, cotton, etc., which is impregnated with the plastics mixture. Should not be used as a synonym for filler.

relative heat gain An energy comparison factor for glass products combining the radiant and conductive heat gain under specific conditions (200 BTUs times the shade coefficient + 14 degrees times the summer U-value).

release agent A material used to keep a molding material from adhering to a mold.

resilient tape A preshaped, rubbery sealing material furnished in varying thicknesses and widths, in roll form. May be plain or reinforced with scrim, twine, rubber or other materials.

resin A solid or semi-solid organic material, often of high molecular weight, that has a tendency to flow when subjected to stress. It usually has a softening or melting range and usually fractures cohesively.

resin deposition water repellent products Products that function by depositing a resin on and into a concrete or masonry surface, as opposed to producing a water repellent effect by chemical reaction

resinoid Any of the class of thermosetting synthetic resins, either in their initial, temporarily fusible state, or in their infusible state.

resin rich area In a reinforced plastic part, a space that is filled with a resin and little or no reinforcing material.

resistance to freezing or thawing This can be accomplished by proper air entrapment in the concrete, use of a mix with minimum water content, and proper curing of the concrete.

resite An alternate term for C-stage.

resitol An alternate term for B-stage.

resol An alternate term for A-stage.

resonance The sympathetic vibration of an object when subjected to a vibration of a specific frequency. The object tends to act as a sound source.

restraint Restriction of free movement of hardened concrete. Resistance can be internal or external and may act in one or more directions.

retardation Slowing down the rate of setting or hardening, usually in hot weather, to gain an increase in the time required to reach initial and final set or to develop the early strength of concrete, mortar or plaster.

retarder An admixture that delays the setting of concrete.

retempering Adding water and remixing concrete or mortar that has started to stiffen and becomes harsh.

retrogradation A change of starch pastes from low to high consistency on aging.

reverberation The continuation of sound reflections within a space after the initial sound has ceased.

reverberation time Normally defined as the time that elapses for a sound level to drop 60 decibels after a mean-square sound level is abruptly stopped after being continuously generated at a steady pressure in a room.

reversion The change that occurs in vulcanized rubber as the result of aging or overcuring in the presence of air or oxygen, usually resulting in a semi-plastic mass. It is the basis of rubber reclaiming processes and is aided by the use of swelling solvents, chemical plasticizers, and mechanical disintegration to obtain a workable mass.

Rex hardness The hardness of a *soft* vulcanized rubber or other similar elastic material as measured by the Rex Hardness Gage.

rheology Science of deformation and flow of matter. Deals with laws of plasticity, elasticity, viscosity, and their connection with paints, plastics, rubber, oils, glass, cement, and so forth.

rigidity The property of bodies by which they can resist an instantaneous change of shape. The opposite of elasticity.

rising damp Upward movement of ground water through capillary action.

rolled glass Glass formed by rolling, including patterned and wired glass. As the glass is drawn horizontally from the tank, figured, engraved, or etched, machine rolls impress a pattern on the surface of the glass, varying from almost smooth to deeply geometric, fluted, or random overall designs. It is made in ⅛ inch to ⅜ inch thickness.

room temperature setting adhesive Adhesives that set at 65° F (18.33° C) or below. This is the limit for room temperature specified by ASTM standards.

room temperature vulcanizing (RTV) Vulcanization or curing at room temperature by chemical reaction, particularly of silicones and other rubbers.

rosin A resin obtained as a residue in the distillation of crude turpentine from the sap of the pine tree (gum rosin) or from an extract of the stumps and other tree (wood rosin).

rowlock A header unit that is set on its edge.

rubbing stone Abrasive stone that is used to smooth rough edges of tile.

rust-corrosion Rusting or oxidation of ferrous metal. Principally, causing volumetric expansion and fracturing of surrounding concrete or masonry.

R-Value The resistance of conductive heat energy transfer in one hour through one-square-foot of a specific insulating glass unit assembly for each one degree Fahrenheit temperature difference between the indoor and outdoor air. It is the reciprocal of U-value; $R = 1/U$

safing joint The gap between the floor in a high-rise building and the curtain wall.

sag and flow test Involves vertical applications of compounds to specified surfaces and shapes under predetermined conditions of temperature and time intervals. Tendency to sag or run is observed and is reported as none, very slight, slight, etc., or may be reported as a lineal movement.

sagging Running or flowing in the finish of a coating caused by the application of too much material and/or material that is too thin. See *slumping*.
sailor A brick or block standing on end with the broad face showing.
salt spray test A testing method to compare the corrosion resistance of materials, usually coatings. One of the best corrosion test media known is vapor of salt solutions. The most common procedure is to spray a fine mist of a 20 percent by weight solution of sodium chloride (iron-free) in water into a large, closed container in which the test panels are suspended. The opening temperature is about 95° F (35° C).
sandblasted finish A surface treatment for flat glass obtained by spraying the glass with hard particles so as to clip out and roughen one or both surfaces of the glass. The effect is to increase obscurity and diffusion, but it can make the glass fragile and hard to clean.
sand core A shape made of sand, around which strands of molten glass are wound.
sash A frame into which glass products are glazed, i.e. the operating sash of a window.
scarifier A tool with flexible steel tines used to scratch or rake the unset surface of a first coat of plaster. Also a tool used to roughen the surface of concrete.
score side The upper side of glass coming off the float line, sometimes called the air side.
scoring Grooving, usually horizontal, of the scratch coat of stucco to provide a mechanical key with the brown coat.
scratch coat Coarse cementitious mortar coat applied to smooth, fill, and sometimes level a subsurface prior to application of a mortar setting bed. Its surface is usually scratched or roughened to ensure a proper bond with subsequent mortar coats; the first coat of mortar on a wall or ceiling.
screed strip Strips of wood, metal, or other material applied to a surface, at suitable distances from each other, to serve as a guide for a straightedge to obtain a true mortar surface.
screw-on-bead or stop Stop, molding, or bead fastened by screws as compared to a sealer that is a liquid used to seal a porous surface.
sealant A material used to fill a joint, usually for the purpose of weatherproofing or waterproofing. It forms a seal against gas and liquid entry.
sealant (building joints) Compound used to fill and seal a joint or opening, as contrasted to a sealer that is a liquid used to seal a porous surface.
sealant (for insulating glass units) Formulated elastomeric compounds of specific application and vapor transmission properties as well as controlled adhesion, cohesion and resiliency properties.
sealant spacer A permanent adhesive sealant extrusion that may contain a structural metal insert and a pre-compounded desiccant.
sealed insulated glass units Units constructed of two or more lights of glass separated and hermetically sealed to spacer frames at the glass edges with the enclosed air chamber(s) dehydrated at the plant's atmospheric pressure.

sealer (1) A continuous film to prevent the passage of liquids or gaseous media; a high-bodied adhesive generally of low cohesive strength to fill voids of various sizes to prevent passage of liquid or gaseous media. (2) A coating used to seal the sand-scratched surface of a primer to obtain a smooth, uniform paint base over rough metal. Sealants are products of low pigmentation.

seam A line formed by joining material to form a single ply or layer. See also *lap seam*; *transverse seam*.

secondary sealant A sealant applied into the exterior glass-spacer cavity to provide elastic, structural bonding of the assembly. In single-sealed units, this sealant also has low gas and moisture transmission properties to achieve effective unit performance.

selenium A metal used in powdered form as a flux in glass making.

self-vulcanizing Pertaining to an adhesive that undergoes vulcanization without the application of heat.

semi-cure Partially cured. A term frequently used to designate the first cure of an article that is subjected to more than one cure in its manufacture.

semi-rigid plastic For purposes of general classification, a plastic that has a modulus of elasticity either in flexure or in tension greater than 10,000 and 100,000 psi (69 and 690 mPa) at 73° F (23° C) and 50% relative humidity when tested in accordance with appropriate ASTM methods.

service life The period of time over which significant benefits of some type will remain in existence.

set A term used rather loosely to describe the point at which a film has either dried sufficiently (released enough solvent) so that it is tough or hard, or has cured sufficiently after the addition of the accelerator to sustain the required load or pressure. See also *permanent set*.

setting Placement of lights or panels in sashes or frames. Also the action of a compound as it becomes more firm after application.

setting bed A layer of mortar on which the tile is set. The final coat of mortar on a wall or ceiling may also be called a setting bed.

setting block Small blocks of composition, lead, neoprene, wood, etc., placed under bottom edge of the light or panel to prevent its settling down onto the bottom rabbet or channel after setting, thus distorting the sealant.

setting shrinkage A reduction in volume of concrete prior to the final set of cement, caused by settling of the solids and by the decrease in volume due to the chemical combination of water with cement when an external source of curing water is not present.

setting time A term used to describe the amount of time needed for a material to dry sufficiently through solvent release or cure sufficiently through chemical reaction.

S-glass A magnesia-alumina-silicate glass especially designed to provide high tensile strength.

shading coefficient The ratio of the rate of solar heat gain passing through a glazing system to the solar heat gain that occurs under the same conditions if

the window were made of clear, unshaded double-strength window glass. The lower the number the better the solar heating qualities of the glazing system.
shear The progressive relative displacement of adjacent layers because of strain, or a lateral motion.
shearing Breaking caused by the action of equal and opposed forces located in the same place.
shear test A method of separating two materials by forcing (either by compression or tension) the interfaces to slide over each other. The force exerted is distributed over the entire bonded area at the same time. Strengths are recorded in psi (pounds per square inch).
sheeting A form of plastic in which the thickness is very small in proportion to length and width and in which the plastic is present as a continuous phase throughout, with or without filler.
sheet molding compounds (SMC) SMC's are thermoset, generally polyester materials processed in the form of sheet and containing roughly equal volumes of resin, fiberglass and calcium carbonate, together with a catalyst, that can be cured by heat or pressure.
shelf aging A method of determining the resistance of rubber articles to perishing by storing them under atmospheric conditions, either in light or in darkness, and testing them after definite lapses of time. The natural deterioration of rubber articles kept in storage or *on the shelf* under atmospheric conditions.
shelf life The period of time a packaged adhesive or sealer can be stored under specific temperature conditions and remain suitable for use.
Shore A hardness Measure of firmness of a compound by means of a Durometer Hardness Gauge. (Range of 20 to 25 is about the firmness of an art gum eraser, whereas 90 is about the firmness of a rubber heel).
Shore hardness Measurement of the hardness of a cured elastomeric material by means of a durometer.
shortness A qualitative term that describes an adhesive that does not string, cotton, or other wise form filaments or threads during application.
shrinkage The percent loss of volume of a material when put through a particular process, such as the washing and drying of crude rubber. The percent diminution in the area or volume of a piece of processed, unvulcanized rubber compound on cooling. Also, the contraction of molded vulcanized rubber on cooling.
shrinkage cracking Cracking of a structure or member due to failure in tension caused by external or internal restraints as reduction in moisture content develops or as a carbonation occurs, or both.
sight line Imaginary line along a perimeter of lights or panels corresponding to the top edge of stationary and removable stops, and the line to which sealants contacting the lights or panels are sometimes finished off.
silanes/siloxanes Compounds used to produce chemical reaction or bonding type water repellency as opposed to resin deposition method.
silica A type of sand derived from minerals such as quartz.

silica gel   An adsorbent. An amorphous form of silica dioxide having a large internal surface area and range of pore sizes.

silicone rubber   A rubber prepared by the action of moisture on dichlordimethyl-silicon. These rubbers withstand temperatures from 120 °F to 500° F (47° C to 260° C) and are vulcanized with benzoyl peroxide.

silicone sealant   A sealant having as its chemical composition a backbone consisting of alternating silicone oxygen atoms.

single glazing   A single pane of glass.

size   A coating composition for sealing porous surfaces prior to further finishing.

size of bead   Normally refers to the width of the bead, but there are many situations in which both the width and the depth should be taken into account in design, specification and applications.

sizing   The process of applying a material on a surface to fill the pores and thus reduce the absorption of the subsequent applied material or to otherwise modify the surface properties of the substrate to improve adhesion. The latter is sometimes called size.

skylight   A glass and frame assembly installed into the roof of a building.

sleeve   A liner, generally metallic, used to create an annulus for, or around, the pentrants. May be placed into concrete as it is poured or may be placed around the penétrant and inserted into a wall as it is erected.

slippage   The movement of adherends with respect to each other during the bonding process.

sloped glazing   Any installation of glass that is at a slope of 15 degrees or more from the vertical.

slumping   See *sagging*.

snow load   Load force due to snow accumulation.

soda ash (sodium oxide)   A flux used in glass making.

softening point   The temperature at which a prescribed load will cause the failure of a one square-inch shear bond of cloth to steel. Since softening under heat is progressive, increasing with temperature, it is rarely stated that a product is hard or soft at a certain temperature. The preferred method is to report what weight per inch it can support without failure. A bond under high tension, for instance, will not withstand as much heat as one under light tension, and raising or lowering the temperature will, to a point, decrease or increase the apparent strength of an adhesive.

solar   Pertaining to the sun or to energy from the sun.

solar energy   Thermal radiation from the sun, as measured by the short radiation wave lengths, less than three microns long.

solar energy absorption   The percentage of the solar spectrum energy (ultraviolet, visible, and near infrared) from 300 to 3000 nanometers, that is absorbed by the glass product.

solarization   A change in appearance of glass as a result of exposure to sunlight.

solids (plastic and elastic) Solid: A substance that undergoes permanent deformation only when subjected to shearing stress in excess of some finite value characteristic of the substance (yield stress). *Plastic solid*: A substance that does not deform under a shearing stress until the stress attains the yield stress, when the solid deforms permanently. *Elastic solid*: A substance in which, for all values of the shearing stress below the rupture stress (shear strength), the strain is fully determined by the stress regardless of whether the stress is increasing or decreasing.

solids content test A determination of the nonvolatile matter of a compound at a specific temperature and time interval. Usually expressed in percentage by weight. The difference between this figure and 100 percent represents the volatile matter or loss by evaporation.

solubility The degree to which a substance will dissolve in a particular solvent, usually expressed as grams dissolved in 100 grams of solvent.

soluble salts Mineral salts that are solubilized by moisture in concrete or masonry and carried in solution as water migrates through a substrate.

solvent resistance Ability of a sealant or adhesive to withstand exposure to a solvent.

solvents Broadly defined as substances with the ability to dissolve other substances, solvents are used by the adhesives and sealants industry in several ways. As intermediates, solvents are used in the production of many monomers and resins. Solvents also are used in adhesives, sealants, glazing systems and as surface coatings.

sound absorption The property possessed by material and objects, including air, of converting sound energy to heat energy.

sound transmission class (STC) A single-figure rating of standardized test performance according to ASTM E 413-73, for evaluating the effectiveness of assemblies in isolating airborne sound transmission.

spacer corners Specific methods used in joining the spacer lengths into spacer frames including interlocking keys, bending, soldering or welding.

spacer depth That dimension of the spacer that is measured parallel to the glass surface.

spacers (IG) A hollow shape used to provide a fixed air space between two pieces of glass to obtain thermal or acoustical properties. Small blocks of composition, wood, neoprene, aluminum, etc., placed on each side of the light or the panel to center them in the channel and maintain the width of sealant beads. Prevents excess sealant distortion.

spacers (shims) Small blocks of composition, neoprene, etc., placed on each face of light and panel to center them in the channel and maintain uniform width of sealant beads, preventing excessive sealant distortion.

spacer width That dimension of the spacer that is measured perpendicular to the glass surface and establishes the unit's air space.

spalling Rough breakage of the face of concrete, stone, or brick as may be caused by rust-corrosion or freeze-thaw cycles.

spandrel That portion of the exterior wall of a multi-story commercial building that covers the area below the sill of the vision glass installation.

spandrel glass Heat strengthened float glass with a colored ceramic coating adhered to the back by a heat-fusing process. It has double the strength of the same size and thickness, enabling it to withstand greater uniform loads and thermal stresses. Spandrel glasses cannot be recut after heat strengthening. It is used as fixed opaque colored glass on buildings in front of floor slabs and columns. It is available in a wide variety of colors.

SPE An abbreviation for Society of Plastic Engineers.

specific gravity Specific gravity = weight of substance/weight of equal volume of standard. It is the ratio of the weight of any volume of substance to the weight of an equal volume of a standard substance at stated temperatures. For solids or liquids, the standard substance is usually water, and for gases the standard is air or hydrogen.

specific volume This the reciprocal of specific gravity (1 divided by the specific gravity), and represents the volume in liters of one kilogram or the volume in cubic feet of 1000 pounds (998.9 exactly). It also represents the ratio between the volume of one pound of water (27.72 cubic inches) and the volume of one pound of the material in question.

specimen A piece or portion of a sample used to make a test.

SPI An abbreviation for the Society of Plastics Industry.

spread The quantity of material per unit joint area applied to an adherend. It is preferably expressed in pounds of liquid or solid adhesive per thousand square feet of joint area.

> *Single spread* refers to the application of adhesive to only one adherend of a joint.
>
> *Double spread* refers to the application of material to both adherends of a joint.

sprue The primary feed channel that runs from the outer face of the injection or transfer mold, to the mold gate in a single cavity mold.

sputter coating A micro-thin metallic oxide coating vacuum deposited on the surface of glass after manufacturing. Sputter coatings are vulnerable to moisture and abrasion and must be enclosed in sealed insulating glass units.

stain Attack of a glass surface by water or other solutions involving:

    a. Leaching of sodium ions to the surface of the glass.

    b. An increase in pH on the glass surface.

    c. The breaking of silica bonds in the glass structure.

If this process advances to "c," the glass will have a blotched, streaked, cloudy appearance and cannot be restored to pristine condition short of grinding and polishing the damage away, a process that is generally more expensive than replacing the glass. It is also a transparent coloration achieved on glass by the application of metallic ions at an elevated temperature.

starved joint A joint that has an insufficient amount of adhesive to produce a satisfactory bond. See *dry joints* and ASTM D 907.

stationary stops The permanent stop or lip of a rabbet on the side away from the side on which lights or panels are set.

stearates Hydrophobic salt of stearic (fatty) acid. Used in some resin deposition water repellent products.

steel deck assembly Otherwise known as a fluted deck or floor pans, these floor assemblies consist of concrete that is poured into a corrugated steel pan assembly.

stoce A loose pack of glass, weighing from 4,000 to 10,000 pounds.

stones Crystalline contaminations in glass, usually pieces of undissolved or crystallized silica, bits of refractory, or crystals due to devitrification. Stones are detrimental to appearance and may seriously weaken the glass, particularly if present in highly stressed areas.

stop Either the stationary lip at the back of a rabbet or removable molding at the front of the rabbet, either or both serving to hold light or the panel in sashes or frames, with the help of spacers.

storage life The period of time during which a material can be stored under specified conditions and remain suitable for use.

straight chain polymer A polymer containing groups of molecules attached to each other in a straight line, like a string of beads. No cross-linking of groups is present. Examples are vistonex and polybutene.

strain Deformation of a material resulting from external loading.

strength The maximum stress required to overcome the cohesion of a material. Quantitative: A complex property made up of tensile strength and shearing strength. The force required to break a bar of unit cross-section under tension is the tensile strength. Strength involves the idea of resistance to rupture.

strength, dry The strength of an adhesive joint determined immediately after drying under specified conditions or after a period of conditioning in the standard laboratory atmosphere.

strength, wet The strength of an adhesive determined immediately after removal from a liquid in which it has been immersed under specified conditions of time, temperature and pressure.

stress crack An external or internal crack in a plastic caused by tensile stresses less than its short time mechanical strength.

stretcher A masonry unit that is laid parallel with a wall.

striking off The operation of smoothing off excess compound at the sight-line when applying sealant around lights or panels. In masonry work; the cutting away of mortar with the trowel; also tooling of joints.

stringiness The property of an adhesive or sealant that results in the formation of filaments or threads when adhesive transfer surfaces are separated.

structural foam A thermoplastic generally fabricated by injection molding and containing a high void content (many small bubbles), generated either by the use of nitrogen gas or gas-forming blowing agents.

structural glazing Is based on the use of a sealant not only as a weather seal but also for structural transfer of loads from the glazing panel to its perimeter support system. Only certain specific sealant formulations are suitable for this purpose.

structural glazing gaskets Cured elastomeric channel-shaped extrusions used in place of a conventional sash to install glass products onto structurally supporting sub-frames when the pressure of sealing exerted by the insertion of separated lock strip wedging splines.

structural silicone glazing A system in which the glass product is bonded to the framing members of a curtain wall, using a structural silicone adhesive/sealant without the presence of the outdoor retainers or stops.

styrene ($C_6H_5CH: CH_2$) A colorless liquid hydrocarbon used in the making of synthetic rubbers.

subflorescence Migration of soluble salt where moisture vaporizes below the surface, leaving salts to accumulate and crystallize below the surface of concrete or masonry, causing accelerated deteriorations and fracturing of surface. This is recognized as a counter-productive aspect of water repellent usage.

substance A variety of matter, all specimens of which have the same properties.

substrate A material upon the surface of which an adhesive-containing substance is spread for any purpose, such as bonding or coating. A broader term than adherend.

sun effect Heat from the sun that tends to increase the internal temperature of a space in a building.

sunlight The portion of solar energy that is detectable by the human eye. It accounts for about 44 percent of the total radiation wavelength spectrum.

super-cooled Frozen into shape, or cooled below the freezing point without crystallization.

surface mat A thin mat of fine fibers, used primarily to produce a smooth surface on a reinforced plastic.

surface preparation The procedure required with respect to a foundation surface or the materials to be adhered which will promote optimum performance by the adhesive, coating or sealer. For example, if higher bond strength is required, abrading and/or acid etching the surface can be the means of improving the adhesion of the bonding material to the mating surfaces. Common methods of surface preparation are solvent washing, sandblasting, and vapor degreasing.

surfactants Additives in a polyurethane reaction that lower the surface tension in liquids and permit greater component dispersion and more uniform bubbles.

sweating Exudation of small drops of liquid, usually a plasticizer or softener, on a reinforced plastic.

swelling The property in raw or vulcanized rubber of absorbing organic liquids such as benzene, gasoline, etc., and swelling to many times its original volume. The property is also shown by other colloids in contact with other liquids. In a general sense, it may be called any increase in volume of a solid substance caused by the absorption of a liquid.

syneresis The exudation of small amounts of liquid by gels on standing.

synthesis The formation or building up of more complex material, or compound, by the chemical reaction of elements, or simpler compounds, or radicals.

synthetic A compound produced by chemical reactions as opposed to those of natural origins.

tack The property of an adhesive that enables it to form a bond of measurable strength immediately after adhesive and adherend are brought into contact under low pressure.

tack, dry Property of certain adhesives, particularly non-vulcanizing rubber adhesives, to adhere on contact to themselves at a stage in the evaporation of volatile constituents, even though they seem dry to the touch.

tackiness The stickiness of the film while in the stage of drying. For instance, after a paint or varnish sets up, it usually retains a sticky or tacky feel until it is practically dry. Stickiness: A quality possessed by a solid having a low yield value and high mobility by means of which contact readily results in adhesion. For example, adhesive, varnish, printer's ink, and gold size under working conditions are tacky or sticky substances. When most of the above dry out, set, gel, or harden through chemical or other change, they lose tack or stickiness. Those compounds that retain tack long after drying are said to be *permanently tacky* or to possess *after tack*.

tack range The period of time during which an adhesive will remain in the tacky-dry condition after application to an adherend, under specified conditions of temperature and humidity.

tank A glass furnace.

tar base Descriptive of a compound whose liquid content is composed of materials obtained from coal tar as distinguished from compounds whose liquid content is composed of vegetable oils or asphaltic materials.

tear resistance The force required to tear completely across a notched specimen tested according to prescribed procedures, expressed in pounds per inch of specimen thickness.

temperature The degree of heat or cold as measured in degrees Centigrade or Fahrenheit.

temperature, absolute Temperature measured on a scale for which zero is set at $-459.69°$ F ($-273.16°$ C) presumably the temperature at which all molecular motion stops in a gas under constant pressure. The scale is called Kelvin, and $1° K = 1° C = \frac{5}{9}°$ F.

temperature cracking Cracking due to tensile failure, caused by temperature drop in members subjected to external restraints or temperature differential in members subjected to internal restraints.

temperature of deflection under load The temperature at which a simple beam has deflected a given amount under load (formerly called heat distortion temperature).

tempered glass Fully tempered (FT) glass is re-heated to just below the softening point, about $1300°$ F or ($744.4°$ C), and then rapidly cooled. If broken, fully tempered glass always *dices* into a multitude of small particles.

tensile strength The capacity of a material to resist a force tending to stretch it. Ordinarily the term is used to denote the force required to stretch a material to rupture, and is known variously as *breaking load, breaking stress,* or *ultimate* *tensile strength.* In rubber testing, it is the load in pounds per square inch or kilos per square centimeter of original cross-sectional area, supported at the moment of rupture, on being elongated at a constant rate.

tension pull A term of total pull in pounds at the conclusion of a tension test. This test subjects, for example, a hose assembly to increasing tension load in a suitable machine until failure occurs by separation of the specimen from the end fittings or by rupture of the hose structure.

tension-stress-strain testing Determination of stress and strain (tensile strength and elongation) with the use of dumbbell specimens in conformance with the ASTM Method G-412 at 75° F or (24° C). Rings may also be used.

terra cotta Traditional Italian raw material used to produce unglazed red tiles, which can be vitreous, generally extruded and ½ inch (13 mm) thick or more. Surfaces may be rustic or smooth and waxed for luster after installation.

thermal break A material with a low thermal conductance used to separate exterior and interior materials. The thermal break is intended to stop the flow or transfer of heat.

thermal conductivity Ability of a material to conduct heat.

thermal expansion Capable of being repeatedly softened by heat and hardened by cooling.

thermoplastic A product (or chemical) that melts when heated and solidifies when cooled. Thermoplastic products may be composed of either linear or branched polymers.

thermosetting Having the property of undergoing a chemical reaction by the action of heat, catalysts, ultraviolet light, etc., leading to a relatively infusible state.

thick-bed mortar A thick layer of mortar (more than ½ inch (13 mm)) that is used in leveling.

thin-bed or thin-set methods A term referring to all tile setting methods except the full mortar-bed. Used to describe the bonding of the tile with suitable materials applied approximately ⅛ inch (3 mm) or less in thickness.

thinning Addition of a slight amount of unleaded gasoline or solvent to an oleo-resinous glazing by the glazier to soften its consistency.

Thiokol A commonly used name for the first commercial synthetic elastomer, produced by Thiokol Chemical Company in 1930. Thiokol A is produced by the reaction of ethylene dichloride and sodium tetra sulfide, $CL\text{-}CH_2$, $CH_2 CL + Na_2S_4$ $(\text{-}CH_2 CH_2 \text{-}S_4\text{-})_X$. The atoms of sulfur may all be in a straight line. Thiokol is vulcanized by heating with zinc oxide and is important because it is practically insoluble in petroleum oils. Thiokols are also prepared from the other dichlor compounds: di (2-chloroethyl) ether, $CI\text{-}CH_2 CH_2 OCH_2 CH_2 CL$, and di (2-chloroethyl) formal, $CL\text{-}CH_2 CH_2 OCH_2 OCH_2 CH_2 CL$. The latter Thiokols are somewhat more soluble than Thiokol A. See *polysulfide elastomer*.

thixotropic A term used to describe certain colloidal dispersions which, when at rest, assume a gel-like condition but when agitated, stirred or subjected to pressure or other mechanical action at ordinary temperatures, are transformed into a liquid condition. The action is reversible and can be repeated at will. Thixotropic colloidal dispersions occur in nature, the best-known example being bentonite, a colloidal American clay. Rubber dispersions are not thixotropic colloids.

through penetration An opening in a wall or floor that passes completely through the assembly.

tie coat One layer of a coating system used to improve the adhesion of adjacent or succeeding coats.

tile backer board Aggregated Portland cement board with vinyl coated woven, glass-fiber in back and front surfaces. Interior only. Highly water-resistant.

time-dependent deformation The deformation of concrete occurring with appreciable time (as days, weeks or months); includes creep and characteristics affected by age and strength changes such as elasticity, drying, shrinkage, and temperature effects.

tinted glass Body-colored glass of a specific batch ingredient formulation to produce light reducing and/or heat absorbing glass products.

TL Sound transmission loss.

toluene dissocyanate (TDI) An aromatic isocyanate with three substitutions on a benzene ring, two of which are -NCO. It has two isomers (2,4 and 2,6, although 2,4 is the most common). It is used primarily in flexible foams.

tooling Operation of pressing and sticking a compound in a joint to press it against the sides of the joint and secure good adhesion. Also, the finishing off of the surface of a compound in a joint so that it is flush with the surface. A narrow, blunt-bladed tool is used for this purpose.

total heat gain/summer/daytime (BTU per hour, per square foot) The sum of the radiant energy and the conductive energy transmitted into the building. (Shade coefficient times ASHRAE solar heat gain factors + summer U-value (the indoor/outdoor temperature difference.)

total heat gain/summer/nighttime (BTU per hour, per square) The conductive energy transmitted into the building. (Summer U-value (the indoor/outdoor temperature difference.)

total heat loss/winter/nighttime (BTU per hour, per square foot) The conductive energy transmitted to the outdoors. Winter U-value (the outdoor/indoor temperature difference.)

total solar heat absorption The percentage of incident solar radiation that is absorbed by a glazing system.

total solar reflectance The percent of solar radiation that is reflected by a glazing system.

total solar transmittance The percent of incident solar radiation that directly passes through a glazing system.

toxicity A term referring to the physiological effect of absorbing a poisonous substance into the system either through the skin, mucous membranes or respiratory system. When describing their toxic effect, solvents are usually classified as having high, medium or low toxic effect, depending upon whether a solvent

vapor concentration of less than 100, 100 to 400, or more than 400 parts per million respectively, is the maximum amount permissible in the air for safe or healthful working conditions.
transfer molding   Modified compression molding where an auxiliary hydraulic ram is used in a direction opposite to the mold-clamping ram.
transmission   The passage of sound from one location to another through an intervening medium, such as a partition or air space.
transmittance   The ability of the glass to transmit solar energy in the visible light, the ultraviolet, and the infrared ranges, centrally measured in percentages of each.
transverse seam   A seam joining two materials across the width of the finished product.
T-rating   A measure of the thermal conductivity of a firestop system. The time required for various elements on the exposed side of a system to exceed 326° F (163.33° C) over the temperature at the start of the test.
trim units   Various shapes, including bases, caps, corners, moldings, angles, etc., necessary to achieve a finished installation.
triol   A pall with three hydroxyl (-OH) groups.
triple glazing   Windows that are made of three panes of glass with an air space between the lights, and may be filled with argon gas for increased energy efficiency.
trowel   A flat, broad-bladed steel hand tool used to finish concrete or to apply or shape.
tube drawing   Drawing glass tubing into shape.
tuck-pointing   See *pointing*.
two-part compound   A product that is necessarily packaged in two separate containers. It is comprised of a base and the curing agent or accelerator. The two compounds are uniformly mixed just prior to use, since, when mixed, it cures and its useful life is limited from the standpoint of application characteristics.
ultimate elongation   The elongation at the moment of rupture.
ultraviolet light (UV)   A form of luminous energy occupying a position in the spectrum of sunlight beyond the violet, and having wavelengths of less than 3900 Angstrom units, which is the limit of the visible spectrum. Ultraviolet rays affect chemical reactions, exhibit bactericidal action, and cause many substances to fluoresce.
undercut   Any indentation in, or protrusion from, a piece beyond the direct line of flow.
underlayment   Factory proportioned combination of Portland cement, sand and additives used with or without a latex liquid (according to manufacturer's instructions) to level an even substrate and to provide a suitable tile setting surface.
Underwriter's Laboratories (UL)   A non-profit independent laboratory organization that tests a variety of building materials. Products that have been approved carry the *UL* label.

unglazed tile   Fire-hardened clay tile with color that runs throughout the body.

uniform bead   Compound applied to a joint, with uniform width and appearance.

unit   Term normally used to refer to one single light of insulating glass.

united inches   The sum of the dimensions of one length and one width of a light of glass.

urea plastics   Plastics based on resins made by the condensation of urea and aldehydes.

urethane $CO(NH_2)OC_2H_5$. (Urethane, ethyl carbamate, ethyl urethane). Urethane is not used directly in the production of urethane plastics or polyurethane foams.

urethane plastics   Plastics based on polymers in which the repeated structural units in the chain are of the urethane type, or on copolymers in which urethane and other types of repeated structural units are present in the chains.

U-value   The amount of conductive heat energy (BTUs) transferred through a one-square-foot area of a specific insulating glass unit for each one degree Fahrenheit temperature difference between the indoor and outdoor air. It is the inverse of the R value; $U=1/R$.

UV inhibitor   A material that inhibits or blocks ultraviolet light.

vacuum forming   A forming process in which a heated plastic sheet is drawn against the mold surface by evacuating the air between it and the mold.

vapor   A gaseous state of fluid; small particles of matter suspended in the air.

vapor barrier   A layer of material through which water vapor will not pass readily or at all.

vapor transmission   The passage of moisture as a gas or vapor as opposed to liquid form. A wall that can do this is said to *breathe*.

vegetable oil base   Formulated with a vehicle of vegetable oils usually processed with resins by application of heat.

venting   Providing circulation of air or ventilation between two walls or partitions. Venting accomplished by the use of tubes, breather vents or openings left in the wall.

vibrator   A machine used to eliminate trapped air bubbles and consolidate freshly placed concrete.

vinyl   Derived from ethylene (hydrocarbon gas) the compounds of which are polymerized to form high molecular weight plastics and resins such as vinyl acetate, vinyl styrene, etc. It is a base material for plastisols and organisols and also widely used in emulsion form as polyvinyl acetate.

vinyl acetate plastics   Plastics based on polymers of vinyl acetate or copolymers of vinyl acetate with other monomers, the vinyl acetate being in greater amounts by mass.

vinyl chloride plastics   Plastics based on polymers of vinyl chloride or copolymers of vinyl chloride with other monomers, the vinyl chloride being in greatest amount by mass.

vinyl glazing   Holding glass in place with the extruded vinyl channel or roll-in type.

vinylidene chloride plastics   Plastics based on polymer resins made by the polymerization of vinylidene chloride with other unsaturated compounds, the vinylidene chloride being in the greatest amount by weight.

virgin material   A plastic material in the form of pellets, granules, powder, floc or liquid that has not been subjected to use or processing other than required for its initial manufacture.

viscosimeter or viscometer   An instrument used for measuring the viscosity or fluidity of liquids and plastic materials. Various types are used, such as the Baybolt, Redwood, Engler, based on rate of flow through a tube. Some types, (Brookfield, Stormer) are based on the torsion principle, and others on the time taken for a metal ball to fall through a column of light of definite time length. For rubbers, including GO-S, the Mooney viscometer is widely used for both the raw and compound material.

viscosity   A manifestation of internal friction; opposed to mobility. The property of fluids by which they resist an instantaneous change of shape (i.e. resistance to flow). It is measured by the force required to cause two parallel liquid surfaces of unit area and unit distance apart to slide past each other in the liquid with velocity. This is expressed in *poises*, or dyne-seconds per square centimeter. Water at 20.2° C has a viscosity of 1 centipoise and is taken as the standard comparison. A number of terms and materials have been proposed for special applications of viscosity. These are listed below:

Water	1
Kerosene	10
Motor oil, SAE 10	100
Castor oil, glycerine	1000
Corn syrup	10,000
Molasses	100,000

Syrup: Material that slumps under its own weight (will not maintain its shape) when made into a ball with a diameter of one inch or less.

Thin: Any material tested on a Ford Cup or any material testing up to 40 on a #26 MacMichael wire.

Medium: Any material testing 40 to 300 on a #22 MacMichael wire, up to 40 on a #22 MacMichael wire.

Heavy: Any material testing 40 to 100 on a #22 MacMichael wire or from 0 to 65 on a #18 MacMichael wire.

Paste or Mush: Material that will flow or slump under its own weight (not hold its shape) in a diameter greater than one inch. Viscosity is recorded in the range of 400 to 150 cone penetrometer.

Dough: Material that will generally not flow under its own weight. Viscosity is recorded in the range of 150 to 0 cone penetrometer.

viscosity coefficient   The shearing stress necessary to induce a unit velocity flow gradient in a material.

visible light reflectance   The percent of total visible light that is reflected by a glazing system.

visible light transmittance   The percentage of light in the visible spectrum range of 390 to 780 nanometers that is directly transmitted through a glass product.

vitreous   Having a water absorption of more than 0.5 percent but no more than 3 percent.

vitrifying   Those temperatures required to transform sand into glass. Ceramic tile fired at vitrifying temperatures will have the water absorption rate of 0.5 to 3 percent.

volatile   The property of liquids to change into a gas and pass away by evaporation under normal atmospheric conditions.

volume change   This occurs as drying, shrinkage, creep, expansion, or contraction due to external thermal forces.

vulcanization   An irreversible process during which a rubber compound, through a change in its chemical structure (for example, cross-linking), becomes less plastic and more resistant to swelling by organic liquids. Elastic properties are conferred, improved or extended over greater range on temperature during the process.

waisting   Glass narrowing in the middle.

wale   A long horizontal member on the formwork to hold studs in place.

wall tile   Nonvitreous tile with water absorption lower than seven percent, relatively, inexpensive, generally nonloadbearing, ceramic tiles dry-pressed from soft clays under lower temperatures than vitreous tiles. In most cases, wall tiles derive color and surface texture from a glaze or glazes, added in one or more separate additional firing operations.

warehouse set   The partial hardening of bagged cement or prepackaged mixes caused by the absorption of atmospheric moisture during improper or lengthy storage.

warp   Distortion caused by uniform change of internal stresses. The lengthwise direction of glass cloth.

water absorption   The process of assimilating or soaking up water.

waterproofing   A treatment for concrete or other structures to prevent the passage of liquid water under hydrostatic, dynamic, or static pressure.

water-reducing agent   A material that increases the slump and workability of freshly mixed concrete without increasing the amount of water.

water repellent   The material or treatment for concrete surfaces that is intended to provide resistance to penetration by water in the absence of hydrostatic pressure.

water resistance   The ability to withstand swelling by water for a specified time and temperature, usually 48 hours at 212° F (100° C), expressed as a percentage of swelling or volume increase of specimen.

water-resistant   Capillary pores exist that permit the passage of water and water vapor, but there are few or no openings larger than capillaries that permit leakage of significant amounts of water.

water vapor A psychrometric term used to denote the water in air (actually low pressure, superheated steam) that has been evaporated into the air at a temperature corresponding to the boiling point of water at that very low pressure.

water vapor transmission The amount of water vapor passing through a given area of an adhesive or sealant sheet or film in a given time, when the sheet or film is maintained at a constant temperature and when the faces are exposed to certain different relative humidities. The result is usually expressed as grams per-24-hours-per-square-meter.

weather-o-meter An apparatus for estimating the comparative resistance of a soft vulcanized rubber compound to deterioration when exposed to light having a frequency range approximating that of sunlight. The criterion used in estimating resistance to light is the percentage decrease in tensile strength and in elongation at break. A supplementary criterion is the observed extent of surface crazing and cracking. During the test, sprays of clean water are forced on the specimens to simulate the action of rain (ASTM: D 750).

weatherstripping Material used to close off openings around windows and doors.

webbing Filaments or threads that form when adhesive transfer surfaces are separated.

weep hole Opening at the base of a cavity wall to collect moisture and dispense it or a breather installed in sealant to relieve moisture.

weeping Failure of a compound to support its own weight in a joint, but not as pronounced as *sagging*.

wet lay up A method of making a reinforced plastic in which the polymer compound is applied as a liquid and as the reinforcement is put in place.

wet strength Commonly used to designate strength after immersion in water.

white wool A process used in making glass fiber.

wind load Force exerted by winds on building panels and completed structures; may be inward (positive) or outward (negative). Outward forces generally occur on the leeward sides of the buildings.

window Unit with panes of glass encased in a framework.

wired reinforced glass Glass having a layer of meshed wire completely embedded in the glass light. It may have polished or patterned surfaces. Approved wired glass is used as transparent or translucent fire retardant glazing, sometimes as decorative glass, or as security glazing. It breaks more easily than unwired glass of the same thickness, but the wire restrains the fragments from falling out of the frame when broken.

workability (concrete) This is the property most important to the contractor who places the concrete into forms and finishes it. Workability includes the properties of cohesiveness, plasticity, and non-segregation.

working life The period of time during which an adhesive, after mixing with catalyst, solvent, or other compounding ingredients, remains suitable to use. Synonym to *pot life*. ASTM D 907.

workmanship The quality of work put into a product or its installation.

wrinkling The formulating of wrinkles in the skin of a compound during the formation of its surface skin by oxidation after application.
WVT Abbreviation for water vapor transmission.
wythe A continuous vertical section of a wall that is one masonry unit wide.
xylene $C_6H_4(CH_3)_2$ A mixture of three isomers, ortho-, meta-, and para-xylene, used in a solvent for alkyd resins.
yield strength The first stress in an adhesive (or adherent) less than maximum attainable stress, at which an increase in strain occurs without an increase in stress.
yield value The lowest pressure at which a plastic will follow at a specified temperature.
Young's modulus The ratio of normal stress to corresponding strain for tensile or compressive stress less than the proportional limit of the material. ASTM D 1053.
zeolite A class of compounds that are crystalline, alumino silicates of group I and group II elements of the periodic table. They may occur in nature or be synthetically produced. The framework structure and pore size of these compounds is orderly and controlled. They are effective adsorbents with a very high surface area.

Note: These definitions are not designed to have any legal significance or consequences.

REFERENCES

1. Building Research Institute (defunct).
2. The Language of Sealed Insulating Glass Units, SIGMA, Vol. XV, 12/89.
3. Peterson, A.J. 1985. *How Glass is Made,* Facts on File Publications. New York: Threshold Books Ltd.
4. *A Glass Primer*, Glass Digest, 4 and 5/1990.
5. Fenestration for Sound Control, The Construction Specifier, Construction Specifications Institute, 4/1990.
6. Whittington, Lloyd R. 1968. *Whittington's Dictionary of Plastics*, 1st ed. Sponsored by the Society of Plastics Engineers, Inc.
7. *The Homeowner's Guide to Building with Concrete*, Brick & Stone, 1988. The Portland Cement Association. Rodale Press.

APPENDIX A
TEMPERATURE CONVERSION CHART

TABLE A.1. Temperature Conversion Table

°F.	Reading in °F. or °C. to be converted	°C.	°F.	Reading in °F. or °C. to be converted	°C.
....	−458	−272.22	....	−358	−216.67
....	−456	−271.11	....	−356	−215.56
....	−454	−270.00	....	−354	−214.44
....	−452	−268.89	....	−352	−213.33
....	−450	−267.78	....	−350	−212.22
....	−448	−266.67	....	−348	−211.11
....	−446	−265.56	....	−346	−210.00
....	−444	−264.44	....	−344	−208.89
....	−442	−263.33	....	−342	−207.78
....	−440	−262.22	....	−340	−206.67
....	−438	−261.11	....	−338	−205.56
....	−436	−260.00	....	−336	−204.44
....	−434	−258.89	....	−334	−203.33
....	−432	−257.78	....	−332	−202.22
....	−430	−256.67	....	−330	−201.11
....	−428	−255.56	....	−328	−200.00
....	−426	−254.44	....	−326	−198.89
....	−424	−253.33	....	−324	−197.78
....	−422	−252.22	....	−322	−196.67
....	−420	−251.11	....	−320	−195.56
....	−418	−250.00	....	−318	−194.44
....	−416	−248.89	....	−316	−193.33
....	−414	−247.78	....	−314	−192.22
....	−412	−246.67	....	−312	−191.11
....	−410	−245.56	....	−310	−190.00
....	−408	−244.44	....	−308	−188.89
....	−406	−243.33	....	−306	−187.78
....	−404	−242.22	....	−304	−186.67
....	−402	−241.11	....	−302	−185.56
....	−400	−240.00	....	−300	−184.44
....	−398	−238.89	....	−298	−183.33
....	−396	−237.78	....	−296	−182.22
....	−394	−236.67	....	−294	−181.11
....	−392	−235.56	....	−292	−180.00
....	−390	−234.44	....	−290	−178.89
....	−388	−233.33	....	−288	−177.78
....	−386	−232.22	....	−286	−176.67
....	−384	−231.11	....	−284	−175.56
....	−382	−230.00	....	−282	−174.44
....	−380	−228.89	....	−280	−173.33
....	−378	−227.78	....	−278	−172.22
....	−376	−226.67	....	−276	−171.11
....	−374	−225.56	....	−274	−170.00
....	−372	−224.44	−457.6	−272	−168.89
....	−370	−223.33	−454.0	−270	−167.78
....	−368	−222.22	−450.4	−268	−166.67
....	−366	−221.11	−446.8	−266	−165.56
....	−364	−220.00	−443.2	−264	−164.44
....	−362	−218.89	−439.6	−262	−163.33
....	−360	−217.78	−436.0	−260	−162.22

TABLE A.2. Temperature Conversion Table

°F.	Reading in °F. or °C. to be converted	°C.	°F.	Reading in °F. or °C. to be converted	°C.
−432.4	−258	−161.11	−216.4	−138	−94.44
−428.8	−256	−160.00	−212.8	−136	−93.33
−425.2	−254	−158.89	−209.2	−134	−92.22
−421.6	−252	−157.78	−205.6	−132	−91.11
−418.0	−250	−156.67	−202.0	−130	−90.00
−414.4	−248	−155.56	−198.4	−128	−88.89
−410.8	−246	−154.44	−194.8	−126	−87.78
−407.2	−244	−153.33	−191.2	−124	−86.67
−403.6	−242	−152.22	−187.6	−122	−85.56
−400.0	−240	−151.11	−184.0	−120	−84.44
−396.4	−238	−150.00	−180.4	−118	−83.33
−392.8	−236	−148.89	−176.8	−116	−82.22
−389.2	−234	−147.78	−173.2	−114	−81.11
−385.6	−232	−146.67	−169.6	−112	−80.00
−382.0	−230	−145.56	−166.0	−110	−78.89
−378.4	−228	−144.44	−162.4	−108	−77.78
−374.8	−226	−143.33	−158.8	−106	−76.67
−371.2	−224	−142.22	−155.2	−104	−75.56
−367.6	−222	−141.11	−151.6	−102	−74.44
−364.0	−220	−140.00	−148.0	−100	−73.33
−360.4	−218	−138.89	−144.4	−98	−72.22
−356.8	−216	−137.78	−140.8	−96	−71.11
−353.2	−214	−136.67	−137.2	−94	−70.00
−349.6	−212	−135.56	−133.6	−92	−68.89
−346.0	−210	−134.44	−130.0	−90	−67.78
−342.4	−208	−133.33	−126.4	−88	−66.67
−338.8	−206	−132.22	−122.8	−86	−65.56
−335.2	−204	−131.11	−119.2	−84	−64.44
−331.6	−202	−130.00	−115.6	−82	−63.33
−328.0	−200	−128.89	−112.0	−80	−62.22
−324.4	−198	−127.78	−108.4	−78	−61.11
−320.8	−196	−126.67	−104.8	−76	−60.00
−317.2	−194	−125.56	−101.2	−74	−58.89
−313.6	−192	−124.44	−97.6	−72	−57.78
−310.0	−190	−123.33	−94.0	−70	−56.67
−306.4	−188	−122.22	−90.4	−68	−55.56
−302.8	−186	−121.11	−86.8	−66	−54.44
−299.2	−184	−120.00	−83.2	−64	−53.33
−295.6	−182	−118.89	−79.6	−62	−52.22
−292.0	−180	−117.78	−76.0	−60	−51.11
−288.4	−178	−116.67	−72.4	−58	−50.00
−284.8	−176	−115.56	−68.8	−56	−48.89
−281.2	−174	−114.44	−65.2	−54	−47.78
−277.6	−172	−113.33	−61.6	−52	−46.67
−274.0	−170	−112.22	−58.0	−50	−45.56
−270.4	−168	−111.11	−54.4	−48	−44.44
−266.8	−166	−110.00	−50.8	−46	−43.33
−263.2	−164	−108.89	−47.2	−44	−42.22
−259.6	−162	−107.78	−43.6	−42	−41.11
−256.0	−160	−106.67	−40.0	−40	−40.00
−252.4	−158	−105.56	−36.4	−38	−38.89
−248.8	−156	−104.44	−32.8	−36	−37.78
−245.2	−154	−103.33	−29.2	−34	−36.67
−241.6	−152	−102.22	−25.6	−32	−35.56
−238.0	−150	−101.11	−22.0	−30	−34.44
−234.4	−148	−100.00	−18.4	−28	−33.33
−230.8	−146	−98.89	−14.8	−26	−32.22
−227.2	−144	−97.78	−11.2	−24	−31.11
−223.6	−142	−96.67	−7.6	−22	−30.00
−220.0	−140	−95.56	−4.0	−20	−28.89

TABLE A.3. Temperature Conversion Table

°F.	Reading in °F. or °C to be converted	°C.	°F.	Reading in °F. or °C. to be converted	°C.
−0.4	−18	−27.78	+116.6	+47	+8.33
+3.2	−16	−26.67	+118.4	+48	+8.89
+6.8	−14	−25.56	+120.2	+49	+9.44
+10.4	−12	−24.44	+122.0	+50	+10.00
+14.0	−10	−23.33	+123.8	+51	+10.56
+17.6	−8	−22.22	+125.6	+52	+11.11
+19.4	−7	−21.67	+127.4	+53	+11.67
+21.2	−6	−21.11	+129.2	+54	+12.22
+23.0	−5	−20.56	+131.0	+55	+12.78
+24.8	−4	−20.00	+132.8	+56	+13.33
+26.6	−3	−19.44	+134.6	+57	+13.89
+28.4	−2	−18.89	+136.4	+58	+14.44
+30.2	−1	−18.33	+138.2	+59	+15.00
+32.0	±0	−17.78	+140.0	+60	+15.56
+33.8	+1	−17.22	+141.8	+61	+16.11
+35.6	+2	−16.67	+143.6	+62	+16.67
+37.4	+3	−16.11	+145.4	+63	+17.22
+39.2	+4	−15.56	+147.2	+64	+17.78
+41.0	+5	−15.00	+149.0	+65	+18.33
+42.8	+6	−14.44	+150.8	+66	+18.89
+44.6	+7	−13.89	+152.6	+67	+19.44
+46.4	+8	−13.33	+154.4	+68	+20.00
+48.2	+9	−12.78	+156.2	+69	+20.56
+50.0	+10	−12.22	+158.0	+70	+21.11
+51.8	+11	−11.67	+159.8	+71	+21.67
+53.6	+12	−11.11	+161.6	+72	+22.22
+55.4	+13	−10.56	+163.4	+73	+22.78
+57.2	+14	−10.00	+165.2	+74	+23.33
+59.0	+15	−9.44	+167.0	+75	+23.89
+60.8	+16	−8.89	+168.8	+76	+24.44
+62.6	+17	−8.33	+170.6	+77	+25.00
+64.4	+18	−7.78	+172.4	+78	+25.56
+66.2	+19	−7.22	+174.2	+79	+26.11
+68.0	+20	−6.67	+176.0	+80	+26.67
+69.8	+21	−6.11	+177.8	+81	+27.22
+71.6	+22	−5.56	+179.6	+82	+27.78
+73.4	+23	−5.00	+181.4	+83	+28.33
+75.2	+24	−4.44	+183.2	+84	+28.89
+77.0	+25	−3.89	+185.0	+85	+29.44
+78.8	+26	−3.33	+186.8	+86	+30.00
+80.6	+27	−2.78	+188.6	+87	+30.56
+82.4	+28	−2.22	+190.4	+88	+31.11
+84.2	+29	−1.67	+192.2	+89	+31.67
+86.0	+30	−1.11	+194.0	+90	+32.22
+87.8	+31	−0.56	+195.8	+91	+32.78
+89.6	+32	±0.00	+197.6	+92	+33.33
+91.4	+33	+0.56	+199.4	+93	+33.89
+93.2	+34	+1.11	+201.2	+94	+34.44
+95.0	+35	+1.67	+203.0	+95	+35.00
+96.8	+36	+2.22	+204.8	+96	+35.56
+98.6	+37	+2.78	+206.6	+97	+36.11
+100.4	+38	+3.33	+208.4	+98	+36.67
+102.2	+39	+3.89	+210.2	+99	+37.22
+104.0	+40	+4.44	+212.0	+100	+37.78
+105.8	+41	+5.00	+213.8	+101	+38.33
+107.6	+42	+5.56	+215.6	+102	+38.89
+109.4	+43	+6.11	+217.4	+103	+39.44
+111.2	+44	+6.67	+219.2	+104	+40.00
+113.0	+45	+7.22	+221.0	+105	+40.56
+114.8	+46	+7.78	+222.8	+106	+41.11

TABLE A.4. Temperature Conversion Table

°F.	Reading in °F. or °C. to be converted	°C.	°F.	Reading in °F. or °C. to be converted	°C.
+224.6	+107	+41.67	+332.6	+167	+75.00
+226.4	+108	+42.22	+334.4	+168	+75.56
+228.2	+109	+42.78	+336.2	+169	+76.11
+230.0	+110	+43.33	+338.0	+170	+76.67
+231.8	+111	+43.89	+339.8	+171	+77.22
+233.6	+112	+44.44	+341.6	+172	+77.78
+235.4	+113	+45.00	+343.4	+173	+78.33
+237.2	+114	+45.56	+345.2	+174	+78.89
+239.0	+115	+46.11	+347.0	+175	+79.44
+240.8	+116	+46.67	+348.8	+176	+80.00
+242.6	+117	+47.22	+350.6	+177	+80.56
+244.4	+118	+47.78	+352.4	+178	+81.11
+246.2	+119	+48.33	+354.2	+179	+81.67
+248.0	+120	+48.89	+356.0	+180	+82.22
+249.8	+121	+49.44	+357.8	+181	+82.78
+251.6	+122	+50.00	+359.6	+182	+83.33
+253.4	+123	+50.56	+361.4	+183	+83.89
+255.2	+124	+51.11	+363.2	+184	+84.44
+257.0	+125	+51.67	+365.0	+185	+85.00
+258.8	+126	+52.22	+366.8	+186	+85.56
+260.6	+127	+52.78	+368.6	+187	+86.11
+262.4	+128	+53.33	+370.4	+188	+86.67
+264.2	+129	+53.89	+372.2	+189	+87.22
+266.0	+130	+54.44	+374.0	+190	+87.78
+267.8	+131	+55.00	+375.8	+191	+88.33
+269.6	+132	+55.56	+377.6	+192	+88.89
+271.4	+133	+56.11	+379.4	+193	+89.44
+273.2	+134	+56.67	+381.2	+194	+90.00
+275.0	+135	+57.22	+383.0	+195	+90.56
+276.8	+136	+57.78	+384.8	+196	+91.11
+278.6	+137	+58.33	+386.6	+197	+91.67
+280.4	+138	+58.89	+388.4	+198	+92.22
+282.2	+139	+59.44	+390.2	+199	+92.78
+284.0	+140	+60.00	+392.0	+200	+93.33
+285.8	+141	+60.56	+393.8	+201	+93.89
+287.6	+142	+61.11	+395.6	+202	+94.44
+289.4	+143	+61.67	+397.4	+203	+95.00
+291.2	+144	+62.22	+399.2	+204	+95.56
+293.0	+145	+62.78	+401.0	+205	+96.11
+294.8	+146	+63.33	+402.8	+206	+96.67
+296.6	+147	+63.89	+404.6	+207	+97.22
+298.4	+148	+64.44	+406.4	+208	+97.78
+300.2	+149	+65.00	+408.2	+209	+98.33
+302.0	+150	+65.56	+410.0	+210	+98.89
+303.8	+151	+66.11	+411.8	+211	+99.44
+305.6	+152	+66.67	+413.6	+212	+100.00
+307.4	+153	+67.22	+415.4	+213	+100.56
+309.2	+154	+67.78	+417.2	+214	+101.11
+311.0	+155	+68.33	+419.0	+215	+101.67
+312.8	+156	+68.89	+420.8	+216	+102.22
+314.6	+157	+69.44	+422.6	+217	+102.78
+316.4	+158	+70.00	+424.4	+218	+103.33
+318.2	+159	+70.56	+426.2	+219	+103.89
+320.0	+160	+71.11	+428.0	+220	+104.44
+321.8	+161	+71.67	+431.6	+222	+105.56
+323.6	+162	+72.22	+435.2	+224	+106.67
+325.4	+163	+72.78	+438.8	+226	+107.78
+327.2	+164	+73.33	+442.4	+228	+108.89
+329.0	+165	+73.89	+446.0	+230	+110.00
+330.8	+166	+74.44	+449.6	+232	+111.11

TABLE A.5. Temperature Conversion Table

°F.	Reading in °F. or °C. to be converted	°C.	°F.	Reading in °F. or °C. to be converted	°C.
+453.2	+234	+112.22	+669.2	+354	+178.89
+456.8	+236	+113.33	+672.8	+356	+180.00
+460.4	+238	+114.44	+676.4	+358	+181.11
+464.0	+240	+115.56	+680.0	+360	+182.22
+467.6	+242	+116.67	+683.6	+362	+183.33
+471.2	+244	+117.78	+687.2	+364	+184.44
+474.8	+246	+118.89	+690.8	+366	+185.56
+478.4	+248	+120.00	+694.4	+368	+186.67
+482.0	+250	+121.11	+698.0	+370	+187.78
+485.6	+252	+122.22	+701.6	+372	+188.89
+489.2	+254	+123.33	+705.2	+374	+190.00
+492.8	+256	+124.44	+708.8	+376	+191.11
+496.4	+258	+125.56	+712.4	+378	+192.22
+500.0	+260	+126.67	+716.0	+380	+193.33
+503.6	+262	+127.78	+719.6	+382	+194.44
+507.2	+264	+128.89	+723.2	+384	+195.56
+510.8	+266	+130.00	+726.8	+386	+196.67
+514.4	+268	+131.11	+730.4	+388	+197.78
+518.0	+270	+132.22	+734.0	+390	+198.89
+521.6	+272	+133.33	+737.6	+392	+200.00
+525.2	+274	+134.44	+741.2	+394	+201.11
+528.8	+276	+135.56	+744.8	+396	+202.22
+532.4	+278	+136.67	+748.4	+398	+203.33
+536.0	+280	+137.78	+752.0	+400	+204.44
+539.6	+282	+138.89	+755.6	+402	+205.56
+543.2	+284	+140.00	+759.2	+404	+206.67
+546.8	+286	+141.11	+762.8	+406	+207.78
+550.4	+288	+142.22	+766.4	+408	+208.89
+554.0	+290	+143.33	+770.0	+410	+210.00
+557.6	+292	+144.44	+773.6	+412	+211.11
+561.2	+294	+145.56	+777.2	+414	+212.22
+564.8	+296	+146.67	+780.8	+416	+213.33
+568.4	+298	+147.78	+784.4	+418	+214.44
+572.0	+300	+148.89	+788.0	+420	+215.56
+575.6	+302	+150.00	+791.6	+422	+216.67
+579.2	+304	+151.11	+795.2	+424	+217.78
+582.8	+306	+152.22	+798.8	+426	+218.89
+586.4	+308	+153.33	+802.4	+428	+220.00
+590.0	+310	+154.44	+806.0	+430	+221.11
+593.6	+312	+155.56	+809.6	+432	+222.22
+597.2	+314	+156.67	+813.2	+434	+223.33
+600.8	+316	+157.78	+816.8	+436	+224.14
+604.4	+318	+158.89	+820.4	+438	+225.56
+608.0	+320	+160.00	+824.0	+440	+226.67
+611.6	+322	+161.11	+827.6	+442	+227.78
+615.2	+324	+162.22	+831.2	+444	+228.89
+618.8	+326	+163.33	+834.8	+446	+230.00
+622.4	+328	+164.44	+838.4	+448	+231.11
+626.0	+330	+165.56	+842.0	+450	+232.22
+629.6	+332	+166.67	+845.6	+452	+233.33
+633.2	+334	+167.78	+849.2	+454	+234.44
+636.8	+336	+168.89	+852.8	+456	+235.56
+640.4	+338	+170.00	+856.4	+458	+236.67
+644.0	+340	+171.11	+860.0	+460	+237.78
+647.6	+342	+172.22	+863.6	+462	+238.89
+651.2	+344	+173.33	+867.2	+464	+240.00
+654.8	+346	+174.44	+870.8	+466	+241.11
+658.4	+348	+175.56	+874.4	+468	+242.22
+662.0	+350	+176.67	+878.0	+470	+243.33
+665.6	+352	+177.78	+881.6	+472	+244.44

TABLE A.6. Temperature Conversion Table

°F.	Reading in °F. or °C. to be converted	°C.	°F.	Reading in °F. or °C. to be converted	°C.
+885.2	+474	+245.56	+1101.2	+594	+312.22
+888.8	+476	+246.67	+1104.8	+596	+313.33
+892.4	+478	+247.78	+1108.4	+598	+314.44
+896.0	+480	+248.89	+1112.0	+600	+315.56
+899.6	+482	+250.00	+1115.6	+602	+316.67
+903.2	+484	+251.11	+1119.2	+604	+317.78
+906.8	+486	+252.22	+1122.8	+606	+318.89
+910.4	+488	+253.33	+1126.4	+608	+320.00
+914.0	+490	+254.44	+1130.0	+610	+321.11
+917.6	+492	+255.56	+1133.6	+612	+322.22
+921.2	+494	+256.67	+1137.2	+614	+323.33
+924.8	+496	+257.78	+1140.8	+616	+324.44
+928.4	+498	+258.89	+1144.4	+618	+325.56
+932.0	+500	+260.00	+1148.0	+620	+326.67
+935.6	+502	+261.11	+1151.6	+622	+327.78
+939.2	+504	+262.22	+1155.2	+624	+328.89
+942.8	+506	+263.33	+1158.8	+626	+330.00
+946.4	+508	+264.44	+1162.4	+628	+331.11
+950.0	+510	+265.56	+1166.0	+630	+332.22
+953.6	+512	+266.67	+1169.6	+632	+333.33
+957.2	+514	+267.78	+1173.2	+634	+334.44
+960.8	+516	+268.89	+1176.8	+636	+335.56
+964.4	+518	+270.00	+1180.4	+638	+336.67
+968.0	+520	+271.11	+1184.0	+640	+337.78
+971.6	+522	+272.22	+1187.6	+642	+338.89
+975.2	+524	+273.33	+1191.2	+644	+340.00
+978.8	+526	+274.44	+1194.8	+646	+341.11
+982.4	+528	+275.56	+1198.4	+648	+342.22
+986.0	+530	+276.67	+1202.0	+650	+343.33
+989.6	+532	+277.78	+1205.6	+652	+344.44
+993.2	+534	+278.89	+1209.2	+654	+345.56
+996.8	+536	+280.00	+1212.8	+656	+346.67
+1000.4	+538	+281.11	+1216.4	+658	+347.78
+1004.0	+540	+282.22	+1220.0	+660	+348.89
+1007.6	+542	+283.33	+1223.6	+662	+350.00
+1011.2	+544	+284.44	+1227.2	+664	+351.11
+1014.8	+546	+285.56	+1230.8	+666	+352.22
+1018.4	+548	+286.67	+1234.4	+668	+353.33
+1022.0	+550	+287.78	+1238.0	+670	+354.44
+1025.6	+552	+288.89	+1241.6	+672	+355.56
+1029.2	+554	+290.00	+1245.2	+674	+356.67
+1032.8	+556	+291.11	+1248.8	+676	+357.78
+1036.4	+558	+292.22	+1252.4	+678	+358.89
+1040.0	+560	+293.33	+1256.0	+680	+360.00
+1043.6	+562	+294.44	+1259.6	+682	+361.11
+1047.2	+564	+295.56	+1263.2	+684	+362.22
+1050.8	+566	+296.67	+1266.8	+686	+363.33
+1054.4	+568	+297.78	+1270.4	+688	+364.44
+1058.0	+570	+298.89	+1274.0	+690	+365.56
+1061.6	+572	+300.00	+1277.6	+692	+366.67
+1065.2	+574	+301.11	+1281.2	+694	+367.78
+1068.8	+576	+302.22	+1284.8	+696	+368.89
+1072.4	+578	+303.33	+1288.4	+698	+370.00
+1076.0	+580	+304.44	+1292.0	+700	+371.11
+1079.6	+582	+305.56	+1295.6	+702	+372.22
+1083.2	+584	+306.67	+1299.2	+704	+373.33
+1086.8	+586	+307.78	+1302.8	+706	+374.44
+1090.4	+588	+308.89	+1306.4	+708	+375.56
+1094.0	+590	+310.00	+1310.0	+710	+376.67
+1097.6	+592	+311.11	+1313.6	+712	+377.78

TABLE A.7. Temperature Conversion Table

°F.	Reading in °F. or °C. to be converted	°C.	°F.	Reading in °F. or °C. to be converted	°C.
+1317.2	+714	+378.89	+1533.2	+834	+445.56
+1320.8	+716	+380.00	+1536.8	+836	+446.67
+1324.4	+718	+381.11	+1540.4	+838	+447.78
+1328.0	+720	+382.22	+1544.0	+840	+448.89
+1331.6	+722	+383.33	+1547.6	+842	+450.00
+1335.2	+724	+384.44	+1551.2	+844	+451.11
+1338.8	+726	+385.56	+1554.8	+846	+452.22
+1342.4	+728	+386.67	+1558.4	+848	+453.33
+1346.0	+730	+387.78	+1562.0	+850	+454.44
+1349.6	+732	+388.89	+1565.6	+852	+455.56
+1353.2	+734	+390.00	+1569.2	+854	+456.67
+1356.8	+736	+391.11	+1572.8	+856	+457.78
+1360.4	+738	+392.22	+1576.4	+858	+458.89
+1364.0	+740	+393.33	+1580.0	+860	+460.00
+1367.6	+742	+394.44	+1583.6	+862	+461.11
+1371.2	+744	+395.56	+1587.2	+864	+462.22
+1374.8	+746	+396.67	+1590.8	+866	+463.33
+1378.4	+748	+397.78	+1594.4	+868	+464.44
+1382.0	+750	+398.89	+1598.0	+870	+465.56
+1385.6	+752	+400.00	+1601.6	+872	+466.67
+1389.2	+754	+401.11	+1605.2	+874	+467.78
+1392.8	+756	+402.22	+1608.8	+876	+468.89
+1396.4	+758	+403.33	+1612.4	+878	+470.00
+1400.0	+760	+404.44	+1616.0	+880	+471.11
+1403.6	+762	+405.56	+1619.6	+882	+472.22
+1407.2	+764	+406.67	+1623.2	+884	+473.33
+1410.8	+766	+407.78	+1626.8	+886	+474.44
+1414.4	+768	+408.89	+1630.4	+888	+475.56
+1418.0	+770	+410.00	+1634.0	+890	+476.67
+1421.6	+772	+411.11	+1637.6	+892	+477.78
+1425.2	+774	+412.22	+1641.2	+894	+478.89
+1428.8	+776	+413.33	+1644.8	+896	+480.00
+1432.4	+778	+414.44	+1648.4	+898	+481.11
+1436.0	+780	+415.56	+1652.0	+900	+482.22
+1439.6	+782	+416.67	+1655.6	+902	+483.33
+1443.2	+784	+417.78	+1659.2	+904	+484.44
+1446.8	+786	+418.89	+1662.8	+906	+485.56
+1450.4	+788	+420.00	+1666.4	+908	+486.67
+1454.0	+790	+421.11	+1670.0	+910	+487.78
+1457.6	+792	+422.22	+1673.6	+912	+488.89
+1461.2	+794	+423.33	+1677.2	+914	+490.00
+1464.8	+796	+424.44	+1680.8	+916	+491.11
+1468.4	+798	+425.56	+1684.4	+918	+492.22
+1472.0	+800	+426.67	+1688.0	+920	+493.33
+1475.6	+802	+427.78	+1691.6	+922	+494.44
+1479.2	+804	+428.89	+1695.2	+924	+495.56
+1482.8	+806	+430.00	+1698.8	+926	+496.67
+1486.4	+808	+431.11	+1702.4	+928	+497.78
+1490.0	+810	+432.22	+1706.0	+930	+498.89
+1493.6	+812	+433.33	+1709.6	+932	+500.00
+1497.2	+814	+434.44	+1713.2	+934	+501.11
+1500.8	+816	+435.56	+1716.8	+936	+502.22
+1504.4	+818	+436.67	+1720.4	+938	+503.33
+1508.0	+820	+437.78	+1724.0	+940	+504.44
+1511.6	+822	+438.89	+1727.6	+942	+505.56
+1515.2	+824	+440.00	+1731.2	+944	+506.67
+1518.8	+826	+441.11	+1734.8	+946	+507.78
+1522.4	+828	+442.22	+1738.4	+948	+508.89
+1526.0	+830	+443.33	+1742.0	+950	+510.00
+1529.6	+832	+444.44	+1745.6	+952	+511.11

TABLE A.8. Temperature Conversion Table

°F.	Reading in °F. or °C. to be converted	°C.	°F.	Reading in °F. or °C. to be converted	°C.
+1749.2	+954	+512.22	+2498.0	+1370	+743.33
+1752.8	+956	+513.33	+2516.0	+1380	+748.89
+1756.4	+958	+514.44	+2534.0	+1390	+754.44
+1760.0	+960	+515.56	+2552.0	+1400	+760.00
+1763.6	+962	+516.67	+2570.0	+1410	+765.56
+1767.2	+964	+517.78	+2588.0	+1420	+771.11
+1770.8	+966	+518.89	+2606.0	+1430	+776.67
+1774.4	+968	+520.00	+2624.0	+1440	+782.22
+1778.0	+970	+521.11	+2642.0	+1450	+787.78
+1781.6	+972	+522.22	+2660.0	+1460	+793.33
+1785.2	+974	+523.33	+2678.0	+1470	+798.89
+1788.8	+976	+524.44	+2696.0	+1480	+804.44
+1792.4	+978	+525.56	+2714.0	+1490	+810.00
+1796.0	+980	+526.67	+2732.0	+1500	+815.56
+1799.6	+982	+527.78	+2750.0	+1510	+821.11
+1803.2	+984	+528.89	+2768.0	+1520	+826.67
+1806.8	+986	+530.00	+2786.0	+1530	+832.22
+1810.4	+988	+531.11	+2804.0	+1540	+837.78
+1814.0	+990	+532.22	+2822.0	+1550	+843.33
+1817.6	+992	+533.33	+2840.0	+1560	+848.89
+1821.2	+994	+534.44	+2858.0	+1570	+854.44
+1824.8	+996	+535.56	+2876.0	+1580	+860.00
+1828.4	+998	+536.67	+2894.0	+1590	+865.56
+1832.0	+1000	+537.78	+2912.0	+1600	+871.11
+1850.0	+1010	+543.33	+2930.0	+1610	+876.67
+1868.0	+1020	+548.89	+2948.0	+1620	+882.22
+1886.0	+1030	+554.44	+2966.0	+1630	+887.78
+1904.0	+1040	+560.00	+2984.0	+1640	+893.33
+1922.0	+1050	+565.56	+3002.0	+1650	+898.89
+1940.0	+1060	+571.11	+3020.0	+1660	+904.44
+1958.0	+1070	+576.67	+3038.0	+1670	+910.00
+1976.0	+1080	+582.22	+3056.0	+1680	+915.56
+1994.0	+1090	+587.78	+3074.0	+1690	+921.11
+2012.0	+1100	+593.33	+3092.0	+1700	+926.67
+2030.0	+1110	+598.89	+3110.0	+1710	+932.22
+2048.0	+1120	+604.44	+3128.0	+1720	+937.78
+2066.0	+1130	+610.00	+3146.0	+1730	+943.33
+2084.0	+1140	+615.56	+3164.0	+1740	+948.89
+2102.0	+1150	+621.11	+3182.0	+1750	+954.44
+2120.0	+1160	+626.67	+3200.0	+1760	+960.00
+2138.0	+1170	+632.22	+3218.0	+1770	+965.56
+2156.0	+1180	+637.78	+3236.0	+1780	+971.11
+2174.0	+1190	+643.33	+3254.0	+1790	+976.67
+2192.0	+1200	+648.89	+3272.0	+1800	+982.22
+2210.0	+1210	+654.44	+3290.0	+1810	+987.78
+2228.0	+1220	+660.00	+3308.0	+1820	+993.33
+2246.0	+1230	+665.56	+3326.0	+1830	+998.89
+2264.0	+1240	+671.11	+3344.0	+1840	+1004.4
+2282.0	+1250	+676.67	+3362.0	+1850	+1010.0
+2300.0	+1260	+682.22	+3380.0	+1860	+1015.6
+2318.0	+1270	+687.78	+3398.0	+1870	+1021.1
+2336.0	+1280	+693.33	+3416.0	+1880	+1026.7
+2354.0	+1290	+698.89	+3434.0	+1890	+1032.2
+2372.0	+1300	+704.44	+3452.0	+1900	+1037.8
+2390.0	+1310	+710.00	+3470.0	+1910	+1043.3
+2408.0	+1320	+715.56	+3488.0	+1920	+1048.9
+2426.0	+1330	+721.11	+3506.0	+1930	+1054.4
+2444.0	+1340	+726.67	+3524.0	+1940	+1060.0
+2462.0	+1350	+732.22	+3542.0	+1950	+1065.6
+2480.0	+1360	+737.78	+3560.0	+1960	+1071.1

TABLE A.9. Temperature Conversion Table

°F.	Reading in °F. or °C. to be converted	°C.	°F.	Reading in °F. or °C. to be converted	°C.
+3578.0	+1970	+1076.7	+4604.0	+2540	+1393.3
+3596.0	+1980	+1082.2	+4622.0	+2550	+1398.9
+3614.0	+1990	+1087.8	+4640.0	+2560	+1404.4
+3632.0	+2000	+1093.3	+4658.0	+2570	+1410.0
+3650.0	+2010	+1098.9	+4676.0	+2580	+1415.6
+3668.0	+2020	+1104.4	+4694.0	+2590	+1421.1
+3686.0	+2030	+1110.0	+4712.0	+2600	+1426.7
+3704.0	+2040	+1115.6	+4730.0	+2610	+1432.2
+3722.0	+2050	+1121.1	+4748.0	+2620	+1437.8
+3740.0	+2060	+1126.7	+4766.0	+2630	+1443.3
+3758.0	+2070	+1132.2	+4784.0	+2640	+1448.9
+3776.0	+2080	+1137.8	+4802.0	+2650	+1454.4
+3794.0	+2090	+1143.3	+4820.0	+2660	+1460.0
+3812.0	+2100	+1148.9	+4838.0	+2670	+1465.6
+3830.0	+2110	+1154.4	+4856.0	+2680	+1471.1
+3848.0	+2120	+1160.0	+4874.0	+2690	+1476.7
+3866.0	+2130	+1165.6	+4892.0	+2700	+1482.2
+3884.0	+2140	+1171.1	+4910.0	+2710	+1487.8
+3902.0	+2150	+1176.7	+4928.0	+2720	+1493.3
+3920.0	+2160	+1182.2	+4946.0	+2730	+1498.9
+3938.0	+2170	+1187.8	+4964.0	+2740	+1504.4
+3956.0	+2180	+1193.3	+4982.0	+2750	+1510.0
+3974.0	+2190	+1198.9	+5000.0	+2760	+1515.6
+3992.0	+2200	+1204.4	+5018.0	+2770	+1521.1
+4010.0	+2210	+1210.0	+5036.0	+2780	+1526.7
+4028.0	+2220	+1215.6	+5054.0	+2790	+1532.2
+4046.0	+2230	+1221.1	+5072.0	+2800	+1537.8
+4064.0	+2240	+1226.7	+5090.0	+2810	+1543.3
+4082.0	+2250	+1232.2	+5108.0	+2820	+1548.9
+4100.0	+2260	+1237.8	+5126.0	+2830	+1554.4
+4118.0	+2270	+1243.3	+5144.0	+2840	+1560.0
+4136.0	+2280	+1248.9	+5162.0	+2850	+1565.6
+4154.0	+2290	+1254.4	+5180.0	+2860	+1571.1
+4172.0	+2300	+1260.0	+5198.0	+2870	+1576.7
+4190.0	+2310	+1265.6	+5216.0	+2880	+1582.2
+4208.0	+2320	+1271.1	+5234.0	+2890	+1587.8
+4226.0	+2330	+1276.7	+5252.0	+2900	+1593.3
+4244.0	+2340	+1282.2	+5270.0	+2910	+1598.9
+4262.0	+2350	+1287.8	+5288.0	+2920	+1604.4
+4280.0	+2360	+1293.3	+5306.0	+2930	+1610.0
+4298.0	+2370	+1298.9	+5324.0	+2940	+1615.6
+4316.0	+2380	+1304.4	+5342.0	+2950	+1621.1
+4334.0	+2390	+1310.0	+5360.0	+2960	+1626.7
+4352.0	+2400	+1315.6	+5378.0	+2970	+1632.2
+4370.0	+2410	+1321.1	+5396.0	+2980	+1637.8
+4388.0	+2420	+1326.7	+5414.0	+2990	+1643.3
+4406.0	+2430	+1332.2	+5432.0	+3000	+1648.9
+4424.0	+2440	+1337.8	+5450.0	+3010	+1654.4
+4442.0	+2450	+1343.3	+5468.0	+3020	+1660.0
+4460.0	+2460	+1348.9	+5486.0	+3030	+1665.6
+4478.0	+2470	+1354.4	+5504.0	+3040	+1671.1
+4496.0	+2480	+1360.0	+5522.0	+3050	+1676.7
+4514.0	+2490	+1365.6	+5540.0	+3060	+1682.2
+4532.0	+2500	+1371.1	+5558.0	+3070	+1687.8
+4550.0	+2510	+1376.7	+5576.0	+3080	+1693.3
+4568.0	+2520	+1382.2	+5594.0	+3090	+1698.9
+4586.0	+2530	+1387.8	+5612.0	+3100	+1704.4

APPENDIX B
GREEK ALPHABET

TABLE B.1 Greek Letters
Common Uses as Symbols in Glass Technology

English spelling	Greek capital letters	Greek small letters
Alpha	A	α—Coefficient of thermal expansion; natural or low-temperature crystal form
Beta	B	β—Specific heat constant; secondary or high-temperature crystal form; compressibility
Gamma	Γ	γ—Surface tension; higher crystal form
Delta	Δ	δ—Increment or differential; birefringence
Epsilon	E	ϵ—Unit elongation or strain; energy potential
Zeta	Z	ζ—Logarithm of viscosity; deformation at breaking point
Eta	H	η—Viscosity, poises; entropy
Theta	Θ	θ—Degrees of thermal shock; plane angle
Iota	I	ι
Kappa	K	κ—K Dielectric constant; κ conductivity
Lambda	Λ	λ—Wave length
Mu	M	μ—Micron (10^{-4} cm); $\mu\mu$ micromicron (10^{-10} cm); mμ millimicron (10^{-7} cm)
Nu	N	ν—Dispersion
Xi	Ξ	ξ
Omicron	O	o
Pi	Π	π—Circumference $\div$ diameter, 3.1416
Rho	P	ρ—Electrical resistivity
Sigma	Σ	σ—Σ Summation; σ Stefan-Boltzmann constant; interfacial surface tension; Poisson's ratio
Tau	T	τ
Upsilon	Y	υ
Phi	Φ	ϕ—Fluidity or $1/\eta$
Chi	X	χ
Psi	Ψ	ψ
Omega	Ω	ω—Ohms

APPENDIX C
PERIODIC TABLE OF ELEMENTS

Note:
Atomic weights are those of the most commonly available long-lived isotopes on the 1973 IUPAC Atomic Weights of the Elements. A value given in parentheses denotes the mass number of the longest-lived isotope.

APPENDIX D
CONVERSION CHART

Table D.1.

Length:
1 ft = 12 in
1 yd = 3 ft
1 mi = 5280 ft
1 in = 25.4 mm
1 in = 2.54 cm
1 meter = 3.281 ft
1 mi = 1.609 km

Area:
1 ft^2 = 144 in^2
1 yd^2 = 9 ft^2
1 in^2 = 6.452 cm^2
1 mi^2 = 2.590 km^2
1 mi^2 = 640 acres

Volume:
1 ft^3 = 1728 in^3
1 yd^3 = 27 ft^3
1 in^3 = 16.39 cc
1 liter = 61.02 in^3
1 U.S. gal = 231 in^3
1 ft^3 = 7.481 U.S. gal

Velocity:
1 fps = 60 fpm
1 fpm = 60 fph
1 fps = 3600 fph
1 fps = 30.48 cm/s
1 mph = 1.467 fps
1 radian/s = 9.55 rpm

Volume Flow Rate:
1 cfm = 60 cfh
1 cfs = 60 cfm
1 cfm = 7.481 gal pm
1 cfs = 26,930 gal ph

Energy, Heat, and Work:
1 cal = 4.187 joules
1 hp-hr = 2544 Btu
1 kwh = 3412 Btu
1 hp-hr = 1,980,000 ft · lb
1 kg-m = 7.233 ft · lb

Power and Heat Flow:
1 kw = 1.341 hp
1 hp = 550 ft · lb/sec
1 hp = 42.41 Btu/min
1 Btu/s = 1.055 kw
1 kw = 3412 Btu/h
1 hp = 2544 Btu/h

Mass:
1 lb (avdp) = 16 oz (avdp)
1 short ton = 2000 lb (avdp)
1 lb (avdp) = 7000 grains
1 lb (avdp) = 453.6 g
1 kg = 2.205 lb (avdp)

Density and Specific Gravity:
Specific gravity relative to water (abbr. sgw) of 1.00 = 62.43 lb/ft^3
Specific gravity relative to air (abbr. sga) of 1.000 = 0.0763 lb/ft^3
1 g/cc = 62.43 lb/ft^3
1 lb/in^3 = 27.68 g/cc
1 lb/in^3 = 1728 lb/ft^3
1 lb/U.S. gal = 7.481 lb/ft^3
$1 \, \frac{\text{short ton}}{\text{yd}^3} = 74.07 \text{ lb/ft}^3$

Pressure:
1 psi = 144 psf
1 kg/cm^2 = 14.22 psi
1 atm = 14.7 psi
1 psi = 2.309 ft wc (for water at 59°F)
1 psi = 27.71 in wc (for water at 59°F)
1 psi = 2.036 in Hg (for mercury at 32°F)
1 in Hg = 13.61 in wc (for water at 59°F and mercury at 32°F)
1 osi = 1.732 in wc

Energy, Heat, and Work:
1 Btu = 252.0 cal
1 Btu = 0.2520 kg cal
1 therm = 100,000 Btu
1 Btu = 778.2 ft · lb
1 Btu = 1055 joules

Heat Flux:
1 cal/h cm^2 = 3.687 Btu/h/ft^2
1 watt/cm^2 = 3170 Btu/h/ft^2

Thermal Conductivity:
$1 \, \frac{\text{Btu ft}}{\text{h ft}^2 \, °\text{F}} = 12 \, \frac{\text{Btu in}}{\text{h ft}^2 \, °\text{F}}$
$1 \, \frac{\text{Btu ft}}{\text{h ft}^2 \, °\text{F}} = 14.88 \, \frac{\text{cal cm}}{\text{h cm}^2 °\text{C}}$
$1 \, \frac{\text{watt cm}}{\text{cm}^2 °\text{C}} = 57.79 \, \frac{\text{Btu ft}}{\text{h ft}^2 °\text{F}}$

Heat Content:
$1 \, \frac{\text{Btu}}{\text{lb}} = 0.556 \, \frac{\text{cal}}{\text{g}}$
$1 \, \frac{\text{Btu}}{\text{lb °F}} = 1 \, \frac{\text{cal}}{\text{g °C}}$

APPENDIX E
EQUIVALENTS

TABLE E.1 Units of Capacity

Fl oz	Pint	Qt	Gal	British gal	Cc	Liter	Cu in	Cu ft
1	0.625	0.0313	0.0078	0.0065	29.57	0.0296	1.805	0.00105
16	1	0.5	0.125	0.104	473	0.473	28.88	0.0167
32	2	1	0.25	0.2082	946	0.946	57.75	0.0334
128	8	4	1	0.8327	3785	3.785	231	0.1337
153.7	9.608	4.804	1.201	1	4546	4.546	277.4	0.1605
0.338	0.0021	0.00106	0.00026	0.00022	1	0.001	0.0610	0.000035
33.82	2.113	1.057	0.2642	0.2201	1000	1	61.02	0.03531
0.5541	0.0346	0.0173	0.0043	0.0036	16.39	0.0164	1	0.000579
957.4	59.84	29.92	7.480	6.232	28,320	28.32	1728	1

TABLE E.2 Conversion Factors for Concentrations

G/100 cc	G/l	Oz/gal	Grains/gal	Parts per million
1	10	1.335	584	10^4
0.1	1	0.1335	58.4	1000
0.749	7.49	1	437.5	7490
1.712×10^{-3}	1.712×10^{-2}	2.88×10^{-2}	1	17.12
10^{-4}	10^{-3}	1.35×10^{-3}	0.0584	1

TABLE E.3 Metric and Avoirdupois Weights

G	Kg	Metric ton	Grain	Ounce	Pound	Ton
1	0.001	1×10^{-6}	15.43	0.03527	0.0022	1.1×10^{-6}
1000	1	0.001	15.43×10^3	35.27	2.2016	0.0011
10^4	1000	1	15.43×10^6	35.27×10^3	2.2×10^3	1.1023
0.0648	6.48×10^{-5}	6.48×10^{-8}	1	2.285×10^{-3}	1.428×10^{-4}	7.14×10^{-8}
28.35	0.02835	2.835×10^{-5}	437.5	1	0.0625	3.125×10^{-6}
453.6	0.4536	4.536×10^{-4}	7000	16	1	5×10^{-4}
9.072×10^4	907.2	0.9072	14×10^6	32×10^3	2000	1

TABLE E.4 Fluid Measure (Small Units)

Cc	Minim	Dram	Fl oz	Gill	British fl oz	British gill
1	16.23	0.2705	0.0338	0.00845	0.352	0.00704
0.0616	1	0.0167	0.00208	0.000521	0.00217	0.000434
3.697	60	1	0.1250	0.0312	0.1302	0.0260
29.57	480	8	1	0.250	1.041	0.2082
118.3	1920	32	4	1	4.164	0.8328
28.41	461.2	7.686	0.9608	0.2402	1	0.2000
142.0	2306	37.43	4.804	1.201	5	1

Gill = 0.25 pint

TABLE E.5 Electric Heating

Volts × Amperes	= Watts
Watts × 3.41	= Btu/h
1 Amp. at 110 V	= 375 Btu/h
Amperes	= V/R (ohms)
Watt = Volts × Amp.	= V^2/R

For constant voltage, the power consumed and the heat generated in a resistance wire of given diameter vary inversely as its length.

TABLE E.6 Temperature Conversion

To change Centigrade to Fahrenheit, multiply by $9/5$ and add 32.
To change Fahrenheit to Centigrade, subtract 32 and multiply remainder by $5/9$:

$$°F = (°C \times 1.8) + 32$$

and

$$°C = \frac{°F - 32}{1.8}$$

The Centigrade degree is a longer temperature interval than the Fahrenheit degree in the ratio of 9:5.

APPENDIX F
CONVERSIONS (METRIC AND COLOR), PROPERTIES AND COMPARISONS OF BUILDING MATERIALS

Fraction	Decimal	Fraction	Decimal	Millimeters	Decimal	Millimeters	Decimal
1/64	.016	33/64	.516	1.0	.039	13.0	.512
1/32	.031	17/32	.531	1.5	.059	13.1	.516
3/64	.047	35/64	.547	2.0	.079	13.2	.520
1/16	.063	9/16	.563	2.5	.098	13.5	.531
5/64	.078	37/64	.578	3.0	.118	14.0	.551
3/32	.094	19/32	.594	3.5	.138	14.5	.571
7/64	.109	39/64	.609	4.0	.157	15.0	.591
1/8	.125	5/8	.625	4.5	.177	15.5	.610
9/64	.141	41/64	.641	5.0	.197	15.9	.626
5/32	.156	21/32	.656	5.5	.217	16.0	.630
11/64	.172	43/64	.672	6.0	.236	16.5	650
3/16	.188	11/16	.688	6.5	.256	17.0	.669
13/64	.203	45/64	.703	7.0	.276	17.5	.689
7/32	.219	23/32	.719	7.5	.295	18.0	.709
15/64	.234	47/64	.734	8.0	.315	18.5	.728
1/4	.250	3/4	.750	8.5	.335	19.0	.748
17/64	.266	49/64	.766	9.0	.354	19.5	.768
9/32	.281	25/32	.781	9.5	.374	20.0	.787
19/64	.297	51/64	.797	10.0	.394	20.5	.807
5/16	.313	13/16	.813	10.1	.398	21.0	.827
21/64	.328	53/64	.828	10.5	.413	21.5	.846
11/32	.344	27/32	.844	10.8	.425	22.0	.866
23/64	.359	55/64	.859	11.0	.433	22.5	.886
3/8	.375	7/8	.875	11.2	.441	23.0	.906
25/64	.391	57/64	.891	11.5	.453	23.5	.925
13/32	.406	29/32	.906	11.7	.461	24.0	.945
27/64	.422	59/64	.922	12.0	.472	24.5	.965
7/16	.438	15/16	.938	12.5	.492	25.0	.984
29/64	.453	61/64	.953				
15/32	.469	31/32	.969				
31/64	.484	63/64	.984				
1/2	.500	1	1.00				

TABLE F.1 Decimal Equivalencies of Fractions

INCHES	MILLIMETERS	FEET	METERS
1/4	6.35	2500	762
		4000	1219.2
		6400	1950.72
		8000	2438.4
3/8	9.525	1400	426.72
		2100	640.08
		3600	1097.28
		4200	1280.16
1/2	12.7	800	243.84
		1250	381
		2500	762
5/8	15.875	550	167.64
		775	236.22
		1550	472.44
3/4	19.05	400	121.92
		550	167.64
		1100	335.28
7/8	22.225	850	259.08
1	25.4	500	152.4
		600	182.88
1 1/8	28.575	500	152.4
1 1/4	31.75	400	121.92
		864	263.35
1 1/2	38.1	552	168.25
		770	234.7
2	50.8	360	109.73
		525	160.02
2 1/2	63.5	240	73.15
3	76.2	144	43.89
4	101.6	90	27.43

TABLE F.1a   The Above Metric Conversions Refer to all Nomaco Backer Rod Lines

Table F-2 Expansion Properties of Building Materials

MATERIAL	TYPICAL COEFFICIENTS OF EXPANSION IN INCH PER INCH PER °F	EXPANSION IN 64ths OF AN INCH PER 100° F TEMPERATURE RISE PER 10' LENGTH (APPROX.)
STAINLESS STEEL (300 Series)	.0000096	7
110 SOFT COPPER	.0000094	7
110 COLD ROLLED COPPER	.0000098	7
TIN	.0000117	8½
ALUMINUM 3003	.0000129	9½
LEAD	.0000150	11
ZINC, ROLLED	.0000174	13
ZINC ALLOY (With grain)	.0000130	9½
ZINC ALLOY (Across grain)	.0000098	7
MONEL	.0000078	6
GALVANIZED STEEL	.0000067	5
STEEL	.0000067	5
LIMESTONE	.0000044	3
GLASS	.0000047	3½
MARBLE	.0000056	4
SLATE	.0000058	4
BRICK	.0000031	2
CONCRETE	.0000078	6

Table F-3 Color Conversion Chart

Index	PHOTOMETRIC INDEX[2] (A.O.C.S.) 440/550 mmu	LOVIBOND 5-1/4" CELL Y/R	TRANSMISSION % THRU 2.5 cm 440/550 mmu	GARDNER 1953	A.P.H.A.	A.S.T.M. D-15000	Index
	0/0		100/100		0		
	4/0		90/100		50		
—A	7/0		85/100		100		A—
—B	11/0.3	3/0.3		1	200		B—
	14/0.5	4/0.5	72/99		300	1	
—C			65/98	2	400		C—
—D	22/1	7/1	59/97				D—
	25/1.5				450		
—E	28/2	8/1.5	51/95		500		E—
	32/3	10/2		3			
—F		14/3	39/90	4	A.P.H.A. F.A.C. 1		F—
	44/5	16/4					
—G		18/5 (5-1/4" Cell)	28/85	5		1-1/2	G—
		5/1 (1" Cell)					
—H	62/10		21/75	6		3	H—
	75/15					2	
—I	88/20	7/2	12/60	7	5	2-1/2	I—
—J	108/30	10/3	8/50	8, 9	7, 9	3	J—
		16/5		10	13	3-1/2	
—K	150/50	20/10	2/30	11, 12	15	4	K—
—L	200/100	-/20	1/15	13	19	4-1/2	L—
—M	-/200	-/30	0/0	14, 15, 16	31, 35	5, 6, 7	M—
Index	PHOTOMETRIC INDEX[2] (A.O.C.S.) 440/550 mmu	LOVIBOND 1" CELL Y/R	TRANSMISSION % THRU 2.5 cm 440/550 mmu	GARDNER 1953	F.A.C.	A.S.T.M. D-15000	Index

[1] Comparisons of color scales of different systems are very difficult and inaccurate. Thus, this conversion chart should be used only for fatty acids and only to obtain approximate values.

[2] Absorbency readings were taken on Coleman 6A Spectrophotometer using 25 mm Cuvette.

Table F-4   Comparable Thicknesses and Weights
Stainless Steel, Aluminum and Copper

STAINLESS STEEL			ALUMINUM				COPPER		
Thickness (Inch)	Gauge (U.S. Standard)	Lb. sq. ft.	Thickness (Inch)	Gauge (B&S)	Lb. sq. ft.		Thickness (Inch)	Oz. sq. ft.	Lb. sq. ft.
.010	32	.420	.010	30	.141		.0108	8	.500
.0125	30	.525	.0126	28	.177		.0121	9	.563
							.0135	10	.625
.0156	28	.656	.0156	—	.220		.0148	11	.688
			.0179	25	.253		.0175	13	.813
.0187	26	.788	.020	24	.282		.021	16	1.000
.0219	25	.919							
.025	24	1.050	.0253	22	.352				
.031	22	1.313	.0313	—	.441		.027	20	1.250
							.032	24	1.500
.0375	20	1.575	.032	20	.451		.0337	28	1.750
			.0403	18	.563		.0431	32	2.000
			.0453	17	.100				
.050	18	2.100	.0506	16	.126				

Note that U.S. Standard Gauge (stainless sheet) is not directly comparable with the B&S Gauge (aluminum). A 20-gauge stainless averages .0375" thick; while a 20-gauge aluminum averages .032" thick; and 20-ounce copper is .027" thick. The higher strength of stainless steel permits use of thinner gauges than required for aluminum or copper, which makes stainless more competitive with aluminum on a weight-to-coverage basis and provides stainless with a substantial weight saving compared to copper. For example, 100 sq. ft. of .032" aluminum will weigh about 45 pounds, .021" (16-ounce) copper will weigh about 100 pounds, and .015" stainless will weigh about 66 pounds.

APPENDIX G
KEY TO SYMBOLS

Appendix G

Table G-1   Key to Symbols Used in Figures in Chapters 18 thru 22

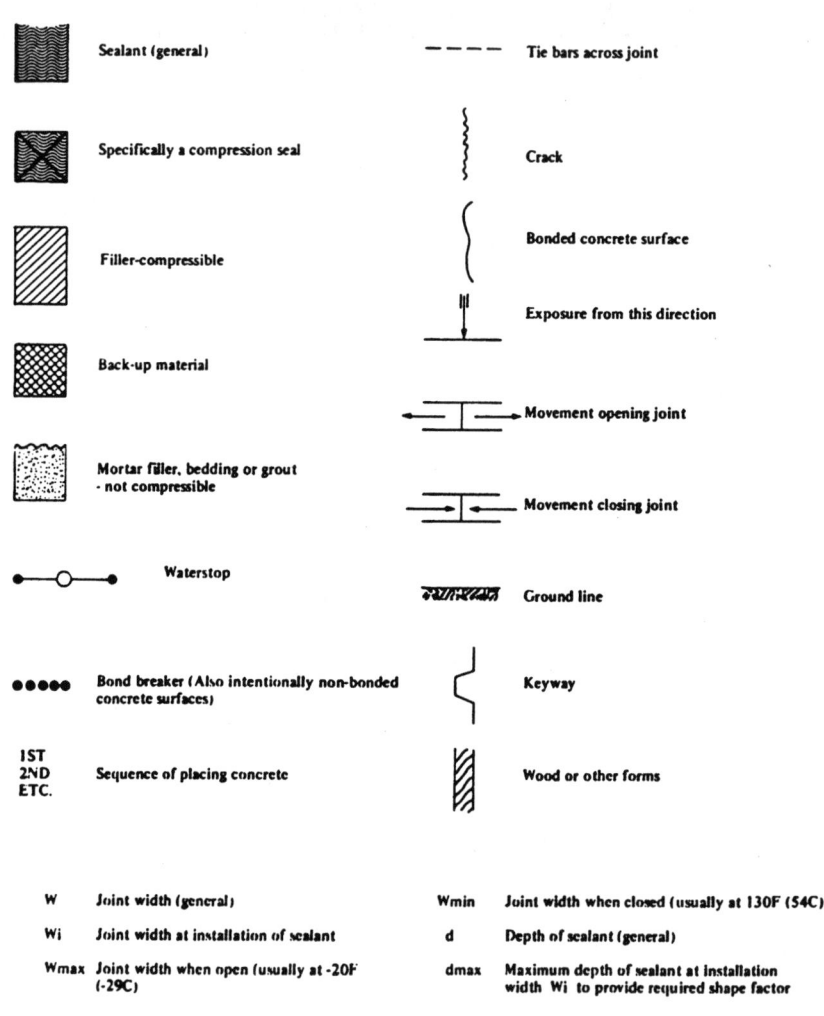

W	Joint width (general)	Wmin	Joint width when closed (usually at 130F (54C)
Wi	Joint width at installation of sealant	d	Depth of sealant (general)
Wmax	Joint width when open (usually at -20F (-29C)	dmax	Maximum depth of sealant at installation width Wi to provide required shape factor

APPENDIX H
ESTIMATES AND REQUIREMENTS

Joint Estimation Table

JOINT WIDTH	Per 1/8" Installed Depth			Per 1/4" Installed Depth			Per 1/2" Installed Depth		
	CU IN. PER FT.	FT/GAL	GAL/ 100 FT	CU IN PER FT	FT/GAL	GAL/ 100 FT	CU IN PER FT	FT/GAL	GAL/ 100 FT
0.5	0.8	308.0	0.3	1.5	154.0	0.6	3.0	77.0	1.3
0.8	1.1	205.3	0.5	2.2	102.7	1.0	4.5	51.3	1.9
1.0	1.5	154.0	0.6	3.0	77.0	1.3	6.0	38.5	2.6
1.5	2.2	102.7	1.0	4.5	51.3	1.9	9.0	25.7	3.9
2.0	3.0	77.0	1.3	6.0	38.5	2.6	12.0	19.2	5.2
2.5	3.8	61.6	1.6	7.5	30.8	3.2	15.0	15.4	6.5
3.0	4.5	51.3	1.9	9.0	25.7	3.9	18.0	12.8	7.8
3.5	5.2	44.0	2.3	10.5	22.0	4.5	21.0	11.0	9.1
4.0	6.0	38.5	2.6	12.0	19.2	5.2	24.0	9.6	10.4
5.0	7.5	30.8	3.2	15.0	15.4	6.5	30.0	7.7	13.0
6.0	9.0	25.7	3.9	18.0	12.8	7.8	36.0	6.4	15.6

CU IN PER FT	FT/GAL	GAL/ 100 FT
6.0	38.5	12.0
9.0	25.7	18.0
12.0	19.2	24.0
18.0	12.8	36.0
24.0	9.6	48.0
30.0	7.7	60.0
36.0	6.4	72.0
42.0	5.5	84.0
48.0	4.8	96.0
60.0	3.9	120.0
72.0	3.2	144.0

Look up the joint width in left hand column. Calculations for depths ranging from 1/8" (0.125") to 1" are provided. Stated values are for caulking. To calculate spray requirements, add actual joint width + over spray to determine JOINT WIDTH. Example: actual joint width =4", over spray =1" on both sides. JOINT WIDTH in table would be 6". For wider joints such as 8" joints, double the values shown in the 4" JOINT WIDTH estimation.

H = Height of Flute

L = Length of Joint

Joint

Amount in Cubic Inches for Each Side of Joint:

L = Length of Joint (ft)   W = Width of Joint (in)
H = Height of flute (in)

Area of Flutes= (L x 12 x H)/2 = _____ sq in
Area of Joint= L x 12 x W= _____ sq in
Area of Overspray L x 12= _____ sq in
Volume= (Area of flutes +
Area of Joint + Area of
Overspray) x .125"= _____ sq in
* Based upon 1/8" (0.125") wet coating thickness
Remember to Multiply by Two for Both Sides
(1) 5 Gallon pail of SpecSeal Elastomeric Spray contains 1155 cubic inches

H-2

SEALANT REQUIREMENTS IN CUBIC INCHES PER HALF INCH OF INSTALLED DEPTH

Pipe Size Trade Size	Pipe O.D.	Size of Hole (Inches) 1-1/2	2	3	4	5	6	7	8	10	12	14	26
1/2"	0.840	1.22	1.29	3.26	6.01								
1"	1.313	0.42	0.89	2.86	5.61	9.14							
1-1/2"	1.900			2.12	4.87	8.40							
2"	2.375			1.32	4.07	7.60	11.92						
2-1/2"	2.875				3.04	6.57	10.89						
3"	3.500				1.47	5.01	9.33	14.43	20.32				
3-1/2"	4.000					3.53	7.85	12.96	18.85				
4"	4.500					1.87	6.19	11.29	17.18	31.32	48.60		
6"	5.563						1.98	7.09	12.98	22.03	39.31		
8"	8.653									10.06	27.34		
10"	10.750										11.75	63.20	
12"	12.750											26.30	
24"	24.000												76.60

IMPORTANT NOTE: This table is for estimation purposes only. Consult UL Fire Resistance Directory or STI Product & Application Guide for specific installation requirements and limitations.

PRODUCT ESTIMATION INFORMATION

NOMINAL CROSS-SECTIONAL AREA OF PILLOWS

CAT. NO.	SSB14	SSB24	SSB26	SSB36
Nom. Dim[A]	1" x 4"	2" x 4"	2" x 6"	3" x 6"
in cm	2.54 x 10.2	5.1 x 10.2	5.1 x 15.24	7.6 x 15.24
Nom. Area[B]	4 in^2	8 in^2	12 in^2	18 in^2
	25.8 cm^2	51.6 cm^2	77.4 cm^2	116 cm^2
Effective Yield[C]	2.9 in^2	5.7 in^2	8.6 in^2	12.9 in^2
	18.7 cm^2	36.8 cm^2	55.5 cm^2	83.2 cm^2

NOTES: PILLOW LENGTH = 9" (22.9cm). A. Nom dimensions (uncompressed). B. Cross-sectional area (uncompressed) C. Cross-sectional area (compressed)

Calculating Pillow Requirement

Measure the size of the opening to be sealed and calculate the total area of the opening in square inches. Measure and calculate the approximate area occupied by the penetrants. Calculate the net area to be sealed by subtracting the area occupied by the penetrants from the total area of the opening. To allow for the required compression of the pillows, multiply the net area by 1.4. This will provide a compression factor of 29%.

In the example shown at left, the opening is 12" x 24" with an 18" wide tray. The cable depth in the tray is about 3". The area of the opening is 12 x 24 = 288 sq. in. The approximate area of the cables is 3 x 18 = 54 sq. in. Subtracting the area of the cables from the total area of the opening yields a net area of 234 sq. in. 234 x 1.4 = approx. 328 sq. in. to be filled by pillows. Using the table at left to determine the nominal area of the various pillows, we can determine that approximately 28 (328 - 12) SSB26 pillows would be required. The number of pillows required will of course vary by the size of the pillow being utilized. Generally, a small percentage of smaller pillows will be required along with the larger ones. A test opening of this size utilized 24 SSB26 pillows, along with 4 SSB24's and 4 SSB14's.

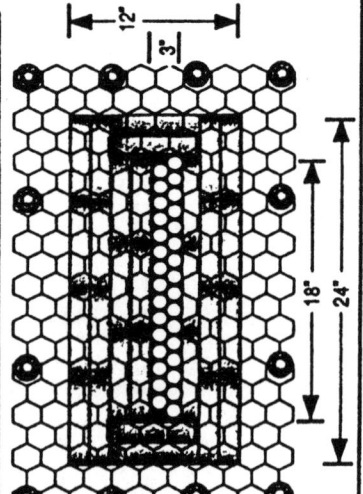

APPENDIX I
CONCRETE COMPOSITION

Estimating Cubic Yards of Concrete for Slabs

Thickness, in.	Area in Square Feet (width × length)					
	10	25	50	100	200	300
4	0.12	0.31	0.62	1.23	2.47	3.70
5	0.15	0.39	0.77	1.54	3.09	4.63
6	0.19	0.46	0.93	1.85	3.70	5.56

NOTE: Does not allow for losses due to uneven subgrade, spillage, etc. Add 5 to 10 percent for such contingencies.

Guide for Ordering Ready-Mix Concrete for Drives, Walks, and Patios

Maximum-Size Aggregate, in.	Minimum Cement Content, lb. per cu. yd.	Typical Slump, in.	Compressive Strength at 28 Days, lb. per sq. in.*	Air Content, percent by volume
3/8	610	3 to 5	3,500	7½ ± 1
1/2	590	3 to 5	3,500	7½ ± 1
3/4	540	3 to 5	3,500	6 ± 1
1	520	3 to 5	3,500	6 ± 1
1½	470	3 to 5	3,500	5 ± 1

NOTE: Slump should neither exceed 5 inches nor fall below 3 inches.
*Use 4,000 psi for freeze-thaw and deicers.

Proportions by Weight to Make 1 Cubic Foot of Concrete

Maximum-Size Coarse Aggregate, in.	Air-Entrained Concrete				Concrete without Air			
	Cement, lb.	Sand, lb.	Coarse Aggregate, lb.*	Water, lb.	Cement, lb.	Sand, lb.	Coarse Aggregate, lb.*	Water, lb.
3/8	29†	53	46	10	29	59	46	11
1/2	27	46	55	10	27	53	55	11
3/4	25	42	65	10	25	47	65	10
1	24	39	70	9	24	45	70	10
1 1/2	23	38	75	9	23	43	75	9

NOTE: Above proportions based on wet sand. See page 27 for adjustments for damp or very wet sand.
*If crushed stone is used, decrease coarse aggregate by 3 pounds and increase sand by 3 pounds.
†Metric conversion: 1 lb. = 0.454 kg; 10 lb. = 4.54 kg; 1 in. = 25 mm; 1 cu. ft. = 0.028 m³.

Proportions by Volume

Maximum-Size Coarse Aggregate, in.	Air-Entrained Concrete				Concrete without Air			
	Cement	Sand	Coarse Aggregate	Water	Cement	Sand	Coarse Aggregate	Water
³⁄₈	1	2¼	1½	½	1	2½	1½	½
½	1	2¼	2	½	1	2½	2	½
¾	1	2¼	2½	½	1	2½	2½	½
1	1	2¼	2¾	½	1	2½	2¾	½
1½	1	2¼	3	½	1	2½	3	½

NOTES:
The combined volume is approximately two-thirds of the sum of the original bulk volumes.
Above proportions based on wet sand.

INDEX

Acoustical sealants
   ASTM standard, 18.21
   tapes, 16.4
Acrylics
   anaerobic acrylic adhesives, 3.14
   anaerobics compared, 3.3
   application tables, 3.20
   binders, 3.4
   butyls compared, 8.11
   caulks
      binder types, 3.3
      component materials, 3.4
      latex-based, 3.5
      oil-based, 3.23
      Rhoplex, 3.6
      solvent-release types, 3.6
      vertical applications, 3.22
   curing processes, 3.17
   development, 3.2
   emulsions, properties, 3.25
   first- and second-generation, 3.3
   flammability, 3.16
   formulations, 3.19
   health and safety considerations, 3.3
   latex polymers, 3.24
   mixing equipment, 3.4
   modified acrylic adhesives, 3.14
   movement capabilities of joint sealants, 13.19
   non-volatile modified acrylic adhesives, 3.16
   oil and resin-based materials, displacement of, 6.2
   open joint sealants, comparison table, 1.4, 21.5
   PMMA types, 3.1
   polymer selection, 3.3
   restrictions on use, 3.1, 3.3
   Rhoplex as binder, 3.6
   sealant backbone comparison table, 11.6
   sealant materials comparison table, 22.15
   sealant types distinguished, 1.2
   silicone unpigmented sealants compared, 3.7
   solvent release caulks, 3.6
   solvent release system formula, 17.4
   solvent release tables, 3.20
   tensile shear values for modified acrylics, 3.18
   vertical caulks, use in, 3.22
   vertical mastic applications, 3.20
   water- vs solvent-based systems, 3.1
Adhesion standards
   (*See also* specifications)
   ASTM peel test, 18.16
   SWRI Validation Program, 18.10
Adhesive properties and uses
   classification system, 1.17
   comparison table, 1.19
   definitions, 1.1
   resin types, 1.1
   sealants distinguished, 1.17, 4.1
   solvent-based systems, 3.18
Airfield runway joint sealants
   (*See also* jet fuels, resistance to)
   designs, joint, 19.4
   markets, 17.2
   sealant comparison table, 21.12
   silicones, 13.8
   skewed joints, 22.28
   slabs on grade, 19.15, 19.17
American Architectural Manufacturers Association (AAMA) specification and test standards, 18.31
American Association of State Highway and Transportation Officials standards, 18.30

American Concrete Institute (ACI)
  crack repair studies, 22.8
  joint design studies, 19.18, 22.28
  standards, 18.28
  waterstop design, 19.17
American Society of Testing and Materials (ASTM)
  cement classifications, 7.7
  construction standards, 1.7
  firestop product standards, 15.7
  grout specifications, 7.38
  National Bureau of Standards, replacement of, 18.12
  sealant standards, 18.12-18.25
  smoke emission properties of firestop products, 15.8
  specifications, use of standards in, 18.10
Anaerobics and cyanoacrylates
  acrylic adhesives
    anaerobic types, 3.14
    compared, 3.3
  advantages of, 1.18
  application types, 4.4
  bonding action of cyanoacrylates, 4.10
  chemical composition, 4.1
  competing adhesives, comparison table, 4.5
  cure rates, 4.2
  curing processes, 3.17
  cyanoacrylates
    bonding action, 4.10
    properties, 4.6
    storage, 4.11
  formaldehyde, 4.11
  nonporous materials, use on, 4.1
  oxygen deprivation, effect of, 4.1
  pot life, 4.6
  primers, 4.2
  stabilizers, 4.4
  storage, 4.2, 4.6
    cyanoacrylates, 4.11
  superglues, 4.1
  threaded fasteners, use on, 3.16
  types, 4.4
  uncured properties, 4.5
Animal glues, 1.2

Application conditions (*see* installation conditions)
Application methods (*see* installation methods)
Asbestos
  (*See also* toxicity)
  prohibitions on uses of, 15.28, 16.3
  roofing materials, 3.26
Asphalts
  advantages, 6.2
  applications, 6.4
  ASTM standards, 18.14
  concrete joints, 7.11
  concrete, compatibility with, 6.3
  disadvantages, 1.12
  failures caused by use of low grade sealants, 22.1
  field molded rubber types, 1.15
  history of oil- and resin-based materials, 6.1
  hot-poured bituminous materials, 6.2, 19.22
    deck-sealing standards, 18.21
  limitations, 6.2
  low grade sealants resulting in failures, 22.1
  neoprene compatibility, 10.6
  open joint sealants, comparison table, 1.4, 21.5
  pour points, 19.22
  preformed foam sealants impregnated with, 16.4
  reclaimed rubber in, 10.10
  rubber types, 1.13
  sagging problems, 1.11, 21.13
  sealant materials comparison table, 22.15
  tapes, 16.5
  tar-based compounds, 19.22
  thermoplastic properties, 1.12, 5.1
  types, 6.3
  waterstops, preformed materials used for, 1.9
ASTM (*see* American Society of Testing and Materials)
Backup materials (*see* joints)
Bituminous sealants
  (*See also* asphalts)
  generally, 19.22
  deck-sealing standards, 18.21
  hot-poured, 6.2, 19.22

Bond breakers (*see* joints)
Bridge joint sealants
    armoring of joint faces, 22.18
    compression seals, 22.18, 22.21
    deck joints, 19.12
    expansion joints, 19.5
    gland seals, 22.24
    illustrations, 22.25
    joint design, 19.12
    markets, 17.2
    modular systems, 22.21
    sealant comparison table, 21.12
    seals, 19.25
    silicones, 13.8
    specification standards, 18.29
    strip seals, 22.24
    suspension bridges, joint movement in, 21.11
    tension-compression seals, 22.24
Bubble testing, ASTM standard, 18.20
Butyls
    acrylics compared, 8.11
    advantages, 8.4
    ASTM standards, 18.13, 18.22
    bulk sealant formulations, 8.6
    caulks formula, 17.8
    chemical composition, 8.2
    curable and noncurable compounds, 8.1
    flexible waterstop seals, use as, 1.15
    gaskets, 8.1
    glazing applications, 8.7
    glazing formulas, 17.6
    hot melts
        combinations, 8.6
        formulas, 17.6
    insulating glass sealing formulas, 17.6
    isobutylene and isoprene components, 8.1
    latex paint over materials containing, 3.23
    limitations, 8.4
    movement capabilities of joint sealants, 13.19
    neoprene blending, 8.3
    pipe tape wrap, 8.7
    polyisobutylene, chemical composition, 8.2
    polysulfides compared, 8.11
    polyurethane materials compared, 8.11
    preformed joint fillers, use in, 1.6
    preparation, 8.2, 8.5
    properties, 8.4
    roof repair applications, 8.3, 8.6
    rubber, butyl, 3.19, 8.1
    sealant backbone comparison chart, 11.6
    sealant formula, 17.8
    silicones compared, 8.11
    single- and two-component formulations, 8.11
    skinning and nonskinning formulations, 8.4
    solvent release formulation, 8.8
    tapes, 8.1, 8.6, 16.2
        formula, 17.10
    thixotropic formulations, 8.10
    vertical caulks, use in, 3.21
    waterstops, preformed, 1.8
    weather resistance, 16.3
Canadian sealant standards, 18.25
Caulking (*see* installation methods)
Caulks
    (*See also* gun grade sealants)
    acrylic vs oil-base, 3.23
    asphalts, 6.3
    band aid repairs, 13.22
    butyls, 17.8
    coloring of, 3.7
    cord, caulking, 16.1
    failure analysis, 13.18
    firestop, 15.10
    Hypalon, 10.14
    latex-based, 3.5
    oil-base, 6.1
        acrylics compared, 3.23
    oleoresinous, 6.1
    polybutenes
        formula, 17.12
        rope caulk, 16.3
    polyisobutylene, 8.1
    polyvinyl, 3.24
    remedial, 13.17
    removal of old materials, 13.21
    sealant types used for, 1.2
    silicone, 13.17
    translucent and transparent types, 3.6
    vertical joints, 3.21

IN.4 INDEX

Cellulosic resins, 1.1
Cement
    ASTM type classifications, 7.7
    composition, 7.6
    definition, 7.5
    masonry cement, 7.8
    plaster, 7.8
    plaster added to, 7.35
    portland cement invented, 7.1
    preformed joint fillers, use in, 1.7
    properties, 7.1
    terrazzo, 17.14
    tile mortar, thin bed, 17.14
    topping mix, 17.13
    variations in composition, 7.7
Chemical curing (*see* curing processes)
Chemical elements table, Appx C
Chemical fasteners, 13.6
Clear sealants
    advantages of, 3.6
    formulation table, 3.8
Coal tar polyvinyl chlorides
    development of, 1.12, 22.14
    shape factor considerations, 22.28
    weather resistance, 1.13
Cohesion standards, SWRI Validation Program, 18.10
Compression seals
    (*See also* preformed sealants)
    adhesion lubricants, 22.21
    application methods, 19.23
    ASTM standards, 18.25
    bridges, use in, 22.21
    compartmentalized, 21.10, 22.18
    composition, 1.16
    developments of new types, 22.18
    epoxy primers, 19.25
    field splices, 19.25
    foam types, 19.25
    installation, 22.19
    large movement joints, use in, 19.25
    low-temperature recovery, 21.10
    mechanical interlocking, 22.18
    modular compression systems, 21.14
    modular systems, 22.21
    movement of joint, 19.24, 21.10
    neoprene-based lubricants, 19.23
    over compression, 21.10
    size selection, 21.10
    splices, 19.25
    tension-compression types
        composition, 1.17
        developments, 22.18
        illustration, 22.26
        improvements, 22.24
    under compression, 21.10
    width of joint, 19.24, 21.10
Concrete
    airfield runways (*see* airfield runway joint sealants)
    American Concrete Institute standards, 18.28
    application conditions, 19.3
    armored joints, 22.18
    articulated joints, 19.6
    asphalt joint fillers, 7.11
    asphalts, compatibility with, 6.3
    ASTM joint sealer standards, 18.19
    backup joint filler materials, 19.10
    bond breakers, 19.10
    bridges (*see* bridge joint sealants)
    butt joints, 19.7
    combined joints, 19.6
    construction joints, 7.11, 7.15, 7.28
        multiple functions, 19.6
    continuous joints, 7.28
    contraction joints, 7.29, 19.5
    control joints, 19.5
        creation, 7.14
        plane of movement, 7.11
        spacing, 7.19
    cracks
        causes, 7.10
        fatigue, 22.9
        joints distinguished, 19.7
        preparing for sealing, 22.9
        probability, 22.10
        problems caused by, 7.10
        repairs, 7.15, 7.19
        routing, 22.10
        sawed joints to retard, 22.2

shrinkage compensating formulations, 7.25
  thermal, 22.9
  uncontrolled, 22.5
curling of slabs, 7.10
defective joint designs, 19.20
definition, 7.5
design of joints generally, 19.12
deteriorating joints, 7.10
dimensions of floor panels, 22.3
doweled joints, 7.11, 7.15, 7.16
elasticity, modulus of, 7.19
elastomeric joint fillers, 7.32
elongated panels, 22.3
epoxies, use in concrete construction, 2.1
epoxy joint fillers, 7.30
epoxy joint repairs, 22.8
expansion joints, 7.11, 19.5
exposed reinforced designs, 7.4
fillers, joint, 7.11
filling joints, 7.29
fire protection advantages of, 7.1
forklift traffic and floor joint design, 22.3
formation of joints, 7.11
functions of joints, 7.10
gypsum concrete, 7.35
highways (*see* highway joint sealants)
hinge joints, 19.6
history of, 7.2
importance of, 7.1
intermediate joints, 22.3
isolation joints, 7.11, 7.29, 19.5
keyed joints, 7.11, 7.15, 7.17
lap joints, 19.7,
  sealing, 19.9
layout of joints, 7.15, 7.18
leave-in-place forms, 7.33
mixture classifications, 7.7, 7.25
modulus of elasticity, 7.19
monolithic slabs, 19.6
mortar, use as, 7.5
movement of slabs, research as to, 22.13
necessity of joints in, 19.2
one- and two-stage joints, 19.16
panel dimensions and joint design, 22.3

plastic ribbon method of sealing joints, 7.14
preparations for sealing joints, 19.19
problems caused by joints, 7.29
random spacing of joints, 19.17
raveling, 22.4
reinforcement bars, joint placement and, 22.3
relief joints, 22.8
repairs
  cracking of, 7.20
  finite element model, 7.24
  joint seals, 22.2
  modelling studies, 7.24
  spalling, 22.8
  strength of, 7.23
rigid and semi-rigid fillers, 7.30, 7.33
roads (*see* highway joint sealants)
runways (*see* airfield runway joint sealants)
sawed joints, 7.14, 22.2
  curing cut joints, 22.8
  depth of cut, 22.7, 22.10
  development of practice, 22.2
  precutting fresh concrete, 7.14
  purpose, 22.2
  raveling, 22.4
  safety considerations, 22.7
  sequencing cuts, 22.7
  technique, 22.7
  timing, 22.3, 22.5
sealant reservoirs, 19.16
sealing of joints, 7.15
  necessity of, 19.3
shape factor for sealants, 19.18
shrinkage, 7.25
silicones sealants, 13.8
skewed joints
  lateral movement, 21.17
  purpose, 19.18
  seal improvement, 22.28
slabs on grade, 19.15, 19.17
sliding joints, 19.6
soft joint fillers, 7.32
solvents, use in preparation for joint sealing, 19.19
spacing of joints, 7.15, 7.18, 22.3

Concrete (*continued*)
  spalling, 7.10, 7.29,
    joints as cause of, 22.3
    repair, 22.8
  special-purpose joints, 19.6
  square panel rule, 22.3
  steel protected joints, 7.33
  stepped joints, 19.7
  strengthening slab edges, 7.33
  tongue and groove joints, 7.15, 7.17
  types of joints, 7.10, 7.28
  uncontrolled cracking, 22.5
  underwater joints, 19.13
  walkways, 19.15
  waterstops, 19.14
  wear resistant seal systems, 22.24
Construction adhesive formula, 17.5
Construction Document Technologists (CDT), 18.11
Construction Specifications Institute (CSI) (*see* specifications)
Contact adhesives
  formulas, 17.5
  neoprene, 10.10
Contraction joints (*see* joints)
Conversion tables, Appx D
Copper waterstops, 1.15
Corp of Engineers specification standards, 18.31
Corrosion
  acrylics, 3.3
  concrete reinforcement, 13.8
  galvanic, silicones used to prevent, 13.7
  gypsum systems, 7.36
  salts, 7.36
  silicones in curtainwall construction methods, 13.7
  solvents causing, 15.11
  water-based sealants, 15.11
Curing processes
  acrylics, 3.17
  anaerobics, 3.17, 3.17
    rates of curing, 4.2
    uncured properties, 4.5
  anaerobics, cure rates, 4.2

butyls, curable and noncurable compounds, 8.1
chemical curing generally, 1.13
concrete, cut joints in, 22.8
epoxies, 1.13
  agents, 2.1
  ambient cured systems, 2.6
  rapid cure formulations, 2.6
  temperature and curing process, 2.5
exothermic reactions, 2.6,
modified acrylics, 3.17
neoprenes, curing and noncuring types, 10.6
polysulfides, 1.13, 11.3
polyurethanes, 1.13, 12.2
  moisture cure formulations, 12.16
rapid cure epoxies, 2.6
ropes, precured and partially cured, 16.1
silicones, 13.5
Cyanoacrylates (*see* anaerobics and cyanoacrylates)
Defoamers, 3.5, 3.11, 12.9
Density conversion tables, Appx D
Design (*see* joints)
Elastomer adhesive classifications, 1.1
Elastomeric sealants
  (*See also* synthetic rubbers)
  advantages of, 1.9
  extrusion standards, 18.15, 18.17
Electric heating conversion table, Appx E
Elements table, Appx C
Elongation, testing of, 5.10
Environmental conditions (*see* Installation conditions)
Environmental protection
  (*See also* solvents; toxicity)
  polyurethanes, 12.21
  water-based technologies, growth in, 15.10
Epoxides, 2.1
Epoxies
  aluminum, bonding of, 11.13
  ambient cured systems, 2.6
  applications, range of, 2.9
  ASTM standards, 18.22
  brittleness, 2.1, 2.2
  chemical curing, 1.13

concrete construction, use in, 2.1
concrete joint fillers, 7.30, 7.33
concrete joint repairs, 22.8
curing agents, 2.1
definition, 2.1
end-use formulations, 2.10
EPI resins, 2.8
flooring
    sand-filled, 17.12
    seamless, 17.13
intermediate resins, 2.7
Kneadatite, 14.4
movement capabilities of joint sealants, 13.19
on-site applications, 2.9
open joint sealants, comparison table, 1.4, 21.5
polyesters compared, 2.1, 2.12
properties table, 2.2
rapid cure formulations, 2.6
resins, 2.4
sand-filled flooring, 17.12
sealant backbone comparison chart, 11.6
sealant materials comparison table, 22.15
seamless flooring, 17.13
shelf life restrictions, 2.3
specialized uses, 2.9
temperature and curing process, 2.5
trowel coating, 17.13
Equivalents tables, Appx E
Ethyl vinyl acetate formulations, 5.2
Ethylene propylene diene monomer
    compression seals, 1.16
    foam tapes based on, 16.5
    hot melt applications, 5.3
    strip gland seals, 1.16
    tension compression seal systems, 1.17
Ethylene vinyl acetate (EVA)
    blending qualities, 8.5
    chemical structure, 5.4
    heat resistance, 5.2
    hot melts, development of, 3.19
Ethylene-acrylic ionomer, 14.2
Exothermic curing reactions, 2.6
Expansion joints (*see* joints)

Exterior insulation finish systems (EIFS), 13.14
Fasteners
    anaerobic thread locks, 3.16
    chemical, development of, 13.6
Field-molded sealants
    bond-breakers and backup materials, 20.9
    butt-joint widths, selecting, 21.2
    cold-applied, 19.22
    compression seals, size selection criteria, 21.10
    defects illustrated, 20.5
    developments, new, 22.14
    hot applied, 19.22
    joint design requirements, 1.3
    masking tape, use of, 19.27
    material comparison table, 21.5
    preformed materials distinguished, 20.1
    preparation for installation, 19.19
    priming requirements, 20.12
    repairs, 22.2
    research, 22.14
    sealant materials comparison table, 22.15
    shape factor requirements, 20.9
        ACI studies, 19.18
        importance of, 20.9
        ratios, 22.14
        short slab designs, 22.28
    temperature, effect on, 20.8
    uses, comparison table, 1.10, 22.30
Fire hazards
    (*See also* solvents)
    acrylics, 3.16
    adhesives comparison table, 1.19
    ASTM surface burning standards, 18.24
    ASTM volatility standards, 18.20, 18.22
    liquid oxygen, exposure of sealants to, 19.28
    non-volatile modified acrylic adhesives, 3.16
    primers, 19.27
    water-based technologies, growth in, 15.10
Fire protection and firestops
    active/passive system, 15.7
    adhesion requirements for firestop materials, 15.8
    annular space defined, 15.9
    application comparisons, 15.6

Fire protection and firestops (*continued*)
  application depth requirements of fire stop products, 15.8
  asbestos materials, 15.28
  ASTM standards
    firestop products, 15.7
    surface burning, 18.24
  backup materials, volatility, 1.14
  blow by, dangers of, 15.8
  cable penetrations, 15.23
  caulks, firestopping, 15.10
  complex penetrations, sealing, 15.26
  concrete, advantages of, 7.1
  development, 15.1
  elastomeric sealants, 15.14
  electric conduit penetrations, 15.15
  electrical boxes, putty bag protection, 15.24
  field assistance by firestop manufacturers, 15.9
  FlameSafe products, 15.12, 15.19
  flamesafe selector guide, 15.4
  gypsum concrete, 7.35
  heat transfer paths, 15.7
  insulation, fireproof, 15.1
  intumescents, firestopping, 15.7, 15.10
  metal pipe penetrations, 15.15, 15.26
  mortars, firestopping, 15.10, 15.25
  multiple penetrations, sealing, 15.26
  Nelson FSC coatings, 15.12
  nonmetalic conduit penetrations, 15.16
  penetrants, partition, defined, 15.9
  Pensil sealants, 15.13, 15.17
  product firestop ratings, 15.3
  putty, firestop, 15.22
  shrinkage of firestop materials, 15.8, 15.9
  silicones
    fire resistance properties, 13.4
    firestop properties, 15.8
  smoke
    emission properties of firestop products, 15.8
    protection of building inhabitants, 15.1
  specification, firestopping, 15.27
  SpecSeal elastomeric sealant, 15.12
  technologies, fire stop, 15.2
  toxicity of firestop sealants, 15.8
  water-based technologies, growth in, 15.10
Flashings
  copper, waterstops composed of, 1.15
  two-stage building joints, 22.32
Flexible foam (*see* foams)
Fluid measure conversion table, Appx E
Fluoroalkoxy-polyphosphazene elastomers, 14.3
Foams
  backup materials, comparison table, 1.6
  chemically resistant, 2.9
  compression seals, use as, 1.16
  double-coated tapes, 16.4
  epoxies, 2.9
  ethylene propylene diene monomer based tapes, 16.5
  flexible types, 1.16
  impregnation, 1.16
  mobile home construction, 16.4
  polyurethanes, 12.9, 16.5
  preformed materials and mastics distinguished, 16.1
  resiliency grading, 16.2
  tapes, foam, 16.1, 16.4
  TDI, 12.6
  waterstops, comparison table, 1.8
Functions of sealants, 20.1
Galvanic corrosion, silicones used to prevent, 13.7
Gaskets
  (*See also* preformed sealants)
  anchoring methods, 16.11, 16.13
  building types, 16.10
  butyl, 8.1
  classifications, 16.8
  fixing, 16.11
  glazing, 16.14
  installation methods, 19.26
  lock strip types, 16.11, 16.15
  mobile home construction, 16.4
  polychlorotrifluoroethylene, 14.2
  polyesters, 16.16
  polyurethanes, 16.16
  preformed materials and mastics distinguished, 16.1

preformed materials used for, 1.8
pressure requirements, 16.9
properties, 16.9
ribbon types, 1.15
self-adhesive types, 16.12
shapes, 21.16
silicones, 16.16
sizes, 21.16
structural, 16.11, 16.14
tapes, 1.15
types, 16.8
zipper types, 16.11, 16.15
Gland seals (*see* strip gland seals)
Glazing
    acrylics, 3.1
    architectural glazing tape, 17.10
    ASTM standards, 18.17
        water penetration, 18.24, 18.25
    butyls
        formulas, 17.6
        tapes, 16.2
        sealants, 8.7
    double pane sealants, 8.7
    extruded mastic insulation, 8.8
    extruded mastic sealant, 17.7
    failure analysis, 13.18
    Greek symbols used in glass technology, Appx B
    hot flow sealants, 5.2
    insulated windows
        ASTM standard, 18.19
        butyls, 8.1, 8.7, 17.6
        comparison tables, 17.6
        design system, 13.11
        extruded mastic, 17.7
        polysulfides, 11.4
        polyurethane, 12.10
        retrofit systems, 13.22
        silicone, 17.7
        two-component insulating glass formula, 12.11
        Vamac VMR, 14.2
        weep systems, 13.18
    markets, 17.2

mastics used in, 1.12
neoprene gaskets, 16.14
nitrile sealants, 10.10
oil-based materials, 6.2
polysulfides, 11.4
polyurethanes, 12.11
preformed materials and mastics distinguished, 16.1
remedial glazing, 13.22
ribbon glazing, 13.13
ribbon type seals, 1.15
sealing compounds compared, 1.8
silicones
    development, 13.1, 13.7
    formula, 17.7
    structural glazing, 16.14
    tapes, application techniques, 16.6
    two- and four-sided, 13.8
    Vamac VMR, 14.2
Glossary, 24.1-24.76
Glue
    animal glues, history of, 1.2
    definition of term, 1.1
Government agency listings, 23.1
Grouts
    ASTM specifications, 7.38
    Portland cement, use in, 7.35
    preformed joint fillers, use in, 1.7
Gun grade sealants
    neoprene types, 10.1
    oil based, 6.5
    polybutene caulks, 17.12
    polyurethane types, 12.15
    vertical joint sealants, 21.13
Hardness
    formulations, 5.11
    measurement, 5.10
    softening point, 5.11
    SWRI Validation Program, 18.10
Heat
    adhesive classifications based on, 1.1
    conversion tables, Appx D
Heat-welding flexible foam compression seals, 1.17

High-strength cement formula, 17.7
Highway joint sealants
  curb joints, 22.24
  gutter joints, 22.24
  joint design, 19.4, 19.12
  markets, 17.2
  random spacing, 19.17
  sealant comparison table, 21.12
  silicones, 13.8
  skewed joints, 19.18, 22.28
  slabs on grade, 19.15, 19.17
  specification standards, 18.29
  standards, 18.30
  wear resistant seal systems, 22.24
Hockman Cycle test, 18.16
Hot melt materials
  (*See also* thermoplastics)
  advantages, 5.2
  antioxidants in, 5.4
  application areas, 5.5
  application methods, 19.22
  ASTM standards, 18.23
  butyls, 8.6
  density, measurement, 5.10
  development of techniques, need for, 22.18
  elongation testing, 5.10
  ethyl vinyl acetate, 5.2
  glazing, hot flow sealants used in, 5.2
  hardness, 5.10
  limitations of, 5.2
  low-density polyethylene, 5.3
  melt index, 5.5, 5.9
  polymers used in, 5.3
  tensile modulus measurement, 5.10
  tensile strength, 5.10
  thermoplastic properties, 5.1
  Vamac VMR, 14.2
  viscosity determination and control, 5.7
Hot-poured bituminous sealants (*See* bituminous sealants)
Hypalon (*See* neoprenes)
Impact resistance
  adhesives comparison table, 1.19
Impregnated flexible foam (*see* foams)

Installation conditions
  access limitations, 1.2
  acrylics, temperature requirements, 3.5, 3.6
  below-grade joints, sealing, 11.11
  butyls, 18.13
  cold storage warehouses, 14.3
  concrete, 19.3
  environment, material choices dictated by, 1.2
  epoxies, temperature requirements, 2.5
  extrusion pump installation of silicones, 13.14
  field-molded sealants, effects of temperature on, 20.8
  joint sealants, 19.18
  lubricants, weather conditions affecting, 19.21
  material choices dictated by environment, 1.2
  oil and resin-based materials, 18.14
  pot life (*see* pot life)
  sealant applications, weather considerations, 13.20
  silicones requirements, 13.4, 18.13
  specialized sealant formulas for unusual conditions, 2.10
  specification requirements, 18.1
  temperature, 13.20
    conversion table, Appx A, Appx E
  thermoplastic sealants, temperature requirements, 1.13
  underwater joints, sealing, 11.11
Installation methods
  backup materials and bond-breakers, 19.21
  caulking
    guns, operation, 19.22
    remedial, 13.17
    skill required in applications, 19.23
  cold-applied sealants, 19.22
  compression seals, 19.23, 22.19
  failures caused by poor workmanship, 22.1
  field-molded sealants, 19.19
  fillers, 19.27
  gaskets, 19.26
  hot-poured sealants, 19.22
  joint preparation, 19.2
  joint sealants, 19.18

INDEX

large movement joints, use of compression seals in, 19.25
mixing-machines, custom, 19.23
neoprene sealants, 19.21
pavement joints, machine installation of, 19.23
pre-mixed sealants, 19.23
premixing of sealants offsite, 19.23
preparations for sealing, 19.19
spraying, 19.21
tapes, 16.6
toxic elastomeric sealants, handling, 19.28
waterstops, 19.26
Insulation
   acoustical sealants, ASTM standard, 18.21
   caulking, 3.21
   electrical, 2.7
   exterior insulation finish systems, 13.14
   fireproof, 15.1
   joint sealant requirements, 19.3
   windows (*see* glazing)
Intumescents (*see* fire protection)
Isobutylene and isoprene (*see* butyls)
Jet fuels, resistance to
   asphalts, 1.12
   ASTM standards, 18.12
   coal tar polyvinyl chlorides, 1.13, 22.14
   epoxies, 2.2
Joints
   adhesion lubricants, 22.21
   airfield runways (*see* airfield runway joint sealants)
   application of sealants, 19.18
   armoring of joint faces, 22.18
   ASTM sealant standards, 18.18
   backup materials, 19.10
      bond breakers, 1.3, 1.14, 11.12
      comparison table, 1.6
      compressibility, 20.9
      design considerations, 11.10
      functions, 11.10
      illustration of application, 11.11
      installation, 19.21
      sealant application requirements, 1.14
      volatility, 1.14
      water absorption, 18.21

below-grade, 11.11
best sealant decisions, 19.2
bond breakers
   adherence requirements, 20.9
   backup materials, 11.12
   function, 1.3
   illustration, 20.11
   installation, 19.21
   purpose, 11.12
   repairs, use in, 13.23
   sealants not requiring, 1.14
   selection criteria, 11.10
   strains on, 19.10
bridge (*see* bridge joint sealants)
butt-joint width limitations, 21.10, 21.12
calculating movement, 21.1
cleaning, 19.2
cleanup of sealed joints, 19.27
cohesive stresses, shape factor and, 22.14
compression seals, 1.16
concrete slabs (*see* concrete)
container seals, 19.13, 19.14
damp
   priming, 1.14
   sealant choices, 1.8
depth of sealant, limitation of, 20.10
depth-width curve ratios, 21.4
design, 19.4
   backup materials, 11.10
   cost considerations, 22.12
   defects in, 19.20
   failures caused by poor design, 22.1
   movement considerations, 21.1
   silicone sealing, 13.12
dimension selection graph, 21.4
eddying, dissipation of water through, 22.28
elastomeric fillers, 7.32
expansion
   backup materials, 1.6
   bridges, 19.5
   concrete slabs, 7.11, 19.5
   fillers, 1.6, 20.10
   specifications, 7.27
failures, new sealant types and, 22.1

INDEX

fillers
- expansion joints, 20.10
- installation, 19.27
- types, 1.6

fluid pressure, 19.4, 19.14, 19.17
heat conditions and sealant choices, 1.11
highway (*see* highway joint sealants)
horizontal and vertical
- sealant choices dictated by, 1.11
- self-leveling sealants, 21.13

hot-poured sealants, 19.22
illustrations of joint types, 11.11
inspection prior to sealing, 19.21
isolation, 7.11
lap joint sealant thicknesses, 21.11
lap shear joint stresses, 13.21
maintenance, 22.12
mechanical interlocking, 22.18
monolithic, 19.6
movement
- accommodation of, 19.3
- calculating, 21.1
- compression seals, 19.24, 21.10
- estimates, 11.9
- estimating, 21.1
- extremes of, determining, 21.16
- lateral movement, 21.17
- maximum sealable, 21.1
- measurement of, 21.16
- moisture control and, 21.17
- multiple joints, 20.7
- planes of, 7.11
- sealant capabilities compared, 13.19
- sealant thicknesses in lap joints, 21.11
- temperature, effects of, 21.17
- unanticipated service conditions, 22.1
- width and, filled-applied sealant chart, 21.9
- working and nonworking, 19.28

neatness of sealed joints, 19.27
nonworking and working, 19.28
one- and two-stage, 19.16
open joint sealant types comparison table, 1.4, 21.5
pavement, 19.4
polysulfide sealants, 11.1
polyurethane pourable sealer, 12.13
preparation for sealant application, 19.2
rigid and semi-rigid fillers, 7.30, 7.33
roads (*see* highway joint sealants)
runways (*see* airfield runway joint sealants)
sagging problems, 1.11
sandblasting, 19.2
sealant choice and configuration of, 1.2
sealant failure analysis, 13.18
sealant reservoirs, 19.16
sealing, necessity of, 19.3
self-leveling sealants, 21.13
Sil-Span repair process, 13.14
silicone sealants, 13.8
skewed
- lateral movement, 21.17
- purpose, 19.18
- seal improvement, 22.28

special purpose joints, 22.33
temperature, effect on movement, 21.17
tension-compression seals
- developments, 22.24
- illustration, 22.26

two-stage building joints, 22.28, 22.32
types, 11.11
underwater, 11.11
- ASTM standards, 18.18

vertical
- sagging problems, 21.13
- sealant choices for, 1.11

walkways, 19.15
waterstop, 19.14
width
- butt-joints, limitations, 21.10, 21.12
- compression seals, 19.24, 21.10
- movement and, filled-applied sealant chart, 21.9
- sealant recommendations, 13.20

width-depth ratios, 11.8
working and nonworking, 19.28

Latex-based caulks, 3.5
Lead
(*See also* toxicity)
- waterstops composed of, 1.15

Low-density polyethylene (LDPE)
  (*See also* polyethylene)
  hot melt polymer, use as, 5.3
  production of, 5.6
Low-temperature flexibility testing
  ASTM standard, 18.20
LP polymers (*see* polysulfides)
Lubricants
  adhesion types, 19.25, 22.21
  neoprenes, use of, 19.23
  polytetrafluoroethylene, 4.4
  thixotropic, 1.16, 19.25
  weather conditions affecting sealant applications, 19.21
Maintenance of sealed joints, 22.12
  (*See also* repairs of sealed joints)
  inspection programs, 22.12
  preformed sealants, 22.12
  renewal, sealant, 22.12
Markets, 17.1-17.3
Masonry, 7.5
Mastic sealant properties, 1.12
Material Safety Data Sheets (MSDS)
  (*See also* toxicity)
  firestops, 15.8, 15.28
  joint sealant choices, 19.27
Mildew-resistance
  Canadian sealant standards, 18.28
  silicones, 13.19
Modular compression seals (*see* compression seals)
Modulus of elasticity, 2.6
Moisture resistance, adhesives comparison table, 1.19
Mortars
  concrete crack repairs, 7.16
  definition, 7.5
  development, 7.7
  firestop, 15.10, 15.25
  preformed joint fillers, use in, 1.7
  thin bed tile mortar, 17.14
National Bureau of Standards, 18.12
Natural resins, 1.1
Neoprenes
  anionic neoprene latex contact adhesive, 10.10
  application methods, 19.21
  asphalt compatibility, 10.6
  butyl blending, 8.3
  compression seals, 1.16
    installation with neoprene-based lubricants, 19.23
  curing and noncuring types, 10.6
  development, 10.1
  flexible waterstop seals, use as, 1.15
  formulas for construction applications, 10.7
  glazing applications, 10.10
  glazing gaskets, 16.14
  Hypalon
    advantages and limitations, 10.14
    caulk, 10.14
    compounding ingredients, 10.13
    development, 10.7
    types, 10.11
  mastic adhesive formulation, 10.8
  natural rubber, displacement of, 10.2
  nitriles
    advantages and limitations, 10.15
    development, 10.9
    formulation, 10.15
    types, 10.10
  open joint sealants, comparison table, 1.4, 21.5
  pot life, 10.4, 10.9
  preformed joint fillers, use in, 1.6
  sealant backbone comparison chart, 11.6
  sealant materials comparison table, 22.15
  sealant uses, 10.6
  shear strength variations, 10.7
  structural glazing gaskets, 16.14
  tension/compression seal systems, 1.17
  two-component sealants, 10.9
  types, 10.3
  viscosity variations, 10.7
  waterstops, preformed, 1.8
Nitriles (*see* neoprenes)
Nitrite butadiene rubber (NBR) waterstops, 1.8
Odors, comparison table, 1.19
Oil- and resin-based materials, 6.1
  advantages, 6.2
  application requirements, 6.4

ASTM standards, 18.14, 18.23
caulks, disadvantages of, 3.23
costs, 6.4
limitations, 6.2
Oleoresinous caulks and sealants, 6.1
Outgassing, ASTM standards, 18.19
Ozone, sealants resistant to, 1.13
Pavement (*see* airfield runway joint sealants; highway joint sealants)
Peel strength
    ASTM standards, 18.22
    comparison table, 1.19
Permapol products
    development, 9.1
    polysulfides compared, 9.2
Pitch, 1.12
Plaster, 7.8
Plastic type sealants, 6.2
Plywood, adhesive use in, 1.2
PMMA (*see* acrylics)
Polybutenes
    caulk, 17.12
    clear sealant formula, 17.11
    curtainwall sealants, use as, 16.4
    flexibility, 6.2
    gun grade caulk, 17.12
    joint mastics, 1.12
    knifing ability, 6.3
    open joint sealants, comparison table, 1.4
    preformed foam sealants impregnated with, 16.4
    rope caulk, 16.3
    solvent-based clear sealant formula, 17.11
    tapes
        formula, 17.9
        use in, 16.2
Polychlorotrifluoroethylene (PCTFE), 14.1
Polyesters
    epoxies compared, 2.1, 2.12
    foam tapes, 16.5
    formed-in-place gaskets, 16.16
    polyols, 12.7
Polyether silicones, 14.2
Polyethylene
    (*See also* low-density polyethylene)
    anaerobics containers, 4.2, 4.6, 4.11
    ASTM standard, 5.10
    bond breaker tapes, 1.14, 13.22
    chemical composition, 5.6
    density measurement, 5.10
    double-coated foam tapes, 16.4
    foam backer rods, 13.13
    open joint sealants, comparison table, 1.4, 21.5
    sealant materials comparison table, 22.15
    thermoplastic properties, 5.1
Polyisobutylene (*see* butyls)
Polymercaptans
    development, 9.1
    properties compared, 11.6
Polymerization
    acetylene experiments, 10.1
    defined, 1.2
    energy release, 2.6
    oxygen inhibition, 1.18
    polyurethane development, 12.1
Polyols
    polyether, 12.8
    two types, 12.7
    urethane reaction, 12.2
Polysulfides
    advantages, 11.4
    ASTM standards, 18.17
    butyls compared, 8.11
    chemical curing, 1.13
    curing mechanisms, 11.3
    development, 11.1
    glazing applications, 11.4
    joint widths and sealant recommendations, 13.20
    latex paint over materials containing, 3.23
    limitations, 11.4
    LP polymers, 11.1
        aluminum, bonding of, 11.13
        design characteristics, 11.8
    movement capabilities of joint sealants, 13.19
    oil and resin-based materials, displacement of, 6.1

one- and two-component types
    formulations, 11.15
        properties, 11.7
    open joint sealants, comparison table, 1.4, 21.5
    Permapol products compared, 9.2
    pour points, 11.2
    properties, 11.2
    sealant backbone comparison chart, 11.6
    sealant materials comparison table, 22.15
    sealant types distinguished, 1.2
    silicones compared, 11.4, 13.19
    single-component sealant formulas, 11.15
    Thiokol LP, 2.11, 9.1, 11.13, 19.1
    two-component sealant formulas, 11.14
    types, 11.5
    urethanes compared, 11.4
    uses, 11.5
    UV resistance, 11.3
Polytetrafluoroethylene (PTFE), 14.1
Polythioethers, 9.2
Polyurethanes
    adhesion lubricants, 22.21
    advantages, 12.10
    ASTM standards, 18.17
    butyls compared, 8.11
    chemical curing, 1.13
    chemistry of, 12.3
    concrete joints, sealing, 7.15
    curing, 12.2
    development, 12.1
    durability, 12.2
    environmental considerations, 12.21
    foams, 12.9, 16.5
    formed-in-place gaskets, 16.16
    gaskets, use in, 1.8
    glazing applications, 12.11
    gun-grade sealant, 12.15
    high-traffic joint sealant, 12.14
    industrial membranes
        spray applied, 12.19
        trowelable, 12.20
    isocyanates, 12.5
    joint widths and sealant recommendations, 13.20
    latex paint over materials containing, 3.23
    limitations, 12.10
    manufacture, 12.1
    metal adhesive, 12.20
    moisture cure formulations, 12.16
    movement capabilities of joint sealants, 13.19
    one-component sealants, 12.3
    one-part pourable joint sealer, 12.13
    open joint sealants, comparison table, 1.4, 21.5
    polyols, 12.7
    polysulfides compared, 11.4
    preformed joint fillers, use in, 1.6
    prepolymers, use of, 12.7
    reaction basics, 12.4
    sealant backbone comparison chart, 11.6
    sealant types distinguished, 1.2
    silicones compared, 13.19
    single component sealant, 12.15, 12.17
    solvent-free formulations, 12.13, 12.21
    spray applied industrial membrane, 12.19
    TDI, 12.6
    trowelable industrial membrane, 12.20
    two-component insulating glass formula, 12.11
    two-component joint sealant, 12.14
    urethane formulations, 12.16
    urethane reaction, 12.4
    urethane technology, versatility of, 12.1
Polyvinyl chlorides (PVC)
    coal tar sealants, 1.12
    coal tar types, development of, 22.14
    field handling characteristics, 1.15
    flexibility, 1.15
    flexible waterstop seals, use as, 1.15
    gaskets, use in, 1.8
    sealant types distinguished, 1.2
    waterstop sealants, 19.26
    waterstops, preformed, 1.8
Portland cement (*see* cement)
Pot life
    anaerobics, 4.6
    neoprenes, 10.4, 10.9
    polysulfides, 11.4

Pot life (*continued*)
   sealant comparison tables, 21.13, 22.31
   sealant type comparisons, 1.11
   super rapid cure formulations, 2.6
   two-component vs one-component sealants, 19.23
Pour points
   asphalts, 19.22
   overheating, 19.22
   polysulfides, 11.2
   sealant comparison tables, 22.31
Pre-mixed sealants, 19.23
Preformed backup materials, comparison table, 1.6
Preformed sealants
   ASTM standards, 18.29
   butyls, 8.1
   comparison table, 1.8
   compression seals (*see* compression seals)
   concrete joint seals, 18.14
   defects in, 20.6
   developments of new types, 22.18
   elastomeric, 1.9
   failures, 20.2
   field-molded materials distinguished, 20.1
   firestops, 15.31
   flexible types, 1.15
   foam types, 16.4
   function illustrated, 20.3
   gaskets (*see* gaskets)
   maintenance, 22.12
   malfunctions, 20.2
   mastics distinguished, 16.1
   material comparison table, 21.5
   preparation of joint surfaces, 19.19
   rigid types, 1.15
   sealant materials comparison table, 22.15
   silicone profiles, 13.14, 13.16
   strip gland seals (*see* strip gland seals)
   tapes (*see* tapes)
   uses, comparison table, 1.10, 22.30
   Vamac VMR, 14.2
   varieties of, 16.1
Pressure conversion tables, Appx D

Primers
   anaerobics, 4.2
   application, 19.21
   butyls, 8.4
   defined, 19.2
   epoxy adhesive, priming with, 19.25
   field-molded sealants, 20.12
   flammable primers, 19.27
   functions of primers, 20.12
   nitriles, 10.15
   polysulfides, 11.4
   polyurethanes, 12.2
   sealant application requirements, 1.14
   silicones, 13.4, 13.12
Publication listings, 23.1
Putty, firestop, 15.22
PVC (*see* polyvinyl chlorides)
Quality assurance, 18.33
Radseal, 14.3
Renewal, sealant, 22.12
Repairs of sealed joints
   (*See also* maintenance of sealed joints)
   band aid repairs, 13.22
   concrete cracks, sealing, 22.9
   concrete joint seals, 22.2, 22.8
   concrete spalling, 22.8
   field-molded sealants, 22.2
   removal of failed sealants, 22.2
   Sil-Span, 13.14
   spalling in concrete, 22.8
   specifications for, 18.8
   widening of concrete joints, 22.2
Resins
   (*See also* epoxies)
   acrylics, displacement of oil and resin-based materials by, 6.2
   caulks, oleoresinous, 6.1
   cellulosic resins, 1.1
   classifications, 1.1
   epoxies, 2.1, 2.4
      EPI resins, 2.8
      intermediate resins, 2.7
   history of oil- and resin-based materials, 6.1
   natural, 1.1

oil and resin-based materials, 6.1
oleoresinous caulks and sealants, 6.1
polysulfides compared to oil and resin-based materials, 6.1
silicones, displacement of oil and resin-based material by, 6.2
storage, 6.6
synthetic type classifications, 1.1
tackifying resins, 5.3
Rheology test, ASTM standards, 18.19
Rhoplex, use in acrylic caulks, 3.6
Ribbons (*see* tapes)
Roads (*see* highway joint sealants)
Room temperature vulcanizing (*see* silicones)
Ropes (*see* tapes)
Rubber
(*See also* butyls; neoprenes)
adhesive uses, development of, 1.2
butyl, 3.19, 8.1
flexible waterstop seals, use as, 1.15
gaskets, 1.6
limitations of natural rubber, 10.2
preformed joint fillers, use in, 1.6
tapes, 3.18
tear strength, ASTM standard, 18.19
traditional construction uses, 10.2
waterstops, preformed materials used for, 1.9
Safety considerations
acrylics, 3.3
concrete joint sawing, 22.7
joint sealant material handling, 19.27
Sagging of sealants
asphalts, 1.11, 21.13
support materials, 1.3, 1.6
vertical joints, 21.13
SBR (*see* styrene butadiene rubber)
Sealant properties and uses
adhesives and sealants distinguished, 1.17, 4.1
classifications, 16.1, 20.1
definitions, 1.2
functions of sealants, 20.1
materials comparison table, 22.15
purposes of sealing joints, 12.1
solvent-based systems, 3.18
use table, 1.10

Sealant, Waterproofing & Restoration Institute (SWRI) validation program, 18.10
Sheer strength, adhesives comparison table, 1.19
Shrinkage testing, ASTM standard, 18.20
Silicones
acrylic unpigmented sealants compared, 3.7
airfield runway joint sealants, 13.8
ASTM standards, 18.13, 18.17, 18.18
bridge joint sealants, 13.8
butyls compared, 8.11
Canadian sealant standards, 18.26, 18.27
chemical curing, 1.13
chemistry of, 13.2
cured properties, 13.5
curtainwall construction methods, 13.7
development, 13.1
extrusion pump installation, 13.14
fasteners, use as, 13.6
fire resistance, 13.4
firestop properties, 15.8
formed-in-place gaskets, 16.16
galvanic corrosion, 13.7
glazing, 13.1
development of new applications, 13.7
formula, 17.7
ribbon glazing, 13.13
two- and four-sided, 13.8
highway joint sealants, 13.8
intermediates, 13.2
joint widths and sealant recommendations, 13.20
latex paint over materials containing, 3.23
mildew-resistance, 13.19
modulus type formulations, 13.6, 13.23
movement capabilities of joint sealants, 13.19
oil and resin-based materials, displacement of, 6.2
open joint sealants, comparison table, 1.4, 21.5
paintable formulations, 13.19
polyether compounds, 14.2
polysulfides compared, 11.4, 13.19
polyurethanes compared, 13.19
preformed silicone profiles, 13.14

Silicones (*continued*)
    primer requirements, 13.12
    remedial caulking, 13.17
    room temperature vulcanizing
        cured properties, 13.5
        defined, 13.1
    sealant backbone comparison chart, 11.6
    sealant materials comparison table, 22.15
    sealant types distinguished, 1.2
    Sil-Span, 13.14
    starting formulations, 13.23
Slabs (*see* concrete)
Slump testing, ASTM standards, 18.19
Solvents
    (*See also* fire hazards)
    acrylic solvent release tables, 3.20
    caulks, solvent releasing, 3.6
    clean up solvents, toxicity concerns, 19.28
    corrosion problems, 15.11
    development of solvent based systems, 3.18
    joints, preparation for sealing, 19.19
    latex paint over materials containing, 3.23
    pollution problems, 1.7, 3.19
    polyurethanes, 12.13, 12.21
    release system formula, 17.4
    release, comparison table, 1.4, 1.10
    resistance of adhesives, comparison table, 1.19
    water-based technologies, shift to, 15.10
Specifications
    general requirements, 18.1
    AAMA test methods, 18.31
    American Association of State Highway and Transportation Officials standards, 18.30
    American Concrete Institute standards, 18.28
    application condition requirements, 18.1
    architects' uses of, 18.5
    ASTM standards, 18.12-18.25
        use in specs, 18.10
    Canadian sealant standards, 18.25
    canned specs, 18.8
    computers, use of, 18.4
    Construction Document Technologists, 18.11
    Construction Specifications Institute, 18.33
    consultants for drawing of, 18.3
    contractors' uses of, 18.6
    Corp of Engineer standards, 18.31
    CSI Manual of Practice, 18.9
    CSI standard forms, 18.2
    engineers' uses of, 18.5
    expansion requirements, 7.27
    firestopping, 15.27
    general directions, dangers of, 18.1
    grout, 7.38
    joint sealant installation, 19.18
    long-term usage, testing and, 18.34
    manufacturers' uses of, 18.7
    modifying, 18.4
    National Bureau of Standards, 18.12
    qualify assurance, 18.33
    repair specs, 18.8
    sealant validation, 18.10
    specificity, 18.1
    teamwork approach, 18.4
    testing, sampling standards, 18.10
Staining of sealants, ASTM standards, 18.18
Standards (*see* specifications)
Steel
    concrete joints, steel protected, 7.33
    waterstops composed of, 1.15
Storage
    anaerobics and cyanoacrylates, 4.2, 4.6
    cyanoacrylates, 4.11
    deterioration during, 1.7
    frozen storage requirements, 19.23
    hot melt sealants, 5.2
    neoprenes, 10.6
    oil and resin-based materials, 6.6
    polyurethanes, 12.7
    separation of components during, 3.25
Strip gland seals
    bridge systems, use in, 22.24
    development of, 1.15, 1.16
    illustration, 22.29
    prefabrication, 22.28
Stripping strength of adhesives, 18.22
Styrene butadiene rubber (SBR) waterstops, 1.8
Sunlight (*see* ultraviolet resistance)
Superglues (*see* anaerobics and cyanoacrylates)
Supplier listings, 23.1

Symbols used in figures, key to, Appx G
Syndiotactic polybutadiene, 14.2
Synthetic adhesives
    development, 1.2
    polymerization, 1.2
Synthetic resins, 1.1
Synthetic rubbers
    (*See also* butyls; neoprenes; polysulfides)
    elastomers, development of generally, 19.1
    sealant materials comparison table, 22.15
    tear strength, ASTM standard, 18.19
Tapes
    (*See also* preformed sealants)
    application techniques, 16.6
    architectural glazing tape, 17.10
    asphalt coatings, 16.5
    bonding, 16.1
    butyls, 8.1, 16.2
        formula, 17.10
        pipe tape wrap, 8.7
    cellular preformed, 16.5
    double-coated foam, 16.4
    ethylene propylene diene monomer based foam tapes, 16.5
    foamed and solid, 16.1, 16.4
    Kneadatite, 14.4
    mending, 16.1
    mobile home construction, 16.4
    pipe tape wrap, 8.7
    polybutene formula, 17.9
    polybutene base rope caulk, 16.3
    prefabricated housing construction, use in, 16.7
    preformed materials and mastics distinguished, 16.1
    preformed, comparison table, 16.2
    PVC, 16.4
    resiliency grading, 16.2
    ribbons
        elastomeric, 16.1
        glazing applications, 1.15, 13.13
        Kneadatite epoxy, 14.4
        seals, ribbon types, 1.15
    ropes
        butyls, 8.1
        polybutene base, 16.3

    precured and partially cured, 16.1
    weather resistance qualities of preformed tapes compared, 16.2
Temperature
    (*See also* installation conditions)
    conversion table, Appx A, Appx E
    sealant application considerations, 13.20
Tensile adhesion, ASTM standard, 18.17, 18.19
Tensile modulus measurement, 5.10
Tensile sheer strength adhesives comparison table, 1.19
Tension-compression seals (*see* compression seals)
Testing
    (*See also* American Society of Testing and Materials; specifications)
    laboratories, listings, 23.1
    AAMA standards, 18.31
    sampling standards, 18.10
Thermoplastics
    (*See also* hot melt materials)
    adhesive classifications, 1.1
    advantages and disadvantages, 1.12
    sealants, chemical composition, 1.13
Thermosetting, 1.13
Thiokol LP (*see* polysulfides)
Thixotropic formulations
    butyl sealants, 8.10
    compression seal installations, 1.16
    pavement joint sealing, 19.25
Toxicity
    (*See also* MSDS safety data sheets)
    acrylics, 3.3
    adhesives comparison table, 1.19
    asbestos, 16.3
    ASTM lead-based paint abatement standards, 18.25
    clean up solvents, toxicity concerns, 19.28
    elastomeric sealants, 19.28
    firestop sealants, 15.8
    formaldehyde in anaerobic formulations, 4.11
    joint sealants generally, 19.27
    lead, 18.25
    lead dioxide, 19.28
    modified acrylic adhesives, 3.16

Toxicity (*continued*)
   MSDS safety data sheets, 15.8
   sealant choices generally, 1.7
   water-based technologies, growth in, 15.10
Trade organization listings, 23.1
Two-stage building joints, 22.32
   description, 22.28
   flashings, 22.32
   illustration, 22.33
Ultraviolet resistance
   (*See also* weather resistance)
   acrylics, 3.6, 3.10, 3.22
   ASTM testing, 18.16, 18.17
   butyls, 8.4
   glazing adhesion failures, 13.18
   Hypalon, 10.14
   impregnated foam tape, 16.6
   neoprenes, 10.10
   nitriles, 10.15
   Permapol products, 9.2
   polysulfides, 11.3
   polyurethanes, 12.10, 12.17
      foam, 16.4
   preformed tapes compared, 16.2
   Sil-Span, 13.15
   silicones, 13.2, 13.4, 13.19
Underwater joints, ASTM sealant standards, 18.18
Unsaturated polyether, 14.3
Urethane (*see* polyurethanes)
Vamac VMR, 14.2
Vinyl acetate ethylene (VAE), 5.4
Viscosity
   hot melt materials, 5.7
   measurement, 5.9
Volatility (*see* fire protection)
Vulcanizing, room temperature (*see* silicones)
Warranties
   insulating glass assemblies, 13.18
   sealants, 22.14
Washers, 16.1
Water absorption, ASTM standards, 18.21
Water emulsion sealant types, 6.2
Waterstops
   flexible waterstop seals, 1.15
   installation methods, 19.26

   polyvinyl chloride sealants, 19.26
   preformed materials used for, 1.8
   rigid waterstop seals, 1.15
   sealant choices, comparison table, 1.8
   shape and size choices, 21.11
   splicing, 19.26
Weather conditions
   (*See also* installation conditions)
   sealant application considerations, 13.20
Weather resistance
   (*See also* jet fuels, resistance to; ultraviolet resistance)
   acrylics, 3.1, 3.23
   ASTM testing, 18.16
   butyls, 8.1
      tapes, 16.3
   cellular preformed tapes, 16.5
   coal tar polyvinyl chlorides, 22.14
   epoxies, 2.2
   glazing adhesion failures, 13.18
   Hypalon, 10.11
   joint movement, temperature and, 21.17
   joint sealant characteristics, 1.3
   natural rubber, 10.2
   nitrile, 10.15
   open joint sealants, comparison table, 21.5
   polyurethane foams, 1.8, 16.5
   polyurethanes, 12.5, 12.10
   polyvinyl chloride coal tars, 1.13
   preformed tapes compared, 16.2
   sealants resistant to, 1.13
   Sil-Span, 13.15
   silicones
      advantages generally, 13.2, 13.12
      ASTM standard, 18.18
      corrosion resistance, 13.7
      curtainwall construction methods, 13.7
      relative strength, 13.19
   syndiotactic polybutadiene, 14.2
   Vamac VMR, 14.2
Weathersealing
   sealant types used for, 1.2
Weep systems, 13.18
Weights
   conversion table, Appx E
Windows (*see* glazing)